Probability and Statistics by Example: II

Probability and statistics are as much about intuition and problem solving, as they are about theorem proving. Because of this, students can find it very difficult to make a successful transition from lectures to examinations to practice, since the problems involved can vary so much in nature. Since the subject is critical in many modern applications such as mathematical finance, quantitative management, telecommunications, signal processing, bioinformatics, as well as traditional ones such as insurance, social science and engineering, the authors have rectified deficiencies in traditional lecture-based methods by collecting together a wealth of exercises for which they have supplied complete solutions. These solutions are adapted to needs and skills of students.

Following on from the success of *Probability and Statistics by Example: Basic Probability and Statistics*, the authors here concentrate on random processes, particularly Markov processes, emphasising models rather than general constructions. Basic mathematical facts are supplied as and when they are needed and historical information is sprinkled throughout.

Probability and Statistics by Example: II
Markov Chains: a Primer in Random Processes
and their Applications

Yuri Suhov
University of Cambridge

Mark Kelbert
University of Wales–Swansea

CAMBRIDGE
UNIVERSITY PRESS

Shaftesbury Road, Cambridge CB2 8EA, United Kingdom

One Liberty Plaza, 20th Floor, New York, NY 10006, USA

477 Williamstown Road, Port Melbourne, VIC 3207, Australia

314–321, 3rd Floor, Plot 3, Splendor Forum, Jasola District Centre, New Delhi – 110025, India

103 Penang Road, #05–06/07, Visioncrest Commercial, Singapore 238467

Cambridge University Press is part of Cambridge University Press & Assessment,
a department of the University of Cambridge.

We share the University's mission to contribute to society through the pursuit of
education, learning and research at the highest international levels of excellence.

www.cambridge.org
Information on this title: www.cambridge.org/9780521612340

First published 2008

A catalogue record for this publication is available from the British Library

ISBN   978-0-521-84767-4   Hardback
ISBN   978-0-521-61234-0   Paperback

Contents

Preface

This volume, like its predecessor, *Probability and Statistics by Example*, Vol. 1, was initially conceived with the intention of giving Cambridge students an opportunity to check their level of preparation for Mathematical Tripos examinations. And, as with the first volume, in the course of the preparation, another goal became important: to give the general public a clearer picture of how probability- and statistics-related courses are taught in a place like the University of Cambridge, and what level of knowledge is achieved (or aimed for) by the end of these courses. In addition, the specific topic of this volume, Markov chains and their applications, has in recent years undergone a real surge. A number of remarkable theoretical results were obtained in this field which only twenty years or so ago was considered by many probabilists as a 'dead' zone. Even more surprisingly, an active part in this exciting development was played by applied research. Motivated by a dramatically increasing number of problems emerging in such diverse areas as computer science, biology and finance, applied people boldly invaded the territory traditionally reserved for the few hardened enthusiasts who until then had continued to improve old results by removing one or another condition in theorems which became increasingly difficult to read, let alone apply. We thus felt compelled to include some of these relatively recent ideas in our book, although the corresponding sections have little to do with current Cambridge courses. However, we have tried to follow a distinctively Cambridge approach (as we see it) throughout the whole volume.

On the whole, our feeling is that the modern theory of Markov chains can be compared with a huge and complex living organism which has suddenly woken from a period of hibernation and is now in a state of active consumption and digestion of fresh foodstuff produced by fertile lands around it, flourishing under blissful conditions. As often happens in nature, some parts of this living organism go through vast changes: they become less or more important compared with the previous state. In addition, some parts, like an old skin, may be sloughed off and replaced by new, better adapted to new realities. Our book then can be compared with a photographic snapshot of this giant from a certain distance and angle. We are not able to feature the whole animal (it is simply too big and fast-moving for us), and many details of the picture within the frame of our snapshot are blurred. However, we hope that the overall image is somewhat new and fresh.

At the same time, our goal was to treat those topics that are particularly important, especially in the course of learning the basic concepts of Markov chains. These are the aspects and issues that are particularly thought-provoking for a newcomer and, not surprisingly, usually provide the most fertile grounds for setting up problems suitable for exams. Roughly speaking, all the material from the theory of Markov chains which proved to be useful in examinations in Cambridge during the period 1991–2003 is included in the book.

It has to be said that studying via (or supporting the learning process by going through) a large number of homogeneous problems (with or without solutions) can be rather painstaking. A view popular among the mathematically-minded section of the academic community could be that the most productive way of learning a mathematical subject is to digest proofs of a collection of theorems general enough to serve many particular cases and then treat various questions as examples illustrating such theorems (the present authors were educated in precisely this fashion). The problem is that it ideally suits the mathematically-minded section of the academic community, but perhaps not the rest...

On the other hand, an increasing number of students (mainly, but not always, with a non-mathematical background) strongly oppose (at least psychologically) any attempt at a 'decent' proof of even basic theorems. Moreover, the manual calculations often required in examples whose tailor-made background is obvious also became increasingly unpopular with generations of students for whom computers have become as ordinary as toothbrushes. The authors can refer to their own experience as lecturers when audiences have been convinced more by computer evidence than by a formal proof. There is clearly a problem about how to teach an originally pure mathematics subject to a wider audience. There is some basis for the above unpopularity, although we personally still believe that learning the proof of convergence to an equilibrium distribution of a Markov chain is more productive than seeing twenty or so numerical examples confirming this fact. But an artificial example where, say, a four by four transition matrix is constructed so that its eigenvalues are of a 'nice' form (a particular value 1, easy to find from symmetry or another 'educated guess', and the remaining two from a quadratic equation), may mis- or even back-fire, whereas an efficient modern package could do the job without much fuss. However, our presentation disregards these aspects; we consider it as first step towards a future style of book-writing.

A particular feature of the book is the presence of what we have called 'Worked Examples', along with 'Examples'. The former show readers how to go about solving specific problems; in other words, give explicit guidance about how to make the transition from the theory to the practical issue of solving problems. The end of a worked example is marked by a symbol □. The latter are illustrative, and are intended to reveal more about the underlying ideas. We must note that we have been particularly influenced by books Norris, 1997, and Stroock, 2005. In addition, a number of past and present members of the Statistical Laboratory, DPMMS, University of Cambridge, contributed to creating a particular style of presentation (we wrote about it in the preface to the first volume). It is the pleasure to name here David Williams, Frank Kelly, Geoffrey Grimmett, Douglas Kennedy, James Norris, Gareth Roberts and Colin Sparrow whose lectures we attended, whose lecture notes we read and whose example sheets we worked on. In Swansea, great help and encouragement came from Alan Hawkes, Aubrey Truman and again David Williams. We are particularly grateful to Elie Bassouls who read the early version of the book and made numerous suggestions for improving the presentation. His help extended beyond the usual level of involvement of a careful reader into preparation of a mathematical text and rendered the great service to the authors.

We would like to thank David Trarah for the efforts he made to clarify and strengthen the structure of the book and for his careful editing work which went much further than the usual copyediting. We also thank Sarah Shea-Simonds and Eugenia Kelbert for checking the style of presentation.

The book comprises three chapters divided into sections. Chapters 1 and 2 include material from Cambridge undergraduate courses but go far beyond in various aspects of Markov chain theory. In Chapter 3 we address selected topics from Statistics where the structure of a Markov chain clarifies problems and answers. Typically, these topics become straightforward for independent samples but are technically involved in a general set-up.

The bibliography includes a list of monographs illustrating the dynamics of development of the theory of random processes, particularly Markov chains, and parallel progress in Statistics. References to relevant papers are given in the body of the text. References to Vol. 1 mean *Probability and Statistics by Example*, Volume 1.

1
Discrete-time Markov chains

1.1 The Markov property and its immediate consequences

*Mathematics cannot be learned by lectures alone, anymore
than piano playing can be learned by listening to a player.*
C. Runge (1856–1927), German applied mathematician

Typically, the subject of Markov chains represents a logical continuation from a basic course of probability. We will study a class of *random processes* describing a wide variety of systems of theoretical and practical interest (and sometimes simply amusing). The fact that deep insight into the subject is possible without using sophisticated mathematical tools may also be an explanation of why Markov chains are popular in so many different disciplines which are seemingly remote from pure mathematics.

The basic model for the first half of the book will be a system which changes state in *discrete* time, according to some random mechanism. The collection of states is called a *state space* and throughout the whole book will be assumed finite or countable; we will denote it by I. Each $i \in I$ is called a state; our system will always be in one of these states. Sometimes we will know what state the system occupies and sometimes only that the system is in state i with some probability. Therefore it makes sense to introduce a *probability measure* or *probability distribution* (or, more simply, a distribution) on I. A probability distribution λ on I is simply a countable collection $(\lambda_i, i \in I)$ of non-negative numbers of total sum 1:

$$\lambda_i \geq 0, \ \sum_{i \in I} \lambda_i = 1. \tag{1.1}$$

We can think of a unit 'mass' spread over the set I where point i has mass λ_i. For that reason it is sometimes convenient to speak of a probability mass function $i \in I \mapsto \lambda_i$. Then the probability of a set $J \subseteq I$ is $\lambda(J) = \sum_{j \in J} \lambda_j$.

If $\lambda_i = 1$ for some $i \in I$ and $\lambda_j = 0$ when $j \neq i$, the distribution is 'concentrated' at point i. Then the state of our system becomes 'deterministic'. We will denote such a distribution by δ_i (the Dirac measure being an extreme case).

Sometimes the condition $\sum_{i \in I} \lambda_i = 1$ is not fulfilled; then we simply say that λ is a *measure* on I. If the total mass $\sum_{i \in I} \lambda_i < \infty$, the measure is called finite and can be transformed into a probability distribution by the normalisation: $\widetilde{\lambda}_i = \lambda_i / \sum_{j \in I} \lambda_j$ which defines a probability measure on I, since $\sum_{i \in I} \widetilde{\lambda}_i = \sum_{i \in I} \lambda_i / \sum_{j \in I} \lambda_j = 1$. But even if $\sum_{i \in I} \lambda_i = \infty$ (i.e. the total mass is infinite), we still can assign a finite value $\lambda(J) = \sum_{i \in J} \lambda_i$ to finite subsets $J \subset I$.

The random mechanism through which a change of state occurs is described by a *transition matrix* P, with entries p_{ij}, $i, j \in I$. Entry p_{ij} gives the probability that the system will change state i to j in a unit of time. That is, p_{ij} is the conditional probability that the system will occupy state j at the next time-step given that it is currently in state i. Hence, we have that each entry in P is non-negative but not greater than 1, and the sum of entries along every row equals 1:

$$0 \leq p_{ij} \leq 1 \text{ for all } i, j \in I \text{ and } \sum_{j \in I} p_{ij} = 1 \text{ for all } i \in I. \tag{1.2}$$

A matrix P with these properties is called *stochastic*. By analogy, a probability distribution (λ_i) on I is often called a *stochastic vector*. Then a stochastic matrix is one where every row is a stochastic vector.

Example 1.1.1 The simplest case is 2×2 (a two-state space). Without loss of generality, we may think that the states are 0 and 1: then the entries will be p_{ij}, $i, j = 0, 1$. Here, the stochastic matrix has the form

$$\begin{pmatrix} 1 - \alpha & \alpha \\ \beta & 1 - \beta \end{pmatrix}$$

where $0 \leq \alpha, \beta \leq 1$. In particular, $\alpha = \beta = 0$ gives the identity matrix $\mathbf{I}$ and $\alpha = \beta = 1$ the anti-diagonal matrix:

$$\begin{pmatrix} 1 & 0 \\ 0 & 1 \end{pmatrix}, \begin{pmatrix} 0 & 1 \\ 1 & 0 \end{pmatrix}.$$

A system with the identity transition matrix remains in the initial state forever; in the anti-diagonal case it flips state every time, from 0 to 1 and *vice versa*.

On the other hand, $\alpha = \beta = 1/2$ gives the matrix

$$\begin{pmatrix} 1/2 & 1/2 \\ 1/2 & 1/2 \end{pmatrix}.$$

In this case the system may stay in its state or change it with equal probabilities.

It is convenient to represent the transition matrix by a diagram where arrows show possible transitions and are labelled with the corresponding transition probabilities (arrows leading back to their own origin are often omitted as well as labels for deterministic transitions). See Figure 1.1, top.

La Dolce Beta

(From the series *'Movies that never made it to the Big Screen'*.)

Example 1.1.2 The 4×4 matrix

$$\begin{pmatrix} 0 & 1/3 & 1/3 & 1/3 \\ 1/4 & 1/4 & 1/4 & 1/4 \\ 1/2 & 1/2 & 0 & 0 \\ 0 & 0 & 0 & 1 \end{pmatrix}$$

is represented in Figure 1.1, bottom.

The time will take values $n = 0, 1, 2, \ldots$. To complete the picture, we have to specify in what state our system is at the initial time $n = 0$. Typically, we will assume that the system at time $n = 0$ is in state i with probability λ_i for some given 'initial' distribution λ on I.

Denote by X_n the state of our system at time n. The rules specifying a Markov chain with initial distribution λ and transition matrix P are that

(i) X_0 has distribution λ:

$$\mathbb{P}(X_0 = i) = \lambda_i, \text{ for all } i \in I,$$

(ii) more generally, for all n and $i_0, \ldots, i_n \in I$, the probabilities $\mathbb{P}(X_0 = i_0, X_1 = i_1, \ldots, X_n = i_n)$ that the system occupies states $i_0, i_1, \ldots, i_n$ at times $0, 1, \ldots, n$ is written as a product

$$\mathbb{P}(X_0 = i_0, X_1 = i_1, \ldots, X_n = i_n) = \lambda_{i_0} p_{i_0 i_1} \cdots p_{i_{n-1} i_n}. \tag{1.3}$$

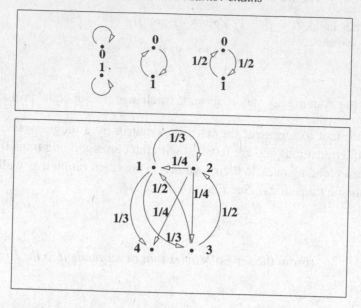

Fig. 1.1

Of course, (i) is a particular case of (ii), with $n = 0$.

An important corollary of (1.3) is the equation for the conditional probability $\mathbb{P}(X_{n+1} = j | X_0 = i_0, \ldots, X_{n-1} = i_{n-1}, X_n = i)$ that the state at time $n+1$ is j, given states $i_0, \ldots, i_{n-1}$ and $i_n = i$ at times $0, \ldots, n-1, n$:

$$\mathbb{P}(X_{n+1} = j | X_0 = i_0, \ldots, X_{n-1} = i_{n-1}, X_n = i)$$
$$= \frac{\mathbb{P}(X_0 = i_0, \ldots, X_{n-1} = i_{n-1}, X_n = i, X_{n+1} = j)}{\mathbb{P}(X_0 = i_0, \ldots, X_{n-1} = i_{n-1}, X_n = i)}$$
$$= \frac{\lambda_{i_0} p_{i_0 i_1} \cdots p_{i_{n-1} i} p_{ij}}{\lambda_{i_0} p_{i_0 i_1} \cdots p_{i_{n-1} i}} = p_{ij}. \tag{1.4}$$

That is, conditional on $X_0 = i_0, \ldots, X_{n-1} = i_{n-1}$ and $X_n = i$, we see X_{n+1} has the distribution $(p_{ij}, j \in I)$. In particular, the conditional distribution of X_{n+1} does not depend on $i_0, \ldots, i_{n-1}$, i.e., depends only on the state i at the last preceding time n.

Formula (1.4) illustrates the 'no memory' property of a Markov chain (only the current state counts for determining probabilities of future states).

Another consequence of (1.3) is an elegant formula involving matrix multiplication for the marginal probability distribution of X_n. Here we ask the question: what is the probability $\mathbb{P}(X_n = j)$ that at time n our system is in state j? For example, for $n = 1$ we can write:

$$\mathbb{P}(X_1 = j) = \sum_{i \in I} \mathbb{P}(X_0 = i, X_1 = j),$$

by considering all possible initial states i. In fact, the events

$$\{\text{state } i \text{ at time } 0, \text{ state } j \text{ at time } 1\}$$

do not intersect for different $i \in I$ and their union gives the event

$$\{\text{state } i \text{ at time } 1\},$$

Now use (1.3) and recall the rules of matrix algebra:

$$\sum_{i \in I} \mathbb{P}(X_0 = i, X_1 = j) = \sum_{i \in I} \lambda_i p_{ij} = (\lambda P)_j.$$

By a direct calculation, this formula is extended to a general n:

$$
\begin{aligned}
\mathbb{P}(X_n = j) &= \sum_{i_0, \ldots, i_{n-1}} \mathbb{P}(X_0 = i_0, \ldots, X_{n-1} = i_{n-1}, X_n = j) \\
&= \sum_{i_0, \ldots, i_{n-1}} \lambda_{i_0} p_{i_0 i_1} \cdots p_{i_{n-1} j} = (\lambda P^n)_j,
\end{aligned}
\tag{1.5}
$$

where P^n is the nth power of the matrix P. That is, the stochastic vector describing the distribution of X_n is obtained by applying the matrix P^n to the initial stochastic vector λ.

Then, similarly,

$$
\begin{aligned}
&\mathbb{P}(X_n = i, X_{n+1} = j) \\
&= \sum_{i_0, \ldots, i_{n-1}} \mathbb{P}(X_0 = i_0, \ldots, X_{n-1} = i_{n-1}, X_n = i, X_{n+1} = j) \\
&= \sum_{i_0, \ldots, i_{n-1}} \lambda_{i_0} p_{i_0 i_1} \cdots p_{i_{n-1} i} p_{ij} = (\lambda P^n)_i p_{ij},
\end{aligned}
$$

and, hence

$$\mathbb{P}(X_{n+1} = j | X_n = i) = \frac{\mathbb{P}(X_n = i, X_{n+1} = j)}{\mathbb{P}(X_n = i)} = \frac{(\lambda P^n)_i p_{ij}}{(\lambda P^n)_i} = p_{ij}. \tag{1.6}$$

In other words, the entry p_{ij} is the conditional probability that the state at the next time-step is j given that at the preceding one it is i.

Moreover,

$$
\begin{aligned}
&\mathbb{P}(X_0 = i, X_n = j) \\
&= \sum_{i_1, \ldots, i_{n-1}} \mathbb{P}(X_0 = i, X_1 = i_1, \ldots, X_{n-1} = i_{n-1}, X_n = j) \\
&= \sum_{i_1, \ldots, i_{n-1}} \lambda_i p_{i i_1} \cdots p_{i_{n-1} j} = \lambda_i (P^n)_{ij},
\end{aligned}
$$

and

$$\mathbb{P}(X_n = j | X_0 = i) = \frac{\mathbb{P}(X_0 = i, X_n = j)}{\mathbb{P}(X_0 = i)} = \frac{\lambda_i (P^n)_{ij}}{\lambda_i} = (P^n)_{ij}. \tag{1.7}$$

That is, the entry $(P^n)_{ij}$ of matrix P^n gives the n-step transition probability from state i to j. We also denote it sometimes by $p_{ij}^{(n)}$.

More generally,

$$\mathbb{P}(X_k = i, X_{n+k} = j) = (\lambda P^k)_i (P^n)_{ij}$$

and

$$\mathbb{P}(X_{k+n} = j | X_k = i) = \frac{\mathbb{P}(X_k = i, X_{k+n} = j)}{\mathbb{P}(X_k = i)} = \frac{(\lambda P^k)_i (P^n)_{ij}}{(\lambda P^k)_i} = (P^n)_{ij}. \qquad (1.8)$$

A corollary of this observation is that the power P^n of a stochastic matrix is again stochastic, viz. $\sum_{j \in I} p_{ij}^{(n)} = 1$ for all $i \in I$. Of course, this fact can be verified directly:

$$\sum_{j \in I} p_{ij}^{(n)} = \sum_{i_1, \ldots, i_{n-1}, j} p_{i i_1} \cdots p_{i_{n-1} j} = \sum_{i_1} p_{i i_1} \cdots \sum_j p_{i_{n-1} j} = 1$$

as at each step (beginning with $\sum_j$) we get the sum 1, owing to (1.2).

Another consequence is that if we apply to a stochastic vector a stochastic matrix (P or more generally P^n), we obtain another stochastic vector. Again, direct calculation confirms this:

$$\sum_j (\lambda P^n)_j = \sum_{i,j} \lambda_i (P^n)_{ij} = \sum_i \lambda_i \sum_j (P^n)_{ij} = \sum_i \lambda_i = 1.$$

An ultimate generalisation of (1.3) is the formula

$$\mathbb{P}(X_{k_1} = i_1, X_{k_2} = i_2, \ldots, X_{k_n} = i_n)$$
$$= (\lambda P^{k_1})_{i_1} (P^{k_2 - k_1})_{i_1 i_2} \cdots (P^{k_n - k_{n-1}})_{i_{n-1} i_n} \qquad (1.9)$$

valid for all times $0 \leq k_1 < k_2 < \cdots < k_n$ and states $i_1, \ldots, i_n \in I$.

It is now time to summarise our findings. Suppose that $\lambda = (\lambda_i)$ is a stochastic vector and $P = (p_{ij})$ a transition matrix on I. The random state X_n at time n is considered as a random variable with values in I.

Definition 1.1.3 A sequence of random variables X_n with values in a finite or countable set I is a *discrete-time Markov chain* (DTMC), or a *Markov chain* for short, with the initial distribution λ and transition matrix P if, for all $i_0, \ldots, i_n \in I$, the joint probability $\mathbb{P}(X_0 = i_0, \ldots, X_n = i_n)$ is given by formula (1.3). In this case we also say that (X_n) is Markov (λ, P) or call it a (λ, P) Markov chain.

Theorem 1.1.4 *If (X_n) is Markov (λ, P), then:*

(i) *the conditional probability*

$$\mathbb{P}(X_{n+1} = j | X_0 = i_0, \ldots, X_{n-1} = i_{n-1}, X_n = i)$$

is equal to the conditional probability $\mathbb{P}(X_{n+1} = j | X_n = i)$ and coincides with p_{ij}. In particular, the conditional distribution of X_{n+1} given that $X_0 = i_0, \dots, X_{n-1} = i_{n-1}, X_n = i$ does not depend on $i_0, \dots, i_{n-1}$ and coincides with $(p_{ij}, j \in I)$, i.e. with row i of P;

(ii) the probability $\mathbb{P}(X_n = i)$ that the state at time n is i equals $(\lambda P^n)_i$;

(iii) the entry $p_{ij}^{(n)}$ of matrix P^n corresponds to the conditional probability $\mathbb{P}(X_{k+n} = j | X_k = i)$, i.e. gives the n-step transition probability from i to j;

(iv) the general probability

$$\mathbb{P}(X_{k_1} = i_1, X_{k_2} = i_2, \dots, X_{k_n} = i_n)$$

is given by (1.9).

Example 1.1.5 Suppose that all rows of P are the same, i.e. $p_{ij} = p_j$ does not depend on i. In addition, suppose that $\lambda_j = p_j$, i.e. λ coincides with the row of P. Then, by (1.3)

$$\mathbb{P}(X_0 = i_0, X_1 = i_1, \dots, X_n = i_n) = p_{i_0} p_{i_1} \cdots p_{i_n}.$$

Also, in this example $P^n = P$, as

$$p_{ij}^{(n)} = \sum_{i_1, \dots, i_{n-1}} p_{i_1} \cdots p_{i_{n-1}} p_j = \sum_{i_1} p_{i_1} \sum_{i_2} p_{i_2} \cdots \sum_{i_{n-1}} p_{i_{n-1}} p_j = p_j,$$

owing to the fact that $\sum_{l \in I} p_l = 1$. Hence,

$$\mathbb{P}(X_n = j) = (\lambda P^n)_j = \sum_{i \in I} p_i p_{ij}^{(n)} = \sum_{i \in I} p_i p_j = p_j.$$

We see that

$$\mathbb{P}(X_0 = i_0, X_1 = i_1, \dots, X_n = i_n) = \mathbb{P}(X_0 = i_0) \mathbb{P}(X_1 = i_1) \cdots \mathbb{P}(X_n = i_n).$$

That is (X_n) is a sequence of independent, identically distributed (IID) random variables.

Example 1.1.6 If P is diagonal then it must coincide with the identity matrix $\mathbf{I}$ where row i is given by the stochastic vector δ_i:

$$\begin{pmatrix} 1 & 0 & 0 & \cdots & 0 \\ 0 & 1 & 0 & \cdots & 0 \\ 0 & 0 & 1 & \cdots & 0 \\ 0 & 0 & 0 & \cdots & 1 \end{pmatrix}.$$

In this case, every power P^n again equals the identity matrix (this property is called idempotency; correspondingly, such a matrix P is called idempotent). Hence, by (1.5), $\mathbb{P}(X_n = i) = \lambda_i$. That is, the distribution of X_n is the same as X_0. In other words, the initial distribution is preserved in time.

Example 1.1.7 For a two-state DTMC, $P = \begin{pmatrix} 1-\alpha & \alpha \\ \beta & 1-\beta \end{pmatrix}$, the entries of P^n can be found by a straightforward calculation. In fact, $P^n = P^{n-1}P$, which for entry $p_{00}^{(n)}$ yields

$$
\begin{aligned}
p_{00}^{(n)} &= p_{00}^{(n-1)}(1-\alpha) + p_{01}^{(n-1)}\beta \\
&= p_{00}^{(n-1)}(1-\alpha) + \left(1 - p_{00}^{(n-1)}\right)\beta = \beta + (1-\alpha-\beta)p_{00}^{(n-1)}.
\end{aligned}
$$

This is a recursion in n, with $p_{00}^{(0)} = 1$ and $p_{00}^{(1)} = 1-\alpha$. Hence,

$$
p_{00}^{(n)} = A + B(1-\alpha-\beta)^n,
$$

with

$$
A + B = 1, \; A + B(1-\alpha-\beta) = 1-\alpha,
$$

and, clearly,

$$
p_{00}^{(n)} = \begin{cases} \dfrac{\beta}{\alpha+\beta} + \dfrac{\alpha}{\alpha+\beta}(1-\alpha-\beta)^n, & \text{if } \alpha+\beta > 0, \\ 1, & \text{if } \alpha = \beta = 0. \end{cases}
$$

Entry $p_{11}^{(n)}$ is obtained by swapping α and β, and entries $p_{01}^{(n)}$ and $p_{10}^{(n)}$ as complements to 1.

Example 1.1.8 In the general case, we can use the eigenvalues and eigenvectors of P to find elements of P^n. Consider a 3×3 example

$$
P = \begin{pmatrix} 0 & 1 & 0 \\ 0 & 2/3 & 1/3 \\ 1/3 & 0 & 2/3 \end{pmatrix}.
$$

The eigenvalues are solutions to the characteristic equation:

$$
\begin{aligned}
\det \begin{pmatrix} -\mu & 1 & 0 \\ 0 & 2/3-\mu & 1/3 \\ 1/3 & 0 & 2/3-\mu \end{pmatrix} &= -\mu^3 + \frac{4}{3}\mu^2 - \frac{4}{9}\mu + \frac{1}{9} \\
&= -(\mu-1)\left(\mu^2 - \frac{1}{3}\mu + \frac{1}{9}\right) = 0,
\end{aligned}
$$

whence

$$\mu_0 = 1, \quad \mu_\pm = \frac{1 \pm i\sqrt{3}}{6}.$$

As the eigenvalues are distinct, matrix P is diagonalisable: there exists an invertible matrix D such that

$$D^{-1}PD = \begin{pmatrix} 1 & 0 & 0 \\ 0 & (1+i\sqrt{3})/6 & 0 \\ 0 & 0 & (1-i\sqrt{3})/6 \end{pmatrix},$$

i.e.

$$P = D \begin{pmatrix} 1 & 0 & 0 \\ 0 & (1+i\sqrt{3})/6 & 0 \\ 0 & 0 & (1-i\sqrt{3})/6 \end{pmatrix} D^{-1}.$$

Then

$$P^n = D \begin{pmatrix} 1 & 0 & 0 \\ 0 & [(1+i\sqrt{3})/6]^n & 0 \\ 0 & 0 & [(1-i\sqrt{3})/6]^n \end{pmatrix} D^{-1},$$

and each entry of P^n is a sum of the form

$$A + B \left(\frac{1+i\sqrt{3}}{6} \right)^n + C \left(\frac{1-i\sqrt{3}}{6} \right)^n.$$

The coefficients A, B and C may be complex; they vary from entry to entry and are found from the initial values $n = 0, 1, 2$. For $n = 0$, P^0 is the identity matrix (just as in the scalar case $p^0 = 1$ for any p ($p = 0$ included!)); for $n = 1$, we use the matrix P and for $n = 2$ we have to square it, to obtain P^2. For instance, suppose that the states are 1, 2 and 3; then the entries are $p_{ij}^{(n)}$, $i, j = 1, 2, 3$. Then, for $p_{12}^{(n)}$:

$$p_{12}^{(0)} = A + B + C = 0, \quad p_{12}^{(1)} = A + B\frac{1+i\sqrt{3}}{6} + C\frac{1-i\sqrt{3}}{6} = 1,$$

and

$$p_{12}^{(2)} = A + B \left(\frac{1+i\sqrt{3}}{6} \right)^2 + C \left(\frac{1-i\sqrt{3}}{6} \right)^2 = \frac{2}{3}.$$

The calculations may be simplified if we get rid of imaginary parts (as all entries $p_{ij}^{(n)}$ of P^n are real non-negative). To this end, observe that $\mu_\pm$ are complex conjugate roots and write

$$\frac{1 \pm i\sqrt{3}}{6} = \frac{1}{3} \left(\frac{1 \pm i\sqrt{3}}{2} \right) = \frac{1}{3} e^{\pm i\pi/3} = \frac{1}{3} \left(\cos\frac{\pi}{3} \pm i\sin\frac{\pi}{3} \right).$$

Then

$$\left(\frac{1\pm i\sqrt{3}}{6}\right)^n = \left(\frac{1}{3}\right)^n e^{\pm in\pi/3} = \left(\frac{1}{3}\right)^n \left(\cos\frac{\pi n}{3} \pm i\sin\frac{\pi n}{3}\right),$$

and

$$p_{ij}^{(n)} = \alpha + \left(\frac{1}{3}\right)^n \left(\beta\cos\frac{\pi n}{3} + \gamma\sin\frac{\pi n}{3}\right),$$

where $\alpha = A$, $\beta = (B+C)$ and $\gamma = i(B-C)$ must be real. Again, we have the equations for $n = 0,1,2$; for $p_{12}^{(n)}$ they are

$$\alpha + \beta = 0, \quad \alpha + \frac{1}{3}\left(\frac{1}{2}\beta + \frac{\sqrt{3}}{2}\gamma\right) = 1, \quad \alpha + \frac{1}{9}\left(-\frac{1}{2}\beta + \frac{\sqrt{3}}{2}\gamma\right) = \frac{2}{3},$$

whence

$$\alpha = \frac{3}{7}, \quad \beta = -\frac{3}{7}, \quad \gamma = \frac{9}{7}\sqrt{3}.$$

In particular, $\lim\limits_{n\to\infty} p_{12}^{(n)} = 3/7$.

Example 1.1.9 Consider another 3×3 matrix

$$P = \begin{pmatrix} 1/3 & 0 & 2/3 \\ 1/3 & 2/3 & 0 \\ 1/3 & 1/3 & 1/3 \end{pmatrix}.$$

Here the characteristic equation is:

$$-\mu^3 + \frac{4}{3}\mu^2 - \frac{1}{3}\mu = -(\mu - 1)\left(\mu - \frac{1}{3}\right)\mu = 0,$$

with the eigenvalues

$$\mu_0 = 1, \quad \mu_1 = \frac{1}{3}, \quad \mu_2 = 0.$$

Hence, the entries $p_{ij}^{(n)}$ have a simple form

$$p_{ij}^{(n)} = A + B\left(\frac{1}{3}\right)^n + C\cdot 0^n.$$

Again we use three initial conditions, with P^0, P and P^2. For instance, for $p_{11}^{(n)}$:

$$A + B + C = 1, \quad A + \frac{1}{3}B = \frac{1}{3}, \quad A + \left(\frac{1}{3}\right)^2 B = \frac{1}{3},$$

whence $A = 1/3$, $B = 0$, $C = 2/3$ and $p_{11}^{(n)} \equiv 1/3$. Similarly, $p_{12}^{(n)} = 1/3 - (1/3)^n$ and $p_{13}^{(n)} = 1/3 + (1/3)^n$. As $n \to \infty$, all entries of the first row of P^n approach $1/3$ (in fact, the same is true for all 9 entries in P^n).

Example 1.1.10 We can make a number of observations. First, 1 is always an eigenvalue of any stochastic matrix P. In fact: (i) the eigenvalue equation reads $\det(\mu I - P) = \det(\mu I - P)^T = \det(\mu I - P^T) = 0$, i.e. the eigenvalues of P and its transpose P^T coincide; (ii) 1 is always an eigenvalue of P^T: the corresponding eigenvector is the row $\mathbf{1} = (1, \ldots, 1)$ of 1s. Formally, $\mathbf{1}P^T = \mathbf{1}$, or equivalently, $P\mathbf{1}^T = \mathbf{1}^T$ for the column $\mathbf{1}^T$. To check the last equation, observe that every entry of the column $P\mathbf{1}^T$ is 1

$$\left(P\mathbf{1}^T\right)_i = \sum_{j \in I} p_{ij} = 1,$$

because P is stochastic.

Therefore, the characteristic polynomial of a stochastic matrix is divisible by $(\mu - 1)$; in the 3×3 case this leads to a quadratic quotient polynomial, and all eigenvalues can be found.

Second, if there is a complex eigenvalue μ_+ of P then the complex conjugate $\mu_- = \overline{\mu}_+$ is also an eigenvalue, as this is the only way of producing a real characteristic polynomial from the product of linear monomials, $(\mu - \mu_+)(\mu - \mu_-) = \mu^2 - (\mu_+ + \mu_-)\mu + \mu_+\mu_-$, with real coefficients $\mu_+ + \mu_-$ and $\mu_+\mu_- = |\mu_\pm|^2$. Then, writing

$$\mu_\pm = |\mu_\pm|e^{\pm i\phi} = |\mu_\pm|(\cos\phi \pm i\sin\phi),$$

we can work with real summands only, of the form $\beta\cos(n\phi)$ and $\gamma\sin(n\phi)$.

Third, the coefficient A (in front of 1) in the equation for $p_{ij}^{(n)}$ typically identifies the limit $\lim_{n \to \infty} p_{ij}^{(n)}$. This is because the modulus $|\mu| \leq 1$ for any eigenvalue μ of P, and 'generically' (although not always), any eigenvalue $\mu \neq 1$ has $|\mu| < 1$. This fact is more delicate and will be commented on in subsequent sections. Then in the decomposition

$$p_{ij}^{(n)} = A + \sum_{\text{eigenvalues } \mu_s \neq 1} B_s \mu_s^n$$

all terms except for A are suppressed as $n \to \infty$. (In the case of $P = \begin{pmatrix} 0 & 1 \\ 1 & 0 \end{pmatrix}$ this is not true: the eigenvalues are 1 and -1 and there is no limit $\lim_{n \to \infty} p_{ij}^{(n)}$ as P^n oscillates between I for n even and P for n odd.)

It has to be said that many (even very simple) examples may lead to rather cumbersome computations for entries $p_{ij}^{(n)}$. For example, the matrix

$$P = \begin{pmatrix} 1/3 & 1/3 & 1/3 \\ 0 & 1/2 & 1/2 \\ 1/3 & 2/3 & 0 \end{pmatrix}$$

has the characteristic equation

$$-\mu^3 + \frac{5}{6}\mu^2 + \frac{5}{18}\mu - \frac{1}{9} = -(\mu - 1)\left(\mu^2 + \frac{1}{6}\mu - \frac{1}{9}\right) = 0,$$

with eigenvalues

$$\mu_0 = 1, \quad \mu_{\pm} = \frac{-1 \pm \sqrt{17}}{12}.$$

This leads to the equation

$$p_{ij}^{(n)} = A + B\left(\frac{-1+\sqrt{17}}{12}\right)^n + C\left(\frac{-1-\sqrt{17}}{12}\right)^n,$$

with

$$A + B + C = \delta_{ij}, \quad A + B\left(\frac{-1+\sqrt{17}}{12}\right) + C\left(\frac{-1+\sqrt{17}}{12}\right) = p_{ij},$$

and

$$A + B\left(\frac{-1+\sqrt{17}}{12}\right)^2 + C\left(\frac{-1-\sqrt{17}}{12}\right)^2 = p_{ij}^{(2)}.$$

For instance, for $p_{21}^{(n)}$ the final expression is

$$\frac{4}{19} + \frac{2}{19}\left(\frac{6}{\sqrt{17}} - 1\right)\left(\frac{-1+\sqrt{17}}{12}\right)^n - \frac{2}{19}\left(\frac{6}{\sqrt{17}} + 1\right)\left(\frac{-1-\sqrt{17}}{12}\right)^n.$$

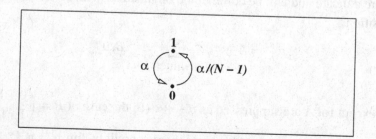

Fig. 1.2

Example 1.1.11 A helpful property is the presence of symmetries in P: it may reduce the number of states in the Markov chain. For example, the $N \times N$ matrix

$$P = \begin{pmatrix} 1-\alpha & \alpha/(N-1) & \cdots & \alpha/(N-1) \\ \alpha/(N-1) & 1-\alpha & \cdots & \alpha/(N-1) \\ \vdots & \vdots & \cdots & \vdots \\ \alpha/(N-1) & \alpha/(N-1) & \cdots & 1-\alpha \end{pmatrix}$$

describes a model of a virus mutation where a virus retains its genotype or changes to one of $(N-1)$ other types with equal probabilities.

To calculate $p_{11}^{(n)}$, we reduce the number of states to two (say, 1 and 0 (another)), by considering original transitions from a state 1 to itself or to another state, and backwards, without further specification (as for our problem all other states are indistinguishable). The reduced two-state chain has the 2×2 transition matrix

$$\begin{pmatrix} 1-\alpha & \alpha \\ \alpha/(N-1) & 1-\alpha/(N-1) \end{pmatrix}.$$

We can apply the formulas of Example 1.1.7 (with $\beta = \alpha/(N-1)$)

$$p_{11}^{(n)} = \frac{\alpha/(N-1)}{\alpha + \alpha/(N-1)} + \frac{\alpha}{\alpha + \alpha/(N-1)} \left(1 - \alpha - \frac{\alpha}{N-1} \right)^n$$

$$= \frac{1}{N} + \frac{N-1}{N} \left(1 - \frac{\alpha N}{N-1} \right)^n.$$

Also, by symmetry,

$$p_{ij}^{(n)} = \frac{1 - p_{11}^{(n)}}{N-1}$$

for $i \neq j$.

We are now in a position to establish the famous *Markov property* of a DTMC. It asserts that the Markov chain begins afresh after any given time n (from its current state).

Theorem 1.1.12 Let (X_n) be Markov (λ, P). Then, for all $m \geq 1$ and $i \in I$, conditional on $X_m = i$, $(X_{m+n}, n \geq 0)$ is Markov (δ_i, P). In particular, conditional on $X_m = i$, the random variables $X_{m+1}, X_{m+2}, \ldots$ are independent of the variables $X_0, \ldots, X_{m-1}$.

In other words, in a DTMC, the past states $(X_0, \ldots, X_{m-1})$ and the future states $(X_{m+1}, X_{m+2}, \ldots)$ are conditionally independent, given the present $(X_m = i)$.

Proof Recall that the stochastic vector δ_i has entries δ_{ij}, $j \in I$. We want to check that for any event A determined by $X_0, \ldots, X_{m-1}$, and B determined by $X_{m+1}, \ldots, X_{m+1+n}$ for some n, (i) the conditional probability $\mathbb{P}(A \cap B | X_m = i)$ decouples:

$$\mathbb{P}(A \cap B | X_m = i) = \mathbb{P}(A | X_m = i)\mathbb{P}(B | X_m = i), \tag{1.10}$$

and (ii) the conditional probability $\mathbb{P}(B | X_m = i)$ is calculated as in the Markov chain (δ_i, P):

$$\mathbb{P}(B | X_m = i) = \sum_{(j_1, \ldots, j_n) \in B} p_{ij_1} \cdots p_{j_{n-1} j_n}. \tag{1.11}$$

First, let A and B be of the form

$$A = \{X_0 = i_0, \ldots, X_{m-1} = i_{m-1}\}, \ B = \{X_{m+1} = j_1, \ldots, X_{m+n} = j_n\}$$

for some sequence of states $i_0, \ldots, i_{m-1}, j_1, \ldots, j_n \in I$. Generally, A and B are disjoint unions of such 'elementary' events.

For A and B as above,

$$\mathbb{P}\big(A \cap B \cap \{X_m = i\}\big)$$
$$= \mathbb{P}\big(X_0 = i_0, \ldots, X_{m-1} = i_{m-1}, X_m = i, X_{m+1} = j_1, \ldots, X_{m+n} = j_n\big)$$
$$= \lambda_{i_0} p_{i_0 i_1} \cdots p_{i_{m-1} i} p_{i j_1} \cdots p_{j_{n-1} j_n}.$$

For a general B we have to sum over $(j_1, \ldots, j_n) \in B$:

$$\lambda_{i_0} p_{i_0 i_1} \cdots p_{i_{m-1} i} \sum_{(j_1, \ldots, j_n) \in B} p_{ij_1} \cdots p_{j_{n-1} j_n}.$$

The sum $\sum_{(j_1, \ldots, j_n) \in B}$ gives the conditional probability $\mathbb{P}(B | X_m = i)$, and it is calculated as in the (δ_i, P) Markov chain.

Next, for a general A we sum over $(i_0, \ldots, i_{m-1}) \in A$:

$$\mathbb{P}\big(A \cap B \cap \{X_m = i\}\big) = \sum_{(i_0, \ldots, i_{m-1}) \in A} \lambda_{i_0} p_{i_0 i_1} \cdots p_{i_{m-1} i} \mathbb{P}(B | X_m = i)$$
$$= \mathbb{P}(A \cap \{X_m = i\})\mathbb{P}(B | X_m = i).$$

Finally, to produce the conditional probability $\mathbb{P}(A \cap B | X_m = i)$, we divide by $\mathbb{P}(X_m = i)$:

$$\mathbb{P}(A \cap B | X_m = i) = \frac{\mathbb{P}(A \cap B \cap \{X_m = i\})}{\mathbb{P}(X_m = i)}$$
$$= \frac{\mathbb{P}(A \cap \{X_m = i\})}{\mathbb{P}(X_m = i)} \mathbb{P}(B | X_m = i)$$
$$= \mathbb{P}(A | X_m = i)\,\mathbb{P}(B | X_m = i),$$

as required.

$\square$

In future we will write $\mathbb{P}_i$ for the conditional probabilities $\mathbb{P}(\,\cdot\,|X_0 = i)$ given that the state at time 0 is i. Similarly, $\mathbb{E}_i$ stands for expectation under distribution $\mathbb{P}_i$.

Worked Example 1.1.13 Three girls A, B and C are playing table tennis. In each game, two of the girls play against each other and the third girl does not play. The winner of any given game n plays again in game $n + 1$. The probability that girl x will beat girl y in any game that they play against each other is $s_x/(s_x + s_y)$ for $x, y \in \{A, B, C\}$, $x \neq y$, where s_A, s_B, s_C represent the playing strengths of the three girls.

   (a) Represent this process as a DTMC by defining the possible states and constructing the transition matrix.

   (b) Determine the probability that the two girls who play each other in the first game will play each other again in the fourth game. Show that this probability does not depend on which two girls play in the first game.

Solution (a) Label states by A, B, C indicating which player is *not* playing in a given game. Then the transition matrix is $\{A, B, C\} \times \{A, B, C\}$:

$$\begin{pmatrix} 0 & s_C/(s_B+s_C) & s_B/(s_B+s_C) \\ s_C/(s_A+s_C) & 0 & s_A/(s_A+s_C) \\ s_B/(s_A+s_B) & s_A/(s_A+s_B) & 0 \end{pmatrix}.$$

The process is a Markov chain because the results of the subsequent games are independent.

   (b) Here, we look for the probability that after three steps the chain returns to a given initial state.

From the symmetry, this probability is the same for any choice of the initial state and is equal to

$$p_{AB}p_{BC}p_{CA} + p_{AC}p_{CB}p_{BA} = \frac{2s_A s_B s_C}{(s_A + s_B)(s_B + s_C)(s_C + s_A)}.$$

                                                                         □

Worked Example 1.1.14 A rock concert held in a hall with N numbered seats attracted a huge crowd of spectators. The lights have been dimmed and $N - 1$ seats have already been taken, and now the last spectator enters the hall. The first $N - 1$ spectators were advised by the ushers, rather imprudently, to take their seats completely at random, but the last spectator is determined to sit in the place indicated on her ticket. If her place is free, she takes it, and the concert is ready to begin. However, if her seat is taken, she loudly insists that the occupier vacates it. In this case the occupier decides to follow the same rule: if the free seat is his, he takes it, otherwise he insists on his place being vacated. The same policy is then adopted by the next unfortunate spectator, and so on. Each move takes 45 seconds. What is the expected duration of the delay caused by these displacements?

Solution (sketch) It is important to keep in mind that, initially, the $N-1$ spectators are distributed so that (a) the probability that seat j is free is $1/N$, $j = 1, \ldots, N$, (b) given that seat j is free, the probability that the first spectator entering the hall takes seat i_1, the second spectator takes seat i_2, ..., the $(N-1)$st takes seat i_{N-1}, equals $1/(N-1)!$, for any sequence $i_1, \ldots, i_{N-1}$ covering the set $\{1, \ldots, N\} \setminus \{j\}$. Consider a DTMC with states $N, N-1, \ldots, 0$. Here, state 0 means that the spectator attempting the free seat 'succeeds' (i.e. the available seat is indeed her correct place), state $1 \leq n \leq N-1$ means that the $(N-n)$th move is 'unsuccessful', and N is the initial state. Then the probability of transition $n \to 0$ is equal to $1/n$ and the probability of transition from $n \to (n-1)$ is $(n-1)/n$; the probability of transition $0 \to 0$ equals 1. Let $E(n)$ denote the expected number of transitions (displacements) until the Markov chain enters state 0 from state n; we are interested in the quantity $E(N)$. A useful remark is that $E(n)$ is the expected number of displacements for the hall with n seats.

The key fact is the following recursion

$$
\begin{aligned}
E(n) &= \frac{n-1}{n} \times [1 + E(n-1)] + \frac{1}{n} \times 0 \\
&= \frac{n-1}{n} [1 + E(n-1)], \quad n = 1, 2, \ldots, N
\end{aligned}
$$

with

$$E(1) = 0.$$

The solution is

$$E(N) = \frac{1}{N}(N - 1 + N - 2 + \cdots + 1) = \frac{N-1}{2}.$$

If $N = 121$, the expected delay will be $45 \times \dfrac{120}{2}$ secs $= 45$ min. $\square$

Worked Example 1.1.15 The point about this problem is that it is often useful to introduce probability where originally it was not present. Assume that the circle of unit perimeter $\mathscr{C}_1 = \left\{ z : |z| = \dfrac{1}{2\pi} \right\}$ has been partitioned into two disjoint (measurable) sets, one, called red, of length $2/3$ and the other, called blue, of length $1/3$. Prove that it is always possible to inscribe in the circle a square such that at least three of its four vertices have the red colour.

Solution Given such a partition, consider a random inscribed square where we choose an anchor point on $\mathscr{C}_1$ uniformly; this determines the square uniquely. Let us number the vertices 1, 2, 3, 4, say clock-wise, beginning with the anchor. Set

$$X_i = \mathbf{1}(\text{vertex } i \text{ falls in the red set}), \quad i = 1, 2, 3, 4.$$

Then

$$\mathbb{E}(\text{the number of red vertices}) = \mathbb{E}X_1 + \mathbb{E}X_2 + \mathbb{E}X_3 + \mathbb{E}X_4 = 4\mathbb{E}X_1 = 8/3 > 2.$$

Hence, the sum $X_1 + X_2 + X_3 + X_4$ must take values 3 or 4 with positive probability, and therefore the inscription in question is always possible.

We can actually assess the probability P that at least three vertices will be red. In fact, the following bound holds: $P \geq P_0$ where $P_0 = \frac{1}{3}$ is found from the equation

$$(1 - P_0)2 + 4P_0 = \frac{8}{3}$$

(which corresponds to the situation where with probability P_0 we have four red vertices and with the complementary probability $1 - P_0$ just two). □

Concluding this section, we would like to note that Definition 1.1.3 above introduces a class of so-called *homogeneous*, or *time-homogeneous* Markov chains. We omit the term 'homogeneous' except for a few cases when we consider 'inhomogeneous' chains (which will occur with continuous time Markov chains; see Section 2.4). We only mention that in an inhomogeneous Markov chain, the transition probability from state i to j depends on the time of transition. Consequently, instead of a single transition matrix P, we have to introduce a family of matrices P_n where $n = 0, 1, 2, \ldots$, describing probabilities of transition from state i at time n to state j at time $n + 1$.

Dial M For Markov
(From the series *'Movies that never made it to the Big Screen'.*)

1.2 Class division

Communicating class struggle
(From the series *'When they go political'.*)

Class division is a natural partition of the state space I, generated by the transition matrix P. We still work with finite state spaces and finite matrices, unless otherwise stated.

Definition 1.2.1 States i and j belong to the same *communicating class* if $p_{ij}^{(n)} > 0$ and $p_{ji}^{(n')} > 0$ for some $n, n' \geq 0$. (Recall, for $n = 0$, P^0 is the identity matrix $\mathbf{I}$. Therefore, the diagonal entry $p_{ii}^{(0)}$ in matrix P^0 is always equal to 1.) The fact that $p_{ij}^{(n)} > 0$ and $p_{ji}^{(n')} > 0$ for some $n, n' \geq 0$ is denoted by $i \leftrightarrow j$. When one of these conditions holds, we write $i \to j$ or $j \to i$, and if we want to stress that some of them do not, we write $i \nrightarrow j$ or $j \nrightarrow i$.

To check that we have a correctly defined partition, observe that: (i) each state communicates with itself, i.e. $i \leftrightarrow i$, as $p_{ii}^{(0)} = 1$ (communication is a reflexive relation); (ii) the relation $i \leftrightarrow j$ is symmetric, i.e. holds or does not hold regardless of the order within the pair i, j (this is obvious from the definition); (iii) if $i \leftrightarrow j$ and $j \leftrightarrow k$ then $i \leftrightarrow k$ (communication is a transitive relation). Indeed, as $p_{ik}^{(n+n')} = \sum_{l \in I} p_{il}^{(n)} p_{lk}^{(n')} \geq p_{ij}^{(n)} p_{jk}^{(n')}$, and similarly for $p_{ki}^{(n+n')}$. Then, because of (i), each state i belongs to some class, because of (ii) each class is correctly defined as an (unordered) subset of I, and because of (iii) any state j falls in no more than one class. States from different classes, of course, do not communicate.

A useful fact is that $i \to j$ if and only if there exists a sequence of states

$$i_0 = i, i_1, \ldots, i_{n-1}, i_n = j$$

such that $p_{i_l i_{l+1}} > 0$ for each pair (i_l, i_{l+1}), $0 \leq l < n$. In fact,

$$p_{ij}^{(n)} = \sum_{i_1, \ldots i_{n-1}} p_{i i_1} \cdots p_{i_{n-1} j},$$

and the whole sum is > 0 if and only if there exists at least one non-zero summand.

Definition 1.2.2 A communicating class C is called *closed* if for all $i \in C$ and $j \in I$ such that $i \to j$, state $j \in C$. In other words, a state cannot escape from a closed communicating class. Otherwise, i.e. when a state (and indeed, all states) can escape from a class, it is called *non-closed or open*. States forming non-closed communicating classes are often called *non-essential*: they indeed are not essential in the long run. A state i is called *absorbing* if $p_{ii} = 1$. Equivalently, the communicating class of an absorbing state i consists solely of i (and is closed). An open class consisting of a single state j occurs when this state can be visited only once, after which the chain never returns. Some authors restrict their attention exclusively to closed communicating classes and do not consider other types as classes.

> *What I did that was new was to prove . . .*
> *that the class struggle necessarily leads*
> *to the dictatorship of the proletariat.*
> K. Marx (1818–1883), German philosopher

Example 1.2.3 A particle moves from state $i = 1, \ldots, N-1$ to state $i+1$ with probability p and state $i-1$ with probability $1-p$ where $0 < p < 1$.

From states 0 and N it cannot move, i.e. once it reaches one of them, it stays there forever. This example describes a match between two players where the winner of a given game gets from the loser one 'score unit' and the match continues until one of the players is left with no units. Another interpretation is a walk of a drunken person from a pub, where he makes a step towards home (state 0) or a lake (state N). In the first case, the value N is the total number of units of both players before the match; it is obviously preserved in the course of the game. In the second case, it is the distance in steps between the home and the lake (the pub is somewhere in between). The state $i = 0, 1, \ldots, N$ is the number of units in possession of player 1 or the distance from home; p and $1-p$ are the probabilities that player 1 wins or loses a game, or that the drunkard makes a step towards home or the lake. Results of different games or directions of different moves are independent.

Here, the transition matrix is $(N+1) \times (N+1)$:

$$P = \begin{pmatrix} 1 & 0 & 0 & 0 & \cdots & 0 & 0 & 0 \\ 1-p & 0 & p & 0 & \cdots & 0 & 0 & 0 \\ 0 & 1-p & 0 & p & \cdots & 0 & 0 & 0 \\ \vdots & \vdots & \vdots & \vdots & \cdots & \vdots & \vdots & \vdots \\ 0 & 0 & 0 & 0 & \cdots & 0 & p & 0 \\ 0 & 0 & 0 & 0 & \cdots & 1-p & 0 & p \\ 0 & 0 & 0 & 0 & \cdots & 0 & 0 & 1 \end{pmatrix}.$$

Communicating classes are $\{0\}$, $\{1, \ldots, N-1\}$ and $\{N\}$, and classes $\{0\}$ and $\{N\}$ are closed (i.e. states 0 and N are absorbing). Thus, states $1, \ldots, N-1$ are non-essential, and the game will ultimately end at one of the border states.

Example 1.2.4 Consider a 6×6 transition matrix, on states $\{1,2,3,4,5,6\}$, of the form

$$P = \begin{pmatrix} * & 0 & 0 & 0 & * & 0 \\ 0 & 0 & * & 0 & 0 & 0 \\ 0 & 0 & * & 0 & 0 & * \\ 0 & * & 0 & 0 & * & 0 \\ * & 0 & 0 & 0 & 0 & * \\ 0 & * & 0 & 0 & 0 & 0 \end{pmatrix}$$

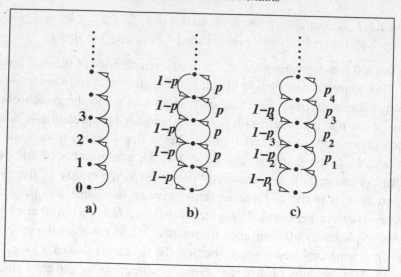

Fig. 1.3

where $*$ stands for a non-zero entry. The communicating classes are $C_1 = \{1,5\}$, $C_2 = \{2,3,6\}$ and $C_3 = \{4\}$, of which only C_2 is closed. If we start in class C_2, we remain in C_2 forever. If we start in C_3 (i.e. at state 4), we will enter C_2 (and then stay in C_2 forever) or C_1. Intuitively, after spending some time in C_1, we must leave it, i.e. enter C_2. This is what happens in reality, as we will soon discover.

A simple but useful fact is

Theorem 1.2.5 *A Markov chain with a finite state space always has at least one closed communicating class.*

To prove Theorem 1.2.5, consider any class, say C_1. If it is not closed, take the next class you can reach from C_1. If it is not closed, you continue. You should end up this process with reaching a closed class.

Remark 1.2.6 The situation with a *countable*, or *denumerable*, Markov chains, where the state space I is countably infinite, is more complicated. Here, you may have no closed class. In addition, states from infinite closed classes may also be non-essential, in the sense that the chain may visit each of them only finitely many times before being driven 'to infinity' (although still within the same closed class).

The simplest examples of a DTMC with countably many states are those where the space I is the set of non-negative integers $\mathbb{Z}_+ = \{0, 1, 2, \cdots\}$. Three examples are shown in Figure 1.3; the corresponding transition matrices are:

$$\text{(a)} \begin{pmatrix} 0 & 1 & 0 & 0 & \cdots & 0 & \cdots \\ 0 & 0 & 1 & 0 & \cdots & 0 & \cdots \\ 0 & 0 & 0 & 1 & \cdots & 0 & \cdots \\ \vdots & \vdots & \vdots & \vdots & \cdots & \vdots & \cdots \end{pmatrix}, \quad \text{(b)} \begin{pmatrix} 0 & 1 & 0 & 0 & \cdots & 0 & \cdots \\ 1-p & 0 & p & 0 & \cdots & 0 & \cdots \\ 0 & 1-p & 0 & p & \cdots & 0 & \cdots \\ \vdots & \vdots & \vdots & \vdots & \cdots & \vdots & \cdots \end{pmatrix}$$

and

$$\text{(c)} \begin{pmatrix} 0 & 1 & 0 & 0 & \cdots & 0 & \cdots \\ 1-p_1 & 0 & p_1 & 0 & \cdots & 0 & \cdots \\ 0 & 1-p_2 & 0 & p_2 & \cdots & 0 & \cdots \\ \vdots & \vdots & \vdots & \vdots & \cdots & \vdots & \cdots \end{pmatrix}.$$

These models describe so-called *birth-and-death processes*, or *birth-death processes*, where state i represents the size of the population, and during a transition a member of the population may die or a new member may be born. In case (a) only births are allowed, and the chain is deterministic. Here, every state i forms a non-closed class and is non-essential. In model (b) a 'death' occurs with the same chance $1 - p$ and a birth with the same chance p, regardless of the size i of the population at the given time (unless $i = 0$ of course). In real life, it may be a queue of 'tasks' served by a 'server' (e.g. clients waiting in a barber shop with a single seat, or computer programs subsequently executed by a processor). Then i is the number of tasks in the queue. If, before the hairdresser finishes with a current client, a new client comes, we have a jump $i \rightarrow i+1$; otherwise a jump $i \rightarrow i-1$ occurs. From 0 we can only jump to 1 (although in the 'real time' the hairdresser may be waiting for a while for this to happen). There are two situations: $p \geq 1/2$ and $0 < p < 1/2$. Intuitively, if $p \geq 1/2$, tasks will arrive at least as often as they are served, and the queue will become eventually infinite (which may rather please our hairdresser). In this situation, as we shall see, each state i will be visited finitely many times and X_n (the size of the queue at time n) will grow indefinitely with n. If $0 < p < 1/2$, the tasks will arrive less often, and the system will be able to reach an 'equilibrium', with some stationary distribution of the queue size.

An often used modification of model (b) is where 0 is made an absorbing state, with $p_{00} = 1$.

In the more general case (c), the rules of the population dynamics may include chances for every member to die (but only one at a time); e.g. $p_n = \lambda / (\lambda + n\mu)$, $1 - p_n = n\mu / (\lambda + n\mu)$ where $\lambda > 0$ and $\mu > 0$ are 'immigration' and death 'rates'; the whole picture becomes more complicated. We will be able to analyse some of these models in detail in Sections 1.5–1.7.

Discrete-time Markov chains

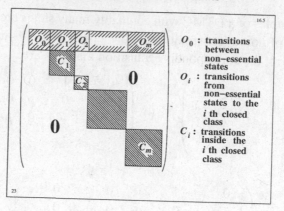

O_0 : transitions between non–essential states

O_i : transitions from non–essential states to the i th closed class

C_i : transitions inside the i th closed class

Fig. 1.4

Let us now go back to finite-state Markov chains In general, after a re-numeration of states, the finite transition matrix P acquires a particular structure, see Figure 1.4. Traditionally, the top left corner is occupied by a square block O_0 formed by probabilities of possible transitions between non-essential states (i.e. between and inside non-closed communicating classes). This block can be zero, if such transitions are not present. Next, square blocks $C_1, \ldots, C_m$ are centred on the main diagonal. They represent (and denote) closed communicating classes of various size; these blocks form stochastic submatrices. The latter means that for all $i = 1, \ldots, m$ and for all states $j \in C_i$, the sum $\sum_{k \in C_i} p_{jk}$ of the entries inside class C_i along row j equals 1. The Markov chains corresponding to individual blocks $C_1, \ldots, C_m$ may be studied separately from each other (which is easier, owing to their lesser size). Blocks $O_1, \ldots, O_m$, to the right of O_0 show transitions from non-essential states to closed communicating classes. These blocks are non-negative rectangular submatrices; some of blocks $O_1, \ldots, O_m$ (but not all) may be zero. (We should not forget that summing the entries along a row of P always gives 1.) If the chain does not have non-essential states, then blocks $O_0, O_1, \ldots, O_m$ are simply absent. The space outside blocks $O_0, O_1, \ldots, O_m$ and $C_1, \ldots, C_m$ is filled with zeros. (There may also be plenty of zeros inside these blocks; see below.)

We call a finite DTMC (X_n) (or equivalently, its transition matrix P) *irreducible* if it has a single communicating class C (which is then automatically closed). In other words, a finite transition matrix is irreducible if any pair of states $i, j \in I$ communicate; equivalently, the whole state space is one (closed) communicating class: $I = C$. Pictorially, the matrix P is reduced in this case to a single block C; see Figure 1.5a). A characteristic feature here is that, for any pair of states $i, j \in I$, the entry $p_{ij}^{(n)}$ of matrix P^n (i.e. the transition probability from i to j in

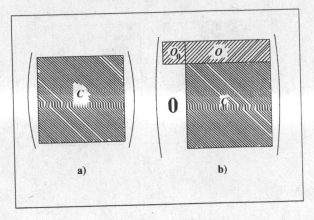

Fig. 1.5

n steps) is strictly greater than 0 for some $n \geq 1$ (depending, in general, on i and j).

Some authors allow a more complicated situation, and call a finite DTMC irreducible if it has a unique closed communicating class and a number of non-closed classes. (The reason is that in this case a finite chain has a unique invariant, or equilibrium distribution; see Section 1.7). The corresponding matrix is shown in Figure 1.5b): it has a single square block C forming a stochastic submatrix plus a top left square block O_0 and a single rectangular block O_1 (or simply O). When you iterate such a matrix P, raising it to a power n, block O_0 will tend to 0 as $n \to \infty$. The behaviour of block O is more complicated: we will analyse it later. As to block C, it will be simply raised to power n (which is handy).

This gives us an idea of what happens when we iterate a general finite transition matrix P with several closed communicating classes. Again, block O_0 in the matrix P^n will tend to 0 as $n \to \infty$. And as before, blocks $C_1, \ldots, C_m$ in P^n will simply be raised to the power n. The last remark illustrates the view that if we have a *reducible* chain, with more than one closed class, we may study separately its 'restrictions' to various closed classes.

For simplicity, let us now assume that a finite matrix P is irreducible. It is not hard to guess that inside block C we may have a 'periodic' picture, with a number v of smaller square 'cells' of equal size which are cyclically permuted by P: cell 1 is taken to cell 2, and so forth, cell v to cell 1. See Figure 1.6. The space outside cells is again filled with zeros.

Such a picture corresponds to a partition of the space I (which under our assumption forms a single (and closed) communicating class) into *periodic subclasses* $W_1, \ldots, W_v$ such that a one-step transition is possible only from a state $j \in W_i$ to a

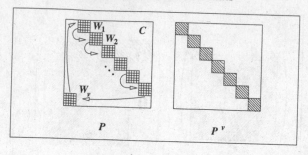

Fig. 1.6

state $k \in W_{i+1}$, $i = 1, \ldots, v$. (Here the sum $i+1$ is understood modulo v, so that $W_{v+1} = W_1$, v being called the *period* of class C.)

Worked Example 1.2.7 Given a state j, define the *period* $v(j)$ of this state as the greatest common divisor of numbers n such that $p_{jj}^{(n)} > 0$. Prove that if states i and j are from the same communicating class then $v(i) = v(j)$. (This justifies the term the 'period of a communicating class'.)

Solution Let i and j be two distinct communicating states. Then $p_{ij}^{(k)} > 0$ for some $k \geq 1$ and $p_{ji}^{(l)} > 0$ for some $l \geq 1$. Assume that $p_{jj}^{(n)} > 0$, then $p_{ii}^{(n+k+l)} \geq p_{ij}^{(k)} p_{jj}^{(n)} p_{ji}^{(l)} > 0$. Therefore, $v(i)$ divides $n + k + l$. Next, $v(i)$ divides $2n + k + l$ as $p_{ii}^{(2n+k+l)} \geq p_{ij}^{(k)} \left(p_{jj}^{(n)}\right)^2 p_{ji}^{(l)} > 0$. Thus, $v(i)$ divides the difference $(2n+k+l) - (n+k+l) = n$. This is true for all n with $p_{jj}^{(n)} > 0$. Then $v(i)$ must divide $v(j)$, as $v(j)$ is the greatest common divisor. A similar argument leads to the conclusion that $v(j)$ divides $v(i)$. Therefore, $v(i) = v(j)$. $\qquad\square$

In the large majority of our examples the period of a closed communicating class equals 1. Such a class (or, equivalently, its transition matrix) is called *aperiodic*. When all communicating classes are aperiodic, the whole Markov chain (or its transition matrix) is called aperiodic.

In general, if you raise the transition matrix P corresponding to a closed communicating class C of period v to the power v, then matrix P^v will decompose into stochastic square submatrices centred on the main diagonal. Pictorially speaking, periodic subclasses $W_1, \ldots, W_v$ will play a rôle of closed communicating classes for matrix P^v. (It has to be said that, formally, the last statement is not correct: some of the W_is may comprise several disjoint closed communicating classes for P^v.)

We see that the whole structure of a finite transition matrix P can be rather intricate. Luckily, most applications and interesting examples do not need an excessive level of generality, and we will be able to make simplifying assumptions.

Worked Example 1.2.8 Consider a stochastic 7×7 matrix

$$P = \begin{pmatrix} 0 & 1/2 & 0 & 0 & 0 & 0 & 1/2 \\ 1/3 & 0 & 0 & 0 & 0 & 1/3 & 1/3 \\ 0 & 1/2 & 0 & 1/2 & 0 & 0 & 0 \\ 0 & 0 & 0 & 0 & 1 & 0 & 0 \\ 0 & 0 & 0 & 1 & 0 & 0 & 0 \\ 0 & 1/2 & 0 & 0 & 0 & 0 & 1/2 \\ 1/3 & 1/3 & 0 & 0 & 0 & 1/3 & 0 \end{pmatrix}.$$

Find all communicating classes of the associated DTMC.

Solution See Figure 1.7. The communicating classes are $\{1,2,6,7\}$, $\{3\}$ and $\{4,5\}$. The closed classes are $\{1,2,6,7\}$ and $\{4,5\}$; 3 is a non-essential state. Class $\{4,5\}$ has a periodic structure. Thus, the limit $p_{ij}^{(n)}$ does not exist (because of oscillations) for $i = 3,4,5$ and $j = 4,5$. For class $\{1,2,6,7\}$ we have a transition submatrix

$$\begin{pmatrix} 0 & 1/2 & 0 & 1/2 \\ 1/3 & 0 & 1/3 & 1/3 \\ 0 & 1/2 & 0 & 1/2 \\ 1/3 & 1/3 & 1/3 & 0 \end{pmatrix},$$

with the symmetry $1 \leftrightarrow 6$ and $2 \leftrightarrow 7$. That is, if we merge 1 with 6 into state I and 2 with 7 into II, we get a two-state chain with the matrix

$$\Pi = \begin{pmatrix} 0 & 1 \\ 2/3 & 1/3 \end{pmatrix}.$$

The last matrix has the characteristic equation

$$\mu^2 - \frac{1}{3}\mu - \frac{2}{3} = 0,$$

with the roots $\mu_0 = 1$, $\mu_1 = -2/3$. Hence, the entries of Π^n have the form $A + B(-2/3)^n$. Adjusting the constants yields

$$\Pi^n = \begin{pmatrix} 2/5 + 3/5\,(-2/3)^n & 3/5 - 3/5\,(-2/3)^n \\ 2/5 - 2/5\,(-2/3)^n & 3/5 + 2/5\,(-2/3)^n \end{pmatrix}.$$

Hence, as $n \to \infty$

$$\Pi^n \to \begin{pmatrix} 2/5 & 3/5 \\ 2/5 & 3/5 \end{pmatrix}.$$

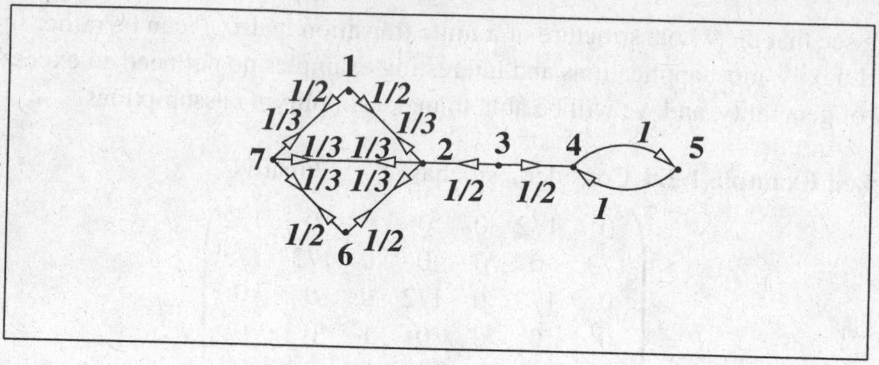

Fig. 1.7

Then, by symmetry, for the original $\{1,2,6,7\}$-block, the limiting matrix is

$$\begin{pmatrix} 1/5 & 3/10 & 1/5 & 3/10 \\ 1/5 & 3/10 & 1/5 & 3/10 \\ 1/5 & 3/10 & 1/5 & 3/10 \\ 1/5 & 3/10 & 1/5 & 3/10 \end{pmatrix}.$$

That is, $p_{i1}^{(n)}, p_{i6}^{(n)} \to 1/5$ and $p_{i2}^{(n)}, p_{i7}^{(n)} \to 3/10$, $i = 1,2,6,7$. Probabilities $p_{3j}^{(n)}$ converge to $(1/2) \lim\limits_{n\to\infty} p_{2j}^{(n)}$, $j = 1,2,6,7$.   □

It has to be stressed that this is not the optimal way of calculating the limits $\lim\limits_{n\to\infty} p_{ij}^{(n)}$. Later on, we will learn about much more efficient ways of doing it.

1.3 Hitting times and probabilities

A hit, a very palpable hit.
W. Shakespeare (1564–1616), English playwright and poet

From now on we denote by $\mathbb{P}_i$ the distribution of a DTMC (X_n) starting from the state $i \in I$. Similarly, $\mathbb{E}_i$ stands for the expectation relative to $\mathbb{P}_i$. Let $A \subset I$ be a set of states. The *hitting time* H^A (of set A in the chain (X_n)) is the first time the Markov chain enters A

$$H^A = \inf \{n \geq 0 : X_n \in A\}. \tag{1.12}$$

The *hitting probability* h_i^A (of set A from state i in chain (X_n)) is the probability that the chain starting from state i will ever hit A

$$h_i^A = \mathbb{P}_i(H^A < \infty) \tag{1.13}$$

When A is a closed class, h_i^A is called the *absorption probability*. The expected value of H^A is denoted by k_i^A:

$$k_i^A = \mathbb{E}_i(H^A) = \sum_{0 < n < \infty} n\mathbb{P}_i(H^A = n) + \infty \cdot \mathbb{P}_i(H^A = \infty), \tag{1.14}$$

so that if $\mathbb{P}_i(H^A = \infty) > 0$, then $k_i^A = \infty$. In other words, $\mathbb{E}_i(H^A) = \infty$ when there is a positive chance the chain starting from i never enters A.

The basis for calculating the hitting probabilities is provided by

Theorem 1.3.1 *Given $A \subset I$, the hitting probabilities h_i^A give the minimal non-negative solutions to the following linear system*

$$h_i^A = \begin{cases} 1, & i \in A, \\ \sum_{j \in I} p_{ij} h_j^A, & \text{otherwise.} \end{cases} \tag{1.15}$$

That is, if $g_i \geq 0$ is any solution of (1.15), then $g_i \geq h_i^A$, $i \in I$.

Proof Recall that h_i^A is calculated for $X_0 = i$. When $i \in A$, $H^A = 0$, so $h_i^A = 1$. If $i \notin A$, we have $H^A \geq 1$, and

$$\begin{aligned} h_i^A &= \sum_{j \in I} \mathbb{P}_i(H^A < \infty, X_1 = j) = \sum_{j \in I} \mathbb{P}_i(X_1 = j)\mathbb{P}_i(H^A < \infty | X_1 = j) \\ &= \sum_j p_{ij} \mathbb{P}_j(H^A < \infty) = \sum_j p_{ij} h_j^A, \end{aligned}$$

by the Markov property.

Now take any non-negative solution g_i. For $i \in A$, $g_i = h_i^A = 1$. For $i \notin A$,

$$\begin{aligned} g_i &= \sum_j p_{ij} g_j = \sum_{j \in A} p_{ij} + \sum_{j \notin A} p_{ij} g_j \\ &= \sum_{j \in A} p_{ij} + \sum_{j \notin A} p_{ij} \left(\sum_{k \in A} p_{jk} + \sum_{k \notin A} p_{jk} g_k \right) \\ &= \mathbb{P}_i(X_1 \in A) + \mathbb{P}_i(X_1 \notin A, X_2 \in A) + \sum_{j \notin A, k \notin A} p_{ij} p_{jk} g_k. \end{aligned}$$

By repeated substitution, for all n,

$$\begin{aligned} g_i &= \mathbb{P}_i(X_1 \in A) + \cdots + \mathbb{P}_i(X_1 \notin A, \ldots, X_{n-1} \notin A, X_n \in A) \\ &+ \sum_{j_1 \notin A} \cdots \sum_{j_n \notin A} p_{ij_1} p_{j_1 j_2} \cdots p_{j_{n-1} j_n} g_{j_n}. \end{aligned}$$

As $g_i \geq 0$, omitting the last sum makes the right-hand side smaller. The first n summands give $\mathbb{P}_i(H^A \leq n)$. Hence

$$g_i \geq \mathbb{P}_i(H^A \leq n), \quad \text{for all } n \geq 0.$$

Then

$$g_i \geq \lim_{n \to \infty} \mathbb{P}_i(H^A \leq n) = \mathbb{P}_i(H^A < \infty) = h_i^A.$$

$\square$

In general, equations, even for more intricate hitting probabilities, give us a powerful tool, especially when a symmetry of a DTMC can be used, as we shall now see.

Worked Example 1.3.2 Construct a graph on seven vertices as follows: take a regular hexagon and join opposite corners by a straight line; let the vertices be the corners of the hexagon together with the point at the centre; let the edges be the perimeter of the hexagon together with the lines joining the corners to the centre. At discrete intervals a particle moves from one vertex of this graph to one of the adjacent vertices at random, and independently of past moves. Suppose the particle starts at a corner A. Find the probability that the particle will return to A without hitting the central vertex C.

Solution See Figure 1.8. Set

$$h_i = \mathbb{P}_i(\text{hit A before C}).$$

Then the probability in question is h_A, and by the symmetry of paths to A,

$$h_A = \frac{2}{3} h_B.$$

Now,

$$h_B = \frac{1}{3} + \frac{1}{3} h_D, \quad h_D = \frac{1}{3} h_B + \frac{1}{3} h_E,$$

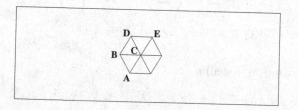

Fig. 1.8

and again by symmetry,

$$h_E = \frac{2}{3} h_D.$$

Then

$$h_D = \frac{1}{3} h_B + \frac{2}{9} h_D, \text{ i.e. } h_D \quad \frac{3}{7} h_D$$

Next,

$$h_B = \frac{1}{3} + \frac{1}{7} h_B, \text{ i.e. } h_B = \frac{7}{18}.$$

Hence, $h_A = 7/27$.   □

Example 1.3.3 Consider the birth-and-death process in Figure 1.3b. Set $h_i = \mathbb{P}_i(\text{hit } 0)$. Then h_i is the minimal non-negative solution to

$$h_0 = 1, \quad h_i = ph_{i+1} + (1-p)h_{i-1}, \quad i \ge 1.$$

For $p \ne 1/2$, this is solved by

$$h_i = A + B \left(\frac{1-p}{p} \right)^i.$$

If $p < 1/2$, minimality and non-negativity imply that $B = 0$ and $A = 1$, with $h_i \equiv 1$.
If $p > 1/2$, the conclusion is that $A = 0$ and $B = 1$, with

$$h_i = \left(\frac{1-p}{p} \right)^i.$$

For $p = 1/2$, the solution has the form

$$h_i = A + Bi,$$

and again the minimality and non-negativity imply that $B = 0$ and $A = 1$, with $h_i \equiv 1$.

Note that h_i is the extinction and $1 - h_i$ the survival probability (conditional on $X_0 = i$). Therefore, the survival probabilities are

$$\begin{cases} 1 - \left(\frac{1-p}{p} \right)^i, \, i \ge 0, & \text{for } p \in (1/2, 1] \\ 0, & \text{for } p \in [0, 1/2]. \end{cases}$$

Every moment dies a man,
Every moment $1\frac{1}{16}$ *is born*
C. Babbage (1792–1871), English mathematician

If we move to the process in Figure 1.3c, the equations become state-dependent:

$$h_0 = 1, \quad h_i = p_i h_{i+1} + (1 - p_i) h_{i-1}, \quad i \geq 1.$$

We solve them by considering the differences

$$u_i = h_{i-1} - h_i, \quad \text{with } p_i u_{i+1} = (1 - p_i) u_i,$$

and

$$u_{i+1} = \frac{1 - p_i}{p_i} u_i = \frac{1 - p_i}{p_i} \frac{1 - p_{i-1}}{p_{i-1}} \cdots \frac{1 - p_1}{p_1} u_1.$$

Set $\gamma_i = ((1 - p_{i-1})/p_{i-1}) \cdots ((1 - p_1)/p_1)$; then, as,

$$u_1 + \cdots + u_i = h_0 - h_i,$$

we obtain

$$h_i = 1 - A(\gamma_0 + \cdots + \gamma_{i-1}).$$

Here $\gamma_0 = 1$ and $A = u_1$. The constant A has to be determined from the condition of non-negative minimality:

$$A = \left(\sum_{i \geq 0} \gamma_i \right)^{-1}.$$

That is,

$$h_i = \begin{cases} 1, & \text{if } \sum\limits_{j \geq 0} \gamma_j = \infty, \\ \left(\sum\limits_{j \geq i} \gamma_j \Big/ \sum\limits_{j \geq 0} \gamma_j \right), & \text{if } \sum\limits_{j \geq 0} \gamma_j < \infty. \end{cases}$$

In particular, in the second case, $h_{i+1} \leq h_i$ and $\lim_{i \to \infty} h_i = 0$. The survival probabilities become

$$\begin{cases} 1 - \left(\sum_{j=i}^{\infty} \gamma_j \Big/ \sum_{j=0}^{\infty} \gamma_j \right), & \text{if } \sum_{j=0}^{\infty} \gamma_j < \infty, \\ 0, & \text{if } \sum_{j=0}^{\infty} \gamma_j = \infty. \end{cases}$$

We proceed to consider the mean hitting times k_i^A. We have

Theorem 1.3.4 *Given* $A \subset I$, *the mean hitting times* k_i^A *give the minimal non-negative solutions to the following linear system*

$$k_i^A = \begin{cases} 0, & i \in A, \\ 1 + \sum_{j \notin A} n_{ij} k_j^A, & i \notin A. \end{cases} \qquad (1.16)$$

That is, if $g_i \geq 0$ *is any solution of* (1.16), *then* $g_i \geq k_i^A$ *for* $i \in I$.

Proof Like h_i^A before, the expected hitting time k_i^A is calculated for $X_0 = i$. When $i \in A$, we have $H^A = 0$, so $k_i^A = 0$. If $i \notin A$, then $H^A \geq 1$, and

$$\mathbb{E}_i(H^A | X_1 = j) = 1 + \mathbb{E}_j H^A$$

by the Markov property. Thus,

$$k_i^A = \mathbb{E}_i(H^A) = \sum_{j \in I} \mathbb{E}_i(H^A \mathbf{1}(X_1 = j)) = \sum_j \mathbb{P}_i(X_1 = j) \mathbb{E}_i(H^A | X_1 = j)$$

$$= 1 + \sum_{j \notin A} \mathbb{P}_i(X_1 = j) \mathbb{E}_j(H^A) = 1 + \sum_{j \notin A} p_{ij} k_j^A.$$

Now let g_i be any non-negative solution. Then $g_i = k_i^A = 0$ for $i \in A$. If $i \notin A$,

$$g_i = 1 + \sum_{j \notin A} p_{ij} g_j = 1 + \sum_{j \notin A} p_{ij} \left(1 + \sum_{k \notin A} p_{jk} g_k \right).$$

Writing 1 as $\mathbb{P}_i(H^A \geq 1)$ and $\sum_{j \notin A} p_{ij}$ as $\mathbb{P}(H^A \geq 2)$, obtain

$$g_i = \mathbb{P}_i(H^A \geq 1) + \mathbb{P}_i(H^A \geq 2) + \sum_{j \notin A} p_{ij} \sum_{k \notin A} p_{jk} g_k.$$

By repeated substitution, for all n,

$$g_i = \mathbb{P}_i(H^A \geq 1) + \cdots + \mathbb{P}_i(H^A \geq n) + \sum_{j_1 \notin A} \cdots \sum_{j_n \notin A} p_{ij_1} p_{j_1 j_2} \cdots p_{j_{n-1} j_n} g_{j_n}$$

$$\geq \mathbb{P}_i(H^A \geq 1) + \cdots + \mathbb{P}_i(H^A \geq n)$$

since $g_i \geq 0$. Then, as $n \to \infty$,

$$g_i \geq \sum_{n \geq 1} \mathbb{P}_i(H^A \geq n) = \mathbb{E}_i H^A = k_i^A.$$

$\square$

Note that in some cases the only non-negative solution to (1.16) is $k_i^A \equiv \infty$, $i \notin A$. As in the case of the h_i^As, (1.16) can be efficiently used, especially when the system has symmetries.

Example 1.3.5 For the birth-and-death process featured in Figure 1.3b, set $k_i = \mathbb{E}_i(H^0)$, the expected time of hitting 0. Then k_i is the minimal non-negative solution to

$$k_0 = 0, \quad k_i = 1 + pk_{i+1} + (1-p)k_{i-1}, \quad i \geq 1.$$

The general solution here is of the form $k_i = A + Bi$; the constants A and B are given by $A = 0$, $B = 1/(1-2p)$. However, for $p \geq 1/2$, there is no finite non-negative solution. Hence, for $i \geq 1$:

$$k_i = \begin{cases} i/(1-2p), & \text{for } 0 \leq p < 1/2, \\ \infty, & \text{for } 1/2 \leq p \leq 1. \end{cases}$$

Worked Example 1.3.6 A flight of stairs has N steps. A frog starts at the bottom of the stairs and tries to jump to the top, making a series of independent jumps as follows. When the frog is on the ith step ($0 < i < N$) it succeeds in jumping up to step $i+1$ with probability α ($0 < \alpha < 1/2$), but with probability α it falls down to step $i-1$ and with probability $1-2\alpha$ it lands again on the ith step. When the frog is at the bottom of the stairs (on step 0) it succeeds in jumping up to step 1 with probability β, $0 < \beta < 1$, but with probability $1-\beta$ it remains where it is. What is the expected number of jumps before the frog reaches the top of the stairs?

Suppose that the same frog starts N steps below the top of an infinite flight of descending stairs. What now is the expected number of jumps before the frog reaches the top of the stairs?

Solution The system of equations for the $[0,N]$ flight is

$$k_N = 0,$$
$$k_i = 1 + \alpha k_{i-1} + (1-2\alpha)k_i + \alpha k_{i+1}, \quad 1 \leq i \leq N-1,$$
$$k_0 = 1 + (1-\beta)k_0 + \beta k_1.$$

Here, the general solution is

$$k_i = A + Bi - \frac{1}{2\alpha} i^2,$$

and the boundary conditions at $i = 0$ and N yield

$$k_i = \frac{N^2 - i^2}{2\alpha} - \frac{N-i}{2\alpha} + \frac{N-i}{\beta},$$

with

$$k_0 = \frac{N(N-1)}{2\alpha} + \frac{N}{\beta}.$$

For infinite stairs, $A + Bi - (i^2/2\alpha)$ cannot be maintained non-negative. Hence, $k_i \equiv \infty$.   □

Worked Example 1.3.7   Consider the Markov chain with state space $\{1, 2, 3, 4, 5, 6\}$ and transition matrix

$$
\begin{pmatrix}
0 & 0 & 1/2 & 0 & 0 & 1/2 \\
1/5 & 1/5 & 1/5 & 1/5 & 1/5 & 0 \\
1/3 & 0 & 1/3 & 0 & 0 & 1/3 \\
1/6 & 1/6 & 1/6 & 1/6 & 1/6 & 1/6 \\
0 & 0 & 0 & 0 & 1 & 0 \\
1/4 & 0 & 1/2 & 0 & 0 & 1/4
\end{pmatrix}.
$$

Determine the communicating classes of the chain, and for each class indicate whether it is closed or not.

Suppose that the chain starts in state 2; determine the probability that it ever reaches state 6.

Suppose that the chain starts in state 3; determine the probability that it is in state 6 after exactly n transitions, $n \geq 1$.

Solution The chain structure is represented in Figure 1.9.

States $1, 3, 6$ form a closed class, $2, 4$ a non-closed class, and state 5 is absorbing (and forms a closed class). If $h_i = \mathbb{P}_i(\text{hit } 6)$ then $h_1 = h_3 = 1$, and

$$
h_2 = \frac{1}{5} h_2 + \frac{1}{5} h_4 + \frac{2}{5},
$$

$$
h_4 = \frac{1}{6} h_4 + \frac{1}{6} h_2 + \frac{1}{2},
$$

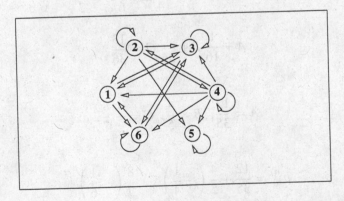

Fig. 1.9

whence $h_2 = 13/19$, $h_4 = 14/19$. Hence, the answer

$$\mathbb{P}_2(\text{hit } 6) = \frac{13}{19}.$$

Now, on class $\{1,3,6\}$, the transition matrix is

$$\begin{pmatrix} 0 & 1/2 & 1/2 \\ 1/3 & 1/3 & 1/3 \\ 1/4 & 1/2 & 1/4 \end{pmatrix}.$$

To find its eigenvalues solve

$$\det \begin{pmatrix} -\mu & (1/2) & (1/2) \\ (1/3) & (1/3)-\mu & (1/3) \\ (1/4) & (1/2) & (1/4)-\mu \end{pmatrix} = 0,$$

i.e.

$$\mu^3 - \frac{7}{12}\mu^2 - \frac{9}{24}\mu - \frac{1}{24} = (\mu-1)\left(\mu^2 + \frac{5}{12}\mu + \frac{1}{24}\right) = 0,$$

with

$$\mu_1 = 1, \quad \mu_2 = -\frac{1}{4}, \quad \mu_3 = -\frac{1}{6}.$$

This yields

$$p_{36}^{(n)} = A + B\left(-\frac{1}{4}\right)^n + C\left(-\frac{1}{6}\right)^n, \quad n = 0,1,\ldots.$$

At $n = 0,1,2$

$$A + B + C = 0, \quad A - \frac{1}{4}B - \frac{1}{6}C = \frac{1}{3},$$

$$A + \frac{1}{16}B + \frac{1}{36}C = \frac{13}{36},$$

giving

$$A = \frac{12}{35}, \quad B = \frac{4}{5}, \quad C = -\frac{8}{7},$$

with

$$p_{36}^{(n)} = \frac{12}{35} + \frac{4}{5}\left(-\frac{1}{4}\right)^n - \frac{8}{7}\left(-\frac{1}{6}\right)^n.$$

$\square$

1.4 Strong Markov property

The strong Markov property asserts that the process begins afresh not only after any given time n but also after a randomly chosen time. An example of such a time is H^i, the time the chain hits a given state $i \in I$. More generally,

Definition 1.4.1 A random variable T depending on $X_0, X_1, \ldots$ and taking values $0, 1, 2, \ldots, \infty$ is called a *stopping time* if the event $\{T = n\}$ is described in terms of random variables $X_1, \ldots, X_n$ only, without involving $X_{n+1}, X_{n+2}, \ldots$.

Pictorially, by watching the chain, you know when you should stop without anticipating future states. The hitting time H^A is an example of a stopping time as for $n = 0$: $\{H^A = 0\} = \{X_0 \in A\}$, and for $n \geq 1$

$$\{H^A = n\} = \{X_0 \notin A, \ldots, X_{n-1} \notin A, X_n \in A\}.$$

When A is reduced to a single state i, the hitting time is often called the *passage time*:

$$H^j = \inf\,[n \geq 0 : X_n = j].$$

On the other hand, the last exit time

$$L^A = \sup\,[n : X_n \in A]$$

is in general not a stopping time as the event $\{L^A = n\}$ requires knowledge of $X_{n+1}, X_{n+2}, \ldots$.

Theorem 1.4.2 *Let* $(X_n, n \geq 0)$ *be Markov* (λ, P) *and assume that* T *is a stopping time. Then, conditional on* $T < \infty$ *and* $X_T = i$, $(X_{T+n}, n \geq 0)$ *is Markov* (δ_i, P). *In particular, conditional on* $T < \infty$ *and* $X_T = i$, *the random variables* $X_{T+1}, X_{T+2}, \ldots$ *are independent of* $X_0, \ldots, X_{T-1}$.

Proof Let A be an event determined by the chain before time T, i.e. by $X_0, \ldots, X_{T-1}$, and B by the chain after time T, i.e. by $X_{T+1}, \ldots, X_{T+n}$ for some n. We want to check that for all $n \geq 1$ and $i \in I$: (i)

$$\mathbb{P}(A \cap B \mid T < \infty, X_T = i) = \mathbb{P}(A \mid T < \infty, X_T = i)\, \mathbb{P}(B \mid T < \infty, X_T = i)$$

and (ii) the conditional probability $\mathbb{P}(B \mid T < \infty, X_T = i)$ is calculated as in the (δ_i, P) Markov chain:

$$\mathbb{P}(B \mid T < \infty, X_T = i) = \sum_{(j_1, \ldots, j_n) \in B} p_{ij_1} \cdots p_{j_{n-1}j_n}.$$

As in the proof of the Markov property, we first assume that A is of the form $\{X_0 = i_0, \ldots, X_{T-1} = i_{T-1}\}$ and B is of the form $\{X_{T+1} = j_1, \ldots, X_{T+n} = j_n\}$ for some $i_0, \ldots, i_{T-1}, j_1, \ldots, j_n \in I$. Given m, the event

$$A \cap \{T = m\} \cap \{X_T = i\} = A \cap \{T = m, X_m = i\}$$

is simply

$$\{X_0 = i_0, \ldots, X_{m-1} = i_{m-1}, X_m = i\}$$

if $T(i_0, \ldots, i_{m-1}, i) = m$; it is empty if $T(i_0, \ldots, i_{m-1}, i) \neq m$. Then the event $A \cap B \cap \{T = m, X_T = i\} = A \cap \{T = m, X_m = i\} \cap B$ has probability

$$\lambda_{i_0} p_{i_0 i_1} \cdots p_{i_{m-1}i} p_{ij_1} \cdots p_{j_{n-1}j_n}\, \mathbf{1}\big(T(i_0, \ldots, i_{m-1}, i) = m\big).$$

For a general B we have to sum over $(j_1, \ldots, j_n) \in B$:

$$\lambda_{i_0} p_{i_0 i_1} \cdots p_{i_{m-1}i}\, \mathbf{1}\big(T(i_0, \ldots, i_{m-1}, i) = m\big) \sum_{(j_1, \ldots, j_n) \in B} p_{ij_1} \cdots p_{j_{n-1}j_n}.$$

The sum $\sum_{(j_1, \ldots, j_n) \in B}$ does not depend on m; it gives the conditional probability $\mathbb{P}(B | T < \infty, X_T = i)$ and is indeed calculated as in the (δ_i, P) Markov chain.

For a general A we now sum over $(i_0, \ldots, i_{m-1}) \in A$:

$$\mathbb{P}(A \cap B \cap \{T = m, X_T = i\}) \times$$
$$\sum_{(i_0, \ldots, i_{m-1}) \in A} \lambda_{i_0} p_{i_0 i_1} \cdots p_{i_{m-1}i}\, \mathbf{1}\big(T(i_0, \ldots, i_{m-1}, i) = m\big)\, \mathbb{P}(B | T < \infty, X_T = i)$$
$$= \mathbb{P}(A \cap \{T = m, X_T = i\})\mathbb{P}(B | T < \infty, X_T = i).$$

Summing over m then gives

$$\mathbb{P}(A\cap B\cap\{T<\infty,X_T=i\})=\mathbb{P}(A\cap\{T<\infty,X_T=i\})\,\mathbb{P}(B|T<\infty,X_T=i).$$

Finally, dividing by $\mathbb{P}(T<\infty,X_T=i)$ yields that the conditional probability $\mathbb{P}(A\cap B\mid T<\infty,X_T=i)$ equals

$$\frac{\mathbb{P}(A\cap\{T<\infty,X_T=i\})}{\mathbb{P}(T<\infty,X_T=i)}\,\mathbb{P}(B|T<\infty,X_T=i)$$
$$=\mathbb{P}(A\mid T<\infty,X_T=i)\,\mathbb{P}(B\mid T<\infty,X_T=i)$$

as required.

The conditional probability $\mathbb{P}(A\cap\{T=m,X_T=i\}\cap B\mid X_m=i)$, given that $X_m=i$, is obtained after division by $\mathbb{P}(X_m=i)=(\lambda P^m)_i$: the ratio is determined by $X_0,\dots,X_m$, and the conditional probability

$$\mathbb{P}\big((A\cap\{T=m\})\cap\{X_{T+1}=j_1,\dots,X_{T+n}=j_n\}\mid X_m=i\big)$$
$$=\mathbb{P}\big((A\cap\{T=m\})\cap\{X_{m+1}=j_1,\dots,X_{m+n}=j_n\}\mid X_m=i\big).$$

By the Markov property we have the decomposition:

$$\mathbb{P}\big((A\cap\{T=m\})\cap\{X_{m+1}=j_1,\dots,X_{m+n}=j_n\}\mid X_m=i\big)$$
$$=\mathbb{P}(A\cap\{T=m\}\mid X_m=i)\,\mathbb{P}(X_{m+1}=j_1,\dots,X_{m+n}=j_n\mid X_m=i)$$
$$=\mathbb{P}(A\cap\{T=m\}\mid X_m=i)\,p_{ij_1}\cdots p_{j_{n-1}j_n}.$$

Hence, the unconditional probability

$$\mathbb{P}\big((A\cap\{T=m\})\cap\{X_{m+1}=j_1,\dots,X_{m+n}=j_n\}\cap\{X_m=i\}\big)$$
$$=\mathbb{P}\big((A\cap\{T=m,X_m=i\})\cap\{X_{m+1}=j_1,\dots,X_{m+n}=j_n\}\big)$$
$$=\mathbb{P}(A\cap\{T=m\}\mid X_m=i)\,\mathbb{P}(X_m=i)p_{ij_1}\cdots p_{j_{n-1}j_n}$$
$$=\mathbb{P}\big(A\cap\{T=m,X_m=i\}\big)p_{ij_1}\cdots p_{j_{n-1}j_n}.$$

Summing over m yields

$$\mathbb{P}\big((A\cap\{T<\infty,X_T=i\}\cap\{X_{T+1}=j_1,\dots,X_{T+n}=j_n\}\big)$$
$$=\mathbb{P}(A\cap\{T<\infty,X_T=i\})\,p_{ij_1}\cdots p_{j_{n-1}j_n}$$

and, dividing by $\mathbb{P}(T<\infty,X_T=i)$,

$$\mathbb{P}\big(A\cap\{X_{T+1}=j_1,\dots,X_{T+n}=j_n\}\mid T<\infty,X_T=i\big)$$
$$=\mathbb{P}\big(A\mid\{T<\infty,X_T=i\}\big)p_{ij_1}\cdots p_{j_{n-1}j_n}.$$

Finally, for a general event B determined by $X_{T+1},\dots,X_{T+n}$ we sum over $(j_1,\dots,j_n)\in B$. $\square$

Worked Example 1.4.3 In the homogeneous birth-and-death process (see Example 1.3.5), what is the distribution of the hitting time $H^0 = \inf \{n \geq 0 : X_n = 0\}$ (the time to extinction)? In other words, what are the probabilities $\mathbb{P}_i(H^0 = k)$ for given i and k?

Solution This can be found by calculating the probability-generating function $\phi_i(s) = \mathbb{E}_i\left(s^{H^0}\right) = \sum_{0 \leq n < \infty} s^n \mathbb{P}_i(H^0 = n)$.
By the strong Markov property,

$$\phi_i(s) = \big(\phi(s)\big)^i, \ \ i \geq 1,$$

where $\phi(s) = \phi_1(s)$. Thus it suffices to analyse the case $i = 1$. Then, given that $X_0 = 1$, we see that $\phi(s)$ is a root of the quadratic equation

$$ps\phi^2 - \phi + qs = 0,$$

given by

$$\phi(s) = \frac{1}{2ps}\left(1 - \sqrt{1 - 4pqs^2}\right), \ \ 0 < s < 1.$$

$\square$

Example 1.4.4 The strong Markov property is very useful when you observe the chain (X_n) only at certain times, for example, when it changes its states (i.e., when $X_{n+1} \neq X_n$) or enters a subset $J \subset I$ (i.e., when $X_n \in J$). The new chain is formally described by introducing the sequence of observation times $T_0, T_1, \ldots$, viz.

$$T_0 = \inf \{n > 0 : X_n \neq X_{n-1}\}, \ \ \text{or} \ \ T_0 = \inf \{n \geq 0 : X_n \in J\},$$

and

$$T_{m+1} = \inf \{n > T_m : X_n \neq X_{n-1}\}, \ \ \text{or} \ \ T_{m+1} = \inf \{n > T_m : X_n \in J\}.$$

Then the chain $(Y_n, n \geq 0)$ is defined by $Y_m = X_{T_m}$.

In either case, each T_m is a stopping time. Assuming that $T_m < \infty$ for all m, the strong Markov property guarantees that (Y_n) is indeed a DTMC. The transition probabilities p_{ij}^Y for the new chain are straightforward: in the first model

$$p_{ij}^Y = \begin{cases} \dfrac{p_{ij}}{1 - p_{ii}}, & i \neq j, \\ 0, & i = j, \end{cases} \qquad i, j \in I, \tag{1.17}$$

and in the second model

$$p_{ij}^Y = p_{ij} + \sum_{k \geq 1} \sum_{j_1, \ldots, j_k \in I \setminus J} p_{ij_1} \cdots p_{j_k j}, \ \ \text{for } i, j \in J. \tag{1.18}$$

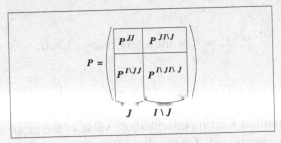

Fig. 1.10

Here $P = (p_{ij})$ is the transition matrix of the original chain (X_n).

The first model, with $Y_{n+1} \neq Y_n$, is called the *jump chain* (for the original DTMC (X_n)); this model will play an important rôle in the analysis of continuous-time Markov chains in Chapter 2. The second model, with $Y_n \in J$, is called a *partially observed* chain. For the partially observed Markov chain, the transition probabilities p_{ij}^Y can be written in terms of matrix blocks P^{JJ}, $P^{JI \setminus J}$, $P^{I \setminus JJ}$ and $P^{I \setminus JI \setminus J}$ extracted from the transition matrix P:

$$p_{ij}^Y = \left(P^{JJ}\right)_{ij} + \left[P^{JI \setminus J}\left(\mathbf{I}_{I \setminus J} - P^{I \setminus JI \setminus J}\right)^{-1} P^{I \setminus JJ}\right]_{ij}, \quad i, j \in J. \qquad (1.19)$$

Here $\mathbf{I}_{I \setminus J}$ stands for the identity matrix over $I \setminus J$. See Figure 1.10.

1.5 Recurrence and transience: definitions and basic facts

> *The eternal silence of these infinite spaces terrifies me.*
> B. Pascal (1623–1662), French mathematician and philosopher

Recurrence and transience are important properties of DTMCs with countably infinite state spaces. In this book, we prefer to pass from a finite to a countable case in a rather casual way: we just extend basic definitions to the case of a countable state space I. Of course, this requires infinite transition matrices $P = (p_{ij}, i, j \in I)$; we have seen such matrices before (see page 21). The theory of infinite matrices is more subtle than the theory of finite matrices; some of its important aspects require working with infinite-dimensional spaces. We will not sail too far in these directions, focussing on properties that are either direct generalisations of their finite-dimensional counterparts or have an intuitively clear probabilistic meaning.

Definition 1.5.1 Given a state $i \in I$, we call it *recurrent* if

$$\mathbb{P}_i(X_n = i \text{ for infinitely many } n) = 1, \qquad (1.20)$$

and *transient* if

$$\mathbb{P}_i(X_n = i \text{ for infinitely many } n) = 0, \tag{1.21}$$

i.e.

$$\mathbb{P}_i(X_n = i \text{ for finitely many } n) = 1.$$

Note that in Definition 1.5.1 no intermediate value of the probability (i.e. strictly between 0 and 1) is mentioned. This is clarified in Theorem 1.5.2 below. Set

$$f_i := \mathbb{P}_i(X_n = i \text{ for some } n \geq 1) \tag{1.22}$$

Theorem 1.5.2 *State i is recurrent if $f_i = 1$ and transient if $f_i < 1$. Therefore, every state is either recurrent or transient.*

Proof A useful random variable is the hitting/passage time of state i (in our context it could also be called the *return time* to state i):

$$T_i = \inf [n \geq 1 : X_n = i], \tag{1.23}$$

with

$$f_i = \mathbb{P}_i(T_i < \infty). \tag{1.24}$$

Then, as was noted, the random variable T_i is a stopping time. By the strong Markov property,

$$\mathbb{P}_i(X_n = i \text{ for at least two values of } n \geq 1) = f_i^2,$$

and more generally, for all k

$$\mathbb{P}_i(X_n = i \text{ for at least } k \text{ values of } n \geq 1) = f_i^k. \tag{1.25}$$

Denote by $B_k^{(i)}$ the event that $X_n = i$ for at least k values of $n \geq 1$. Then, obviously, events $B_k^{(i)}$ are decreasing with k: $B_1^{(i)} \supseteq B_2^{(i)} \supseteq \dots$, and the event that $X_n = i$ for infinitely many values of n is the intersection $\bigcap_{k \geq 1} B_k^{(i)}$. Hence,

$$\mathbb{P}_i(X_n = i \text{ for infinitely many } n) = \lim_{k \to \infty} \mathbb{P}(B_k^{(i)}), \tag{1.26}$$

which equals 1 when $f_i = 1$ and 0 when $f_i < 1$. $\qquad\qquad\square$

O the heavy change, …
Now thou art gone,
and never must return!
J. Milton (1608–1674), English poet

It is worthwhile to introduce yet another random variable (which will be used quite often)

$$V_i = \text{number of visits to } i = \sum_{n \geq 0} \mathbf{1}(X_n = i). \qquad (1.27)$$

Equivalently, V_i counts the total time spent at state i (including initial time 0 when appropriate). Equation (1.25) can be re-written as

$$f_i^k = \mathbb{P}_i(V_i \geq k). \qquad (1.28)$$

An important parameter is the expected value $\mathbb{E}_i V_i$; more precisely, the formula

$$\mathbb{E}_i V_i = \sum_{n \geq 1} \mathbb{P}_i (V_i \geq n) = \sum_{n \geq 0} f_i^n. \qquad (1.29)$$

On the other hand,

$$\mathbb{E}_i V_i = \sum_{n \geq 0} \mathbb{E}_i \mathbf{1}(X_n = i) = \sum_{n \geq 0} p_{ii}^{(n)}; \qquad (1.30)$$

here, $p_{ii}^{(0)} = 1$ as $P^0 = \mathbf{I}$, the identity matrix. From (1.29), (1.30) we see that the following assertion holds true.

Theorem 1.5.3 *The state i is recurrent if*

$$\sum_{n \geq 0} p_{ii}^{(n)} = \infty, \qquad (1.31)$$

and transient if

$$\sum_{n \geq 0} p_{ii}^{(n)} < \infty. \qquad (1.32)$$

Proof According to (1.29), (1.30), the sum $\sum_{n \geq 0} p_{ii}^{(n)}$ coincides with the sum of the geometric progression $\sum_{n \geq 0} f_i^n$. The latter is finite when $f_i < 1$ (and equals $1/(1 - f_i)$), and infinite when $f_i = 1$. $\qquad \square$

Theorem 1.5.3 will be repeatedly used in the analysis of recurrence and transience of states of various chains.

An alternative proof of Theorem 1.5.3 exploits the probability-generating functions of a random variable T_i. Set

$$f_i(n) = \mathbb{P}_i(T_i = n) = \mathbb{P}_i (X_n = i \text{ but } X_l \neq i \text{ for } l = 1, \ldots, n-1), \, n \geq 1, \qquad (1.33)$$

and

$$F(z)(= F_i(z)) = \mathbb{E}z^{T_i} = \sum_{n\geq 1} z^n f_i(n), \quad |z| < 1; \tag{1.34}$$

then $f_i = \lim_{z\to 1} F(z)$.

On the other hand,

$$p_{ii}^{(n)} = \mathbb{P}_i(X_n = i) = f_i(n) + f_i(n-1)p_{ii} + \cdots + f_i(1)p_{ii}^{(n-1)}; \tag{1.35}$$

this implies that if

$$U(z)(= U_i(z)) = \sum_{n\geq 1} p_{ii}^{(n)} z^n, \quad |z| < 1, \tag{1.36}$$

then

$$U(z) = F(z) + F(z)U(z), \quad \text{i.e. } U(z) = \frac{F(z)}{1-F(z)}.$$

Hence, the limiting value $\lim_{z\to 1} U(z)$ is finite if and only if $\lim_{z\to 1} F(z) < 1$. That is, (1.32) holds true if and only if $f_i < 1$.

We conclude this section with the following remark. Equation (1.21) can be written as

$$\mathbb{P}_i(V_i < \infty) = 1, \tag{1.37}$$

whereas (1.32) is

$$\mathbb{E}_i V_i < \infty. \tag{1.38}$$

We see that if the random variable V_i (the total number of visits to state i) is finite with probability 1, then it must have a finite mean; the (more precisely, strong) Markov property excludes an intermediate possibility where $\mathbb{P}_i(V_i < \infty) = 1$ but $\mathbb{E}_i V_i = \infty$.

However, the situation is more subtle when we turn to the random variable T_i (the passage, or return time to state i). We noticed that state i is recurrent if and only if $\mathbb{P}_i(T_i < \infty) = 1$, i.e. the return time to i is finite with probability 1. However, the mean $\mathbb{E}_i T_i$ (or equivalently, $\lim_{z\to 1} F'(z)$) can be finite or infinite. This divides recurrent states into two distinct categories: positive recurrent and null recurrent (see Section 1.7).

Communicating classes for countable DTMCs are defined in the same way as for finite chains. For convenience we repeat the definition:

Definition 1.5.4 States $i, j \in I$ belong to the same *communicating class* if $p_{ij}^{(n)} > 0$ and $p_{ji}^{(n')} > 0$ for some $n, n' \geq 0$. Again, the communicating classes form a partition

of the state space I, and, as some of them may be infinite, the number of communicating classes can also be infinite. Next, as in the finite case, a communicating
class C is called *closed* if $i \to j$ then $j \in C$, for all $i \in C$. Finally, we say that the
chain is *irreducible* if it has a unique communicating class (automatically closed).
In other words, in an irreducible DTMC, the whole of the state space I is a single
(closed) communicating class.

Remark 1.5.5 Observe that if the state space I is finite, the definition of a transient
state coincides with that of a non-essential state (i.e., a state from a non-closed
communicating class). In other words, in the finite case every state from a non-
closed class is transient, and every state from a closed class is recurrent. However,
as we noted in Remark 1.2.6, in the case of a countable DTMC a closed class
can consist entirely of transient states, which are, from a 'physical' point of view,
non-essential. It shows that in the countable case the concept of transience is more
relevant than that of a closed communicating class.

Our aim now is to prove that recurrence and transience are class properties. This
means that if states i, j lie in the same communicating class then they are either
both recurrent or both transient. We therefore could use

Definition 1.5.6 A communicating class is called *recurrent* (resp. *transient*) if all
its states are recurrent (resp., transient).

Theorem 1.5.7 *Within the same communicating class, all states are of the same
type. Every finite closed communicating class is recurrent.*

Proof Let C be a communicating class. Then, for all distinct $i, j \in C$, $p_{ij}^{(m)} > 0$ and
$p_{ji}^{(n)} > 0$ for some $m, n \geq 1$. Then for all $r \geq 0$:

$$p_{ii}^{(n+m+r)} \geq p_{ij}^{(m)} p_{jj}^{(r)} p_{ji}^{(n)} \quad \text{and} \quad p_{jj}^{(n+m+r)} \geq p_{ji}^{(n)} p_{ii}^{(r)} p_{ij}^{(m)},$$

as the RHS in each inequality takes into account only a part of the possibilities of
return.

Hence

$$p_{jj}^{(r)} \leq \frac{p_{ii}^{(n+m+r)}}{p_{ij}^{(m)} p_{ji}^{(n)}}$$

and, for $r \geq n + m$,

$$p_{jj}^{(r)} \geq p_{ji}^{(n)} p_{ii}^{(r-n-m)} p_{ij}^{(m)}.$$

Then the series $\sum_r p_{ii}^{(r)}$ and $\sum_r p_{jj}^{(r)}$ converge or diverge together.

Now let C be a finite closed communicating class, and $j \in C$. Then, with $X_0 = j \in C$, $X_n \in C$ for all n. Hence, there exists a state $i \in C$ visited infinitely often:

$$0 < \mathbb{P}_j(V_i = \infty) = \mathbb{P}_j(T_i < \infty)\, \mathbb{P}_i(V_i = \infty).$$

Then $\mathbb{P}_i(V_i = \infty) > 0$, i.e. state i is recurrent. Then every state from C is recurrent.

$\square$

Definition 1.5.8 A transition matrix P (and a (λ, P) Markov chain) is called *recurrent* (resp. *transient*) if every state i is recurrent (respectively, *transient*).

We conclude this section with one more statement involving passage, or return, times.

Theorem 1.5.9 *If P is irreducible and recurrent then each random variable T_j (the passage time to state j) is finite with probability 1. That is, $\mathbb{P}(T_j < \infty) = 1$ for all j and initial distributions λ.*

Proof By the Markov property

$$\mathbb{P}(T_j < \infty) = \sum_i \lambda_i \mathbb{P}_i(T_j < \infty).$$

Given i, take m with $p_{ji}^{(m)} > 0$. Write

$$1 = \mathbb{P}_j(V_j = \infty) \leq \mathbb{P}_j(X_n = j \text{ for some } n \geq m)$$

(obviously, there is equality here, but the inequality will also do). Further,

$$\mathbb{P}_j(X_n = j \text{ for some } n \geq m)$$
$$= \sum_k p_{jk}^{(m)} \mathbb{P}_j(X_n = j \text{ for some } n \geq m \mid X_m = k)$$
$$= \sum_k p_{jk}^{(m)} \mathbb{P}_k(T_j < \infty) \leq \sum_k p_{jk}^{(m)} = 1.$$

We see that each summand $p_{jk}^{(m)} \mathbb{P}_k(T_j < \infty)$ must be equal to $p_{jk}^{(m)}$; otherwise we would have that $1 < 1$. Therefore,

$$\mathbb{P}_i(T_j < \infty) p_{ji}^{(m)} = p_{ji}^{(m)}, \text{ i.e. } \mathbb{P}_i(T_j < \infty) = 1.$$

This is true for all i, hence for all initial distributions λ. Also, it is true for all j. $\square$

Worked Example 1.5.10 Suppose that P is irreducible and recurrent and that the state space contains at least two states. Define a new transition matrix $\widehat{P} = (\widehat{p}_{ij})$ by

$$\widehat{p}_{ij} = \begin{cases} 0 & \text{if } i = j, \\ (1 - p_{ii})^{-1} p_{ij} & \text{if } i \neq j. \end{cases}$$

Prove that $\widehat{P}$ is also irreducible and recurrent.

Solution If $P = (p_{ij})$ is irreducible then $p_{ii} < 1$ for all state i (unless the total number of states is 1). The matrix $\widehat{P}$ describes the Markov chain obtained from the original DTMC by recording the jumps to the new state only; clearly it is irreducible. Formally, take the sequence $i_0, \ldots, i_m$ as above, then $\widehat{p}_{i_l i_{l+1}} > 0$. Now check the recurrence of $\widehat{P}$: if in the original chain $p_{ii} = 0$ then the return to state i occurs in both chains on the same event, hence the return probability to state i will be the same. If $p_{ii} > 0$ then in the new chain, the return probability is equal to

$$\frac{1}{1 - p_{ii}} \times \mathbb{P}_i(\text{return to } i \text{ after time 1 in the original chain})$$

$$= \frac{1}{1 - p_{ii}}(1 - p_{ii})$$

which is 1. Alternatively, $h\widehat{P} = h$ if and only if $hP = h$, i.e. the solutions to both equations are the same. Hence, the minimal solution to $h\widehat{P} = h$ with $h_i = 1$ is the same as that to $hP = h$. Therefore, it identically equals 1, and the new chain is recurrent if and only if the original one is. $\square$

1.6 Recurrence and transience: random walks on lattices

The only reason for time is so that everything doesn't happen at once.
A. Einstein (1879–1955), German physicist

Random walks on cubic lattices are popular and interesting models of countable Markov chains. Here we have a 'particle' that jumps at times $n = 1, 2, \ldots$ from its current position $\underline{i} \in \mathbb{Z}^d$ to another site $\underline{j} \in \mathbb{Z}^d$ with probability $p_{\underline{i}\underline{j}}$, regardless of the past sample trajectory. We will mostly focus on homogeneous nearest-neighbour random walks where the probabilities $p_{\underline{i}\underline{j}}$ are greater than 0 only when $\underline{i}$ and $\underline{j}$ are neighbouring sites and depend only on the direction from $\underline{i}$ to $\underline{j}$ (i.e. are determined by $p_{\underline{0},\underline{j}}$ where $\underline{j}$ is a neighbour of the origin $\underline{0} = (0, \ldots, 0)$). For $d = 1$ the lattice $\mathbb{Z}^d$ is simply the set of integers; here a random walk (RW) is specified by the probabilities p and $q = 1 - p$ of jumps to the right and the left.

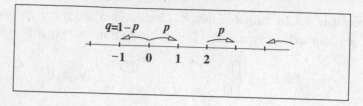

Fig. 1.11

This is an intuitively appealing extended version of the drunkard model (or birth-and-death process); see Example 1.2.3. Here, the state space is $I = \mathbb{Z}(= \mathbb{Z}^1)$, and the transition probability matrix is infinite and has a distinct 'diagonal' structure

$$
P = \begin{pmatrix}
\ddots & \ddots & \ddots & \ddots & \ddots & \ddots & \ddots \\
\ddots & q & 0 & p & 0 & \dots & \ddots \\
\ddots & 0 & q & 0 & p & 0 & \ddots \\
\ddots & \ddots & 0 & q & 0 & p & \ddots \\
\ddots & \ddots & \ddots & \ddots & \ddots & \ddots & \ddots
\end{pmatrix},
\tag{1.39}
$$

with entries p above and q below the main diagonal, and the rest filled with zeros.

If $d = 2$, then $\mathbb{Z}^2$ is a plane square lattice; here we will consider the symmetric nearest-neighbour RW where the probabilities of jumping in any direction are the same and equal $1/4$.

This is an infinitely extended two-dimensional version of the drunkard model.

If $d = 3$, then $\mathbb{Z}^3$ is the three-dimensional cubic lattice; we may think of it as an infinitely extended crystal. Then our walking particle may model a solitary quantum electron moving between heavy ions or atoms fixed at the sites of the lattice. The probability of moving to one of the six neighbours equals $1/6$.

One can also imagine a higher-dimensional model for any given d. Here, the probability of jump equals $1/(2d)$.

Theorem 1.6.1 *For $d = 1$, the nearest-neighbour random walk on $\mathbb{Z}$ is transient, unless $p = q = 1/2$, in which case it is recurrent.*

Proof The DTMC in question is obviously irreducible, so it is enough to check that the origin 0 is a recurrent state. We want to assess $\sum_n p_{00}^{(n)}$. Observe that

$$
p_{00}^{(n)} = \begin{cases}
0, & n \text{ odd}, \\
\dfrac{(2k)!}{k!k!}\, p^k q^k, & n = 2k \text{ even},
\end{cases}
\tag{1.40}
$$

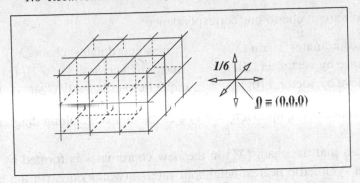

Fig. 1.12

as we need to make an equal number of steps to the right and the left. By using Stirling's formula,

$$n! \approx \sqrt{2\pi} n^{n+1/2} e^{-n}, \text{ as } n \to \infty,$$

we have

$$p_{00}^{(2k)} \approx \frac{(2k)^{2k+1/2}}{\sqrt{2\pi} k^{2k+1}} p^k q^k = \frac{1}{\sqrt{\pi k}} 2^{2k} (pq)^k. \tag{1.41}$$

Now,

$$pq = p(1-p) \le \frac{1}{4}, \ 0 \le p \le 1,$$

and the only point of equality is $p = q = 1/2$. In other words, $\rho := 4pq < 1$ for $p \ne 1/2$ and $\rho = 1$ for $p = 1/2$. Consequently, with $p_{00}^{(2k)} \approx \frac{1}{\sqrt{\pi k}} \rho^k$,

$$\sum_n p_{00}^{(n)} \begin{cases} < \infty, & p \ne 1/2, \\ = \infty, & p = 1/2. \end{cases} \tag{1.42}$$

$\square$

Theorem 1.6.2 *The nearest-neighbour symmetric random walk on $\mathbb{Z}^d$ is recurrent for $d = 2$ and transient for $d = 3$ (and also for $d > 3$).*

Proof $d = 2$: again consider a fixed state, say $\underline{0} = (0,0)$. Every closed path on $\mathbb{Z}^2$ must have equally many jumps to the left and the right and equally many jumps up and down.

Hence again $p_{00}^{(n)} = 0$ when n is odd.

A useful idea is to project the random walk onto orthogonal axes rotated by $\pi/4$.

The moves are in one-to-one correspondence:

old coordinates: chain (X_n) new coordinates: chain (X_n')

move by vector $\pm(1;0)$ $\leftrightarrow$ move by vector $\pm(1/\sqrt{2})(1;1)$

move by vector $\pm(0;1)$ move by vector $\pm(1/\sqrt{2})(-1;1)$.

In the new coordinates, up to a factor $1/\sqrt{2}$, the jumps are along diagonals of the unit square.

This means that the chain (X_n') in the new coordinates is formed by a pair of independent symmetric nearest-neighbour random walks on $\mathbb{Z}$ (in the horizontal and vertical directions). Return to $\underline{0} = (0,0)$ means return to 0 in each of them. Therefore, for $n = 2k$,

$$p_{\underline{00}}^{(2k)} = \left(\frac{(2k)!}{k!k!}\frac{1}{2^{2k}}\right)^2 \approx \frac{1}{\pi k}. \tag{1.43}$$

Hence, $\sum_k p_{\underline{00}}^{(2k)} = \infty$, and the random walk is recurrent.

For $d = 3$, we still have $p_{\underline{00}}^{(n)} = 0$ when n is odd. If n is even, a path returns to $\underline{0} = (0,0,0)$ if and only if it makes equal numbers of jumps in each of three pairs of opposite directions (up/down, east/west, north/south). So,

$$p_{\underline{00}}^{(2k)} = \sum_{\substack{i,j,l \geq 0:\\ i+j+l=k}} \frac{(2k)!}{(i!)^2(j!)^2(l!)^2}\left(\frac{1}{6}\right)^{2k}$$

$$= \frac{(2k)!}{(k!)^2}\sum_{\substack{i,j,l \geq 0:\\ i+j+l=k}} \left(\frac{k!}{i!j!l!}\right)^2\left(\frac{1}{6}\right)^{2k}$$

$$\leq \frac{(2k)!}{(k!)^2}\left(\max\frac{k!}{i!j!l!}\right)\frac{1}{3^k}\frac{1}{2^{2k}}\sum_{\substack{i,j,l \geq 0:\\ i+j+l=k}} \frac{k!}{i!j!l!}\frac{1}{3^k}.$$

Now, the sum

$$\sum_{\substack{i,j,l \geq 0:\\ i+j+l=k}} \frac{k!}{i!j!l!} = 3^k \tag{1.44}$$

is the number of ways placing k balls into 3 boxes. Also, for $k = 3m$,

$$\frac{(3m)!}{m!m!m!} \geq \frac{(3m)!}{i!j!l!} \quad \text{whenever } i+j+l = 3m. \tag{1.45}$$

In fact, suppose that $i < m < l$. Then when you pass to $i!j!l!$ from $(m!)^3$, you either (a) replace the 'tails' $(i+1)\cdots m$ and $(j+1)\cdots m$ of $m!$ by the product $(m+1)\cdots(m+2m-i-j)$, i.e., $(m+1)\cdots l$ when $j < m$; or (b) replace the tail

$(i + 1)\cdots m$ by the product $(m + 1)\cdots j(m + 1)\cdots(3m - i - j)$; that is, $(m+1)\cdots j(m+1)\cdots l$ when $j > m$. Either way you increase the denominator, hence decrease the ratio.

Then, for $n = 2k = 6m$,

$$p_{00}^{(6m)} \le \frac{(6m)!}{((3m)!)^2}\left(\frac{1}{2}\right)^{6m}\frac{(3m)!}{(m!)^3}\left(\frac{1}{3}\right)^{3m} \tag{1.46}$$

which, by Stirling, is

$$\approx \sqrt{2}\left(\frac{1}{\sqrt{2\pi}}\right)^3\frac{1}{m^{3/2}}. \tag{1.47}$$

Hence, $\sum_m p_{00}^{(6m)} < \infty$. But for $m \ge 1$ we have $p_{00}^{(6m)} \ge (1/6)^2 p_{00}^{(6m-2)}$ and $p_{00}^{(6m)} \ge (1/6)^4 p_{00}^{(6m-4)}$, i.e.

$$p_{00}^{(6m-2)} \le 6^2 p_{00}^{(6m)} \quad \text{and} \quad p_{00}^{(6m-4)} \le 6^4 p_{00}^{(6m)}.$$

Thus,

$$\sum_k p_{00}^{(2k)} \le \sum_m p_{00}^{(6m)}(1 + 6^2 + 6^4) < \infty, \tag{1.48}$$

and the walk is transient. $\qquad\square$

A similar approach can be used in higher dimensions. But there is another way to establish transience in all dimensions $d > 3$. Namely, project the random walk (X_n^d) on $\mathbb{Z}^d$ to three dimensions by discarding all coordinates but the first three. The projected chain $\left(X_n^{\text{proj}}\right)$ on $\mathbb{Z}^3$ stays where it is with probability $(d-3)/d$ (when the original walk jumps in one of the discarded directions), but when it jumps, it behaves as the nearest-neighbour symmetric walk in dimension 3:

$$\mathbb{P}\left(X_{n+1}^{\text{proj}} = \underline{i} \pm \underline{e}^\alpha \,\middle|\, X_n^{\text{proj}} = \underline{i}\right) = \frac{1/(2d)}{1 - (d-3)/d} = \frac{1}{6}, \; \alpha = 1, 2, 3, \tag{1.49}$$

with

$$\underline{e}^1 = (1; 0; 0), \; \underline{e}^2 = (0; 1; 0), \; \underline{e}^3 = (0; 0; 1).$$

Clearly, if the original d-dimensional walk returns to $\underline{0} = (0, \ldots, 0)$, then the projected walk returns to $(0, 0, 0)$. Hence, if the original d-dimensional walk (X_n^d) is recurrent then the projected chain $\left(X_n^{\text{proj}}\right)$ is too. But then consider the random walk on $\mathbb{Z}^3$ obtained from $\left(X_n^{\text{proj}}\right)$ by discarding the stays and recording the jumps only. The latter is the nearest-neighbour symmetric random walk on $\mathbb{Z}^3$ which is transient. By Theorem 1.6.3 below, $\left(X_n^{\text{proj}}\right)$ is also transient. Then so is (X_n^d).

Nearest-neighbour symmetric random walks are often called simple walks. Rephrasing a famous saying, we could state that in two dimensions every road of a simple random walk will lead you to the origin (or any other given site) while in three dimensions and higher it is no longer so. The difference between two and three dimensions emerges in virtually all domains of mathematics.

We conclude this section by analysing the relation between a general DTMC (X_n) and its jump chain (Y_n) obtained when we record only the changes in the state of (X_n). Suppose that the transition matrix of (X_n) is $P = (p_{ij})$. Then for (Y_n) the transition matrix will be $(\widehat{p}_{ij})$ where

$$\widehat{p}_{ij} = \begin{cases} 0, & i = j, \\ \dfrac{p_{ij}}{1 - p_{ii}}, & i \neq j. \end{cases} \tag{1.50}$$

Theorem 1.6.3 *If the jump chain (Y_n) is transient then so is the original chain (X_n).*

Proof If (Y_n) is transient then for all states i

$$\widehat{f}_i = \mathbb{P}_i\big((Y_n) \text{ returns to } i\big) < 1.$$

Now, for (X_n),

$$\begin{aligned} f_i &:= \mathbb{P}_i\big((X_n) \text{ returns to } i\big) = p_{ii} + \sum_{j \neq i} p_{ij}\mathbb{P}_j\big((X_n) \text{ hits } i\big) \\ &= p_{ii} + (1 - p_{ii})\sum_{j \neq i} \frac{p_{ij}}{1 - p_{ii}}\, \mathbb{P}_j\big((X_n) \text{ hits } i\big) \\ &\leq p_{ii} + (1 - p_{ii})\sum_{j \neq i} \widehat{p}_{ij}\, \mathbb{P}_j\big((Y_n) \text{ hits } i\big), \end{aligned}$$

because if (X_n) hits i from j then so does (Y_n). The last expression may be written

$$f_i \leq p_{ii} + (1 - p_{ii})\widehat{f}_i < 1.$$

Hence, $f_i < 1$, and the chain (X_n) is transient. $\qquad\qquad\square$

We will return to this statement later on and give an alternative proof.

Worked Example 1.6.4 (i). Let (X_n, Y_n) be a simple symmetric random walk in $\mathbb{Z}^2$, starting from $(0,0)$, and set $T = \inf\{n \geq 0 : \max\{|X_n|, |Y_n|\} = 2\}$. Determine the quantities $\mathbb{E}T$ and $\mathbb{P}(X_T = 2 \text{ and } Y_T = 0)$.

(ii). Let $(X_n)_{n\geq 0}$ be a DTMC with state space I and transition matrix P. What does it mean to say that a state $i \in I$ is recurrent? Prove that i is recurrent if and only if $\sum_{n=0}^{\infty} p_{ii}^{(n)} = \infty$, where $p_{ii}^{(n)}$ denotes the (i,i) entry in P^n.

Show that the simple symmetric random walk in $\mathbb{Z}^2$ is recurrent.

Solution (i). If $k_i = \mathbb{E}_i T$ and $h_i = \mathbb{P}_i(X_T Y_T = 0)$ then

$$
\begin{aligned}
k_{(0,0)} &= 1 + k_{(-1,0)}, \\
k_{(-1,0)} &= 1 + \frac{k_{(0,0)}}{4} + \frac{k_{(-1,-1)}}{2}, \\
k_{(-1,-1)} &= 1 + \frac{k_{(-1,0)}}{2}, \\
h_{(0,0)} &= h_{(-1,0)}, \\
h_{(-1,0)} &= \frac{1}{4} + \frac{h_{(0,0)}}{4} + \frac{h_{(-1,-1)}}{2}, \\
h_{(-1,-1)} &= \frac{h_{(-1,0)}}{2},
\end{aligned}
$$

by conditioning on the first step, the Markov property and symmetry.

Hence,

$$
\mathbb{E}T = k_{(0,0)} = \frac{9}{2}, \quad h_{(0,0)} = \frac{1}{2}.
$$

By symmetry,

$$
\mathbb{P}(X_T = 2 \text{ and } Y_T = 0) = \frac{1}{4} h_{(0,0)} = \frac{1}{8}.
$$

(ii) The state i is recurrent if $f_i = \mathbb{P}_i(T_i < \infty) = 1$ where $T_i = \inf\{n \geq 1 : X_n = i\}$.

If V_i is the total time spent in i then

$$
\begin{aligned}
\mathbb{P}_i(V_i \geq k+1) &= \mathbb{P}_i(V_i \geq k)\,\mathbb{P}_i(V_i \geq k+1 | V_i \geq k) \\
&= \mathbb{P}_i(V_i \geq k) f_i = \cdots = f_i^{k+1}.
\end{aligned}
$$

Then

$$
\mathbb{E}_i(V_i) = \sum_{k \geq 1} \mathbb{P}(V_i \geq k) = \sum_{k \geq 0} f_i^k.
$$

On the other hand,

$$
\mathbb{E}_i V_i = \mathbb{E}_i \sum_{n \geq 0} \mathbf{1}(X_n = i) = \sum_{n \geq 0} p_{ii}^{(n)}.
$$

Hence, $f_i = 1$ if and only if $\sum_{n \geq 0} p_{ii}^{(n)} = \infty$.

Now let (X_n) be a simple symmetric random walk in $\mathbb{Z}^2$. It is irreducible, hence it suffices to check that $\sum_{n \geq 0} p_{ii}^{(n)} = \infty$ for a single $i \in \mathbb{Z}^2$, say the origin $(0,0)$.

Write $(X_n^{\pm})$ for the projection of (X_n) on the diagonal $\{x = \pm y\}$ in $\mathbb{Z}^2$. Then $(X_n^{\pm})$ are independent simple symmetric random walks on $\frac{1}{\sqrt{2}}\mathbb{Z}$, and return to $(0,0)$ in (X_n) means return to 0 in each of $(X_n^{\pm})$. Next,

$$\mathbb{P}_0(X_{2k}^{\pm} = 0) = \binom{2k}{k}\frac{1}{2^{2k}},$$

and

$$p_{00}^{(2k)} = \mathbb{P}_0(X_{2k}^{+} = 0)\mathbb{P}_0(X_{2k}^{-} = 0).$$

Then Stirling's formula asserts that

$$p_{00}^{(2k)} \approx \left(\frac{\sqrt{2}}{\sqrt{2\pi k}}\frac{(2k)^{2k}}{k^{2k}}\frac{1}{2^{2k}}\right)^2 = \frac{1}{\pi k}, \quad \text{as } k \to \infty.$$

Hence,

$$\sum_{n \geq 0} p_{00}^{(n)} = \sum_{k \geq 0} p_{00}^{(2k)} = \infty.$$

$\square$

Random walks occupy a special place among Markov chains; their (often strikingly beautiful) properties depend on the geometric and algebraic structures (especially symmetries) existing in the state space (in the above examples, the lattice $\mathbb{Z}^d$). In forthcoming sections, we will encounter examples of RWs on other types of graphs and discover more aspects of the related theory.

1.7 Equilibrium distributions: definitions and basic facts

> *Time is a sort of river of passing events, and strong is its current;*
> *no sooner is a thing brought to sight than it is swept by,*
> *and another takes its place, and this too will be swept away.*
> Marcus Aurelius Antoninus (121–80), Roman Emperor

Let (X_n) be a DTMC with transition probability matrix P.

Definition 1.7.1 An initial probability distribution λ is called an *equilibrium* distribution (also a *stationary*, or an *invariant* distribution) if it is preserved in time. That is, for all $j \in I$,

$$\lambda_j = \mathbb{P}(X_0 = j) = \mathbb{P}(X_1 = j) = \cdots = \mathbb{P}(X_n = j) = \cdots. \tag{1.51}$$

As $\mathbb{P}(X_n = j) = \sum_i \lambda_i p_{ij}^{(n)} = (\lambda P^n)_j$, this means that $\lambda = (\lambda_i)$ is an invariant vector for P (that is, an eigenvector with the eigenvalue 1): $\lambda P = \lambda$.

We will denote an equilibrium distribution (ED) by $\pi = (\pi_i)$ and use the equation $\pi P = \pi$ without stressing it every time. Of course, the vector π satisfies two properties: (a) entries $\pi_i \geq 0$ for all $i \in I$ (geometrically, this means that π lies in the non-negative orthant of a Euclidean space); and (b), $\sum_i \pi_i = 1$ (π lies on the hyperplane orthogonal to the vector $\underline{1}$, with all entries 1, that passes through point $(1/|I|,\ldots,1/|I|)$). If we have property (a), but property (b) is not satisfied, we will use the notation μ instead of π and say that μ is an invariant measure (IM): $\mu = (\mu_i)$, $\mu P = \mu$, $\mu_i \geq 0$ for all states i.

One should not confuse two equations $\pi P = \pi$ (invariance) and $Ph = h$ (the hitting time equation).

Example 1.7.2 Consider the 2×2 transition matrix

$$P = \begin{pmatrix} 1-\alpha & \alpha \\ \beta & 1-\beta \end{pmatrix}.$$

Then: (a) if $\alpha + \beta > 0$, it has a unique equilibrium distribution

$$\pi = \left(\frac{\beta}{\alpha+\beta}, \frac{\alpha}{\alpha+\beta} \right);$$

(b) if $\alpha = \beta = 0$ then $P = \begin{pmatrix} 1 & 0 \\ 0 & 1 \end{pmatrix}$, and every vector (x,y) is invariant.

Example 1.7.3 Let $a,b \geq N$, $a,b,N \in \mathbb{Z}_+$. Consider a birth-death Markov chain on $n = 0,1,\ldots,N$ with

$$\lambda_n = (N-n)(a-n), \quad \mu_n = n(b-(N-n)).$$

Show that the equilibrium distribution is hypergeometric

$$\pi_i = \frac{\dbinom{a}{i}\dbinom{b}{N-i}}{\dbinom{a+b}{N}}, \quad i = 0,1,\ldots,N.$$

The non-uniqueness of an ED may occur when the chain has more than one closed communicating class. It may have equilibrium distributions supported by different closed classes. See Figure 1.4. An open communicating class cannot support an equilibrium distribution as π_i always vanishes for states i from open classes.

The multitude of closed communicating classes is the only source of non-uniqueness of an ED, and an irreducible transition matrix P has at most one ED (i.e. one or none). A finite irreducible matrix P always has a unique ED.

Next, if P is (countable) irreducible and transient then it has no ED.

Furthermore, if P is irreducible and recurrent then two cases can occur:

(a) P has a (unique) equilibrium distribution π. Then all probabilities $\pi_i > 0$. In this case we say that P is positive recurrent.

(b) P has no equilibrium distribution. Then we say that P is null recurrent.

More precisely, every irreducible recurrent matrix P has an IM μ, with $\mu P = \mu$ and all entries $\mu_i > 0$. But the series $\sum_{i \in I} \mu_i$ may converge or diverge, and in Definition 1.7.6 below we distinguish two cases:

$$\sum_i \mu_i < \infty : \quad P \text{ positive recurrent,}$$

$$\sum_i \mu_i = \infty : \quad P \text{ null recurrent.}$$

Note the following. Solutions to $\mu P = \mu$ admit addition $((\mu_1 + \mu_2)P = \mu_1 P + \mu_2 P)$ and multiplication by a constant $((c\mu)P = c(\mu P))$. Hence we can compute $\mu_i / \sum_j \mu_j = \pi_i$ to get $\sum_j \pi_j = 1$ (when $\sum_j \mu_j < \infty$). We will see that for an irreducible chain all IMs μ are proportional to each other: $\mu' = c\mu$. In particular, they all have $\mu_i > 0$ for all $i \in I$.

We now turn to the proof of the above properties. The key statement here is Theorem 1.7.4. Set

$$\gamma_i^k = \mathbb{E}_k \sum_{n=0}^{T_k - 1} \mathbf{1}(X_n = i)$$

$$= \begin{cases} \mathbb{E}_k \big(\text{number of visits to } i \text{ before returning to } k\big), \\ \qquad\qquad \text{if } i \neq k \text{ (with } 1 \leq n < T_k), \\ 1, \qquad\qquad \text{if } i = k \text{ (from } n = 0). \end{cases} \quad (1.52)$$

Here, as in (1.23), T_k is the return time to state k:

$$T_k = \inf [n \geq 1 : X_n = k]. \quad (1.53)$$

Then $0 \leq \gamma_i^k \leq \infty$. Consider vectors $\gamma^k = (\gamma_i^k, i \in I)$, parametrised by $k \in I$. Observe that

$$\sum_{i \in I} \gamma_i^k = 1 + \sum_{i \in I : i \neq k} \mathbb{E}_k \big(\text{number of visits to } i \text{ before returning to } k\big)$$

$$= 1 + \mathbb{E}_k(T_k - 1) = \mathbb{E}_k T_k. \quad (1.54)$$

Theorem 1.7.4 (a) *For all states* k

$$\big(\gamma^k P\big)_j := \sum_{i \in I} \gamma_i^k p_{ij} = \gamma_j^k, \quad j \neq k \quad \text{(invariance)}, \quad (1.55)$$

and

$$\big(\gamma^k P\big)_k := \sum_{i \in I} \gamma_i^k p_{ik} \leq 1 = \gamma_k^k \quad \text{(sub-invariance)}. \quad (1.56)$$

(b) We have that $(\gamma^k P)_k = 1$ if and only if k is recurrent. Hence, the vector γ^k is invariant: $\gamma^k = \gamma^k P$ if and only if the state k is recurrent.

(c) If P is irreducible and recurrent then

$$0 < \gamma_i^k < \infty \text{ for all states } i, k \in I.$$

Hence, for an irreducible and recurrent matrix P, the vector γ^k is a 'genuine' invariant vector with strictly positive and finite entries.

Proof (a) By the Markov property for all $m \geq 2$ and states $i \neq k$ and $j \neq k$,

$$\mathbb{P}_k(T_k > m - 1, X_{m-1} = i) p_{ij} = \mathbb{P}_k(T_k > m - 1, X_{m-1} = i, X_m = j), \quad (1.57)$$

and

$$\mathbb{P}_k(T_k > m - 1, X_{m-1} = i) p_{ik} = \mathbb{P}_k(T_k = m, X_{m-1} = i). \quad (1.58)$$

Then, for $j \neq k$,

$$
\begin{aligned}
\gamma_j^k &= \mathbb{E}_k \sum_{0 \leq n \leq T_k - 1} \mathbf{1}(X_n = j) = \sum_{n \geq 1} \mathbb{E}_k \mathbf{1}(X_n = j, T_k > n) \\
&= \sum_{n \geq 1} \mathbb{P}_k(X_n = j, T_k > n) \\
&= p_{kj} + \sum_{n \geq 2} \sum_{i : i \neq k} \mathbb{P}_k(T_k > n - 1, X_{n-1} = i, X_n = j) \\
&= p_{kj} + \sum_{n \geq 2} \sum_{i : i \neq k} \mathbb{P}_k(T_k > n - 1, X_{n-1} = i) p_{ij} \text{ by (1.57)} \\
&= \gamma_k^k p_{kj} + \sum_{i : i \neq k} \sum_{n \geq 1} \mathbb{E}_k \mathbf{1}(T_k > n, X_n = i) p_{ij} = \left(\gamma^k P\right)_j.
\end{aligned}
$$

Further, for $j = k$,

$$
\begin{aligned}
\left(\gamma^k P\right)_k &= \sum_{i \in I} \gamma_i^k p_{ik} = \gamma_k^k p_{kk} + \sum_{i : i \neq k} \gamma_i^k p_{ik} \\
&= p_{kk} + \sum_{i : i \neq k} \mathbb{E}_k \left(\sum_{1 \leq n < T_k} \mathbf{1}(X_n = i) \right) p_{ik} \\
&= p_{kk} + \sum_{i : i \neq k} \sum_{n \geq 1} \mathbb{E}_k \mathbf{1}(T_k > n, X_n = i) p_{ik} \\
&= p_{kk} + \sum_{i : i \neq k} \sum_{n \geq 1} \mathbb{P}_k(T_k = n + 1, X_n = i) \text{ by (1.58)} \\
&= p_{kk} + \sum_{n \geq 2} \mathbb{P}_k(T_k = n) \\
&= \sum_{n \geq 1} \mathbb{P}_k(T_k = n) = \mathbb{P}_k(T_k < \infty) := f_k \leq 1 = \gamma_k^k.
\end{aligned}
$$

Observe that so far we have not used recurrence.

(b) From the last equation, $(\gamma^k P)_k = 1$ if and only if $f_k = 1$, i.e., the state k is recurrent.

(c) If P is irreducible then for all $i, k \in I$ there exist $m, n \geq 0$ such that $p_{ik}^{(n)} > 0$ and $p_{ki}^{(m)} > 0$. Assuming that P is recurrent, the vector γ^k is invariant and hence $\gamma^k P^m = \gamma^k P^n = \gamma^k$. So,

$$\gamma_i^k = \sum_l \gamma_l^k p_{li}^{(m)} \geq \gamma_k^k p_{ki}^{(m)} = p_{ki}^{(m)}.$$

On the other hand

$$1 = \gamma_k^k = \sum_l \gamma_l^k p_{lk}^{(n)} \geq \gamma_i^k p_{ik}^{(n)}, \text{ i.e., } \gamma_i^k \leq \frac{1}{p_{ik}^{(n)}}.$$

$\square$

Theorem 1.7.5 *Suppose that* $\mu = (\mu_i)$ *is an IM: thus* $\mu P = \mu$ *and* $\mu_i \geq 0$ *for all* $i \in I$. *Suppose in addition that* $\mu_k = 1$ *for some given state* k. *Then: (a) for all* $i \in I$,

$$\mu_i \geq \gamma_i^k;$$

(b) for an irreducible and recurrent matrix P, *we have*

$$\mu_i = \gamma_i^k, \text{ for all } i \in I.$$

Proof (a) Invariance plus the fact that $\mu_k = 1$ imply that, for all $j \in I$ and $n \geq 1$,

$$
\begin{aligned}
\mu_j &= \sum_i \mu_i p_{ij} = 1 \cdot p_{kj} + \sum_{i: i \neq k} \mu_i p_{ij} = p_{kj} + \sum_{i \neq k} \sum_l \mu_l p_{li} p_{ij} \\
&= p_{kj} + \sum_{i \neq k} p_{ki} p_{ij} + \sum_{i \neq k} \sum_{l \neq k} \mu_l p_{li} p_{ij} = \cdots \\
&= p_{kj} + \sum_{i \neq k} p_{ki} p_{ij} + \cdots + \sum_{i_1, \ldots, i_{n-1} \neq k} p_{ki_1} \cdots p_{i_{n-1} j} \\
&\quad + \sum_l \sum_{i_1, \ldots, i_{n-1} \neq k} \mu_l p_{li_1} \cdots p_{i_{n-1} j}.
\end{aligned}
$$

Now, the non-negativity implies that the last expression is

$$\geq \mathbb{P}_k (X_1 = j, T_k > 1) + \mathbb{P}_k (X_2 = j, T_k > 2) + \cdots$$
$$+ \mathbb{P}_k (X_n = j, T_k > n),$$

which tends to γ_j^k as $n \to \infty$.

(b) Now let P be irreducible and recurrent. Then γ^k is invariant: $\gamma^k P = \gamma^k$. Then $\tilde{\mu} = \mu - \gamma^k$ is also invariant: $\tilde{\mu} = \tilde{\mu} P$, and, owing to (a), non-negative: $\tilde{\mu}_i \geq 0$ for all $i \in I$. But, for $i = k$, $\tilde{\mu}_k = \mu_k - \gamma_k^k = 1 - 1 = 0$.

Next, given $i \in I$, there exists $n \geq 1$ with $p_{ik}^{(n)} > 0$. Then, as

$$0 = \tilde{\mu}_k = \sum_l \tilde{\mu}_l p_{lk}^{(n)} \geq \tilde{\mu}_i p_{ik}^{(n)},$$

we obtain that $\widetilde{\mu}_i = 0$. Hence, $\widetilde{\mu} = 0$ and $\mu = \gamma^k$.   □

We see that for an irreducible recurrent chain, everything is fixed by the condition $\mu_k = 1$. More precisely, if μ is a non-zero IM, i.e. $\mu P = \mu$, $\mu_i \geq 0$ and $\mu_k > 0$ for some state k, then

$$\mu = \mu_k \gamma^k.$$

This implies that all non-zero IMs are proportional: $\mu' = c\mu$. Next, every non-zero IM has all entries finite and strictly positive. In particular, all vectors γ^k are proportional:

$$\gamma_i^k \gamma^i = \gamma^k, \quad i, k \in I. \tag{1.59}$$

Now, for an irreducible recurrent chain, we have two cases: (i) all non-zero IMs μ have

$$\sum_{j \in I} \mu_j < \infty, \tag{1.60}$$

and (ii) all non-zero IMs μ have

$$\sum_{j \in I} \mu_j = \infty. \tag{1.61}$$

Definition 1.7.6 In the case (i) we call the irreducible Markov chain (or matrix P) *positive recurrent*, and in case (ii) *null recurrent*.

If the number of states $|I| < \infty$ then the case (ii) is impossible. Hence, an irreducible finite DTMC is always positive recurrent and has a (unique) equilibrium distribution $\pi = (\pi_i)$. Furthermore, equilibrium probabilities π_i are strictly positive.

We now see that, in general, when P is positive recurrent then normalising $\mu_j / \sum_i \mu_i = \pi_j$ yields a (unique) equilibrium distribution. It has all $\pi_i > 0$. Then γ^k is recovered by division:

$$\gamma^k = \frac{1}{\pi_k} \pi, \text{ i.e., } \gamma_i^k = \frac{\pi_i}{\pi_k}. \tag{1.62}$$

In other words, we obtain the following

Theorem 1.7.7 *In an irreducible positive recurrent chain with equilibrium distribution π, for all states $k \neq i$*

$$\mathbb{E}_k (\text{the number of visits to } i \text{ before returning to } k) = \frac{\pi_i}{\pi_k}. \tag{1.63}$$

For $i = k$ we obtain

Theorem 1.7.8 *In an irreducible positive recurrent chain with equilibrium distribution π, for all states k,*

$$m_k := \mathbb{E}_k T_k = \text{the mean return time to state } k = \frac{1}{\pi_k} < \infty. \qquad (1.64)$$

Proof In (1.54) we observed that

$$\mathbb{E}_k T_k = 1 + \mathbb{E}_k(T_k - 1) = 1 + \sum_{i:i\neq k} \gamma_i^k = \sum_i \gamma_i^k < \infty.$$

Hence,

$$m_k = \sum_i \frac{\pi_i}{\pi_k} = \frac{1}{\pi_k},$$

implying that $m_k = 1/\pi_k$. $\qquad\qquad\square$

Our results in this section are summarised in Table 1.1.

1.8 Positive and null recurrence

> *Not to know what has been transacted in former times*
> *is to be always a child.*
> *If no use is made of the labours of past ages,*
> *the world must remain always in the infancy of knowledge.*
> Marcus Tullius Cicero (106–43 BC), Roman orator and statesman

Throughout this section we work with initial distributions $\lambda = \delta_i$, i.e. consider DTMCs starting from a particular state, and use the above notations $\mathbb{P}_i$ and $\mathbb{E}_i$. The state space I is assumed to be countably infinite (and further specified in examples below). For simplicity, we omit reference to I: statements of the type 'for all i' mean for all $i \in I$, and we assume that the transition matrix P is irreducible.

We begin by elaborating Definitions 1.5.1 and 1.7.6. Recall

$$T_i = \min[n \geq 1 : X_n = i]$$

stands for the return time to state i.

(I) Irreducible DTMCs with more than one state have transition probabilities $0 < p_{ij} < 1$ for all states $i, j \in I$ (no absorption).

(II) An irreducible DTMC (X_n) can be transient or recurrent:
(i) Transient: $\mathbb{P}_i(\text{return time } T_i < \infty) < 1$, i.e. $\mathbb{P}_i(T_i = \infty) > 0$, for all $i \in I$. Equivalently:
$\mathbb{P}_i(i \text{ is not visited in } (X_n) \text{ after some finite time}) = 1$
and $\sum_{n \geq 0} p_{ii}^{(n)} < \infty$, for all $i \in I$. Equivalently:
$h_j^{\{i\}} = \mathbb{P}_j(\text{hit } i) < 1$, for some states j and i.
(ii) Recurrent: $\mathbb{P}_i(\text{return time } T_i < \infty) = 1$,
i.e., $\mathbb{P}_i(T_i = \infty) = 0$ for all $i \in I$. Equivalently:
$\mathbb{P}_i(i \text{ visited in } (X_n) \text{ at arbitrarily large times}) = 1$
and $\sum_{n \geq 0} p_{ii}^{(n)} = \infty$, for all $i \in I$. Equivalently: $h_j^{\{i\}} = \mathbb{P}_j(\text{hit } i)$
$= 1$, for all states j and i. In this case, for all i, the vector
$\gamma^i = (\gamma_j^i)$ from (1.62) has $0 < \gamma_j^i < \infty$ and gives an IM for (X_n);
all such IMs are of the form $\alpha \gamma^i$. In particular,
$$\text{vector } \gamma^k = (\gamma_k^i)^{-1} \times \text{ vector } \gamma^i, \text{ for all states } i, k.$$

(III) Next, an irreducible recurrent DTMC can be
(i) Null Recurrent: $m_i = \mathbb{E}_i(\text{return time } T_i) = \infty$,
for all $i \in I$; in this case there is no IM $\mu = (\mu_i)$ with $\sum_j \mu_j < \infty$.
Hence, there is no ED.
(ii) Positive Recurrent: $m_i < \infty$, for all $i \in I$; in this case any invariant measure $\mu = (\mu_i)$ has $\sum_j \mu_j < \infty$, and there exists
a unique equilibrium distribution $\pi = (\pi_i)$,
where $\pi_i = (\mu_i / \sum_j \mu_j) > 0$. In this case, $\gamma^k = m_k \pi$. Furthermore,
$$\mathbb{E}_i T_i = \frac{1}{\pi_i}, \quad \text{and } \mathbb{E}_i(\text{time at } k \text{ before } T_i) = \frac{\pi_k}{\pi_i}, \quad \text{for all states } i, k.$$
Finite irreducible DTMCs are always PR.

Table 1.1

Definition 1.8.1 Set $f_i = \mathbb{P}_i(T_i < \infty)$ and $m_i = \mathbb{E}_i T_i$. A state i is called

$$
\begin{aligned}
\textit{recurrent (R), if} \quad & f_i = 1; \text{ equivalently, } \sum_n p_{ii}^{(n)} = \infty, \text{ or} \\
& \mathbb{P}_i(X_n = i \text{ for infinitely many } n) = 1, \\
\textit{positive recurrent (PR), if} \quad & m_i = \mathbb{E}_i T_i < \infty, \\
\textit{null recurrent (NR), if} \quad & m_i = \mathbb{E}_i T_i = \infty, \text{ but } f_i = 1, \\
\textit{transient (T), if} \quad & f_i < 1; \text{ equivalently, } \sum_n p_{ii}^{(n)} < \infty, \text{ or} \\
& \mathbb{P}_i(X_n = i \text{ for infinitely many } n) = 0. \quad (1.65)
\end{aligned}
$$

As these are class properties, in the case of an irreducible matrix P, either all states are PR or all states are NR or all states are T.

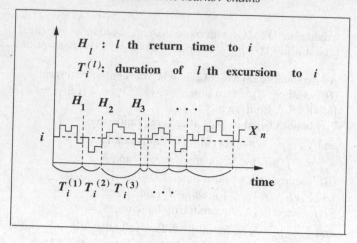

H_l : l th return time to i

$T_i^{(l)}$: duration of l th excursion to i

Fig. 1.13

Definition 1.8.2 Given $l = 0, 1, \ldots$, define the subsequent return (or passage) times $H_l(= H_l^i)$ to state i by $H_0 = 0$, $H_1 = T_i$, and

$$H_l = \inf \left[n \geq H_{l-1} + 1 : X_n = i \right], \ l > 1. \tag{1.66}$$

The difference

$$T_i^{(l)} = \begin{cases} H_l - H_{l-1}, & \text{if } H_{l-1} < \infty, \\ 0, & \text{if } H_{l-1} = \infty, \end{cases} \tag{1.67}$$

gives the time between the $(l-1)$st and lth return times to i, or the duration of the lth excursion to states i, $l = 1, 2, \ldots$. Obviously, $T_i^{(1)} = T_i$. See Figure 1.13.

> *May you always live in interesting return times.*
> (From the series *'Thus spoke Superviser'*.)

The above analysis of positive and null recurrence combined with the strong Markov property leads to the following

Theorem 1.8.3 *Assume that the chain (X_n) is recurrent and let i be any state. Under the distribution $\mathbb{P}_i$, the variables $T_i^{(1)}$, $T_i^{(2)}$, $\ldots$ are independent and identically distributed (IID) random variables, with positive integer values, finite with probability 1. That is, for all $k \geq 1$ and positive integers $t_1, \ldots, t_k$,*

$$\mathbb{P}_i(T_i^{(1)} = t_1, \ldots, T_i^{(k)} = t_k) = \prod_{1 \leq l \leq k} \mathbb{P}_i(T_i = t_l), \text{ and } \sum_{t=1,2,\ldots} \mathbb{P}_i(T_i = t) = 1. \tag{1.68}$$

Furthermore, the expectation

$$m_i := \mathbb{E}_i T_i = \begin{cases} 1/\pi_i, & \text{if the chain } (X_n) \text{ is PR,} \\ \infty, & \text{if the chain } (X_n) \text{ is NR or T.} \end{cases} \quad (1.69)$$

Here π (π_i) is the (unique) equilibrium distribution of the positive recurrent DTMC (X_n).

The example of IID random variables (RVs) $T_i^{(1)}, T_i^{(2)}, \ldots$ is quite intriguing, as their (common) distribution is determined by the transition matrix P and varies in a rather intricate way when we change P. Therefore, to analyse the sequence $(T_i^{(n)})$, one needs to develop a general theory of IID RVs (in particular, it was one of the strong motives for a general theory of summation of IID RVs).

An example of a general statement about IID RVs which we will use in the next section is the following 'strong' Law of Large Numbers (LLN) for the sequence $(T_i^{(n)})$:

Theorem 1.8.4 *Under the assumptions of Theorem 1.8.3, for all states i, with probability 1, the average*

$$\frac{1}{n} \left(T_i^{(1)} + T_i^{(2)} + \cdots + T_i^{(n)} \right)$$

converges, as $n \to \infty$, to the expected value m_i specified in (1.69); symbolically $\left(T_i^{(1)} + T_i^{(2)} + \cdots + T_i^{(n)} \right) / n \overset{\mathbb{P}_i-\text{a.s.}}{\to} m_i$. *That is,*

$$\mathbb{P}_i \left(\lim_{n \to \infty} \frac{1}{n} \sum_{l=1}^{n} T_i^{(l)} = m_i \right) = 1. \quad (1.70)$$

Remark 1.8.5 In previous sections we have already used various properties and facts that hold with probability 1; there will be more examples of this in the forthcoming sections. It has to be said that some of these facts and properties are rather delicate and require careful analysis. An example of such a property is convergence with probability 1 in Theorem 1.8.4. (This property is behind the term 'strong', as opposed to 'weak', LLN; see below.) The alternative term for this form of convergence is 'almost sure convergence with respect to the probability distribution $\mathbb{P}_i$', which is reflected in the notation $\overset{\mathbb{P}_i-\text{a.s.}}{\to}$ that we often use. When the probability distribution in question is specified from the context, we write $\overset{\text{a.s.}}{\to}$. We will discuss properties of convergence with probability 1 in more detail in Chapter 3.

Remark 1.8.6 We want to stress that the statement of Theorem 1.8.4 holds in 'full generality', regardless of whether the value m_i is finite or infinite, let alone existence of a finite *second moment* $\mathbb{E}\big(T_i^{(1)}\big)^2$ or finite *higher moments* $\mathbb{E}\big(T_i^{(1)}\big)^n$, $n \geq 3$. In fact, the assertion of Theorem 1.8.4 holds in a much wider context of *ergodic processes*.

We will not discuss here the proof of Theorem 1.8.4; the interested reader is referred to more advanced books, e.g., Grimmett & Stirzaker, 1982, Stroock, 2005.

Example 1.8.7 Random walks on $\mathbb{Z}^d$

(a) *Symmetric nearest-neighbour random walk.* We know that the symmetric nearest-neighbour RW on $\mathbb{Z}^d$ (also called the *simple* RW) is recurrent for $d = 1$ and $d = 2$ and transient for $d = 3$. First, consider $d = 1$. The invariance equations read

$$\pi_i = \frac{1}{2}\,\pi_{i-1} + \frac{1}{2}\,\pi_{i+1}, \ i \in \mathbb{Z},$$

and have an obvious non-negative solution $\pi_i \equiv 1$ (which is unique, up to a positive factor). As $\sum_{i \in \mathbb{Z}} 1$ diverges, the walk is null recurrent.

Hence, any IM $\lambda \geq \underline{0}$ has $\lambda_i = \text{const} > 0$. Then, for all $i \neq k$,

$$\gamma_i^k = \mathbb{E}_k\big(\text{number of visits to } i \text{ before returning to } k\big) = 1.$$

[You may find this surprising since it might be expected that

$$1 < \gamma_{k+1}^k < \gamma_{k+2}^k < \cdots .]$$

More precisely,

$$\mathbb{P}_k\big(\text{number of visits to } i \text{ before returning to } k \text{ is } n\big)$$
$$= \left(\frac{1}{2|k-i|}\right)^2 \left(1 - \frac{1}{2|k-i|}\right)^{n-1},$$

see Worked Example 1.8.9 below. Also

$$m_k = \mathbb{E}_k(\text{return time to } k) = \infty, \ k \in \mathbb{Z}.$$

For $d = 2$, the invariance equations are similar

$$\pi_{(i_1,i_2)} = \frac{1}{4}\sum\big(\pi_{(i_1\pm 1,i_2)} + \pi_{(i_1,i_2\pm 1)}\big), \ \underline{i} = (i_1,i_2) \in \mathbb{Z}^2,$$

and again have $\pi_{\underline{i}} \equiv 1$ as a solution. Hence, the walk is null recurrent, and as before,

$$\gamma_{\underline{i}}^k \equiv 1.$$

For $d = 3$, $\pi_{\underline{i}} \equiv 1$ is still an IM (this remains true for all d). However, as the walk is transient, the vectors γ^k are sub-invariant, not invariant. Hence, it is no longer true that $\gamma_{\underline{i}}^k \equiv 1$, although $m_{\underline{k}}$ is still $\equiv \infty$.

(b) *Asymmetric nearest-neighbour random walk on* $\mathbb{Z}$. Here the invariance equations are

$$\pi_i = p\pi_{i-1} + (1-p)\pi_{i+1}, \ \ i \in \mathbb{Z},$$

and $p \neq 1/2$. The RW is transient. A general non-negative solution

$$\pi_i = A + B\left(\frac{p}{1-p}\right)^i$$

contains two parameters, $A, B \geq 0$, and violates $\sum_i \pi_i < \infty$. We see that not all IMs λ are proportional. Again, the γ^k are sub-invariant, not invariant. Also, it is not true that γ_i^k is of the form λ_i/λ_k for some IM λ. But again $m_k \equiv \infty$, as

$$1 - f_k = \mathbb{P}_k(\text{no return to } k \text{ in a finite time}) > 0.$$

Example 1.8.8 (Homogeneous birth-and-death process) This is a RW on the state space $\mathbb{Z}_+ = \{0, 1, 2, \ldots\}$, with

$$p_{ii+1} = p, \ p_{ii-1} = 1-p, \ i \geq 1, \ p_{01} = q, \ p_{00} = 1-q,$$

where $0 \leq p, q \leq 1$. Consider the case $0 < q \leq 1$ and $0 < p < 1$, when the chain is irreducible. Then the answer is

$$\begin{aligned} p < 1/2: &\quad \text{positive recurrent,} \\ p = 1/2: &\quad \text{null recurrent,} \\ p > 1/2: &\quad \text{transient,} \end{aligned}$$

regardless of q.

In fact, the invariance equations

$$\begin{aligned} \pi_i &= p\pi_{i-1} + (1-p)\pi_{i+1}, \ i > 1, \\ \pi_1 &= q\pi_0 + (1-p)\pi_2, \\ \pi_0 &= (1-q)\pi_0 + (1-p)\pi_1, \end{aligned}$$

still admit the solution $\pi_i = A + B(p/(1-p))^i, i > 0$.

For $p < 1/2$, a further reduction seems reasonable: $A = 0$. At $i = 0, 1$ we obtain the same equation

$$q\pi_0 = pB.$$

To normalise, write

$$1 = B\left(\frac{p}{q} + \frac{p}{1-p} + \frac{p^2}{(1-p)^2} + \cdots\right) = B\left(\frac{p}{q} + \frac{p/(1-p)}{1 - p/(1-p)}\right)$$

$$= B\frac{p(1-2p+q)}{q(1-2p)}, \quad \text{whence } B = \frac{q(1-2p)}{p(1+q-2p)}.$$

Therefore,

$$\pi_0 = \frac{1-2p}{1+q-2p}, \quad \pi_i = \frac{q}{p}\left(\frac{p}{1-p}\right)^i \pi_0, \quad i \geq 1,$$

and the chain is positive recurrent, as claimed.
Further, for $p < 1/2$,

$$\gamma_i^k = \mathbb{E}_k\left(\text{number of visits to } i \text{ before returning to } k\right) = \frac{\pi_i}{\pi_k}$$

$$= \begin{cases} \left(\dfrac{p}{1-p}\right)^{i-k}, & 0 < i, k < \infty, \ i \neq k, \\[2mm] \dfrac{q}{p}\left(\dfrac{p}{1-p}\right)^i, & 0 = k < i < \infty, \\[2mm] \dfrac{p}{q}\left(\dfrac{1-p}{p}\right)^k, & 0 = i < k < \infty, \end{cases}$$

and

$$m_k = \mathbb{E}_k(\text{return time to } k) = \frac{1}{\pi_k}, \quad k \in \mathbb{Z}_+.$$

For $p \geq 1/2$, we have to consider $f_i = \mathbb{P}_i(T_i < \infty)$. Writing

$$\mathbb{P}_0(T_0 < \infty) = 1 - q + q\,\mathbb{P}_1(\text{hit } 0),$$

we see that if $\mathbb{P}_1(\text{hit } 0) < 1$, the chain is transient. But

$$\mathbb{P}_i(\text{hit } i-1) = \frac{1-p}{p}, \quad i \geq 1;$$

see Section 1.5. Hence, for $p > 1/2$ the chain is transient.
It remains to check the case $p = 1/2$. Here, $f_i = 1$, and the chain is recurrent. The invariance equations

$$\pi_i = \frac{1}{2}\pi_{i-1} + \frac{1}{2}\pi_{i+1}, \quad i > 1,$$

have the general solution $\pi_i = A + Bi$, $i \geq 1$. At $i = 1, 0$ they have the form

$$\pi_1 = q\pi_0 + \frac{1}{2}\pi_2, \quad \pi_0 = (1-q)\pi_0 + \frac{1}{2}\pi_1,$$

which yields $B = 0$ and

$$\pi_i \equiv A, \quad i \geq 1, \quad \pi_0 = \frac{1}{2q}A,$$

and the non-negative IMs correspond to $A \geq 0$. We see that the inequality $\sum_i \pi_i < \infty$ cannot hold unless $A = 0$. Thus, the chain does not have an equilibrium distribution, and hence is null recurrent.

Then, for $p = 1/2$, (i) all non-negative IMs $\lambda = (\lambda_i)$ are proportional to each other, and each such measure different from 0 has $\lambda_i = A > 0$ for $i \geq 1$ and $\lambda_0 = A/(2q) > 0$. Furthermore, (ii) all vectors γ^k, $k \geq 0$, must be invariant and hence proportional to each other. With the normalisation $\gamma_k^k = 1$, the only possibility is that (a) $\gamma_i^k \equiv 1$ and $\gamma_0^k = 1/(2q)$ for all $k, i > 1$ and (b) $\gamma_i^0 = 2a$ for all $i > 1$. (This looks even more surprising, as one might expect that for $k \geq 1$

$$\gamma_0^k < \cdots < \gamma_{k-2}^k < \gamma_{k-1}^k < 1 < \gamma_{k+1}^k < \gamma_{k+2}^k < \cdots,$$

and there is no reason to believe that $\gamma_{k-1}^k = \gamma_{k+1}^k$ because of the asymmetry of the model.)

Finally, it is not difficult to check that for all $i > k \geq 1$

$$\mathbb{P}_k\left(\text{number of visits to } i \text{ before returning to } k \text{ is } n\right)$$

$$= \left(\frac{1}{2(i-k)}\right)^2 \left(1 - \frac{1}{2(i-k)}\right)^{n-1},$$

as in the case of the symmetric RW on $\mathbb{Z}$.

Cherchez la Gamme: a Musical On Vectorial Return Times
(From the series *'Movies that never made it to the Big Screen'*.)

Worked Example 1.8.9

(i) Let $X = (X_n : n \geq 0)$ be a random walk on the integers, which moves one step rightwards or one step leftwards with probability $1/2$, at each time step. Show that

$$\mathbb{P}(X_{2n} = 0 | X_0 = 0) = \binom{2n}{n}\left(\frac{1}{2}\right)^{2n},$$

and deduce that X is recurrent.

(ii) Let X be given as above, and assume that $X_0 = 0$. Let m be a strictly positive integer, and let N be the number of visits to the point m before returning to 0.
Find $\mathbb{P}(N \geq 1)$, and deduce that

$$\mathbb{P}(N = n) = \left(\frac{1}{2m}\right)^2 \left(1 - \frac{1}{2m}\right)^{n-1}, \quad n \geq 1.$$

Solution (i) First, $\mathbb{P}(X_{2n} = 0 | X_0 = 0) = p_{00}^{(2n)}$ is the probability that the sample path of length $2n$ starts at and returns to 0. This is because each such path must have n steps right and n steps left, the total number of such paths is $\binom{2n}{n}$, and each of them has the same probability $(1/2)^{2n}$. Hence the formula for $\mathbb{P}(X_{2n} = 0 | X_0 = 0)$.

The sum $\sum_{n=1}^{\infty} \mathbb{P}(X_n = 0 | X_0 = 0)$ coincides with $\sum_{n=1}^{\infty} \mathbb{P}(X_{2n} = 0 | X_0 = 0)$ (return at odd times is not possible), and is evaluated via Stirling's formula: $n! \approx \sqrt{2\pi} n^{n+1/2} e^{-n}$. This leads to the series $\sum_n 1/(\sqrt{\pi n})$ which diverges. So, by Theorem 1.5.3, the state 0 is recurrent. The same argument works for every state i. Hence, the chain is recurrent. (The same conclusion holds because recurrence is a class property; see Theorem 1.5.7.)

(ii) For the present write $\mathbb{P}$ to mean $\mathbb{P}_0$, the distribution of the (δ_0, P) chain. Then $\mathbb{P}(N \geq 1) = \mathbb{P}_0(\text{hit } m \text{ before returning to } 0)$. By conditioning on the first step, we have

$$\mathbb{P}(N \geq 1) = \frac{1}{2} \mathbb{P}_1(\text{hit } m \text{ before visiting } 0),$$

where $\mathbb{P}_i$ stands for the distribution of the (δ_i, P) chain. Set

$$h_i = \mathbb{P}_i(\text{hit } m \text{ before visiting } 0);$$

then

$$h_i = \frac{1}{2} h_{i-1} + \frac{1}{2} h_{i+1}, \quad 1 \leq i < m.$$

The general solution $h_i = A + Bi$ is specified by $h_0 = 0$, $h_m = 1$: $A = 0$, $B = 1/m$. Hence, $h_1 = 1/m$, and $\mathbb{P}(N \geq 1) = 1/(2m)$.

Clearly, $1 - 1/(2m) = \mathbb{P}(N = 0) = \mathbb{P}_0(\text{hit } 0 \text{ again before visiting } m)$. By symmetry,

$$\mathbb{P}_m(\text{hit } m \text{ again before visiting } 0) = 1 - \frac{1}{2m}.$$

To be in event $\{N = n\}$, a sample path from 0 must hit m before returning to 0, return to m $n-1$ times without visiting 0 and then proceed to 0 without returning to m. By the strong Markov property,

$$\mathbb{P}(N = n) = \frac{1}{2m} \left(1 - \frac{1}{2m}\right)^{n-1} \frac{1}{2m},$$

the last factor being $\mathbb{P}_m(\text{hit } 0 \text{ before returning to } m)$, again by symmetry. Hence the result.

$\square$

Worked Example 1.8.10 Consider a Markov chain on the state space $I = \{0, 1, 2, \ldots\} \cup \{1', 2', 3', \ldots\}$ with transition probabilities as illustrated in Figure 1.14 where $0 < q < 1$ and $p = 1 - q$.

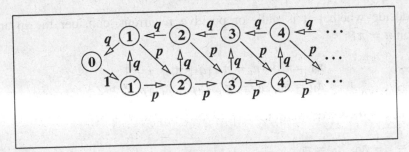

Fig. 1.14

For each value of q, determine whether the chain is transient, null recurrent or positive recurrent.

When the chain is positive recurrent, calculate the invariant distribution.

Solution For $i \geq 1$ set

$$a = \mathbb{P}_i(\text{hit } i-1), \quad b = \mathbb{P}_{i'}(\text{hit } i);$$

these probabilities do not depend on the value of i because of the homogeneous property of the chain. Conditioning on the first jump and using the strong Markov property we get

$$a = q + pba^2, \quad b = q + pba,$$

whence

$$b = \frac{q}{1-pa}, \text{ and } a = q + \frac{pqa^2}{1-pa}.$$

Thus,

$$p(1+q)a^2 - (pq+1)a + q = 0,$$

and the solutions are

$$a = 1 \text{ and } a = \frac{q}{1-q^2}.$$

We are interested in the minimal solution

$$\frac{q}{1-q^2} < 1 \text{ if and only if } q < \frac{\sqrt{5}-1}{2}.$$

Therefore, the chain is recurrent if and only if $q \geq \left(\sqrt{5}-1\right)/2$ and transient if and only if $q < \left(\sqrt{5}-1\right)/2$.

To decide whether it is null- or positive recurrent, consider the invariance equation $\pi = \pi P$:

$$\pi_0 = \pi_1 q, \ \pi_i = \pi_{i+1} q + \pi_{i'} q, \ i \geq 1,$$
$$\pi_{1'} = \pi_0, \ \pi_{i'} = \pi_{(i-1)'} p + \pi_{i-1} p, \ i' \geq 2.$$

This admits a recursive solution:

$$\pi_1 = \frac{1}{q} \pi_0, \ \pi_{1'} = \pi_0,$$

$$\pi_2 = \left(\frac{1}{q^2} - 1 \right) \pi_0 = \frac{1}{q} \frac{1-q^2}{q} \pi_0, \ \pi_{2'} = (1-q) \left(1 + \frac{1}{q} \right) \pi_0 = \frac{1-q^2}{q} \pi_0,$$

$$\pi_3 = \frac{1}{q} \left(\frac{1-q^2}{q} \right)^2 \pi_0, \ \pi_{3'} = \left(\frac{1-q^2}{q} \right)^2 \pi_0.$$

By induction, one gets the general formulas

$$\pi_i = \frac{1}{q} \left(\frac{1-q^2}{q} \right)^{i-1} \pi_0, \ \pi_{i'} = \left(\frac{1-q^2}{q} \right)^{i-1} \pi_0,$$

and the equilibrium distribution will exist if and only if both series converge; that is, $(1-q^2)/q < 1$, i.e. $q > \left(\sqrt{5} - 1 \right)/2$. Hence, the chain is null recurrent when $q = \left(\sqrt{5} - 1 \right)/2$ and positive recurrent when $q > \left(\sqrt{5} - 1 \right)/2$. In the latter case

$$\pi_0 = \left[1 + \sum_{i \geq 1} \left(\frac{1}{q} + 1 \right) \left(\frac{1-q^2}{q} \right)^{i-1} \right]^{-1} = \frac{q^2+q-1}{q^2+2q}.$$

$\square$

Worked Example 1.8.11 Let (W_n) be the birth-and-death process on $\mathbb{Z}_+ = \{0,1,2,\ldots\}$ with the following transition probabilities

$$p_{i,i+1} = p_{i,i-1} = \frac{1}{2}, \ i \geq 1$$
$$p_{01} = 1.$$

By relating (W_n) to the symmetric simple random walk (Y_n) on $\mathbb{Z}$, or otherwise, prove that (W_n) is a recurrent Markov chain. By considering IMs, or otherwise, prove that (W_n) is null recurrent.

Calculate the vectors $\gamma^k = (\gamma_i^k, i \in \mathbb{Z}_+)$ for the chain (W_n), $k \in \mathbb{Z}_+$.

Finally, let $W_0 = 0$ and let N be the number of visits to 1 before returning to 0. Show that $\mathbb{P}_0(N = n) = (1/2)^n$, $n \geq 1$.

Solution Now, (W_n) is an irreducible Markov chain. Also, $W_n = |Y_n|$ where (Y_n) is the nearest-neighbour symmetric random walk on $\mathbb{Z}$. Hence, for all $i \in \mathbb{Z}$,

$$\mathbb{P}_{|i|}((W_n) \text{ returns to } i) \geq \mathbb{P}_i((Y_n) \text{ returns to } i);$$

but the right hand side equals 1 since (Y_n) is recurrent. Hence, the left-hand side equals 1, and (W_n) is recurrent.

To check null recurrence, it suffices to prove that (W_n) has no equilibrium distribution. Consider the invariance equations

$$\pi_0 = \frac{1}{2}\pi_1, \ \pi_1 = \pi_0 + \frac{1}{2}\pi_2,$$

$$\pi_i = \frac{1}{2}\pi_{i-1} + \frac{1}{2}\pi_{i+1}, \ i \geq 2.$$

The second line has a general solution $\pi_i = A + Bi$, $i \geq 1$. From the first line, $B = 0$ and $\pi_0 = A/2$. Hence, any IM π is of the form

$$\pi_i = A, \ i \geq 1, \ \pi_0 = \frac{1}{2}A,$$

where $A \geq 0$. It has $\sum_i \pi_i = \infty$ unless $A = 0$. Thus, no equilibrium distribution can exist, and (W_n) is null recurrent.

Therefore, for the chain (W_n),

$$\gamma_i^k = \frac{\pi_i}{\pi_k} = \begin{cases} 1, & i,k \geq 1 \text{ or } i = k = 0, \\ 1/2, & i = 0, \ k \geq 1, \\ 2, & i \geq 1, \ k = 0. \end{cases}$$

Next, by the strong Markov property,

$$\begin{aligned} \mathbb{P}_0(N = n) &= \mathbb{P}_0(N \geq 1) \\ &\quad \times (\mathbb{P}_1(\text{return to 1 without visiting 0}))^{n-1} \\ &\quad \times \mathbb{P}_1(\text{hit 0 without returning to 1}) \\ &= \frac{1}{2^n}. \end{aligned}$$

In fact,

$$\mathbb{P}_0(N \geq 1) = 1 \qquad (\text{as } p_{01} = 1),$$

$\mathbb{P}_1(\text{return to 1 without visiting 0}) = 1 - p_{10} = \frac{1}{2}$ (as the chain hits 0 from 1 with probability $1/2$ and is recurrent),

and

$$\mathbb{P}_1(\text{hit } 0 \text{ without returning to } 1) = p_{10} = \tfrac{1}{2}.$$

$\square$

> *There are many points in an infinite state space.*
> *More than stars in the sky or grains of sand in Sahara.*
> (From the series *'Thus spoke Superviser'*.)

1.9 Convergence to equilibrium. Long-run proportions

> *Time is the image of eternity.*
> Laertius Diogenes, 2nd century AD, Greek writer

Convergence to equilibrium means that, as the time progresses, the Markov chain 'forgets' about its initial distribution λ. In particular, if $\lambda = \delta^{(i)}$, the Dirac delta concentrated at i, the chain 'forgets' about the initial state i. Clearly, this is related to properties of the n-step matrix P^n as $n \to \infty$. Consider first the case of a finite chain.

Theorem 1.9.1 *Suppose that a finite $m \times m$ transition matrix P^n converges, in each entry, to a limiting matrix $\Pi = (\pi_{ij})$:*

$$\lim_{n \to \infty} p_{ij}^{(n)} = \pi_{ij}, \ \text{ for all } \ i, j \in I. \tag{1.71}$$

Then: (a) every row $\pi^{(i)}$ of Π is an equilibrium distribution

$$\pi^{(i)} P = \pi^{(i)} \text{ or } \pi_{ij} = \sum_{l} \pi_{il} p_{lj}.$$

(b) If P is irreducible then all rows $\pi^{(i)}$ coincide: $\pi^{(1)} = \cdots = \pi^{(m)} = \pi$. In this case,

$$\lim_{n \to \infty} \mathbb{P}(X_n = j) = \pi_j \text{ for all } j \in I \text{ and the initial distribution } \lambda.$$

Proof (a) For all states j we have

$$\begin{aligned}
\left(\pi^{(i)} P\right)_j &= \sum_{l \in I} \pi_{il} p_{lj} = \sum_{l} \lim_{n \to \infty} p_{il}^{(n)} p_{lj} = \lim_{n \to \infty} \sum_{l} p_{il}^{(n)} p_{lj} \\
&= \lim_{n \to \infty} p_{ij}^{(n+1)} = \pi_{ij} = \left(\pi^{(i)}\right)_j.
\end{aligned} \tag{1.72}$$

(b) If P is irreducible then all rows $\pi^{(i)}$ of Π coincide as there is a unique equilibrium distribution. Also,

$$\lim_{n \to \infty} \mathbb{P}(X_n = j) = \lim_{n \to \infty} \sum_i \lambda_i p_{ij}^{(n)} = \sum_i \lambda_i \lim_{n \to \infty} p_{ij}^{(n)} = \pi_j. \tag{1.73}$$

$\square$

For a countable chain, our argument in (1.72) requires a justification of exchanging the order of the limit and summation. We will do this later in this section.

We see from Theorem 1.9.1 that the equilibrium distribution of a chain can be identified from the limit of matrices P^n as $n \to \infty$. More precisely, if we know that P^n converges to a matrix Π whose rows are equal to each other then these rows give the equilibrium distribution π. We see therefore that convergence $P^n \to \Pi$ where

Π has a structure $\pi \begin{pmatrix} \pi \\ \text{---} \\ \text{---} \\ \cdots \\ \text{---} \end{pmatrix}$ is a crucial factor.

So when does $P^n \to \Pi$? A simple counterexample is $P = \begin{pmatrix} 0 & 1 \\ 1 & 0 \end{pmatrix}$. Here,

$$P^n = \begin{pmatrix} 1 & 0 \\ 0 & 1 \end{pmatrix}, \quad n \text{ even};$$
$$P^n = \begin{pmatrix} 0 & 1 \\ 1 & 0 \end{pmatrix}, \quad n \text{ odd}. \tag{1.74}$$

More generally, consider an $m \times m$ matrix $P = \begin{pmatrix} 0 & 1 & 0 & \cdots & 0 \\ 0 & 0 & 1 & \cdots & 0 \\ \cdots & \cdots & \cdots & \cdots & \cdots \\ 1 & 0 & 0 & \cdots & 0. \end{pmatrix}$

Here again, the equilibrium distribution is unique: $\pi = (1/m, \ldots, 1/m)$.

We know that in these examples, the matrix P is periodic. Recall that P is aperiodic if and only if for all $i \in I$

$$p_{ii}^{(n)} > 0 \text{ for all } n \text{ large enough.} \tag{1.75}$$

If in addition, P is irreducible then, for all $i, j \in I$,

$$p_{ij}^{(n)} > 0 \text{ for all } n \text{ large enough.} \tag{1.76}$$

Theorem 1.9.2 *Assume P is irreducible, aperiodic and positive recurrent. Then, as $n \to \infty$,*

$$P^n \to \Pi.$$

The entries of the limiting matrix Π *are constant along columns. In other words the rows of* Π *are repetitions of the same vector* π *which is the (unique) equilibrium distribution for P. Hence, the irreducible aperiodic and positive recurrent Markov chain forgets its initial distribution: for all* λ *and* $j \in I$

$$\lim_{n \to \infty} \mathbb{P}(X_n = j) = \pi_j.$$

Proof Consider two Markov chains: $(X_n^{(i)})$, which is $(\delta^{(i)}, P)$; and (X_n^π), which is (π, P). Then

$$p_{ij}^{(n)} = \mathbb{P}_i(X_n^{(i)} = j), \quad \pi_j = \mathbb{P}(X_n^\pi = j).$$

To evaluate the difference between these probabilities, we will identify their 'common part', by coupling the two Markov chains, i.e. running them together. One way is to run both chains independently. This means that we consider the Markov chain (Y_n) on $I \times I$, with states (k, l) where $k, l \in I$, the transition probabilities

$$p_{(k,l)(u,v)}^Y = p_{ku} p_{lv}, \quad k, l, u, v \in I, \tag{1.77}$$

and the initial distribution

$$\mathbb{P}(Y_0 = (k, l)) = \mathbf{1}(k = i)\pi_l, \quad k, l \in I.$$

But a better way is to run the chain (W_n) where the transition probabilities are

$$p_{(k,l)(u,v)}^W = \begin{cases} p_{ku} p_{lv}, & \text{if } k \neq l, \\ p_{ku} \mathbf{1}(u = v), & \text{if } k = l, \end{cases} \quad k, l, u, v \in I, \tag{1.78}$$

with the same initial distribution

$$\mathbb{P}(W_0 = (k, l)) = \mathbf{1}(k = i)\pi_l, \quad k, l \in I. \tag{1.79}$$

Indeed, (1.78) determines a transition probability matrix on $I \times I$: all entries $p_{(k,l)(u,v)}^W \geq 0$ and the sum along a row equals 1. In fact,

$$\sum_{u,v \in I} p_{(k,l)(u,v)}^W = \begin{cases} \sum_u p_{ku} \sum_v p_{lv}, & \text{if } k \neq l \\ \sum_u p_{ku}, & \text{if } k = l \end{cases} = 1.$$

Further, partial summation gives the original transitional probabilities P:

$$\sum_{v \in I} p_{(k,l)(u,v)}^W = p_{ku}, \quad \sum_{u \in I} p_{(k,l)(u,v)}^W = p_{lv}.$$

Pictorially, the two components of the chain (W_n) behave individually like $(X_n^{(i)})$ and (X_n^π); together they evolve independently (i.e. as in (Y_n)) until the (random) time T when they coincide

$$T = \inf \left[n \geq 1 : X_n^{(i)} = X_n^\pi \right],$$

after which they stay together. Therefore,

$$p_{ij}^{(n)} - \pi_j = \mathbb{P}^W \left(X_n^{(i)} = j \right) - \mathbb{P}^W \left(X_n^\pi = j \right).$$

Writing

$$\mathbb{P}^W \left(X_n^{(i)} = j \right) = \mathbb{P}^W \left(X_n^{(i)} = j, T \leq n \right) + \mathbb{P}^W \left(X_n^{(i)} = j, T > n \right) \quad (1.80)$$

and

$$\mathbb{P}^W \left(X_n^\pi = j \right) = \mathbb{P}^W \left(X_n^\pi = j, T \leq n \right) + \mathbb{P}^W \left(X_n^\pi = j, T > n \right), \quad (1.81)$$

we see that the first summands cancel each other:

$$\mathbb{P}^W \left(X_n^{(i)} = j, T \leq n \right) = \mathbb{P}^W \left(X_n^\pi = j, T \leq n \right),$$

as the events $\left\{ X_n^{(i)} = j, T \leq n \right\}$ and $\{ X_n^\pi = j, T \leq n \}$ coincide. Hence

$$p_{ij}^{(n)} - \pi_j = \mathbb{P}^W \left(X_n^{(i)} = j, T > n \right) - \mathbb{P}^W \left(X_n^\pi = j, T > n \right)$$

and

$$\left| p_{ij}^{(n)} - \pi_j \right| \leq \mathbb{P}^W (T > n) = \mathbb{P}^Y (T > n). \quad (1.82)$$

The last bound is called the coupling inequality.

Thus, it suffices to check that $\mathbb{P}^Y (T > n) \to 0$, i.e. $\mathbb{P}(T < \infty) = 1$. But (Y_n) is an irreducible positive recurrent Markov chain. (Irreducibility follows from the fact that the original matrix P is irreducible and aperiodic (equation (1.76) is helpful here) and positive recurrence from the fact that (Y_n) has the equilibrium distribution $(\pi \times \pi)_{(k,l)} = \pi_k \pi_l$.) Hence, by Theorem 1.5.9, for all states $l \in I$,

$$\mathbb{P}^Y \left(T_{(l,l)} < \infty \right) = 1,$$

where

$$T_{(l,l)} = \inf \left[n \geq 0 : X_n^{(i)} = X_n^\pi = l \right].$$

As $T \leq T_{(l,l)}$, the statement follows. $\qquad\square$

In the case of a finite irreducible aperiodic chain it is possible to establish that the rate (or speed) of convergence of $p_{ij}^{(n)}$ to π_j is geometric. This means that for some $m \geq 1$

$$p_{ij}^{(m)} \geq \rho \quad \text{for all states } i, j. \tag{1.83}$$

Theorem 1.9.3 *If P is finite irreducible and aperiodic then for all states i, j*

$$\left| p_{ij}^{(n)} - \pi_j \right| \leq (1 - \rho)^{n/m - 1}, \tag{1.84}$$

where m and ρ are as in (1.83).

Proof Repeat the scheme of the proof of Theorem 1.9.2: we have to assess $\mathbb{P}^Y(T > n)$. But in the finite case, we can write

$$\mathbb{P}_{(k,l)}^W(T \leq m) \geq \sum_{u \in I} p_{ku}^{(m)} p_{lu}^{(m)} \geq \rho \sum_{u \in I} p_{lu}^{(m)} = \rho,$$

i.e.

$$\mathbb{P}_{(k,l)}^W(T > m) \leq (1 - \rho) \quad \text{for all } k, l \in I.$$

Then, by the strong Markov property,

$$\mathbb{P}^W(T > n) \leq \mathbb{P}^W\left(T > \left\lfloor \frac{n}{m} \right\rfloor m\right) \leq \mathbb{P}^W(T > m)^{\lfloor n/m \rfloor}$$

and the assertion of Theorem 1.9.3 follows.   □

An instructive example is as follows

Worked Example 1.9.4 Consider a pack of cards labelled $1, 2, \ldots, 52$. We repeatedly take the top card and insert it uniformly at random in one of the 52 possible places, that is, either on the top or on the bottom or in one of the 50 places inside the pack. How long on average will it take for the bottom card to reach the top?

Let p_n denote the probability that after n iterations the cards are found in increasing order. Show that, irrespective of the initial ordering, p_n converges as $n \to \infty$, and determine the limit p. You should give precise statements of any general results to which you appeal.

Show that, at least until the bottom card reaches the top, the ordering of cards inserted beneath it, is uniformly random. Hence or otherwise show that, for all n,

$$|p_n - p| \leq 52(1 + \ln 52)/n.$$

Solution Label the places $1, 2, \ldots, 52$ where 1 is bottom. Suppose the bottom card has reached place m. Then the top card is inserted below it with probability $m/52$. The expected time until this happens satisfies

$$k_m = 1 + \left(1 - \frac{m}{52}\right) k_m,$$

with $k_m = 52/m$. Then the total expected time to reach the top equals

$$k_1 + \cdots + k_{51} = 52 \left(1 + \frac{1}{2} + \cdots + \frac{1}{51}\right).$$

The card ordering performs a Markov chain on the set of permutations $\mathscr{S}_{52}$ (the permutation group). The chain is aperiodic, as the top card may be replaced at the top. The chain is also irreducible as it always can be brought to increasing order, by repeatedly inserting the top card at the bottom until the bottom becomes 1, then inserting the top card in place 2, etc. By symmetry, the uniform distribution on $\mathscr{S}_{52}$ is invariant.

Hence, by the theorem that *for an irreducible aperiodic Markov chain (X_n) with equilibrium distribution $\pi = (\pi_i)$, $\lim\limits_{n\to\infty} \mathbb{P}(X_n = j) = \pi_j$ for all j,* we have

$$\lim_{n\to\infty} p_n = p = \frac{1}{(52)!}.$$

Finally, suppose we have inserted k cards beneath the original bottom card, and these are ordered equiprobably at random. When the next card is inserted beneath the bottom card it is equally likely to go in each of the $k+1$ places. That is, the $k+1$ cards will still be ordered randomly. This applies inductively until $k = 51$.

Then let T be the time the bottom card reaches the top. The pack is randomly ordered at time $T + 1$. By the strong Markov property it remains so at time $(T+1) \vee n = \max[T+1, n]$. Therefore,

$$|p_n - p| = |\mathbb{P}(\text{increasing at time } n) - \mathbb{P}(\text{increasing at } (T+1) \vee n)|$$
$$\leq \mathbb{P}(T \geq \hat{n}) \leq \frac{1}{n} \mathbb{E}T = \frac{52}{n} \left(1 + \frac{1}{2} + \cdots + \frac{1}{51}\right) \leq \frac{52}{n}(1 + \ln 52).$$

$\square$

> *What I say is, patience.*
> *And shuffle the cards.*
> M de Cervantes (1547–1661), Spanish writer

Remark 1.9.5 For a transient or null recurrent irreducible aperiodic chain, the matrix P^n converges to a zero matrix:

$$\lim_{n \to \infty} P^n = \mathbf{O}.$$

We will not give here the formal proof of this assertion. (For a transient case the proof is based on the fact that the series $\sum_{n \geq 1} p_{ii}^{(n)} < \infty$.)

The remaining part of this section focusses on *long-run or long-time proportions*. This is a subject of so-called *ergodic theorems* which study *time averages* along trajectories of random processes (in our situation, Markov chains). One of the striking phenomena here is the fact that, under certain irreducibility-type assumptions, limiting time averages coincide with expected values relative to equilibrium distributions. The latter can be considered as *space averages* (i.e. averages over state space I). Thus, the above fact can be phrased as 'the long-run time-average equals the space average'; this is a formal expression of a 'mixing property' of a random process (in fact, there exists an entire hierarchy of such properties). Mixing properties are believed to be behind many phenomena observed in nature and in various aspects of human activities. Historically, these properties are connected with the names of two famous theoretical physicists of the 19th Century, American J.W. Gibbs (1839–1903), and Austrian L. Boltzmann (1844–1906). Ergodic theorems in turn form the basis of the Ergodic Theory, a well-developed mathematical discipline embracing a broad spectrum of concepts and methods.

> *In the long-run proportion we are all dead.*
> J. Maynard Keynes (1883–1946), British economist

A natural example is as follows.

Definition 1.9.6 Consider the number of visits to state i before time n:

$$V_i(n) = \sum_{k=0}^{n-1} \mathbf{1}(X_k = i). \tag{1.85}$$

The limit (if it exists)

$$\lim_{n \to \infty} \frac{V_i(n)}{n} \tag{1.86}$$

is called the *long-run proportion* of the time spent in state i.

More generally, if $f : I \to \mathbb{R}$ is a function on the state space I, then we consider the sum

$$V(f, n) = \sum_{k=0}^{n-1} f(X_k) \tag{1.87}$$

and the limit

$$\lim_{n \to \infty} \frac{V(f,n)}{n}. \tag{1.88}$$

Theorem 1.9.7 *For all states* $i \in I$, *the ratio* $V_i(n)/n$ *converges almost surely:*

$$\mathbb{P}_i\left(\lim_{n \to \infty} \frac{V_i(n)}{n} = r_i\right) = 1, \tag{1.89}$$

where

$$r_i = \begin{cases} \pi_i, & \text{if } i \text{ is positive recurrent,} \\ 0, & \text{if } i \text{ is null recurrent or transient.} \end{cases} \tag{1.90}$$

Proof First, suppose that state i is transient. Then, as we know, the total number V_i of visits to i is finite with probability 1. See (1.27), (1.37). Hence, $V_i/n \to 0$ as $n \to \infty$ with probability 1. As $0 \le V_i(n) \le V_i$, we deduce that $V_i(n)/n \to 0$ as $n \to \infty$ with probability 1.

Now let i be recurrent. Then the times $T_i^{(1)}, T_i^{(2)}, \ldots$ between successive returns to state i are finite with $\mathbb{P}_i$-probability 1. By Theorem 1.8.3, they are IID random variables, with mean value m_i equal to $1/\pi_i$ in the positive recurrent case and to ∞ in the null recurrent case. Obviously,

$$T_i^{(1)} + \cdots + T_i^{(V_i(n))} \ge n,$$

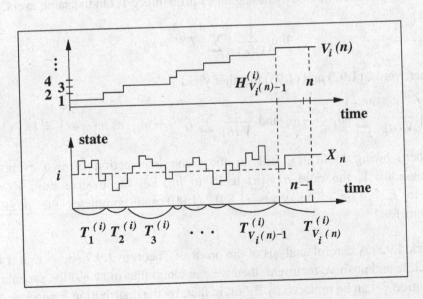

Fig. 1.15

but

$$T_i^{(1)} + \cdots + T_i^{(V_i(n)-1)} \leq n - 1 :$$

see Figure 1.15. So we can write

$$\frac{1}{V_i(n)} \left(T_i^{(1)} + \cdots + T_i^{(V_i(n)-1)} \right) \leq \frac{n}{V_i(n)} \leq \frac{1}{V_i(n)} \left(T_i^{(1)} + \cdots + T_i^{(V_i(n))} \right). \quad (1.91)$$

By Theorem 1.8.4, on an event of $\mathbb{P}_i$-probability 1, the limit $\lim\limits_{n \to \infty} \frac{1}{n} \sum_{l=1}^{n} T_i^{(l)} = m_i$ holds:

$$\mathbb{P}_i \left(\frac{1}{n} \sum_{l=1}^{n} T_i^{(l)} \to m_i, \text{ as } n \to \infty \right) = 1. \quad (1.92)$$

Next, as i is recurrent, the sequence $(V_i(n))$ increases indefinitely, again on an event of $\mathbb{P}_i$-probability 1:

$$\mathbb{P}_i \left(V_i(n) \nearrow \infty, \text{ as } n \to \infty \right) = 1. \quad (1.93)$$

Then we can put in (1.92) a summation up to $V_i(n)$, instead of n and, correspondingly, divide by the factor $V_i(n)$:

$$\lim_{n \to \infty} \frac{1}{V_i(n)} \sum_{l=1}^{V_i(n)} T_i^{(l)} = m_i.$$

This relation holds on the intersection of the two aforementioned events of probability 1, which obviously has again $\mathbb{P}_i$-probability 1. On the same event,

$$\lim_{n \to \infty} \frac{1}{V_i(n)} \sum_{l=1}^{V_i(n)-1} T_i^{(l)} = m_i.$$

In other words, (1.92) and (1.93) together yield

$$\mathbb{P}_i \left(\frac{1}{V_i(n)} \sum_{l=1}^{V_i(n)-1} T_i^{(l)} \to m_i \text{ and } \frac{1}{V_i(n)} \sum_{l=1}^{V_i(n)} T_i^{(l)} \to m_i, \text{ as } n \to \infty \right) = 1. \quad (1.94)$$

But then, owing to (1.91), still on the same intersection of two events of $\mathbb{P}_i$-probability 1, the ratio $n/V_i(n)$ tends to m_i, i.e. the inverse ratio $V_i(n)/n$ tends to $r_i = 1/m_i$. This gives (1.89), (1.90) and completes the proof of Theorem 1.9.7.

$$\square$$

Remark 1.9.8 A careful analysis of the proof of Theorem 1.9.7 shows that if P is irreducible and positive recurrent, then we can claim that in (1.89) the probability distribution $\mathbb{P}_i$ can be replaced by $\mathbb{P}_j$, or, in fact, by the distribution $\mathbb{P}$ generated by an arbitrary initial distribution λ. This is possible because sums $T_i^{(1)} + \cdots + T_i^{(n)}$

still behave asymptotically as if the RVs $T_i^{(l)}$ were IID. (In reality, the distribution of the first RV, $T_i^{(1)} = T_i = H_1^i$, will be different and depend on the choice of the initial state.)

Theorem 1.9.9 *Let P be a finite irreducible transition matrix. Then, for any initial distribution λ and a bounded function f on I,*

$$\mathbb{P}\left(\lim_{n\to\infty} \frac{V(f,n)}{n} = \pi(f)\right) = 1, \qquad (1.95)$$

where

$$\pi(f) = \sum_{i\in I} \pi_i f(i). \qquad (1.96)$$

Proof The proof of Theorem 1.9.9 is a refinement of that of Theorem 1.9.7. More precisely, (1.95) is equivalent to

$$\mathbb{P}\left(\lim_{n\to\infty} \left|\frac{V(f,n)}{n} - \pi(f)\right| = 0\right) = 1.$$

In other words, we have to check that on an event of $\mathbb{P}$-probability 1,

$$\left|\frac{V(f,n)}{n} - \pi(f)\right| \to 0, \quad \text{as } n \to \infty. \qquad (1.97)$$

Writing $V(f,n) = \sum_{i\in I} V_i(n) f(i)$ and $\pi(f) = \sum_{i\in I} \pi_i f(i)$, we can transform and bound the left-hand side in (1.97) as follows

$$\left|\frac{V(f,n)}{n} - \pi(f)\right| = \left|\sum_{i\in I}\left(\frac{V_i(n)}{n} - \pi_i\right) f(i)\right| \le \sum_{i\in I} \left|\frac{V_i(n)}{n} - \pi_i\right| |f(i)|.$$

We know that, for all $i \in I$, on an event of $\mathbb{P}_i$-probability 1, $V_i(n)/n \to \pi_i$. Remark 1.9.5 allows us to claim convergence $V_i(n)/n \to \pi_i$ on an event of $\mathbb{P}_j$-probability 1 (that is, regardless of the choice of the initial state), or, even stronger, on an event of $\mathbb{P}$-probability 1, where $\mathbb{P}$ is the distribution of the (λ, P) Markov chain with any initial distribution λ. Then (1.95) follows, which completes the proof. $\qquad\square$

Worked Example 1.9.10 Describe the long-time behaviour of discrete time Markov chains on a finite state space. What about the convergence of probabilities, or almost-sure behaviour? Explain what happens when the chain is not irreducible.

Solution The state space splits into open classes $O_1,\ldots,O_j$ and closed classes $C_{j+1},\ldots,C_{j+l}$. If $l = 1$ (a unique closed class), it is irreducible. Starting from an open class, say O_i, we end up in closed class C_k with probability h_i^k. These probabilities satisfy

$$h_i^k = \sum_{r=1}^{j+l} \widehat{p}_{ir} h_r^k.$$

Here, $\widehat{p}_{ir}$ is the probability that we exit class O_i to class O_r or C_r, and for $r = j+1$, $\ldots, j+l$: $h_r^k = \delta_{r,k}$.

The chain has a single ED $\pi^{(r)}$ concentrated on C_r, $r = j+1, \ldots, j+l$ (hence, a unique ED when $l = 1$). Furthermore, any ED is a mixture of the EDs $\pi^{(r)}$.

Starting in C_r, we have, for any function f on C_r,

$$\frac{1}{n}\sum_{t=0}^{n} f(X_t) \to \sum_{i \in C_r} \pi_i^{(r)} f(i) \quad \text{almost surely.}$$

Moreover, in the aperiodic case (where $\gcd \{n : p_{aa}^{(n)} > 0\} = 1$ for some $a \in C_r$), for all $i_0 \in C_r$,

$$\mathbb{P}(X_n = i | X_0 = i_0) \to \pi_i^r,$$

and the convergence is with geometric speed.

$\square$

1.10 Detailed balance and reversibility

> *Reversal of Time, Reversal of Fortune*
> (From the series *'Movies that never made it to the Big Screen'*.)

Let $(X_0, X_1, \ldots)$ be a Markov chain and fix $N \geq 1$. What can we say about the time reversal of (X_n), i.e. the family $(X_{N-n}, n = 0, 1, \ldots, N) = (X_N, X_{N-1}, \ldots, X_0)$?

Theorem 1.10.1 *Let (X_n) be a (π, P) Markov chain where $\pi = (\pi_i)$ is an equilibrium distribution for P with $\pi_i > 0$ for all $i \in I$. Then: (a) for all $N \geq 1$, the time reversal $(X_N, X_{N-1}, \ldots, X_0)$ is a $(\pi, \widehat{P})$ Markov chain where $\widehat{P} = (\widehat{p}_{ij})$ has*

$$\widehat{p}_{ij} = \frac{\pi_j}{\pi_i} p_{ji}. \tag{1.98}$$

(b) *If P is irreducible then so is $\widehat{P}$.*

Proof (a) First, observe that $\widehat{P}$ is a stochastic matrix; that is, $\widehat{p}_{ij} \geq 0$ and

$$\sum_j \widehat{p}_{ij} = \frac{1}{\pi_i}\sum_j \pi_j p_{ji} = \frac{1}{\pi_i}\pi_i = 1.$$

Next, π is $\widehat{P}$-invariant:

$$\sum_i \pi_i \widehat{p}_{ij} = \sum_i \pi_j p_{ji} = \pi_j \sum_i p_{ji} = \pi_j.$$

Now pull the factor $\pi_{\bullet}$ through the product

$$
\begin{aligned}
\mathbb{P}(X_N = i_N, \ldots, X_0 = i_0) &= \mathbb{P}(X_0 = i_0, \cdots, X_N = i_N) \\
&= \pi_{i_0} p_{i_0 i_1} \cdots p_{i_{N-1} i_N} \\
&= \widehat{p}_{i_1 i_0} \pi_{i_1} \cdots p_{i_{N-1} i_N} \\
&= \widehat{p}_{i_1 i_0} \widehat{p}_{i_2 i_1} \pi_{i_2} \cdots \\
&= \widehat{p}_{i_1 i_0} \cdots \widehat{p}_{i_N i_{N-1}} \pi_{i_N} \\
&= \pi_{i_N} \widehat{p}_{i_N i_{N-1}} \cdots \widehat{p}_{i_1 i_0}.
\end{aligned}
$$

We see that (X_{N-n}) is a $(\pi, \widehat{P})$ Markov chain.

(b) If P is irreducible then any pair of states i, j is connected; that is, there exists a path $i = i_0, i_1, \ldots, i_n = j$ with

$$
\begin{aligned}
0 < p_{i_0 i_1} \cdots p_{i_{n-1} i_n} &= (1/\pi_{i_0}) \pi_{i_0} p_{i_0 i_1} \cdots p_{i_{n-1} i_n} \\
&= (1/\pi_{i_0}) \widehat{p}_{i_1 i_0} \pi_{i_1} \cdots p_{i_{n-1} i_n} = \cdots \\
&= (1/\pi_{i_0}) \widehat{p}_{i_1 i_0} \cdots \widehat{p}_{i_n i_{n-1}} \pi_{i_n}.
\end{aligned}
$$

So, $\widehat{p}_{i_1 i_0} \cdots \widehat{p}_{i_n i_{n-1}} > 0$, and j, i are connected in $\widehat{P}$. $\qquad \square$

The case where chain (X_{N-n}) has the same distribution as (X_n) is of a particular interest.

Theorem 1.10.2 *Let (X_n) be a Markov chain. The following properties are equivalent:*

(i) *for all $n \geq 1$ and states $i_0, \ldots, i_n$,*

$$\mathbb{P}(X_0 = i_0, \ldots, X_n = i_n) = \mathbb{P}(X_0 = i_n, \ldots, X_n = i_0). \tag{1.99}$$

(ii) *The chain (X_n) is in equilibrium, i.e. $(X_n) \sim (\pi, P)$ where π is an equilibrium distribution for P, and*

$$\pi_i p_{ij} = \pi_j p_{ji} \text{ for all states } i, j \in I. \tag{1.100}$$

Proof (i) $\Rightarrow$ (ii). Take $n = 1$,

$$\mathbb{P}(X_0 = i, X_1 = j) = \mathbb{P}(X_0 = j, X_1 = i),$$

and sum over j

$$\sum_j \mathbb{P}(X_0 = i, X_1 = j) = \mathbb{P}(X_0 = i) = \lambda_i,$$

$$\sum_j \mathbb{P}(X_0 = j, X_1 = i) = \mathbb{P}(X_1 = i) = (\lambda P)_i.$$

So, $\lambda_i = (\lambda P)_i$ for all i, i.e., $\lambda P = \lambda$. Hence, the chain is in equilibrium: $\lambda = \pi$. Next, for all i, j,

$$\mathbb{P}(X_0 = i, X_1 = j) = \pi_i p_{ij} = \mathbb{P}(X_0 = j, X_1 = i) = \pi_j p_{ji}.$$

(ii) $\Rightarrow$ (i). Write

$$\mathbb{P}(X_0 = i_0, \ldots, X_n = i_n) = \pi_{i_0} p_{i_0 i_1} \cdots p_{i_{n-1} i_n}$$

and use (1.100) to pull $\pi_\bullet$ through the product

$$
\begin{aligned}
\pi_{i_0} p_{i_0 i_1} \cdots p_{i_{n-1} i_n} &= p_{i_0 i_1} \pi_{i_1} \cdots p_{i_{n-1} i_n} = \cdots \\
&= p_{i_0 i_1} \cdots p_{i_n i_{n-1}} \pi_{i_n} \\
&= \pi_{i_n} p_{i_n i_{n-1}} \cdots p_{i_0 i_1} \\
&= \mathbb{P}(X_0 = i_n, \ldots, X_n = i_0).
\end{aligned}
$$

$\square$

Definition 1.10.3 A Markov chain (X_n) satisfying (1.99) is called *reversible*. Equations (1.100) are called *detailed balance* equations (DBEs).

So, the assertion of Theorem 1.10.2 reads: a Markov chain is reversible if and only if it is in equilibrium, and the DBEs are satisfied.

DBEs are a powerful tool for identification of an ED.

Theorem 1.10.4 *If λ and P satisfy the DBEs*

$$\lambda_i p_{ij} = \lambda_j p_{ji}, \quad i, j \in I,$$

then λ is an ED for P, that is $\lambda P = \lambda$.

Proof Sum over j:

$$\lambda_i \sum_j p_{ij} = \lambda_i,$$

$$\sum_j \lambda_j p_{ji} = (\lambda P)_i.$$

The two expressions are equal for all i, hence the result.

$\square$

So, for a given matrix P, if the DBEs can be solved (that is, a probability distribution that satisfies them can be found), the solution will give an ED. Furthermore, the corresponding Markov chain will be reversible.

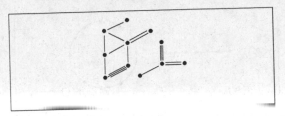

Fig. 1.16

An interesting and important class of Markov chains is formed by *random walks on graphs*. We have seen examples of such chains: a birth-death process (a RW on $\mathbb{Z}^1$ or its subset), a RW on a plane square lattice $\mathbb{Z}^2$ and, more generally, a RW on a d-dimensional cubic lattice $\mathbb{Z}^d$. A feature of these examples is that a wandering particle can jump to any of its neighbouring sites; in the symmetric case, the probability of each jump is the same. This idea can be extended to a general *graph*, with directed or non-directed links (edges). Here, we focus on non-directed graphs; a graph is understood as a collection G of *vertices* some of which are joined by non-directed *edges*, or *links*, possibly several. Non-directed here means that the edges can be traversed in both directions; sometimes it's convenient to think that each edge is formed by a pair of opposite arrows.

A graph is called *connected* if any two distinct vertices are connected with a path formed by edges. The *valency* v_i of a vertex i is defined as the number of edges at i. The *connectedness* v_{ij} is the number of edges joining vertices i and j. These features are illustrated in Figure 1.16.

The RW on a graph has the following transition matrix $P = (p_{ij})$:

$$p_{ij} = \begin{cases} v_{ij}/v_i, & \text{if } i \text{ and } j \text{ are connected,} \\ 0, & \text{otherwise.} \end{cases} \tag{1.101}$$

The matrix P is irreducible if and only if the graph is connected. The vector $v = (v_i)$ satisfies the DBEs. That is, for all vertices i, j,

$$v_i p_{ij} = v_{ij} = v_j p_{ji}, \tag{1.102}$$

and hence v is P-invariant. We obtain the following straightforward result.

Theorem 1.10.5 *The RW on a graph with transition matrix P of the form* (1.101), *is always positive or null recurrent. It is positive recurrent if and only if the total valence $\sum_i v_i < \infty$, in which case $\pi_j = v_j / \sum_i v_i$ is an equilibrium distribution. Furthermore, the chain with equilibrium distribution π is reversible.*

Discrete-time Markov chains

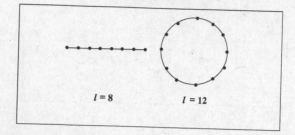

Fig. 1.17

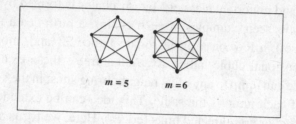

Fig. 1.18

A simple but popular example of a graph is an ℓ-site segment of a one-dimensional lattice: here the valency of every vertex equals 2, except for the endpoints where the valency is 1. See Figure 1.17.

An interesting class is formed by graphs with a constant valency: $v_i \equiv v$; again the simplest case is $v = 2$, where ℓ vertices are placed on a circle (or on a perfect polygon or any closed path). See again Figure 1.17. A popular example of a graph with a constant valency is a fully connected graph with a given number of vertices, say $\{1, \ldots, m\}$: here the valency equals $m - 1$, and the graph has $m(m-1)/2$ (non-directed) edges in total. See Figure 1.18.

Another important example is a regular cube in d dimensions, with 2^d vertices. Here the valency equals d, and the graph has $d2^{d-1}$ (still non-directed) edges joining neigbouring vertices. See Figure 1.19.

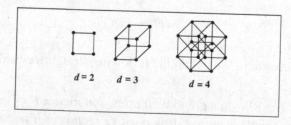

Fig. 1.19

Popular examples of infinite graphs of constant valency are lattices and trees.

In the case of a general finite graph of constant valency $v_i = v$ for any vertex i, the sum $\sum_i v_i$ equals $v \times |G|$ where $|G|$ is the number of vertices. Then probabilities $p_{ij} = p_{ji} = v_{ij}/v$, for all neighbouring pairs i, j. That is, the transition matrix $P = (p_{ij})$ is Hermitian: $P = P^T$. Furthermore, the equilibrium distribution $\pi = (\pi_i)$ is uniform: $\pi_i = 1/|G|$.

In Linear Algebra courses, it is asserted that a (complex) Hermitian matrix has an orthonormal basis of eigenvectors, and its eigenvalues are all real. This handy property is nice to retain whenever possible. For a DTMC, even when P is originally non-Hermitian, it can be 'converted' into a Hermitian matrix by changing the scalar product. We will explore further this avenue in Sections 1.12–1.14.

> *Time present and time past*
> *Are both perhaps present in the future,*
> *And time future contained in time past.*
> T.S. Eliot (1888–1965), American poet

Worked Example 1.10.6 (i) We are given a finite set of airports. Assume that between any two airports, i and j, there are $a_{ij} = a_{ji}$ flights in each direction on every day. A confused traveller takes one flight per day, choosing at random from all available flights. Starting from i, how many days on average will pass until the traveller returns again to i? Be careful to allow for the case where there may be no flights at all between two given airports.

(ii) Consider the infinite tree T with root R, where for all $m \geq 0$, all vertices at distance 2^m from R have degree 3, and where all other vertices (except R) have degree 2. Show that the random walk on T is recurrent.

Solution (i) Let $X_0 = i$ be the starting airport, X_n the destination of the nth flight, and I denote the set of airports reachable from i. Then (X_n) is an irreducible Markov chain on I, so the expected return time to i is given by $(1/\pi_i)$, where π is the unique equilibrium distribution. We will show that $1/\pi_i = \sum_{j,k\in I} a_{jk} / \sum_{k\in I} a_{ik}$.

In fact,

$$p_{jk} = \frac{a_{jk}}{\sum_{l\in I} a_{jl}} \quad \text{and} \quad \left(\sum_{l\in I} a_{jl}\right) p_{jk} = \left(\sum_{l\in I} a_{kl}\right) p_{kj}.$$

So the vector $v = (v_j)$ with $v_j = \sum_{l\in I} a_{jl}$ is in detailed balance with P. Hence

$$\pi_j = \sum_{k\in I} a_{jk} \Big/ \sum_{k,l\in I} a_{kl}.$$

(ii) Consider the distance X_n from the root R at time n. Then $(X_n)_{n \geq 0}$ is a birth-death Markov chain with transition

$$q_i = p_i = 1/2, \text{ if } i \neq 2^m,$$

$$q_i = 1/3, \ p_i = 2/3, \text{ if } i = 2^m.$$

By a standard argument for $h_i = \mathbb{P}_i(\text{hit } 0)$ we have

$$h_0 = 1, \ h_i = p_i h_{i+1} + q_i h_{i-1}, \ i \geq 1,$$

$$p_i u_{i+1} = q_i u_i, \ u_i = h_{i-1} - h_i,$$

$$u_{i+1} = \frac{q_i}{p_i} u_i = \gamma_i u_1, \ \gamma_i = \frac{q_i \cdots q_1}{p_i \cdots p_1},$$

and

$$u_1 + \cdots + u_i = h_0 - h_i, h_i = 1 - A(\gamma_0 + \cdots + \gamma_{i-1}).$$

The condition $\sum_i \gamma_i = \infty$ forces $A = 0$ and hence $h_i = 1$ for all i. Here,

$$\gamma_{2^m - 1} = 2^{-m},$$

so $\sum_i \gamma_i = \infty$ and the walk is recurrent.   □

DBEs are convenient tools for finding an equilibrium distribution: if a measure $\lambda \geq \underline{0}$ is in detailed balance with P and has $\sum_i \lambda_i < \infty$, then $\pi_j = \lambda_j / \sum_i \lambda_i$ is an equilibrium distribution.

Worked Example 1.10.7 Suppose that $\pi = (\pi_i)$ forms an ED for the transition matrix $P = (p_{ij})$, with $\pi P = \pi$, but that the DBEs (1.100) are not satisfied. What is the time reversal of the chain (X_n) in equilibrium?

Solution Assume, for definiteness, that P is irreducible, and $\pi_i > 0$ for all $i \in I$. The answer comes out after we define the transition matrix $P^{\mathrm{TR}} = (p_{ij}^{\mathrm{TR}})$ by

$$\pi_i p_{ij}^{\mathrm{TR}} = \pi_j p_{ji}, \ i, j \in I, \tag{1.103}$$

or

$$p_{ij}^{\mathrm{TR}} = \frac{\pi_j}{\pi_i} p_{ji}, \ i, j \in I. \tag{1.104}$$

Equations (1.103), (1.104) indeed determine a transition matrix, as, for all $i, j \in I$,

$$p_{ij}^{\mathrm{TR}} \geq 0, \text{ and } \sum_{j \in I} p_{ij}^{\mathrm{TR}} = \frac{1}{\pi_i} \sum_{j \in I} \pi_j p_{ji} = \frac{1}{\pi_i} \pi_i = 1.$$

Next, π gives an ED for P^{TR}: for all $j \in I$,

$$\sum_{i \in I} \pi_i p_{ij}^{TR} = \sum_{i \in I} \pi_j p_{ji} = \pi_j.$$

Then, repeating the argument from the proof of Theorem 1.10.1, we obtain that for all $N \geq 1$, the time reversal $(X_{N-n}, \, 0 < n < N)$ is a DTMC in equilibrium, with transition matrix P^{TR} and the same ED π. Symbolically,

$$(X_n^{TR}) \sim (\pi, P^{TR}) - \text{DTMC}, \tag{1.105}$$

where $(X_n^{TR}) = (X_{N-n})$ stands for the time reversal of (X_n). $\square$

It is instructive to remember that P^{TR} was proven to be a stochastic matrix because π is an ED for the original transition matrix P while the proof that π is an ED for P^{TR} used only the fact that P is stochastic.

Example 1.10.8 The detailed balance equations have a useful geometric meaning. Consider the state space $I = \{1, \ldots, s\}$. The matrix P generates a linear transformation $\mathbb{R}^s \to \mathbb{R}^s$, where the vector $\mathbf{x} = \begin{pmatrix} x_1 \\ \vdots \\ x_s \end{pmatrix}$ is taken to be $P\mathbf{x}$. Assuming that P is irreducible, let π be the ED, with $\pi_i > 0$, $i = 1, \ldots, s$. Consider a 'tilted' scalar product $\langle \cdot, \cdot \rangle_\pi$ in $\mathbb{R}^s$, where

$$\langle \mathbf{x}, \mathbf{y} \rangle_\pi = \sum_{i=1}^{s} x_i y_i \pi_i. \tag{1.106}$$

Then the detailed balance equations (1.100) mean that P is self-adjoint (or Hermitian) relative to the scalar product $\langle \cdot, \cdot \rangle_\pi$. That is,

$$\langle \mathbf{x}, P\mathbf{y} \rangle_\pi = \langle P\mathbf{x}, \mathbf{y} \rangle_\pi, \quad \mathbf{x}, \mathbf{y} \in \mathbb{R}^s. \tag{1.107}$$

In fact,

$$\langle \mathbf{x}, P\mathbf{y} \rangle_\pi = \sum_{i,j} x_i p_{ij} y_j \pi_i = \sum_{i,j} x_i p_{ji} y_j \pi_j = \langle P\mathbf{x}, \mathbf{y} \rangle_\pi.$$

The converse is also true: equation (1.107) implies (1.100), since we can take as $\mathbf{x}$ and $\mathbf{y}$ the vectors δ_i and δ_j with the only non-zero entries being 1 at positions i and j, respectively, for all $i, j = 1, \ldots, s$.

This observation yields a benefit, since Hermitian matrices have all eigenvalues real, and their eigenvectors are mutually orthogonal (relative to the scalar product in question – in this instance, $\langle \cdot, \cdot \rangle_\pi$). We will use this in Section 1.12.

Remark 1.10.9 The concept of reversibility and time reversal will be particularly helpful in a continuous-time setting of Chapter 2.

(I.1)   For $h_i^j = \mathbb{P}_i(\text{hit } j)$ the equations are
$$h_j^j = 1, \ h_i^j = \sum_{l \in I} p_{il} h_l^j = (h^j P^T)_i, \ i \neq j,$$
where
$$h^j = (h_i^j, i \in I), \ \text{with} \ h_j^j = 1.$$
Here, $h_i^j \equiv 1$ is always a solution
$$\underline{1} P^T = \underline{1}, \ \text{as} \ (\underline{1} P^T)_i = \sum_l p_{il} = 1 \ \text{for all} \ i \in I.$$

(I.2)   For $k_i^j = \mathbb{E}_i(\text{time to hit } j)$ the equations are
$$k_j^j = 0, \ k_i^j = 1 + \sum_{l \in I, \, l \neq j} p_{il} k_l^j = 1 + (k^j P^T)_i, \ i \neq j.$$
where
$$k^j = (k_i^j, i \in I), \ \text{with} \ k_j^j = 0.$$
Here, taking $0 \cdot \infty = 0$, we have $k_i^j = (1 - \delta_{ij})\infty$ is always a
solution when the chain is irreducible.

These equations are produced by conditioning on the *first* jump.
The vectors h^j and k^j are labelled by the terminal states while
their entries h_i^j and k_i^j indicate the initial states The solution
we look for is identified as a minimal non-negative solution
satisfying the normalisation constraints $h_j^j = 1$ and $k_j^j = 0$.

(II.1)   For
$$\gamma_i^k = \mathbb{E}_k(\text{time spent in } i \text{ before returning to } k)$$
the equations are
$$\gamma_k^k = 1, \ \gamma_i^k = \sum_l \gamma_i^k p_{li}, \ i \neq k,$$
or
$$\gamma^k = \gamma^k P, \ \text{when } k \text{ is recurrent.}$$
Here, conditioning is on the *last* jump, and the vectors γ^k are
labelled by starting states. The identification of the solution
is by the conditions $\gamma_i^k \geq 0$ and $\gamma_k^k = 1$.

(II.2)   Similarly, for an equilibrium distribution
(or more generally, an invariant measure),
$$\pi = \pi P.$$
The identification here is through the condition $\pi_i \geq 0$ and
$\sum_i \pi_i = 1$.

(II.3)   A solution to the detailed balance equations
$$\pi_i p_{ij} = \pi_j p_{ji},$$
always produces an invariant measure. If in addition,
$\sum_i \pi_i = 1$, it gives an equilibrium distribution. As the detailed
balance equations are usually easy to solve (when they have a
solution), they are a powerful tool which is always worth trying
when you need to find an equilibrium distribution.

Table 1.2

It is now time to give a brief summary of essential results established so far about the various equations emerging in the analysis of DTMCs. We have seen two sets of equations: **(I)** for hitting probabilities h_i^A and mean hitting times k_i^A; and **(II)** for equilibrium distributions $\pi = (\pi_i)$ and expected times γ_i^k spent in state i before returning to k. Although they are in a sense similar, there are also differences between them which are important to remember. These are listed in Table 1.2.

1.11 Controlled and partially observed Markov chains

The Crying Control Theory

(From the series *'Movies that never made it to the Big Screen'*.)

We begin this section with a popular example of a *controlled* Markov chain.

Worked Example 1.11.1 Let $m \gg 1$ distinct objects be inspected in a random order, one at a time, without return. One wishes to select the best object but can't take any previously rejected. By introducing a suitable Markov chain, argue that your optimal strategy is to reject the initial k objects then take the first one better than anything seen before and determine $k = k(m)$. Check that $m/k \approx e$ for m large.

Solution Set $X_0 = 1$ and

$$X_1 = \begin{cases} m+1, & \text{if the first object is the best,} \\ i, & \text{if the } i\text{th object is the first one to be better than} \\ & \text{anything before,} \end{cases}$$

$$X_2 = \begin{cases} m+1, & \text{if the first or the } X_1\text{th object is the best,} \\ j, & \text{if the } j\text{th object is the first one after time } X_1 \text{ to be} \\ & \text{better than anything before.} \end{cases}$$

In general,

$$X_r = \begin{cases} m+1, & \text{if } X_{r-1} = m+1 \text{ or the } X_{r-1}\text{st object is the best,} \\ j, & \text{if the } j\text{th object is the first one after } X_{r-1} \text{ to be} \\ & \text{better than anything before.} \end{cases}$$

Then $X_1, X_2, \ldots, X_m = \begin{cases} 2 \\ 3 \\ \vdots \\ m \\ m+1 \end{cases}$ (and $X_n \equiv m+1$ for $n > m$).

Also, $X_{n+1} > X_n \geq n$ if $X_n \leq m$, $1 \leq n \leq m$.

A typical sample trajectory of (X_n) is given in Figure 1.20.

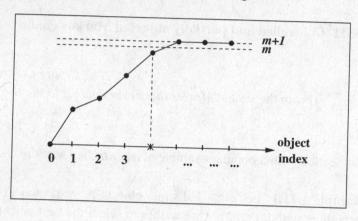

Fig. 1.20

For (X_n) to be a Markov chain, we should have the lack of memory in conditional probabilities

$$\mathbb{P}(X_{n+1} = j | X_n = i, X_{n-1} = i_{n-1}, \ldots, X_1 = i_1, X_0 = 1) = p_{ij}.$$

With the help of some combinatorics,

$$\mathbb{P}\big(i \text{ the (unique) best among } \{1, \ldots, i\}\big) = \frac{(i-1)!}{i!} = \frac{1}{i},$$

and

$$\mathbb{P}\big(j \text{ the best, } i \text{ the second best among } \{1, \ldots, j\}\big) = \frac{(j-2)!}{j!} = \frac{1}{j(j-1)}.$$

Now

$$p_{1j} = \mathbb{P}_1(X_1 = j) = \mathbb{P}\big(j \text{ the best, } 1 \text{ the second best among } \{1, \ldots, j\}\big)$$
$$= \frac{1}{j(j-1)}, \ 1 \leq j \leq m,$$

and

$$p_{1m+1} = \mathbb{P}_1(X_1 = m+1) = \mathbb{P}\big(1 \text{ the overall best}\big) = \frac{1}{m}.$$

Further,

$$\mathbb{P}_1(X_2 = j | X_1 = i) = \frac{\mathbb{P}_1(X_2 = j, X_1 = i)}{\mathbb{P}_1(X_1 = i)}$$

$$= \frac{1}{\mathbb{P}(l \text{ the best, } 1 \text{ the second best, among } \{1, \ldots, i\})}$$

$$\times \mathbb{P}\Big(j \text{ the best, } i \text{ the second best, among} \{1, \ldots, j\};$$

$$1 \text{ the second best among} \{1, \ldots, i\}\Big)$$

$$= \frac{1/j(j-1) \cdot 1/(i-1)}{1/i(i-1)} = \frac{i}{j(j-1)}, \quad 1 \le i < j \le m,$$

and

$$\mathbb{P}_1(X_2 = m+1 | X_1 = i) = \frac{\mathbb{P}_1(X_2 = m+1, X_1 = i)}{\mathbb{P}_1(X_1 = i)}$$

$$= \frac{\mathbb{P}(1 \text{ the second best among } \{1, \ldots, i\}; i \text{ the absolute best})}{\mathbb{P}(i \text{ the best, } 1 \text{ second best, among } \{1, \ldots, i\})}$$

$$= \frac{1/m \cdot 1/(i-1)}{1/i(i-1)} = \frac{i}{m}, \quad 1 < i \le m.$$

In general, for $1 \le i < j \le m$,

$$p_{ij} = \mathbb{P}_1(X_{n+1} = j | X_n = i, X_{n-1} = i_{n-1}, \ldots, X_1 = i_1)$$
$$= \frac{1/j(j-1) \cdot 1/(i-1) \cdot 1/(i_{n-1}-1) \cdots 1/(i_1-1)}{1/i(i-1) \cdot 1/(i_{n-1}-1) \cdots 1/(i_1-1)} = \frac{i}{j(j-1)},$$

for $j = m+1$:

$$p_{im+1} = \mathbb{P}_1(X_{n+1} = m+1 | X_n = i, X_{n-1} = i_{n-1}, \ldots, X_1 = i_1)$$
$$= \frac{1/m \cdot 1/(i-1) \cdot 1/(i_{n-1}-1) \cdots 1/(i_1-1)}{1/i(i-1) \cdot 1/(i_{n-1}-1) \cdots 1/(i_1-1)} = \frac{i}{m}, \quad 1 \le i \le m,$$

and, of course, $p_{m+1m+1} = 1$. The transition matrix is $(m+1) \times (m+1)$:

$$\begin{pmatrix} 0 & 1/1 \cdot 2 & 1/2 \cdot 3 & \cdots & 1/(m-1)m & 1/m \\ 0 & 0 & 2/2 \cdot 3 & \cdots & 2/(m-1)m & 2/m \\ 0 & 0 & 0 & \cdots & 3/(m-1)m & 3/m \\ \cdots & \cdots & \cdots & \cdots & \cdots & \cdots \\ 0 & 0 & 0 & \cdots & 1/m & (m-1)/m \\ 0 & 0 & 0 & \cdots & 0 & 1 \\ 0 & 0 & 0 & \cdots & 0 & 1 \end{pmatrix} \begin{matrix} 1 \\ 2 \\ 3 \\ \vdots \\ m-1 \\ m \\ m+1 \end{matrix}$$

To determine where to stop, consider the decision rule

$$d(j) = \begin{cases} 0 \text{ continue,} \\ 1 \text{ stop,} \end{cases} \quad 1 \le j \le m.$$

We have $d(m) = 1$, trivially. To set up $d(m-1)$, recall that state $m-1$ means the $(m-1)$st object is the best among $\{1,\ldots,m-1\}$. The probability that it is the best overall is $p_{m-1\,m+1} = (m-1)/m$ and is bigger than $(m-1)/m(m-1) = 1/m = p_{m-1\,m}p_{m\,m+1}$, the probability that the mth is the best overall. Hence, $d(m-1) = 1$.

Similarly, to determine $d(m-2)$, we compare $p_{m-2\,m+1} = (m-2)/m$ and $p_{m-2\,m}p_{m\,m+1} + p_{m-2\,m-1}p_{m-1\,m+1}$ which equals

$$\frac{m-2}{m(m-1)} + \frac{m-2}{(m-1)(m-2)} \cdot \frac{m-1}{m} = \frac{m-2}{m(m-1)} + \frac{1}{m}.$$

Equivalently, we compare

$$1 \text{ and } \frac{1}{m-1} + \frac{1}{m-2}.$$

And so on. Clearly,

$$d(m) = d(m-1) = \cdots = d(k+1) = 1, \quad d(k) = \cdots = d(1) = 0,$$

and $k = k(m)$ is determined as the largest value for which

$$\frac{1}{k} + \cdots + \frac{1}{m-1} > 1.$$

For m large, seek k such that

$$\int_k^m dy \, \frac{1}{y} = \ln\left(\frac{m}{k}\right) = 1,$$

i.e. $m/k \approx e$ and $k \approx m/e$.

It is worth noting that the probability of the successful choice under the optimal strategy equals $1/e = 0.3678$ in the limit $m \to \infty$. Indeed, the probability of success,

$$P_{\text{opt}} = \frac{k(m)-1}{m} \sum_{i=k(m)-1}^{m-1} \frac{1}{i} \approx \frac{k(m)-1}{m} \ln\left(\frac{m-1}{k(m)-1}\right) \approx \frac{1}{e} = 0.3678$$

as $k(m) \approx m/e$.

$\square$

Worked Example 1.11.2 Check that the optimal value $k = k(m)$ satisfies the bounds

$$\frac{m-1/2}{e} + \frac{1}{2} - \frac{3e-1}{2(2m+3e-1)} \leq k \leq \frac{m-1/2}{e} + \frac{3}{2}.$$

Solution [Sketch] Use the following inequalities

$$1 \leq \sum_{j=k-1}^{m-1} \frac{1}{j} < \int_{k-3/2}^{m-1/2} \frac{dx}{x} = \ln\left[\frac{m-1/2}{k-3/2}\right], \qquad (1.108)$$

and

$$1 \geq \sum_{j=k}^{m-1} \frac{1}{j} > \int_{k}^{m} \frac{dx}{x} + \frac{1}{2}\left(\frac{1}{k} - \frac{1}{m}\right). \qquad (1.109)$$

Observe that bound (1.108) implies that

$$k < \frac{m-1/2}{e} + \frac{3}{2},$$

whereas (1.109) implies that

$$e > \frac{m}{k}e^{(1/k-1/m)/2} \geq \frac{m}{k}\left[1 + \frac{1}{2}\left(\frac{1}{k} - \frac{1}{m}\right)\right].$$

Substituting for k its upper bound $\dfrac{m-1/2}{e} + \dfrac{3}{2}$ yields the result. $\qquad \square$

Remark 1.11.3 Worked Example 1.11.1 is known in the literature as the Secretary Problem. Other names are also used: a dowry problem, a beauty contest problem and a Googol problem (the name for the number 10^{100}, used long before Google appeared). It has generated a noticeable literature as the problem provides a direct challenge and is easy to state. For the historical background and relation to other known problems, see T.S. Ferguson. "Who solved the secretary problem?", *Statistical Science*, **4** (1989), 282–296; J. Havil. "Optimal Choice"; in *Gamma: Exploring Euler's Constant*, Princeton University Press (2003), 34–138. We mention also the following papers: Y.S. Chow, S. Moriguti, H. Robbins, S.M. Samuel. "Optimal selection based on relative rank (the 'Secretary problem')", *Israel Journ. Math.*, **2** (1964), 81–90; J.P. Gilbert, F. Mosteller. "Recognizing the maximum of a sequence", *Journ. Amer. Statist. Assoc.*, **61** (1966), 35–73. In the last paper an asymptotic formula is derived for the mean number of attempts in the above scheme (that is, mean stopping time). The problem also admits various generalisations, for instance when one takes into account a 'satisfaction' value which attains a maximum when the overall best object had been selected, a value which is somewhat less if the selected object is the second in quality, and so on. See T.J. Stewart. "The secretary problem with an unknown number of options." *Oper. Res.*, **29** (1981), 130–145.

> *I claim not to have controlled events, but*
> *confess plainly that events have controlled me.*
> A. Lincoln (1809–1865), US President

Worked Example 1.11.4 (The Secretary problem with two choices.) Suppose two attempts are allowed, and you are successful if the best candidate is among the two selected. Prove that, asymptotically as $m \to \infty$, the probability of success is 0.5910.

Solution An argument similar to that used above in Worked Example 1.11.1 shows that the optimal strategy lies within the class of strategies indexed by a pair of natural numbers (r,s) (thresholds), where $1 \leqslant r < s \leq m$. These strategies work as follows. First, we reject $r-1$ initial objects. After that, we mark the first object that is better than everyone earlier, and make a 'conditional (or tentative) offer'. This object is called the first candidate, or candidate one. If the first candidate occurred before round $(s-1)$, we reject all objects seen after him but prior to round $(s-1)$ (included). In this case we wait until an object appears, after round $(s-1)$, who is better than everyone before him and select him (of course, he has to be better than the first candidate). This is called the second candidate, or candidate two. If candidate two does not appear, the choice goes to the first candidate. If the first candidate occurred after round s then we simply wait for a better object, and our choice, naturally, would go to him. In the absence of the second candidate we are forced to accept candidate one, and if the first candidate does not occur then we concede a defeat and do not make choice.

To compute the probability of success with the threshold strategy (r,s), the event that the choice is a success is partitioned into three pairwise disjoint events:

(a) the first candidate is the best among all m (in this case the second choice is not made);

(b) the second candidate is the best among all m, and no choice has been made before $s-1$ (included);

(c) the second candidate is the best among all m, and the first choice has been used in one of the rounds $r, r+1, \ldots, s-1$.

It is possible to show that the probability of the events (a), (b) and (c) are

$$P(\mathrm{a}) = \begin{cases} \dfrac{r-1}{m}\left(\dfrac{1}{r-1}+\dfrac{1}{r}+\cdots+\dfrac{1}{m-1}\right), & \text{if } r>1, \\ \dfrac{1}{m}, & \text{if } r=1, \end{cases}$$

$$P(\mathrm{b}) = \frac{r-1}{m}\sum_{v=s+1}^{m}\sum_{u=s}^{v-1}\frac{1}{(u-1)(v-1)},$$

$$P(\mathrm{c}) = \frac{s-r}{m}\sum_{v=s}^{m}\frac{1}{v-1}.$$

Remember too that

$$\mathbb{P}\big(\text{success with }(r,s)\big) = P(\mathrm{a})+P(\mathrm{b})+P(\mathrm{c}).$$

We evaluate in detail only the probability $P(\mathrm{b})$: this case is the most involved. Every term in the sum in the right-hand side for $P(\mathrm{b})$ can be written as

$$\frac{r-1}{u-1}\times\frac{1}{u}\times\frac{u}{v-1}\times\frac{1}{m}=P(\mathrm{I})\times P(\mathrm{II})\times P(\mathrm{III})\times P(\mathrm{IV}).$$

Here

$$P(\mathrm{I}) = \mathbb{P}\big(\text{no candidate appears between rounds } r \text{ and } u-1\big),$$
$$P(\mathrm{II}) = \mathbb{P}\big(\text{a candidate appears in round } u\big),$$
$$P(\mathrm{III}) = \mathbb{P}\big(\text{no candidate appears between } u+1 \text{ and } v-1\big),$$
$$P(\mathrm{IV}) = \mathbb{P}\big(\text{a global leader occurs in round } v\big).$$

The above formulas are computationally rather cumbersome. However, assume that m and r are large (and hence so is s). Then

$$P(\mathrm{a}) \approx \frac{r-1}{m}\ln\left(\frac{m-1}{r-2}\right).$$

Next, applying a similar approximation to the internal sum in the expression for $P(\mathrm{b})$, we get

$$P(\mathrm{b}) \approx \frac{r-1}{m}\sum_{v=s+1}^{m}\frac{1}{v-1}\ln\left(\frac{v-2}{s-2}\right).$$

Replacing the sum by an integral, we get that

$$P(\mathrm{b}) \approx \frac{r}{m}\int_{s}^{m}\frac{1}{v}\ln\left(\frac{v}{s}\right)\,dv \approx \frac{r}{2m}\left(\ln\frac{m}{s}\right)^{2}.$$

The similar approximation for $P(\mathrm{c})$ is that $P(\mathrm{c}) \approx \left(\dfrac{s-r}{m}\right)\ln\dfrac{m}{s}$.

To find the asymptotic optimal value of r and s, set $r = \alpha m$ and $s = \beta m$ and maximise in $0 \leq \alpha \leq \beta \leq 1$. Then

$$\mathbb{P}\big(\text{success with } (r,s)\big) \approx \alpha \ln(1/\alpha) + (\alpha/2)\big[\ln(1/\beta)\big]^2 + (\beta - \alpha)\ln(1/\beta).$$

Differentiating with respect to α and β and equating the derivatives to zero, we get two pairs of roots. One pair gives $\beta^* = e^{-1}$, and $\alpha^* = e^{-3/2}$, with the optimal value $\approx e^{-1} + e^{-3/2} \approx 0.5910$.

The other pair of roots yields $\alpha = \beta = e^{-\sqrt{2}}$, which gives the best probability under a strategy with $(s-r)/m \approx 0$. For m large it yields the value $\approx e^{-\sqrt{2}}(\sqrt{2}+1) \approx 0.5860$. This is only marginally worse than the first pair (which is an overall optimum).

$\square$

For finite m the computations may be done numerically. In the table below the optimal r and s and the corresponding probabilities of success are listed for $m = 5, 10, 20, \ldots, 100, \infty$:

m	r_{opt}	s_{opt}	P_{opt}	m	r_{opt}	s_{opt}	P_{opt}
5	2	3	0.70833	40	10	16	0.60386
10	3	4	0.64632	50	12	19	0.60143
20	5	8	0.61781	100	23	38	0.59617
30	7	12	0.60829	∞	$m/e^{3/2}$	m/e	0.59100

We now pass to partially observed chains.

Worked Example 1.11.5 (i) Let J be a proper subset of the finite state space I of an irreducible Markov chain (X_n), whose transition matrix P is partitioned as

$$P = \begin{pmatrix} P^{JJ} & P^{JI\backslash J} \\ P^{I\backslash JJ} & P^{I\backslash JI\backslash J} \end{pmatrix}.$$

If only visits to states in J are recorded, we see a J-valued Markov chain $(\widetilde{X}_n)$; show that its transition matrix is

$$\widetilde{P} = P^{JJ} + P^{JI\backslash J} \sum_{n \geq 0} \left(P^{I\backslash JI\backslash J}\right)^n P^{I\backslash JJ}$$

$$= P^{JJ} + P^{JI\backslash J} \left(\mathbf{I}_{I\backslash J} - P^{I\backslash JI\backslash J}\right)^{-1} P^{I\backslash JJ},$$

where $\mathbf{I}_{I\backslash J}$ is the unit matrix with rows and columns labeled by $i, j \in I \backslash J$.

(ii) The local MP Bill Sykes spends his time in London in the House of Commons (C), in his flat (F), in the bar (B) or with his girlfriend (G). Each hour, he moves from one to another according to the transition matrix P, though his wife (who knows nothing of his girlfriend) believes that his movements are governed by transition matrix P^W:

$$P = \begin{array}{c} \\ C \\ F \\ B \\ G \end{array} \begin{array}{cccc} C & F & B & G \\ \left(\begin{array}{cccc} 1/3 & 1/3 & 1/3 & 0 \\ 0 & 1/3 & 1/3 & 1/3 \\ 1/3 & 0 & 1/3 & 1/3 \\ 1/3 & 1/3 & 0 & 1/3 \end{array} \right), \end{array} \qquad P^W = \begin{array}{c} \\ C \\ F \\ B \end{array} \begin{array}{ccc} C & F & B \\ \left(\begin{array}{ccc} 1/3 & 1/3 & 1/3 \\ 1/3 & 1/3 & 1/3 \\ 1/3 & 1/3 & 1/3 \end{array} \right). \end{array}$$

The public only sees Bill when he is in $J = \{C, F, B\}$; calculate the transition matrix $\widetilde{P}$ which they believe controls his movements.

Each time Bill moves, in the public eye, to a new location, he calls his wife's mobile phone number; write down the transition matrix that governs the sequence of locations from which the public Bill phones, and calculate its invariant distribution.

Bill's wife notes down the location of each of his calls, and is getting suspicious – he does not come to his flat often enough. Confronted, Bill swears his fidelity and resolves to dump his troublesome transition matrix, choosing instead

$$P^* = \begin{array}{c} \\ C \\ F \\ B \\ G \end{array} \begin{array}{cccc} C & F & B & G \\ \left(\begin{array}{cccc} 1/4 & 1/4 & 1/2 & 0 \\ 1/2 & 1/4 & 1/4 & 0 \\ 0 & 3/8 & 1/8 & 1/2 \\ 2/10 & 1/10 & 1/10 & 6/10 \end{array} \right) \end{array}$$

and still insisting that his moves are governed by P^W. Will this deal with his wife's suspicions? Explain your answer.

Solution (i) Compare with Example 1.4.4. To verify that $\widetilde{P} = P^{JJ} + P^{JI\backslash J}(\mathbf{I} - P^{I\backslash JI\backslash J})^{-1}P^{I\backslash JJ}$, write

$$\mathbb{P}\left(\widetilde{X}_1 = j \,\big|\, \widetilde{X}_0 = i\right)$$
$$= p_{ij} + \sum_{n \geq 2} \mathbb{P}\left(X_n = j, X_r \notin J \text{ for } r = 1, \ldots, n-1 \,\big|\, X_0 = i\right)$$
$$= p_{ij} + \sum_{n \geq 0} \sum_{k \notin J} \sum_{l \notin J} p_{ik}\left[(P^{I\backslash JI\backslash J})^n\right]_{kl} p_{lj}, \quad i, j \in J.$$

(ii) By the first part, with $J = \{C, F, B\}$, we have

$$\widetilde{P} = P^{JJ} + P^{JI\backslash J}(\mathbf{I} - P^{I\backslash JI\backslash J})^{-1}P^{I\backslash JJ} = \begin{pmatrix} 1/3 & 1/3 & 1/3 \\ 1/6 & 1/2 & 1/3 \\ 1/2 & 1/6 & 1/3 \end{pmatrix}.$$

Next, the transition matrix for calls from C, F and B is

$$\begin{pmatrix} 0 & 1/2 & 1/2 \\ 1/3 & 0 & 2/3 \\ 3/4 & 1/4 & 0 \end{pmatrix};$$

its invariant distribution $\pi = (\pi_C, \pi_F, \pi_B)$ satisfies

$$\pi_C = \frac{1}{3}\,\pi_F + \frac{3}{4}\,\pi_B, \quad \pi_F = \frac{1}{2}\,\pi_C + \frac{1}{4}\,\pi_B, \quad \pi_B = \frac{1}{2}\,\pi_C + \frac{2}{3}\,\pi_F,$$

and is uniquely determined

$$\pi = \left(\frac{4}{11}, \frac{3}{11}, \frac{4}{11} \right).$$

Now, with P^*,

$$\widetilde{P^*} = \begin{pmatrix} 1/4 & 1/4 & 1/2 \\ 1/2 & 1/4 & 1/4 \\ 1/4 & 1/2 & 1/4 \end{pmatrix},$$

the invariant distribution for $\widetilde{P^*}$ is

$$\pi^* = \left(\frac{1}{3}, \frac{1}{3}, \frac{1}{3} \right).$$

In other words, on average he is spending equal time in each of public states C, F and B.

However, his wife can observe the following differences from P^W:

(a) calls from B following calls from C are twice as frequent as calls from B following calls from F,

(b) he will phone on average $50/71 > 2/3$ of the time whereas with P^W it would be $2/3$. However, the difference is small and this method is not very practical. Indeed, the invariant distribution $\pi^* = (\pi_C^*, \pi_F^*, \pi_B^*, \pi_G^*)$ for P^* obeys $\pi^* P^* = \pi^*$, i.e.

$$\begin{array}{ll} \pi_C^* = \pi_C^*/4 + \pi_F^*/2 + \pi_G^*/5, & \text{i.e.} \quad \pi_C^*/4 = \pi_F^*/2 + \pi_G^*/5, \\ \pi_F^* = \pi_C^*/4 + \pi_F^*/4 + \pi_B^*/8 + \pi_G^*/10, & \text{i.e.} \quad \pi_F^*/4 = \pi_C^*/4 + \pi_B^*/8 + \pi_G^*/10, \\ \pi_B^* = \pi_C^*/2 + \pi_F^*/4 + \pi_B^*/8 + \pi_G^*/10, & \text{i.e.} \quad \pi_B^*/8 = \pi_C^*/2 + \pi_F^*/8 + \pi_G^*/10, \\ \pi_G^* = \pi_B^*/2 + \pi_G^*/5, & \text{i.e.} \quad \pi_G^*/5 = \pi_B^*/2. \end{array}$$

It is again uniquely determined:

$$\pi_C^* = \frac{28}{71}, \quad \pi_F^* = \frac{34}{71}, \quad \pi_B^* = \frac{4}{71}, \quad \pi_G^* = \frac{5}{71}.$$

Hence, in the long term, the frequency of calls will be

$$\frac{28}{71} \times \frac{3}{4} + \frac{34}{71} \times \frac{3}{4} + \frac{4}{71} \times \frac{3}{8} + \frac{5}{71} \times \frac{6}{10} = \frac{51}{71}.$$

$\square$

> *Who can control his fate?*
>
> W. Shakespeare (1564–1616), English playwright and poet

1.12 Geometric algebra of Markov chains, I. Eigenvalues and spectral gaps

Theorem 1.9.3 provides some bounds for the speed of convergence to an equilibrium distribution. This theorem shows that the convergence happens exponentially (or geometrically) fast in n, which is good news. However, the quantity $(1 - \rho)^{1/m}$ in (1.84) can be pretty close to 1, especially if we consider a natural sequence of Markov chains on increasing state spaces I_ℓ.

An example of a situation where such a problem can arise is as follows. Consider an $\ell \times \ell$ matrix A with entries a_{ij} equal 0 or 1. The *permanent* of A is defined like the determinant, but with signs omitted:

$$\mathrm{per}\, A = \sum_\sigma \prod_{i=1}^\ell a_{i\sigma(i)}$$

where σ is a permutation of order ℓ. Then $\mathrm{per}\, A$ equals the number of 'perfect matches' between points $i \in \{1, \ldots, \ell\}$ labelling the rows and $j \in \{1, \ldots, \ell\}$ labelling the columns. A popular interpretation is that of a group of ℓ boys and ℓ girls; the equation $a_{ij} = 1$ means that girl i and boy j like each other, and $a_{ij} = 0$ that they do not. Then $\mathrm{per}\, A$ counts the number of partnerships where each partner in the pair likes the other. This is a computationally hard problem; the best currently available algorithms for calculating $\mathrm{per}\, A$ take of the order of $\ell 2^\ell$ steps. A stochastic method of computing $\mathrm{per}\, A$ involves an associated Markov chain, and it is important to assess how rapidly it converges to its equilibrium distribution for ℓ large.

Example 1.12.1 Let ℓ be a positive integer and place ℓ points $0, 1, \ldots, \ell - 1$ around a unit circle at the vertices of a regular ℓ-gon. Consider a random walk (X_n) on these points, where a particle jumps to one of its nearest neighbour sites with probability $1/2$.

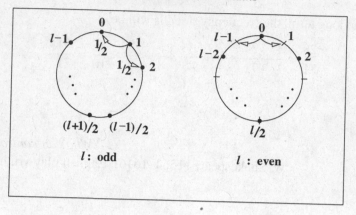

Fig. 1.21

The transition matrix of this Markov chain

$$P = \begin{pmatrix} 0 & 1/2 & 0 & \ldots & 0 & 1/2 \\ 1/2 & 0 & 1/2 & \ldots & 0 & 0 \\ & & & \ldots & & \\ 1/2 & 0 & 0 & \ldots & 1/2 & 0 \end{pmatrix} \qquad (1.110)$$

and has many zeros. In fact, for ℓ even, the whole set of vertices is partitioned into an 'even' subset, $W_e = \{0, 2, \ldots, \ell - 2\}$, and an 'odd', $W_o = \{1, 3, \ldots, \ell - 1\}$. These form periodic subclasses: i.e., if you start at a vertex from the even subset, you will be in the odd subset at all odd times and in the even subset at even times. That is, for ℓ even, the chain (X_n) is periodic, with two subclasses. For ℓ odd, the first power P^m with all positive entries is when $m = \ell - 1$. Further, the minimal entry in $P^{\ell-1}$ is $1/2^{\ell-1}$, which gives the probability $\mathbb{P}_0(X_{\ell-1} = 1) = \mathbb{P}_0(X_{\ell-1} = \ell - 1)$ (as we can get from 0 to 1 or $\ell - 1$ in $\ell - 1$ steps only by travelling all the way along in the corresponding direction).

It is obvious that for all ℓ, the chain is irreducible and has a unique ED

$$\pi = \left(\frac{1}{\ell}, \ldots, \frac{1}{\ell} \right) = \frac{1}{\ell} \underline{1}^{\mathrm{T}}, \quad \text{where } \underline{1} = \begin{pmatrix} 1 \\ \vdots \\ 1 \end{pmatrix}. \qquad (1.111)$$

Moreover, P is reversible with this equilibrium distribution.

Thus, for ℓ odd, the right-hand side of (1.84) will take the form

$$\left(1 - \frac{1}{2^{\ell-1}} \right)^{n/(\ell-1)} \approx \exp\left(-\frac{n}{2^{\ell-1}(\ell-1)} \right). \qquad (1.112)$$

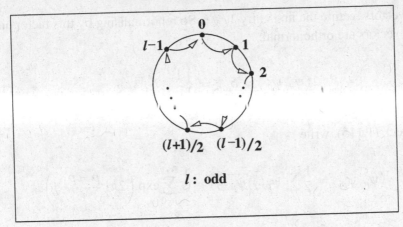

Fig. 1.22

That is, if we want uniformity in convergence when both $\ell, n \to \infty$, we must ensure that $n/(2^{\ell}\ell) \to \infty$, i.e. n must grow faster than $2^{\ell}\ell$. Is it a true bound?

To find out the answer, let us employ some algebra. Matrix (1.110) is Hermitian: $P^{\mathrm{T}} = P$. Hence, it has ℓ orthonormal eigenvectors forming a basis in the ℓ-dimensional real Euclidean space $\mathbb{R}^{\ell}$ (and the ℓ-dimensional complex Euclidean space $\mathbb{C}^{\ell}$), and its eigenvalues are all real. The eigenvectors can be found by using the elegant apparatus of a *discrete Fourier transform*. Namely, consider

$$\psi_p(j) = \frac{1}{\sqrt{\ell}} \exp\left(2\pi i p \, \frac{j}{\ell}\right), \quad j, p = 0, 1, \ldots, \ell - 1. \qquad (1.113)$$

Here $j = 0, 1, \ldots, \ell - 1$ is a discrete argument of these functions, while $p = 0, 1, \ldots, \ell - 1$ is a discrete parameter labeling the functions. Equivalently, we can think of ψ_p as vectors from $\mathbb{C}^{\ell}$, writing $\psi_p = \begin{pmatrix} \psi_p(0) \\ \vdots \\ \psi_p(\ell-1) \end{pmatrix}$ (the entries here are indexed by $0, \ldots, \ell - 1$ instead of the traditional $1, \ldots, \ell$, to make the algebra more transparent). So,

$$\psi_p^{\mathrm{T}} = \left(\frac{1}{\sqrt{\ell}} \overbrace{e^{2\pi i p \cdot 0/\ell}}^{1}, \ldots, \frac{1}{\sqrt{\ell}} e^{2\pi i p(\ell-1)/\ell}\right), \quad p = 0, 1, \ldots, \ell - 1; \quad (1.114)$$

all our vectors feature the first entry $1/\sqrt{\ell}$. So renormalising by this factor ensures that the vectors are orthonormal:

$$\langle \psi_p, \psi_{p'} \rangle = \delta_{p,p'} = \begin{cases} 1, & p = p', \\ 0, & p \neq p' \end{cases}. \tag{1.115}$$

To check (1.115), write

$$\langle \psi_p, \psi_{p'} \rangle = \frac{1}{\ell} \sum_{j=0}^{\ell-1} \psi_p(j) \overline{\psi_{p'}(j)} = \frac{1}{\ell} \sum_{j=0}^{\ell-1} \exp\left(2\pi i \frac{p-p'}{\ell} j \right).$$

When $p = p'$, the right-hand side is equal to $(1/\ell)\sum_{j=0}^{\ell-1} 1 = 1$. Otherwise, i.e. when $p \neq p'$, we have the sum of a geometric progression, with the complex denominator $\exp[2\pi i (p - p')/\ell]$:

$$\langle \psi_p, \psi_{p'} \rangle = \frac{1}{\ell} \frac{\exp[2\pi i(p-p')\ell/\ell]-1}{\exp[2\pi i(p-p')/\ell]-1} = 0,$$

as, in the numerator, $\exp[2\pi i(p-p')\ell/\ell] = \exp[2\pi i(p-p')] = 1$.

Now, we want to verify that the vectors ψ_p are eigenvectors of P:

$$\begin{aligned}
(P\psi_p)(j) &= \frac{1}{2} \psi_p(j-1) + \frac{1}{2} \psi_p(j+1) \\
&= \frac{1}{2\sqrt{\ell}} \left(e^{2\pi i p(j-1)/\ell} + e^{2\pi i p(j+1)/\ell} \right) \\
&= \frac{1}{\sqrt{\ell}} \cos\left(2\pi \frac{p}{\ell} \right) e^{2\pi i p j/\ell} \\
&= \cos\left(2\pi \frac{p}{\ell} \right) \psi_p(j). \tag{1.116}
\end{aligned}$$

Hence, $P\psi_p = \mu_p \psi_p$, and the eigenvalues are

$$\mu_p = \cos\left(2\pi \frac{p}{\ell} \right), \quad p = 0, 1, \dots, \ell - 1. \tag{1.117}$$

The first eigenvalue $\mu_0 = 1$, and the corresponding eigenvector $\psi_0 = \frac{1}{\sqrt{\ell}} \mathbf{1}$ is proportional to the (transpose of the) equilibrium distribution π (compare (1.111)):

$$\pi^{\mathrm{T}} = \frac{1}{\sqrt{\ell}} \psi_0.$$

As mentioned before, the change of the normalisation is explained by differences in requirements: on the one hand, we want $\|\psi_0\|^2 = \langle \psi_0, \psi_0 \rangle = 1$, requiring $\psi_p = \frac{1}{\sqrt{\ell}} \underline{1}$, and on the other, we need $\sum_j \pi_j = \langle \pi^T, \underline{1} \rangle = 1$, requiring $\pi^T = \frac{1}{\ell} \underline{1}$.

Back To The Fourier

(From the series *'Movies that never made it to the Big Screen'*.)

As a matter of fact, one can easily produce real eigenvectors of P (as expected, since P is a real matrix). Note that the complex conjugate $\overline{\psi_p}$ coincides with $\psi_{\ell-p}$, $p = 0, 1, \ldots, \ell - 1$. In fact,

$$\overline{\psi_p(j)} = \frac{1}{\sqrt{\ell}} \exp\left(-2\pi i p \frac{j}{\ell}\right) = \frac{1}{\sqrt{\ell}} \exp\left(2\pi i - 2\pi i p \frac{j}{\ell}\right)$$

$$= \frac{1}{\sqrt{\ell}} \exp\left(2\pi i (\ell - p) \frac{j}{\ell}\right) = \psi_{\ell-p}(j), \quad j = 0, 1, \ldots, \ell - 1.$$

The respective eigenvalues coincide: $\mu_p = \mu_{\ell-p}$, as

$$\cos\left(2\pi \frac{p}{\ell}\right) = \cos\left(2\pi - 2\pi \frac{p}{\ell}\right), \quad p = 0, 1, \ldots, \ell - 1.$$

For $p = 0$ this is trivial, as $\psi_0^T = \frac{1}{\sqrt{\ell}} \underline{1} = \overline{\psi_0}^T$ is real. If ℓ is even and $p = \ell/2$, the vector ψ_p is again real: $\psi_p(j) = \frac{1}{\sqrt{\ell}} e^{\pi i j} = +1$ or -1, depending on the parity of j. In vector notation

$$\psi_{\ell/2} = \frac{1}{\sqrt{\ell}} \underline{1}^a, \quad \text{where } \underline{1}^a = \begin{pmatrix} 1 \\ -1 \\ \vdots \\ 1 \\ -1 \end{pmatrix}.$$

In other words, the vector $\underline{1}^a$ has entries alternating from 1 (for even labels $j = 0, 2, \ldots, \ell - 2$) to -1 (for odd labels $j = 1, 3, \ldots, \ell - 1$). The corresponding eigenvalue $\mu_{\ell/2} = \cos \pi = -1$.

So, apart from $p = 0$ and $p = \ell/2$, for even ℓ, the eigenvectors are grouped into conjugate pairs, with the same eigenvalue. In other words, each of these eigenvalues has multiplicity two. Hence, we can produce the following real orthonormal eigenvectors

$$\frac{1}{\sqrt{2}} (\psi_p + \overline{\psi_p}), \quad \text{with entries} \quad \frac{2}{\sqrt{2\ell}} \cos\left(2\pi p \frac{j}{\ell}\right), \quad j = 0, \ldots, \ell - 1,$$

and

$$\frac{1}{i\sqrt{2}}\,(\psi_p - \overline{\psi_p}), \text{ with entries } \frac{2}{\sqrt{2\ell}}\,\sin\left(2\pi p\,\frac{j}{\ell}\right), \; j = 0,\dots,\ell-1,$$

where $p = 1, 2, \dots, \ell/2 - 1$. For our purposes it matters little whether we use complex or real eigenvectors; what is important is that they form a complete orthonormal system (a basis).

Why such meticulous (although beautiful) algebra? Because we can represent (the transpose of) an initial distribution row-vector $\lambda = (\lambda_i)$ as

$$\lambda^{\mathrm{T}} = \sum_{p=0}^{\ell-1} \langle \lambda^{\mathrm{T}}, \psi_p \rangle \psi_p,$$

and write the row-vector λP^n of probabilities $\mathbb{P}(X_n = i)$ as a linear combination:

$$(\lambda P^n)^{\mathrm{T}} = \sum_{p=0}^{\ell-1} \langle \lambda^{\mathrm{T}}, \psi_p \rangle \left[\cos\left(\frac{2\pi p}{\ell}\right)\right]^n \psi_p. \tag{1.118}$$

The term with $p = 0$ on the right-hand side of (1.118) has the cosine factor 1 and is equal to

$$\langle \lambda^{\mathrm{T}}, \psi_0 \rangle \psi_0 = \frac{1}{\ell}\,\langle \lambda^{\mathrm{T}}, \underline{1} \rangle \underline{1} = \frac{1}{\ell}\,\underline{1} = \pi^{\mathrm{T}}, \text{ as } \langle \lambda^{\mathrm{T}}, \underline{1} \rangle = \sum_i \lambda_i = 1.$$

All other terms comprise factors $\mu_p^n = [\cos(2\pi p/\ell)]^n$; if ℓ is odd, all μ_p with $p \neq 0$ lie strictly between -1 and 1 and hence the rest of the sum on the right-hand side of (1.118) is suppressed as $n \to \infty$:

$$(\lambda P^n)^{\mathrm{T}} \approx \pi^{\mathrm{T}}, \text{ or } \lambda P^n \approx \pi. \tag{1.119}$$

If ℓ is even, we should also count the term with $p = \ell/2$: it comprises the cos factor $(-1)^n$ and equals

$$\frac{1}{\sqrt{\ell}}\,\langle \lambda^{\mathrm{T}}, \psi_{\ell/2} \rangle (-1)^n \psi_{\ell/2} = \frac{1}{\ell}\,\langle \lambda^{\mathrm{T}}, \underline{1}^{\mathrm{a}} \rangle \underline{1}^{\mathrm{a}}.$$

The last expression can be rewritten as

$$(\Lambda^{\mathrm{ev}} - \Lambda^{\mathrm{od}})\alpha^{\mathrm{T}}, \text{ where } \Lambda^{\mathrm{ev}} = \sum_{i \text{ even}} \lambda_i, \; \Lambda^{\mathrm{od}} = \sum_{i \text{ odd}} \lambda_i, \text{ and } \alpha^{\mathrm{T}} = \frac{1}{\ell}\,\underline{1}^{\mathrm{a}}.$$

In this case all μ_p with $p \neq 0, \ell/2$ lie strictly between -1 and 1, and their contribution is suppressed:

$$(\lambda P^n)^{\mathrm{T}} \approx \pi^{\mathrm{T}} + (-1)^n(\Lambda^{\mathrm{ev}} - \Lambda^{\mathrm{od}})\alpha^{\mathrm{T}}, \text{ or } \lambda P^n \approx \pi + (-1)^n(\Lambda^{\mathrm{ev}} - \Lambda^{\mathrm{od}})\alpha. \tag{1.120}$$

Note that $\Lambda^{\text{ev}} + \Lambda^{\text{od}} = \langle \lambda^{\text{T}}, \underline{1} \rangle = 1$, and for $\lambda = \pi$, the invariant distribution, $\Lambda^{\text{ev}} = \Lambda^{\text{od}} = 1/2$ (cancelling the difference $\Lambda^{\text{ev}} - \Lambda^{\text{od}}$). On the other hand, suppose that $\Lambda^{\text{ev}} = 1$ and $\Lambda^{\text{od}} = 0$; i.e., the initial distribution λ is concentrated on the even subclass. Then, for n even,

$$\left(\lambda P^n \right)^{\text{T}} \approx \frac{2}{\ell} \underline{1}_{\text{ev}}, \quad \text{where } \underline{1}_{\text{ev}} = \begin{pmatrix} 1 \\ 0 \\ \vdots \\ 1 \\ 0 \end{pmatrix}.$$

In other words, if ℓ is even and λ is concentrated on the even periodic subclass, then, as $n = 2N \to \infty$, the vector $\left(\lambda P^n \right)^{\text{T}}$ approaches a uniform distribution on the even subclass. Similarly, as $n = 2N + 1 \to \infty$, the vector $\left(\lambda P^n \right)^{\text{T}}$ approaches a uniform distribution on the odd periodic subclass. The picture for ℓ even and λ concentrated on the odd subclass is symmetric.

Now we can assess the speed of convergence of the approximations in (1.119) and (1.120) quite accurately. It is convenient to introduce a 'spectral gap' measuring the distance from the points ± 1 to the absolute values of the μ_ps:

$$\delta^{(\ell)} = \min \left[\left| 1 - |\lambda_p| \right| : \; p = 1, 2, \ldots \text{ with } 2p < \ell \right].$$

Then, for ℓ odd, $\delta^{(\ell)}$ is attained at $p = (\ell \pm 1)/2$:

$$\delta^{(\ell)} = 1 + \cos(\pi(\ell \pm 1)/\ell) = \frac{\pi^2}{2\ell^2} + O(1/\ell^4), \tag{1.121}$$

and

$$\left(\lambda P^n \right)^{\text{T}} = \pi^{\text{T}} + O(\ell e^{-n\delta^{(\ell)}}), \quad \text{or } \lambda P^n = \pi + O(\ell e^{-n\delta^{(\ell)}}). \tag{1.122}$$

In other words, we have convergence to equilibrium as $\ell \to \infty$ if n grows faster than $\ell^2 \ln \ell$. This is much less stringent a restriction, compared with (1.112).

Similarly, for ℓ even, $\delta^{(\ell)}$ is attained at $p = 1$ and $p = \ell/2 \pm 1$:

$$\delta^{(\ell)} = 1 - \cos(2\pi/\ell) = 1 + \cos(\pi \pm 2\pi/\ell) \approx \frac{2\pi^2}{\ell^2} + O(1/\ell^4), \tag{1.123}$$

and

$$\lambda P^n = \pi + (-1)^n (\Lambda^{\text{ev}} - \Lambda^{\text{od}})\alpha + O(\ell e^{-n\delta^{(\ell)}}), \tag{1.124}$$

which requires the same order of growth of n with ℓ.

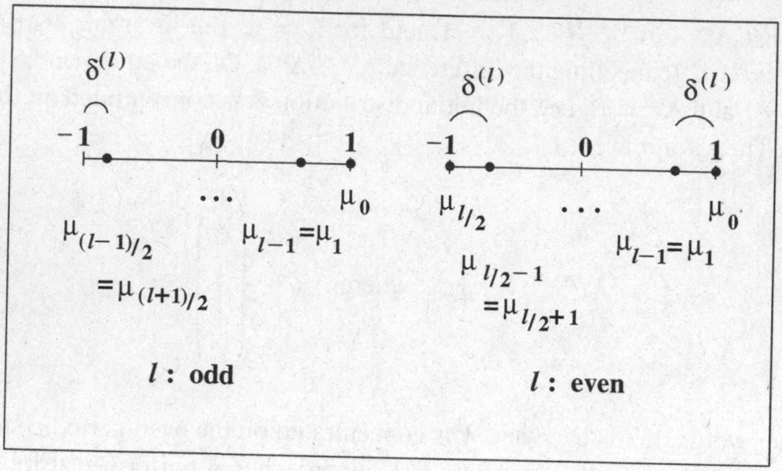

Fig. 1.23

It is instructive to have a look at the matrix $L = \mathbf{I} - P$:

$$L = \mathbf{I} - P = \begin{pmatrix} 1 & -1/2 & 0 & \ldots & 0 & -1/2 \\ -1/2 & 1 & -1/2 & \ldots & 0 & 0 \\ & & & \ldots & & \\ -1/2 & 0 & 0 & \ldots & -1/2 & 1 \end{pmatrix}. \qquad (1.125)$$

This acts on the vector $\psi = \begin{pmatrix} \psi_0 \\ \vdots \\ \psi_{\ell-1} \end{pmatrix}$ via

$$(L\psi)_i = \psi_i - \frac{1}{2}\left(\psi_{i-1 \bmod \ell} + \psi_{i+1 \bmod \ell}\right), \quad i = 0, 1, \ldots, \ell - 1, \qquad (1.126)$$

which can be viewed as a discrete version of the second derivative map (with the minus sign and the coefficient $1/2$ in front):

$$-\frac{1}{2}\frac{d^2}{dx^2} : f(x) \mapsto -(1/2)\,f''(x). \qquad (1.127)$$

The expression (1.127) defines a linear map on the space of twice-differentiable functions $f(x)$, on the real line $\mathbb{R}$ or on an interval, say $[0, 2\pi]$. In the latter case, one usually considers the action of the map on functions satisfying a boundary condition, say, $f(0) = f(2\pi)$ and $f'(0) = f'(2\pi)$; this means that the endpoints 0 and 2π are 'merged', and the interval is turned into a unit circle. In our example, by considering points mod ℓ, we merge vertices 0 and ℓ. In the multi-dimensional case, where functions depend on the variables $x_1, \ldots, x_d$, we replace (1.127) by

$$f(x_1, \ldots, x_d) \mapsto -\frac{1}{2} \sum_{k=1}^{d} \frac{\partial^2}{\partial x_k^2} f(x_1, \ldots, x_d). \tag{1.128}$$

This is called the *Laplace operator*, or the *Laplacian*; a standard notation used for the right-hand side is $-(1/2)\Delta f(x_1, \ldots, x_d)$.

For these reasons, we call the matrix L in (1.125) the *discrete Laplacian* on the unit circle. From the definition we see that L is Hermitian. Moreover, the eigenvectors of L are precisely $\psi_0, \ldots, \psi_{\ell-1}$, and the eigenvalues are of the form $\beta_p = 1 - \mu_p$:

$$\beta_p = 1 - \cos\left(2\pi \frac{p}{\ell}\right), \quad p = 0, 1, \ldots, \ell - 1. \tag{1.129}$$

Note that $\beta_0 = 0$ and $\beta_1 = 1 - \mu_1$, the distance from $\mu_0 = 1$ to the rest of the eigenvalues of P. As all eigenvalues $\beta_p \geq 0$ (and $\beta_p > 0$ for $p \geq 1$), the matrix L is non-negative definite. The discrete Laplacian will be studied in more detail in Section 1.14.

μ *And ℓ's Wedding*

(From the series *'Movies that never made it to the Big Screen'*.)

Example 1.12.2 (The classical Ehrenfest urn problem) Suppose we have d distinct objects (for instance, balls with numbers $1, \ldots, d$) all of which are painted black and white. The balls are put in a box (or an urn). We take a ball from the box at random, change its colour and put it back. The number of states of this system is 2^d (each ball can be black or white). The model can be described as a nearest neighbour random walk on a d-dimensional binary cube, with the number of vertices 2^d and the number of edges $d2^{d-1}$. The vertices (states) are labeled by binary 'strings' $\underline{\alpha}$ of length d. Thus: $\underline{\alpha} = (\alpha_1, \ldots, \alpha_d) \in \{0, 1\}^d$, with $\alpha_1, \ldots, \alpha_d = 0, 1$. The transition matrix $P = (p_{ij})$ has

$$p_{\underline{\alpha},\underline{\alpha}'} = \begin{cases} \dfrac{1}{d}, & \text{if } \underline{\alpha} \text{ and } \underline{\alpha}' \text{ are nearest neighbours,} \\ & \text{i.e. } \underline{\alpha}' \text{ is obtained from } \underline{\alpha} \text{ by changing a single entry } j \\ & \text{(from } \alpha_j = 0 \text{ to } \alpha_j' = 1 \text{ or from } \alpha_j = 1 \text{ to } \alpha_j' = 0), \\ & j = 1, \ldots, d, \\ 0, & \text{otherwise, i.e. when either, } \underline{\alpha} \text{ and } \underline{\alpha}' \text{ differ at more} \\ & \text{than one digit, or coincide.} \end{cases} \tag{1.130}$$

Here the invariant distribution is uniform: $\pi_{\underline{\alpha}} = 1/2^d$, which gives an eigenvector of P, with the eigenvalue $\mu_0 = 1$. Again, all eigenvalues $1 = \mu_0 \geq \cdots \geq \mu_{2^d-1}$ can be computed explicitly (though this is more difficult). The answer is that there are $d+1$ distinct values of the form

$$1 - \frac{2j}{d}, \ 0 \leq j \leq d, \tag{1.131}$$

with geometric multiplicity

$$\binom{d}{j}. \tag{1.132}$$

(The definitions and properties of algebraic and geometric multiplicities of characteristic roots and eigenvalues are discussed below.) To prove the above fact, consider the adjacency matrix A_d of the form $d \times P$. The matrix A_d has all entries $0, 1$ and is a self-adjoint $2^d \times 2^d$ matrix defined, recursively, by

$$A_0 = (0) \quad (1 \times 1),$$

and for $d \geq 1$ by

$$A_d = \begin{pmatrix} A_{d-1} & \mathbf{J}_{d-1} \\ \mathbf{J}_{d-1} & A_{d-1} \end{pmatrix}.$$

Here $\mathbf{J}_{d-1}$ is a $2^{d-1} \times 2^{d-1}$ binary matrix written as the incidence matrix corresponding to the bijective vertex suspensed between two copies of A_{d-1}: see Figure 1.19. We see that A_d is a partitioned matrix (it has non-0 entries whose blocks commute between themselves). Then the characteristic polynomial $D_d(\mu) = \det(\mu \mathbf{I}_d - A_d)$ is $D_0(\mu) = \mu$ and for $d \geq 1$:

$$D_d(\mu) = \det \begin{pmatrix} \mu \mathbf{I}_{d-1} - A_{d-1} & -\mathbf{J}_{d-1} \\ -\mathbf{J}_{d-1} & \mu \mathbf{I}_{d-1} - A_{d-1} \end{pmatrix}.$$

This can be calculated as a determinant of a determinant, bearing in mind that $\mathbf{J}_{d-1}^2 = \mathbf{I}_{d-1}$:

$$\begin{aligned} D_d(\mu) &= \det \left[(\mu \mathbf{I}_{d-1} - A_{d-1})^2 - \mathbf{I}_{d-1} \right] \\ &= \det \left[(\mu - 1)\mathbf{I}_{d-1} - A_{d-1} \right] \left[(\mu + 1)\mathbf{I}_{d-1} - A_{d-1} \right] \\ &= D_{d-1}(\mu - 1) D_{d-1}(\mu + 1). \end{aligned}$$

Iterating yields

$$D_d(\mu) = \prod_{j=0}^{d} (\mu - d + 2j)^{m(d,j)},$$

whence the eigenvalues of A_d are $d - 2j$, $j = 0, \ldots, d$, and those of $P = d^{-1}A_d$ are given by (1.131). To calculate the multiplicities $m(d, j)$, again use the above recursion:

$$\prod_{j=0}^{d} (\mu - d + 2j)^{m(d,j)}$$

$$= \prod_{j=0}^{d-1} (\mu - 1 - d + 1 + 2j)^{m(d-1,j)} \prod_{j=0}^{d-1} (\mu + 1 - d + 1 + 2j)^{m(d-1,j)}$$

$$= \prod_{j=0}^{d-1} (\mu - d + 2j)^{m(d-1,j)} (\mu - d + 2(j+1))^{m(d-1,j)}$$

$$= \prod_{j=0}^{d} (\mu - d + 2j)^{m(d-1,j)+m(d-1,j-1)},$$

where

$$m(d - 1, j) = 0 \text{ if } j < 0 \text{ or } j > d - 1.$$

Therefore,

$$m(d, j) = m(d - 1, j) + m(d - 1, j - 1), \text{ implying that } m(d, j) = \binom{d}{j}.$$

Thus, the algebraic and geometric multiplicity of eigenvalue $d - 2j$ of A_d is as in (1.132). (This proof was supplied by David M.R. Jackson.)

The chain with transition probabilities (1.130) is irreducible and periodic, of period 2, which agrees with the fact that the last eigenvalue $\mu_{2^d-1} = -1$. To obtain an aperiodic chain, it is convenient to change the model by allowing the walk to stay in place with probability $1/(d+1)$, the remaining probability $d/(d+1)$ is again split equally between d nearest neighbours. In this modified form, the example will be commented on in the next section.

Examples 1.12.1 and 1.12.2 prepare us for a discussion of the general spectral properties of a transition matrix. (Many properties stated below hold for a general matrix M with non-negative elements.) Suppose P is an $\ell \times \ell$ transition matrix. One can consider its *right* action, such as $\lambda \mapsto \lambda P$, where P acts on ℓ-dimensional row vectors $\lambda = (\lambda_1, \ldots, \lambda_\ell)$, and the *left* action, $\mathbf{x} \mapsto P\mathbf{x}$, where it acts on ℓ-dimensional column vectors $\mathbf{x} = \begin{pmatrix} x_1 \\ \vdots \\ x_\ell \end{pmatrix}$; in each case the vectors may be real or complex. Of course, the right action of P corresponds to the left action of P^T, the transposed matrix, and vice versa. (But P^T is not necessarily a stochastic matrix.) In what follows, while speaking of eigenvalues and eigenvectors, we mean the right action of the matrix involved. Thus, an *eigenvalue* of P is a number μ such that $\mathbf{y}^{\mathrm{T}}P = \mu\mathbf{y}^{\mathrm{T}}$ for some ℓ-dimensional row-vector $\mathbf{y}^{\mathrm{T}}$. Similarly, an eigenvalue of P^{T} is a number μ such that $\mathbf{y}^{\mathrm{T}}P^{\mathrm{T}} = \mu\mathbf{y}^{\mathrm{T}}$ for some row-vector $\mathbf{y}^{\mathrm{T}}$; evidently, the equality $\mathbf{y}^{\mathrm{T}}P^{\mathrm{T}} = \mu\mathbf{y}^{\mathrm{T}}$ is equivalent to $P\mathbf{y} = \mu\mathbf{y}$. In both cases, μ and $\mathbf{y}$ may be complex.

The *spectrum* of a matrix P is defined as the set of its eigenvalues. Of course, every eigenvalue is a root of the characteristic equation $\det(\mu\mathbf{I} - P) = 0$; moreover, every root is an eigenvalue. The determinant $\det(\mu\mathbf{I} - P) = (-1)^{\ell}\det(P - \mu\mathbf{I})$ is a polynomial of degree ℓ (called the characteristic polynomial of the matrix P), hence it has ℓ roots, some of which may be complex (despite the fact the coefficients of the polynomial are real). We know that every equilibrium distribution π is an eigenvector with the eigenvalue 1, as the invariance equation $\pi P = \pi$ tells us precisely that. Next, if P is irreducible, it has a unique equilibrium distribution; in general, for every closed communicating class, we have a unique equilibrium distribution supported by this class (and any ED is a convex linear combination of those). So, 1 is always an eigenvalue; the tradition is to assign to it the label 0: $\mu_0 = 1$.

Similarly, the *spectrum* of a matrix P^{T} is defined as the set of its eigenvalues. We also know that P^{T} has always eigenvalue 1: the corresponding eigenvector is $\underline{1}^{\mathrm{T}} = (1, \ldots, 1)$. Indeed, $\underline{1}^{\mathrm{T}}P^{\mathrm{T}} = (P\underline{1})^{\mathrm{T}} = \underline{1}^{\mathrm{T}}$, as $P\underline{1} = \underline{1}$ (see (1.3)). In fact, the roots of the characteristic equations for P and P^{T} are the same since the polynomials coincide: $\det(\mu\mathbf{I} - P) = \det(\mu\mathbf{I} - P)^{\mathrm{T}} = \det(\mu\mathbf{I} - P^{\mathrm{T}})$. So, the spectra of P and P^{T} coincide.

A spectre is haunting Europe.
K. Marx (1818–1883), German philosopher

What is the difference between the roots of the characteristic equation (or, equivalently, the roots of the characteristic polynomial) and the eigenvalues? The short answer is: multiplicity. Suppose that the roots of the characteristic polynomial $\det(\mu\mathbf{I} - P)$ are $\mu_0, \ldots, \mu_{\ell-1}$. We then have the product decomposition $\det(\mu\mathbf{I} - P) = \prod\limits_{p=0}^{\ell-1}(\mu - \mu_p)$. But the roots may have multiplicities, so it is also convenient to write $\det(\mu\mathbf{I} - P) = \prod\limits_{p}(\mu - \mu_p)^{\alpha_p}$ where the product is over the distinct roots, or equivalently, over the distinct eigenvalues. Here, one distinguishes between the *algebraic* multiplicity α_p of the root μ_p and the *geometric* multiplicity of the eigenvalue μ_p: the latter equals the number of linearly independent vectors y such that $yP = \mu_p y$ (that is, the dimension of the *eigenspace* $E(\mu_p) \subseteq \mathbb{C}^{\ell}$ of the eigenvalue μ_p). The algebraic multiplicity is always greater than or equal to the geometric one: $\alpha_p \geq \dim E(\mu_p)$. But if μ_p is a root of $\det(\mu\mathbf{I} - P)$ (i.e. has multiplicity $\alpha_p \geq 1$) then $\dim E(\mu_p) \geq 1$, i.e. there exists an eigenvector with the eigenvalue μ_p. Therefore, the matrix P may have less than ℓ linearly independent eigenvectors, but their number is always greater than or equal to the number of distinct roots.

The same is of course true for P^T. Moreover, both the algebraic and geometric multiplicities of the roots μ_p are the same for both matrices P and P^T. As a result, the number of linearly independent eigenvectors for P and P^T is the same. To put it differently, the spectra of P and P^T coincide even when we take into account the multiplicities. Indeed, the last fact holds for any real matrix M.

If P is Hermitian (i.e. $P = P^T$) then the geometric multiplicities coincide with the algebraic ones. In fact, in this case P has an orthonormal basis of ℓ eigenvectors. Furthermore, in this case all the eigenvalues (or, equivalently, all roots of the characteristic equation are real). In other words, the spectrum of P is a subset of real line $\mathbb{R}$.

> *We will do what we always do: raise EVAT, the eigenvalue added tax.*
> (From the series 'When they go political'.)

Also, if $\det(\mu \mathbf{I} - P)$ has pairwise distinct roots then it has ℓ linearly independent eigenvectors.

From now on we will assume that P is irreducible; otherwise we have an invariant subspace for every closed communicating class and can consider the action of P on each subspace. It is also useful to note that the row vector representing the indicator of each closed communicating class (with entries 1 over the class and 0 outside) will be invariant for P^T. The matrix norm $||P||$ mentioned below is defined in a standard fashion:

$$||P|| \;=\; \sup \left[\, ||\mathbf{x}^T P|| : \text{row vector } \mathbf{x}^T \in \mathbb{R}^\ell, \, ||\mathbf{x}|| = 1 \right]$$

$$=\; \sup \left[\frac{||\mathbf{x}^T P||}{||\mathbf{x}||} : \text{row vector } \mathbf{x}^T \in \mathbb{R}^\ell, \, \mathbf{x} \neq \mathbf{0} \right].$$

Here $||\mathbf{x}|| = (\langle \mathbf{x}, \mathbf{x} \rangle)^{1/2} = \left(\sum_{i=1}^\ell |x_i|^2 \right)^{1/2}$, the standard Euclidean norm of a vector, generated by the standard scalar product in $\mathbb{R}^\ell$ or $\mathbb{C}^\ell$.

In Example 1.12.1, we saw that the spectrum of the transition matrix (1.110) lies in the interval $[-1, 1]$, between points $\mu_0 = 1$ and -1. It turns out that this is a general property of transition matrices. More precisely, the eigenvalues of P may be complex, but they must lie within a complex circle of radius 1 centred at the origin. The formal statement is as follows.

Theorem 1.12.3 *Let P be an $\ell \times \ell$ irreducible stochastic matrix. Then:*

(a) *$\mu_0 = 1$ is always an eigenvalue of P and P^{T}, its algebraic and geometric dimensions equal 1, and the corresponding eigenspace of P is generated by the equilibrium distribution π, while the corresponding eigenspace of P^{T} is generated by the vector $\underline{1}^{\mathrm{T}}$. The norm $||P|| = ||P^{\mathrm{T}}||$ equals 1 and all eigenvalues $\mu_p \neq \mu_0$ satisfy $|\mu_p| \leq 1$, i.e. lie within the closed unit circle in the complex plane $\mathbb{C}$.*

(b) *If P is aperiodic, then all the eigenvalues $\mu_p \neq \mu_0$ have $|\mu_p| < 1$, i.e. lie inside the open unit circle in $\mathbb{C}$. Conversely, if all the eigenvalues $\mu_p \neq \mu_0$ have $|\mu_p| < 1$, the chain is aperiodic.*

This is a particular case of the so-called Perron–Frobenius Theorem, which can be stated and proved for general matrices with non-negative entries: see Theorem 1.15.7. We are particularly interested in property (b), so we give a brief proof of this statement. Suppose that P is irreducible and aperiodic. Then, according to Theorem 1.9.2, the vector λP^n converges to the equilibrium distribution π, for any initial distribution λ, in particular, for $\lambda = \delta_i$, $i = 1, \ldots, \ell$, where δ_i is a measure sitting at state i. But the δ_is form an orthonormal basis in spaces of the ℓ-dimensional row vectors $\mathbb{R}^\ell$ and $\mathbb{C}^\ell$:

$$\delta_i = (0, \ldots, 0, 1, 0, \ldots, 0) \text{ (entry 1 in the ith place, zeros elsewhere)}.$$

Hence, the convergence takes place for any ℓ-dimensional row-vector $\mathbf{x}^{\mathrm{T}} = (x_1, \ldots, x_\ell)$, real or complex:

$$\lim_{n \to \infty} \mathbf{x}^{\mathrm{T}} P^n = \lim_{n \to \infty} \sum_{i=1}^{\ell} x_i \delta_i P^n = \sum_{i=1}^{\ell} x_i \pi = \langle \mathbf{x}, \underline{1} \rangle \pi. \tag{1.133}$$

Now, suppose that there exists a row eigenvector ψ^{T}, with an eigenvalue μ where $|\mu| \geq 1$ and $\mu \neq \mu_0 = 1$. Then

$$\psi^T P^n = \mu^n \psi^T \not\to \langle \psi, \underline{1} \rangle \pi,$$

which contradicts (1.133). Hence, there is no such eigenvector, and all eigenvalues $\mu \neq \mu_0$ must have $|\mu| < 1$.

Conversely, assume that for any eigenvalue $\mu \neq \mu_0$, the modulus $|\mu| < 1$, but P is periodic, i.e. has a periodic subclass, $\mathscr{S}_1, \ldots, \mathscr{S}_{k-1}$, of period k. Then, under the matrix P^k, for all $j = 1, \ldots, k-1$, states from $\mathscr{S}_j$ do not communicate with states from outside $\mathscr{S}_j$. That is, under the k-step transition matrix P^k, each subclass $\mathscr{S}_j$ contains (possibly, coincides with) a closed communicating class, $\mathscr{C}_j$. Naturally, for all $j = 1, \ldots, k$, the matrix P^k has an equilibrium distribution: that is, an invariant stochastic row vector, concentrated on $\mathscr{C}_j$. Let us focus on $j = 1$: suppose that

$\pi^{(1)} = (\pi_i^{(1)})$ is an equilibrium distribution for P^k concentrated on $\mathscr{C}_1$. That is, $\pi^{(1)}P^k = \pi^{(1)}$, and entries $\pi_i^{(1)} = 0$ for all $i \notin \mathscr{C}_1$.

Next, the row vector $\pi^{(1)}$ is moved cyclically under the original matrix P and its powers $P^2, \ldots, P^{k-1}$: the vector $\pi^{(j)} = \pi^{(1)}P^{j-1}$ is supported by $\mathscr{C}_j$, and is again an equilibrium distribution for P^k, $j = 1, \ldots, k-1$. But then take a kth root of unity, κ (with $\kappa^k = 1$ but $\kappa \neq 1$), and form the row vector

$$\Pi = \sum_{j=1}^{k} \kappa^{-j} \pi^{(j)}. \tag{1.134}$$

With

$$\Pi P = \sum_{j=1}^{k-1} \kappa^{-j} \pi^{(j+1)} + \pi^{(1)} \kappa^{-k} = \pi^{(1)} + \kappa \sum_{j=2}^{k} \kappa^{-j} \pi^{(j)} = \kappa \Pi,$$

we obtain an eigenvector of P, with the eigenvalue κ where $\kappa \neq 1$, but $|\kappa| = 1$. (And we could repeat the same procedure with all non-trivial kth roots of unity.) But this contradicts the above assumption that all the eigenvalues μ of P, different from $\mu_0 = 1$, have $|\mu| < 1$. This completes the proof of property (b).

The minimal circle in the complex plane $\mathbb{C}$ centred at the origin and containing all the eigenvalues of a matrix is called the spectral circle and its radius is called the *spectral radius*. In other words, the spectral radius $\rho(M)$ of a matrix M is equal to the maximal modulus of an eigenvalue of M. So, according to statement (a) above, the spectral radius $\rho(P)$ of a transition matrix P equals 1 and coincides with the norm $\|P\|$ (in general, one can only say that the norm $\|M\| \geq \rho(M)$).

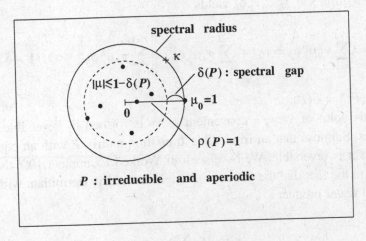

Fig. 1.24

Next, the *spectral gap* of M is defined as the minimal value $\delta(M)$ of the difference $\rho(M) - |\mu|$ for the eigenvalues μ with $|\mu| < \rho(M)$:

$$\delta(M) = \min\left[1 - |\mu| : \mu \text{ eigenvalue with } |\mu| < \rho(M)\right]. \qquad (1.135)$$

According to statement (b) above, the spectral radius of an irreducible aperiodic transition matrix P is attained by a sole eigenvalue $\mu_0 = 1$ with geometric multiplicity 1, excluding the possibility of there being an eigenvalue $\mu \neq 1$ with $|\mu| = 1$.

For an irreducible and aperiodic matrix P, the spectral gap gives the speed of convergence of the row vectors λP^n to the equilibrium distribution π and the speed of convergence of the column vectors $P^n \mathbf{x}$ to the vector $\langle \underline{x}, \pi^T \rangle \underline{1}^T$. Here, the coefficient $\langle \underline{x}, \pi^T \rangle = \sum_{i=1}^{\ell} \pi_i x_i$ plays the same role as $\langle \underline{x}, \underline{1} \rangle$ in (1.133). This is particularly evident when P has ℓ linearly independent row eigenvectors $\psi_0^T = \pi, \psi_1^T, \ldots, \psi_{\ell-1}^T$. Then every ℓ-dimensional row vector $\mathbf{x}^T$ (real or complex) is written as a linear combination $\sum_{p=0}^{\ell-1} u_p \psi_p^T$. Hence,

$$\mathbf{x}^T P^n = \sum_{p=0}^{\ell-1} u_p \mu_p^n \psi_p^T = u_0 \pi + \sum_{p=1}^{\ell-1} u_p \mu_p^n \psi_p^T, \qquad (1.136)$$

and the remaining sum in the right-hand side of (1.136) has the norm

$$\left\| \sum_{p=1}^{\ell-1} u_p \mu_p^n \psi_p^T \right\| \leq (1 - \delta)^n \sum_{p=1}^{\ell-1} |u_p| \|\psi_p\| = O\left((1 - \delta(P))^n\right). \qquad (1.137)$$

Similarly, if P^T has ℓ linearly independent row eigenvectors $\phi_0^T = \underline{1}^T, \phi_1^T, \ldots, \phi_{\ell-1}^T$, then writing $\mathbf{x} = \sum_{p=0}^{\ell-1} v_p \phi_p$ yields

$$P^n \mathbf{x} = \sum_{p=0}^{\ell-1} v_p \mu_p^n \phi_p = v_0 \underline{1} + \sum_{p=1}^{\ell-1} v_p \mu_p^n \phi_p, \quad \left\| \sum_{p=1}^{\ell-1} v_p \mu_p^n \phi_p \right\| \leq O\left((1 - \delta)^n\right),$$

and $\delta = \delta(P^T) = \delta(P)$.

From this point of view, a convenient class is formed by reversible stochastic matrices. Suppose that an irreducible transition matrix P with an equilibrium distribution π is reversible. We have seen in Worked Example 1.10.7 that this is equivalent to the fact that the action of P (right or left) is Hermitian, with respect to the tilted scalar product $\langle \cdot, \cdot \rangle_\pi$:

$$\langle \mathbf{x}, \mathbf{y} \rangle_\pi = \sum_{i=1}^{\ell} x_i \bar{y}_i \pi_i. \qquad (1.138)$$

Moreover, the transposed matrix P^T has the same property: its action (right or left) is Hermitian with respect to $\langle \cdot, \cdot \rangle_\pi$. In fact, let $\mathbf{x} = \begin{pmatrix} x_1 \\ \vdots \\ x_\ell \end{pmatrix}, \mathbf{y} = \begin{pmatrix} y_1 \\ \vdots \\ y_\ell \end{pmatrix}$ be a pair of ℓ-dimensional column vectors (real or complex). Then

$$
\begin{aligned}
\langle P^T \mathbf{x}, \mathbf{y} \rangle_\pi &= \left\langle (\mathbf{x}^T P)^T, \mathbf{y} \right\rangle_\pi = \sum_{i,j=1}^{\ell} x_i p_{ji} \overline{y_j} \pi_j \\
&= \sum_{i,j=1}^{\ell} x_i p_{ij} \overline{y_j} \pi_i \text{ (by reversibility)} \\
&= \sum_i x_i \pi_i \sum_j p_{ij} \overline{y_j} = \left\langle \mathbf{x}, (\mathbf{y}^T P)^T \right\rangle_\pi \\
&= \langle \mathbf{x}, P^T \mathbf{y} \rangle_\pi.
\end{aligned}
\tag{1.139}
$$

In a similar fashion, one shows that

$$
\langle P\mathbf{x}, \mathbf{y} \rangle_\pi = \left\langle (\mathbf{x}^T P^T)^T, \mathbf{y} \right\rangle_\pi = \left\langle \mathbf{x}, (\mathbf{y}^T P^T)^T \right\rangle_\pi = \langle \mathbf{x}, P\mathbf{y} \rangle_\pi.
\tag{1.140}
$$

But this means that, for the both the right and left action of P, the adjoint (or conjugate) matrix P^*, relative to $\langle \cdot, \cdot \rangle_\pi$, coincides with P, i.e. P is Hermitian (i.e. is self-adjoint) relative to $\langle \cdot, \cdot \rangle_\pi$. The same holds also for P^T, as claimed.

The tilted scalar product is non-degenerate (in the sense that $\langle \mathbf{x}, \mathbf{x} \rangle_\pi = 0$ if and only if $\mathbf{x} = \mathbf{0}$ because all the entries $\pi_i > 0$). Here, the standard theorem is applicable, that every Hermitian matrix has an orthonormal eigenbasis, and its spectrum is real (i.e. all eigenvalues are real). In our case it will imply that the eigenvectors $\psi_0, \psi_1, \ldots, \psi_{\ell-1}$ can be made real (as $\psi_0 \propto \pi^T$, the vector ψ_0 can be made positive). And although the orthogonality is meant relative to a scalar product $\langle \cdot, \cdot \rangle_\pi$ and is lost when we return to the standard scalar product $\langle \mathbf{x}, \mathbf{y} \rangle = \sum_{i=1}^{\ell} x_i \overline{y_i}$, linear independence still holds, and (1.136) and (1.137) remain valid (with $u_p = \langle \mathbf{x}, \psi_p \rangle_\pi$).

Summarising:

Theorem 1.12.4 *Let P be an irreducible $\ell \times \ell$ stochastic matrix reversible relative to its equilibrium distribution π. Then both P and its transpose P^T are Hermitian relative to the tilted scalar product $\langle \cdot, \cdot \rangle_\pi$. Thus each of P and P^T has ℓ real eigenvalues, and their spectra coincide. Moreover, the eigenvalues of P and P^T counted with their multiplicities coincide. Arranging the eigenvalues μ_p in decreasing order, we have:*

$$
\mu_0 = 1 > \mu_1 \geq \mu_2 \geq \cdots \geq \mu_{\ell-1} \geq -1.
\tag{1.141}
$$

If P is aperiodic then $\mu_{\ell-1}(P) > -1$, and the spectral gap $\delta = \delta(P) = \delta(P^{\mathrm{T}})$ is given by

$$\delta = \min\left[1 - \mu_1, 1 - |\mu_{\ell-1}|\right].\qquad(1.142)$$

In this case, (1.136), (1.137) imply that

$$\lambda P^n = \pi + O\big((1-\delta)^n\big),\qquad(1.143)$$

for any initial distribution λ.

If P is periodic then $\mu_{\ell-1} = -1$. In this case, P^2 is reducible, with two closed communicating classes, say C_1 and C_2, and and has equilibrium distributions $\pi^{(1)}$ and $\pi^{(2)} = \pi^{(1)}P$ concentrated on a single communicating class. The geometric multiplicity of $\mu_{\ell-1}$ is 1, and the corresponding eigenvector is proportional to the vector

$$\Pi = \pi^{(1)} - \pi^{(2)},\qquad(1.144)$$

cf. (1.120). Then, for all initial distributions $\lambda = (\lambda_1, \ldots, \lambda_\ell)$,

$$\lambda P^n = \pi + (-1)^n(\Lambda^{(1)} - \Lambda^{(2)})\Pi + O\big((1-\delta)^n\big),\qquad(1.145)$$

and $\Lambda^{(1)} = \sum_{i \in C_1}\lambda_i$, $\Lambda^{(2)} = \sum_{i \in C_2}\lambda_i$.

> *Markov processes specialists do it openly in communicating classes.*
>
> (From the series *'How they do it'*.)

1.13 Geometric algebra of Markov chains, II. Random walks on graphs

Many ideas become more transparent when we restrict the presentation to the class of random walks on graphs. The definition was given in Section 1.10; here we focus on finite non-directed graphs without multiple edges and loops. In other words, the connectedness v_{ij} takes values 0 or 1; in the former case we find no edge joining i and j, while in latter case a single edge is associated with two opposite arrows: $(i \to j)$ and $(j \to i)$. Examples 1.12.1 and 1.12.2 fall in this category. In addition, the graphs in these examples feature constant valency: $v_i \equiv \sum_j v_{ij} \equiv \sum_i v_{ij} \equiv v \geq 1$, i.e., for every vertex i, there are v arrows going from i to the neighbouring vertices, and v arrows issuing from these vertices and returning to i ($v = 2$ in Example 1.12.1 and $v = d$ in Example 1.12.2). In general, the valency v_i may depend on i.

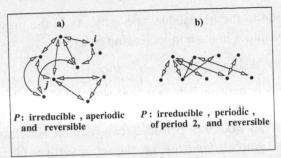

Fig. 1.25

The graph may be drawn with non-directed links or two-way arrows. The random walk on the graph was defined as a DTMC with (finite) state space G, the set of vertices of the graph, and transition matrix P having entries

$$p_{ij} = p_{ji} = v_{ij}/v_i. \tag{1.146}$$

We checked that

$$\pi_i = v_i \bigg/ \sum_{j \in G} v_j \tag{1.147}$$

gives equilibrium probabilities (in other words, the vector π^{T} is an eigenvector for P in $\mathbb{R}^{|G|}$ or $\mathbb{C}^{|G|}$, with the eigenvalue $\mu_0 = 1$). Here $|G|$ stands for the total number of vertices. Moreover, P is reversible relative to the equilibrium distribution $\pi = (\pi_i)$: $\pi_i p_{ij} = \pi_j p_{ji}$, $i, j \in G$. If the valency v_i is constant, the transition matrix $P = (p_{ij})$ is Hermitian, and the above equilibrium distribution $\pi = (\pi_i, i \in G)$ is uniform: $\pi_i = 1/|G|$. It was observed that in this case P has an orthonormal basis of eigenvectors in $\mathbb{R}^{|G|}$ and all its eigenvalues are real. Next, according to Theorem 1.12.3, all the eigenvalues of P lie in the closed interval $[-1, 1]$. In other words, the spectrum of the matrix P is a subset of $[-1, 1]$, and $\mu_0 = 1$ is the right-most point of the spectrum.

In general, the matrix P can be converted into Hermitian, by changing the standard scalar product $\langle \cdot, \cdot \rangle$ in $\mathbb{R}^{|G|}$ into the tilted one:

$$\langle \mathbf{x}, \mathbf{y} \rangle = \sum_{i \in G} x_i \bar{y}_i \pi_i.$$

So, for a general RW on a graph, the spectrum of P is still a subset of the interval $[-1, 1]$, and includes $\mu_0 = 1$ as its right-most point.

If the graph under consideration is connected, the RW is irreducible, and vice versa. See Figure 1.25a. In this case, the chain has a unique equilibrium distribution, and the geometric multiplicity of the eigenvalue $\mu_0 = 1$ is 1. In what follows restrict our attention to the irreducible case only. As in the previous section, we will write the eigenvalues in the non-increasing order:

$$1 = \mu_0 > \mu_1 \geq \cdots \geq \mu_{|G|-1} \geq -1. \tag{1.148}$$

The point -1 may or may not belong to the spectrum of P: it depends on whether the chain is periodic or not. It is possible to check that in our setting, the chain may only have period 1 (aperiodic) or period 2 (two periodic subclasses). If -1 is an eigenvalue of P, the graph is *bipartite*, i.e. the set of vertices G can be partitioned into two disjoint subsets, $G^{(1)}$ and $G^{(2)}$, such that every edge of the graph joins a vertex from $G^{(1)}$ and a vertex from $G^{(2)}$. In this case, the chain is periodic, and the period is 2. See Figure 1.25b. Conversely, if the chain is periodic, of period 2, then -1 is an eigenvalue. If the periodic subclasses are W_1, and W_2, then the eigenvector with eigenvalue -1 is proportional to

$$1_{\mathscr{G}_1} - 1_{\mathscr{G}_2}.$$

Here $1_{\mathscr{G}_l}$ is the vector whose entries are equal to 1 for the states from $\mathscr{G}_l$ and to 0 for the states from the other class.

Therefore, if P is aperiodic, then the point -1 is not an eigenvalue, i.e. it does not belong to the spectrum of P. Consequently, the spectral gap δ is calculated as

$$\min \left[\delta_1, \delta_{-1} \right], \text{ where } \delta_1 = 1 - \mu_1, \ \delta_{-1} = 1 - \left| \mu_{|G|-1} \right|. \tag{1.149}$$

Let us go back to Example 1.12.1 and assume that ℓ is odd; in this case P is aperiodic, and

$$\delta_1 = 1 - \mu_1, \ \delta_{-1} = 1 + \mu_{\ell-1}. \tag{1.150}$$

For any pair i, j of vertices of a regular ℓ-gon, we can identify a *geodesic* from i to j, i.e. the shortest paths going from i to j and formed by the arrows; because ℓ is odd, the geodesic is uniquely defined. For a RW on a general graph, again, a geodesic Γ between two vertices (states) i and j is a path starting at i and ending at j, of shortest length, i.e. with a minimal number of arrows in it. It may be non-unique, but we select one such geodesic for any pair $i, j \in G$ and denote it by $\Gamma_{ij} \sim (i_0, i_1, \ldots, i_L)$; here $i_0 = i$, $i_L = j$, and L is the length of Γ_{ij}. Observe that the geodesic Γ_{ji} does not necessarily coincide with the geodesic Γ_{ij} traversed in the reverse order: the choice of geodesics has to be judicious. The total collection of selected geodesics is denoted by $\mathscr{G}$.

It turns out that, for an irreducible Hermitian stochastic matrix P of the form (1.146), reversible relative to the equilibrium distribution π of the form (1.147), there is a useful inequality for δ_1, called *Poincaré's inequality* (or *Poincaré's bound*):

$$\delta_1 \geq \frac{2E}{D_*^2 \gamma_* b}. \tag{1.151}$$

See Diaconis, P., Stroock, D. "Geometric bounds for eigenvalues of Markov chains". *Ann. Appl. Probab.*, **1** (1991), 36–61. Here E is the total number of (non-directed) edges in the graph. Next, D_* is the maximal valency of any vertex. Furthermore, γ_* is the maximal number of edges in a geodesic across the diagram (the diameter of the directed graph formed by the arrows). Finally, b is the maximal cardinality of the bunch of geodesics including a given arrow

$$b = \max_{e=(i\to j)} \left[\text{number of geodesics } \Gamma \sim (i_0, i_1, \ldots, i_L) \text{ containing arrow } e\right].$$
(1.152)

In Example 1.12.1, with ℓ odd,

$$E = \ell, \quad D_* = 2, \quad \gamma_* = \frac{\ell-1}{2}.$$
(1.153)

To calculate b, we use the fact that the diagram is symmetric, and it does not matter which arrow e one chooses. So, let e be the arrow $0 \to \ell-1$. Suppose a geodesic containing e starts at a vertex i to the right of 0. Then $i \leq (\ell-3)/2$, as the total length of the geodesic could not exceed $(\ell-1)/2$. The geodesic could end in any of the $(\ell-1)/2 - i$ points beyond (and including) $\ell-1$. Hence,

$$b = \sum_{0 \leq i \leq (\ell-3)/2} \left(\frac{\ell-1}{2} - i\right) = \frac{(\ell-1)^2}{4} - \frac{(\ell-1)(\ell-3)}{8} = \frac{\ell^2-1}{8}.$$
(1.154)

Now the bound (1.151) becomes

$$\delta_1 \geq \frac{8\ell}{(\ell-1)^2(\ell+1)},$$
(1.155)

or $\delta_1 \geq 8/\ell^2$ as $\ell \to \infty$. This is less accurate an estimate than (1.121), and the error factor is $2\pi^2/8 \approx 2$.

Further, a useful bound for δ_{-1} is as follows. Define a (directed) *cycle* as a closed path, along (some) edges on our graph, such that it visits each of its vertices exactly once, before returning to the starting point, with a fixed direction of inspection (there are precisely two directions for a given collection of edges). In other words, a cycle passes through a given edge at most once. It is convenient to fix a direction, i.e. to distinguish between the clockwise and anti-clockwise inspections. Consider a collection $\mathscr{S}$ of cycles, of odd length, one for each vertex $i \in G$, and that the cycle $\Sigma = \Sigma_i$ from $\mathscr{S}$ starts and ends at i. Denote by σ_* the maximal length of (i.e. the maximal number of links in) the cycle Σ. Next, let b_* stand for the maximum number of cycles from $\mathscr{S}$ containing a given edge:

$$b_* = \max_{e=(i\to j)} \left[\text{number of cycles from } \mathscr{S} \text{ containing arrow } e\right].$$
(1.156)

Then

$$\delta_{-1} \geq \frac{2}{D_* \sigma_* b_*}. \tag{1.157}$$

In Example 1.12.1, with ℓ odd, $\sigma_* = b_* = \ell$. From (1.156) we obtain

$$\delta_{-1} \geq \frac{1}{\ell^2}, \tag{1.158}$$

which gives the right order in ℓ but an incorrect constant.

In Example 1.12.2, in the original (periodic) setting, the last eigenvalue $\mu_{2^d-1} = -1$. Here, we identify a path $\gamma_{\underline{\alpha},\underline{\alpha}'}$ from $\underline{\alpha}$ to $\underline{\alpha}'$ by changing the entries where $\underline{\alpha}$ differs from $\underline{\alpha}'$ to the complement mod 2, moving from the left to the right, one step at a time. (This choice makes life easier compared with the study of geodesics.) Clearly, $E = d2^{d-1}$, $\gamma_* = d$, $D_* = d$, and for this choice of paths, $b = 2^{d-1}$. To see this, consider an edge (w, z) with w, z differing in only one coordinate, say the jth. A path γ_{xy} crossing over this edge can begin at any x which coincides with w in coordinates after the $(j-1)$st (so 2^{j-1} choices) and ends in any y which coincides with z in coordinates up to the jth (so 2^{d-j} choices). Thus, there are 2^{d-1} paths γ_{xy} crossing an edge. Estimate (1.151) gives

$$\delta_1 \geq \frac{2}{d^2}. \tag{1.159}$$

i.e. is off by a factor of d. A better bound is produced by Cheeger's inequality, see (1.163).

In the modified setting, the last eigenvalue $\mu_{2^d-1} = -1 + 2/(d+1)$. Further, the result of 'geometric evaluation' of the gap from -1 is sharp:

$$\delta_{-1}^{\text{modified}} \geq \frac{2}{d+1}. \tag{1.160}$$

We now pass to *Cheeger's inequality*. Again consider the RW on a general non-directed graph on the vertex set G, without multiple edges. Given a set $S \subset G$ define a *flow in the graph* from S to its complement $S^c = \{1, \ldots, \ell\} \setminus S$ as the fraction of the arrows leading from S to S^c:

$$Q(S, S^c) = \frac{1}{2E} \sum_{(i,j):\, i \in S,\, j \in S^c} \mathbf{1}(p_{ij} > 0). \tag{1.161}$$

Here, as before, E is the number of (non-directed) edges in the graph, and $2E$ is the total number of arrows. Next, set

$$h = \min_{S:\, 1 \le |S| \le \ell/2} \left[\frac{\ell}{|S|} Q(S, S^c) \right]. \tag{1.162}$$

Cheeger's inequality asserts that

$$\frac{h^2}{2} \le \delta_1 \le 2h. \tag{1.163}$$

In Example 1.12.1 with ℓ odd, the minimum is achieved when S is a collection of $(\ell - 1)/2$ subsequent vertices of the ℓ-gon, and

$$h = \frac{2}{\ell - 1}. \tag{1.164}$$

We see that in the Cheeger inequality, the lower bound gives the right order but the constant is off by the factor π^2.

In Example 1.12.2, in the original (periodic) setting, the minimum is achieved by taking S to be the face of the cube, say $S = \{\underline{x}^T : x_1 = 0\}$. This gives

$$h = \frac{1}{d}, \tag{1.165}$$

and the Cheeger inequality becomes

$$\frac{1}{2d^2} \le \delta_1 \le \frac{2}{d}. \tag{1.166}$$

As $\lambda_1 = 1 - 2/d$ in this case, the upper bound is sharp and the lower bound is of the same order as in the Poincaré inequality, with a slightly overrated constant.

The constant in the Poincaré inequality is sensitive to changes in the structure of the graph; this fact can be exploited in order to produce useful bounds. Below we discuss one such bound based on a decomposition of the state space into 'rarely communicating' subsets.

Let $\phi : I \to \mathbb{R}$ be an arbitrary test function. The expectation and variance of ϕ with respect to the invariant distribution π are of course given by

$$\mathbb{E}_\pi \phi = \sum_{i \in I} \pi(i) \phi(i)$$

and

$$\mathrm{Var}_\pi \phi = \sum_{i \in I} \pi(i) (\phi(i) - \mathbb{E}_\pi \phi)^2.$$

The crucial rôle in the proof of the Poincaré and Cheeger bounds is played by the so-called Dirichlet form (see (1.201) below) associated with ϕ and the transition matrix P and defined as

$$\mathscr{E}_\pi(\phi) = \frac{1}{2} \sum_{i,j \in I} (\phi(i) - \phi(j))^2 \pi(i) p(i,j).$$

The Poincaré inequality takes the form

$$\mathscr{E}_\pi(\phi) \geq \lambda \operatorname{Var}_\pi \phi \qquad (1.167)$$

and holds uniformly over all $\phi : I \to \mathbb{R}$. The main point to know is that the constant λ controls the rate of convergence of a Markov chain to the invariant distribution π. To avoid technical problems associated with nearly periodic chains, assume that loop probabilities are uniformly bounded from 0. Denote by $\mathbf{p}^{(n)}(i)$ the row i of the n-step transition matrix $P^n = (p_{ij}^{(n)})$, $i \in I$ (which gives the distribution of the chain when its initial state is i). Define the *total variation distance*

$$\operatorname{dist}_{\mathrm{TV}}\left(\mathbf{p}^{(n)}(i), \pi\right) := \sum_{j \in I} \left| p_{ij}^{(n)} - \pi_j \right|.$$

Assume that $n \geq n(i, \varepsilon)$ where $n(i, \varepsilon)$ has a specific magnitude:

$$n(i, \varepsilon) = O\left(\frac{1}{\lambda}\left(\ln\left(\frac{1}{\pi(i)}\right) + \ln\frac{1}{\varepsilon}\right)\right).$$

Then

$$\operatorname{dist}_{\mathrm{TV}}\left(\mathbf{p}^{(n)}(i), \pi\right) \leq \varepsilon.$$

Here λ is a constant from (1.167). See Jerrum M., Son J-B., Tetali P., Vigoda E. "Elementary bounds on Poincaré and log-Sobolev constants for decomposable Markov chains", *Annals Appl. Probab.*, **14** (2004), 1741–1763.

In many natural cases the state space can be naturally split into several blocks in such a way that transitions between blocks are rare compared with the transitions inside these blocks. This simplifies the study of convergence to equilibrium. Let $I = I_0 \cup \cdots \cup I_{m-1}$ be a decomposition of the state space into m disjoint sets. We use the notation $\mathbb{Z}_m = (0, 1, \ldots, m-1)$. Next, define $\overline{\pi}(i) : \mathbb{Z}_m \to [0, 1]$ by

$$\overline{\pi}(i) = \sum_{j \in I_i} \pi(j) \qquad (1.168)$$

and introduce a new transition matrix $\overline{P} : \mathbb{Z}_m \times \mathbb{Z}_m \to [0, 1]$ by

$$\overline{p}(i,j) = \overline{\pi}(i)^{-1} \sum_{k \in I_i, l \in I_j} \pi(k) p(k,l). \qquad (1.169)$$

The Markov chain on the state space $\mathbb{Z}_m$ and with transition probabilities $\overline{P}$ is the *projection* Markov chain induced by the partition (I_i).

Example 1.13.1 Check that the projection chain has $\overline{\pi}$ as a stationary distribution.

This follows from an obvious equality

$$\sum_i \overline{\pi}(i) \frac{1}{\overline{\pi}(i)} \sum_{k \in I_i, l \in I_j} \pi(k) n(k, l) = \sum_{l \in I_j} \pi(l).$$

For each $k \in \mathbb{Z}_m$ the *restriction* Markov chain on I_k has transition probabilities $P_k : I_k \times I_k \to [0, 1]$ defined by

$$p_k(i, j) = \begin{cases} p(i, j), & \text{if } i \neq j, \\ 1 - \sum_{l \in I_k \setminus i} p(i, l), & \text{if } i = j, \end{cases} \quad i, j \in I_k. \tag{1.170}$$

Example 1.13.2 Prove that both the projection and restriction chains inherit time-reversibility from the original chain. Moreover, $\pi_k(i) = \pi(i)/\overline{\pi}(k)$ is the stationary distribution of the restriction chain.

Due to reversibility, for all $i, j \in I_k$

$$\pi_k(i) p_k(i, j) = \pi_k(j) p_k(j, i).$$

We require both the projection and the restriction chains to be irreducible. Hence, the various stationary distributions $\overline{\pi}$ and $\pi_0, \ldots, \pi_{m-1}$ are unique.

Suppose that the projection chain and the various restriction chains satisfy Poincaré inequalities with constant $\overline{\lambda}$ and $\lambda_0, \ldots, \lambda_{m-1}$, respectively. Define $\lambda_{\min} = \min_i \lambda_i$. Our goal is to obtain a Poincaré inequality for the original Markov chain, with Poincaré constant $\lambda = \lambda(\overline{\lambda}, \lambda_{\min}, \gamma)$, where γ is a further parameter

$$\gamma = \max_{i \in \mathbb{Z}_m} \max_{k \in I_i} \sum_{l \in I \setminus I_i} p(k, l). \tag{1.171}$$

Informally, γ is the probability of escape in one step from a current block of the partition, maximized over all states.

Theorem 1.13.3 *Consider a finite-state time-reversible Markov chain decomposed into a projection chain and m restriction chains as above. Suppose the projection chain satisfies a Poincaré inequality with constant $\overline{\lambda}$, and the restriction chains satisfy inequalities with uniform constant $\lambda_{\min}$. Let γ be defined as in (1.171). Then the original Markov chain satisfies a Poincaré inequality with*

$$\lambda = \min \left[\frac{\overline{\lambda}}{3}, \frac{\overline{\lambda} \lambda_{\min}}{3\gamma + \overline{\lambda}} \right]. \tag{1.172}$$

Proof Consider an arbitrary test function ϕ. Our starting point is the decomposition of $\text{Var}_\pi \phi$ with respect to the partition

$$\text{Var}_\pi \phi = \sum_{i \in \mathbb{Z}_m} \overline{\pi}(i) \text{Var}_{\pi_i} \phi + \sum_{i \in \mathbb{Z}_m} \overline{\pi}(i) \left(\mathbb{E}_{\pi_i} \phi - \mathbb{E}_\pi \phi \right)^2. \tag{1.173}$$

Similarly, for the Dirichlet form,

$$\mathcal{E}_\pi(\phi) = \sum_{i \in \mathbb{Z}_m} \overline{\pi}(i) \mathcal{E}_{\pi_i}(\phi) + \frac{1}{2} \sum_{i,j \in \mathbb{Z}_m, i \neq j} \mathscr{C}_{ij}, \tag{1.174}$$

where

$$\mathscr{C}_{ij} = \sum_{k \in I_i, l \in I_j} \pi(k) p(k,l) (\phi(k) - \phi(l))^2. \tag{1.175}$$

In summations and so forth, the variables i and j always range over $\mathbb{Z}_m$. For all i, j with $i \neq j$ and $\overline{p}(i,j) > 0$, define $\widehat{\pi}_i^j : I_i \to [0,1]$ by

$$\widehat{\pi}_i^j(k) = \frac{\pi_i(k) \sum_{l \in I_j} p(k,l)}{\overline{p}(i,j)}. \tag{1.176}$$

Note that $\widehat{\pi}_i^j$ is a probability distribution on I_i.

The first term on the RHS of (1.173) we simply bound as

$$\sum_i \overline{\pi}(i) \text{Var}_{\pi_i} \phi \;\leq\; \sum_i \frac{1}{\lambda_i} \overline{\pi}(i) \mathcal{E}_{\pi_i}(\phi)$$

$$\leq\; \frac{1}{\lambda_{\min}} \sum_i \overline{\pi}(i) \mathcal{E}_{\pi_i}(\phi). \tag{1.177}$$

The second term we transform, starting with an application of the Poincaré inequality for the projection chain,

$$\sum_i \overline{\pi}(i) \left(\mathbb{E}_{\pi_i} \phi - \mathbb{E}_\pi \phi \right)^2 \leq \frac{1}{2\overline{\lambda}} \sum_{i \neq j} \overline{\pi}(i) \overline{p}(i,j) \left(\mathbb{E}_{\pi_i} \phi - \mathbb{E}_{\pi_j} \phi \right)^2$$

$$\leq \frac{3}{2\overline{\lambda}} \sum_{i \neq j} \overline{\pi}(i) \overline{p}(i,j)$$

$$\times \left[\left(\mathbb{E}_{\pi_i} \phi - \mathbb{E}_{\widehat{\pi}_i^j} \phi \right)^2 + \left(\mathbb{E}_{\widehat{\pi}_i^j} \phi - \mathbb{E}_{\widehat{\pi}_j^i} \phi \right)^2 + \left(\mathbb{E}_{\widehat{\pi}_j^i} \phi - \mathbb{E}_{\pi_j} \phi \right)^2 \right]$$

$$= \frac{3}{2\overline{\lambda}} \left[\Sigma_1 + \Sigma_2 + \Sigma_3 \right], \tag{1.178}$$

where the terms Σ_1, Σ_2 and Σ_3 are associated with corresponding sums, viz.,

$$\Sigma_1 = \sum_{i \neq j} \overline{\pi}(i) \overline{p}(i,j) \left(\mathbb{E}_{\pi_i} \phi - \mathbb{E}_{\widehat{\pi}_i^j} \phi \right)^2.$$

We proceed to bound Σ_1, Σ_2 and Σ_3 separately, noting that Σ_1 and Σ_3 are equal, owing to reversibility. For the second of these terms we have

$$\Sigma_2 = \sum_{i,j} \overline{\pi}(i)\overline{p}(i,j)\Big[\sum_{k\in I_i, l\in I_j} \frac{\pi(k)p(k,l)}{\overline{\pi}(i)\overline{p}(i,j)}(\phi(k)-\phi(l)) \Big]^2 \qquad (1.179)$$

$$\leq \sum_{i\neq j} \overline{\pi}(j)\overline{p}(i,j) \sum_{k\in I_i, l\in I_j} \frac{\pi(k)p(k,l)}{\overline{\pi}(i)\overline{p}(i,j)}(\phi(k)-\phi(l))^2 \qquad (1.180)$$

$$= \sum_{i\neq j}\sum_{k\in I_i, l\in I_j} \pi(k)p(k,l)(\phi(k)-\phi(l))^2 = \sum_{i\neq j} \mathscr{C}_{ij}, \qquad (1.181)$$

where $\mathscr{C}_{ij} = \sum_{k\in I_i, l\in I_j} \pi(k)p(k,l)(\phi(k)-\phi(l))^2$. Here, bound (1.179) uses the fact that $\dfrac{\pi(i)p(i,j)}{\overline{\pi}(i)\overline{p}(i,j)}$ is a joint distribution on $I_i \times I_j$ whose marginals are $\widehat{\pi}_i^j$ and $\widehat{\pi}_j^i$, and (1.180) is just a Cauchy–Schwarz inequality, once we have observed that

$$\sum_{k\in I_i, l\in I_j} \frac{\pi(k)p(k,l)}{\overline{\pi}(i)\overline{p}(i,j)} = 1$$

by definition.

To estimate Σ_1 we use the facts that $\text{Var}\,\xi = \text{Var}\,(\xi-c)$ for any random variable ξ and constant c, and write

$$\text{Var}_{\widehat{\pi}_i^j}\phi = \sum_{k\in I_i} \widehat{\pi}_i^j(k)\Big(\phi(k)-\mathbb{E}_{\pi_i}\phi\Big)^2 - \Big(\mathbb{E}_{\widehat{\pi}_i^j}\phi - \mathbb{E}_{\pi_i}\phi\Big)^2, \qquad (1.182)$$

so that certainly

$$\Big(\mathbb{E}_{\widehat{\pi}_i^j}\phi - \mathbb{E}_{\pi_i}\phi\Big)^2 \leq \sum_{k\in I_i} \widehat{\pi}_i^j(k)\Big(\phi(k)-\mathbb{E}_{\pi_i}\phi\Big)^2. \qquad (1.183)$$

Thus we have the bound

$$\Sigma_1 \leq \sum_{i\neq j} \overline{\pi}(i)\overline{p}(i,j)\sum_{k\in I_i} \widehat{\pi}_i^j(k)\Big(\phi(k)-\mathbb{E}_{\pi_i}\phi\Big)^2$$

$$\overset{!}{=} \sum_i \overline{\pi}(i)\sum_{k\in I_i}\pi_i(k)\Big(\phi(k)-\mathbb{E}_{\pi_i}\phi\Big)^2 \sum_{j\neq i}\frac{\widehat{\pi}_i^j(k)\overline{p}(i,j)}{\pi_i(k)} \qquad (1.184)$$

$$= \sum_i \overline{\pi}(i)\sum_{k\in I_i}\pi_i(k)\Big(\phi(k)-\mathbb{E}_{\pi_i}\phi\Big)^2 \sum_{j\neq i}p(k,I_j)$$

$$\leq \gamma\sum_i \overline{\pi}(i)\text{Var}_{\pi_i}\phi \qquad (1.185)$$

$$\leq \frac{\gamma}{\lambda_{\min}}\sum_i \overline{\pi}(i)\mathscr{E}_{\pi_i}(\phi). \qquad (1.186)$$

where $p(k,I_j) = \sum_{l \in I_j} p(k,l)$. Note that (1.184) is based on the definition of $\hat{\pi}_i^j$, (1.185) uses the definition of γ, and (1.186) is based on the Poincaré inequality for the restriction chains.

Substituting (1.181) and (1.186) into (1.178), and recalling that $\Sigma_1 = \Sigma_3$, we have

$$\sum_i \overline{\pi}(i)\left(\mathbb{E}_{\pi_i}\phi - \mathbb{E}_\pi\phi\right)^2 \leq \frac{3}{2\overline{\lambda}}\sum_{i \neq j}\mathscr{C}_{ij} + \frac{3\gamma}{\overline{\lambda}\lambda_{\min}}\sum_i \overline{\pi}(i)\mathscr{E}_{\pi_i}(\phi). \tag{1.187}$$

Then substituting (1.173) and (1.187) into (1.175) yields

$$\mathrm{Var}_\pi\phi \leq \frac{3}{2\overline{\lambda}}\sum_{i \neq j}\mathscr{C}_{ij} + \frac{3\gamma + \overline{\lambda}}{\overline{\lambda}\lambda_{\min}}\sum_i \overline{\pi}(i)\mathscr{E}_{\pi_i}(\phi). \tag{1.188}$$

Finally, comparing (1.187) with (1.174), we see that

$$\mathscr{E}_\pi(\phi) \geq \lambda\,\mathrm{Var}_\pi\phi,$$

where λ is as in the statement of the theorem.

$\square$

Example 1.13.4 Consider the symmetric random walk on the $2n$ vertex 'pince-nez' graph in Figure 1.26 obtained by joining two disjoint n cycles by a single edge. Suppose transitions within cycles occur with probability 1/3, while the unique transition between cycles happens with probability $p \leq 1/3$. Loop probabilities are symmetric, so the random walk is time-reversible and its stationary distribution is uniform.

Now decompose the set of vertices (states) into two disjoint subsets, I_0 and I_1, where I_0 contains the n vertices in the first cycle and I_1 contains the n vertices in the second cycle. The spectral gap for each cycle considered in isolation is $\frac{2}{3}(1 - \cos(2\pi/n))$. Since $1 - \cos x \geq 2x^2/5$ for $0 \leq x \leq \pi/2$, we have that the spectral gap for each restriction chain is atleast $16\pi^2/15n^2$ (assuming $n \geq 4$), so we may

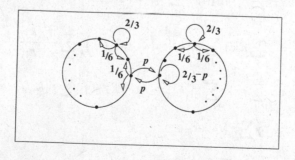

Fig. 1.26

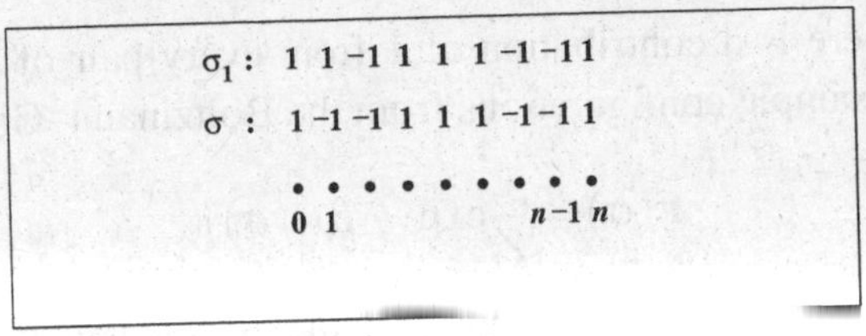

$$
\begin{array}{llllllllll}
\sigma_1: & 1 & 1 & -1 & 1 & 1 & 1 & -1 & -1 & 1 \\
\sigma\ : & 1 & -1 & -1 & 1 & 1 & 1 & -1 & -1 & 1
\end{array}
$$

$$
\begin{array}{cccccccc}
\bullet & \bullet & \bullet & \bullet & \bullet & \bullet & \bullet & \bullet \\
0 & 1 & & & & & n{-}1 & n
\end{array}
$$

Fig. 1.27

take $\lambda_{\min} = 10n^{-2}$. The projection chain in this example is the symmetric two-state chain with transition probability p/n between states, so we take $\bar{\lambda} = 2p/n$. Finally, $\gamma = p$. Hence, the Poincaré constant for the random walk on the pince-nez equals

$$
\lambda = \min \left[\frac{2p}{3n}, \frac{20}{3n^3 + 2n^2} \right]. \tag{1.189}
$$

Hence, $\lambda = O(n^{-3})$.

Example 1.13.5 (The one-dimensional *Ising model*) This example has origins in lattice field theory (where it generated a substantial literature). Recently, it has attracted considerable interest among computer scientists. The model can be set on a general non-directed graph and covers a range of interesting (and complicating) phenomena including phase transitions and irreversibility. One considers spin *configurations* obtained by assigning values ± 1 to each vertex; in the case of a finite graph, the total number of configurations equals $2^{|G|}$ where $|G|$ stands for the cardinality of the vertex set. Here we will focus on a simple case where the graph is a segment of a one-dimensional lattice $\{1, \ldots, n-1\}$, with $|G| = n-1$. It is convenient to attach two additional endpoints 0 and n where the values are kept constant and equal to 1. The space of configurations will consist of 2^{n-1} 'strings' $(\sigma(1), \ldots, \sigma(n-1))$ where $\sigma(i) = \pm 1$, $i = 1, \ldots, n-1$. We also use the values $\sigma(0) = \sigma(n) \equiv 1$. The set of 'extended' strings $\sigma = (\sigma(0), \ldots, \sigma(n))$ still has cardinality 2^{n-1} and will play the rôle of the state space of a Markov chain under consideration. In other words, in the current example, $I = \{\sigma\}$. See Figure 1.27.

The *Hamiltonian* of the Ising system on the path is defined by

$$
H(\sigma) = \sum_{i=0}^{n-1} [1 - \sigma(i)\sigma(i+1)]/2.
$$

In other words, there is a contribution of 1 from every pair of adjacent opposite spins. We wish to sample configurations from the Boltzmann–Gibbs distribution

$$\pi(\sigma) = \frac{1}{Z}\exp(-\beta H(\sigma)) \tag{1.190}$$

on I where Z is the *partition function* and β is the *inverse temperature*, T^{-1}.

One standard way to construct a Markov chain on I with stationary distribution π is through *Glauber dynamics*. For $i \in \{1,\ldots,n-1\}$ and $\sigma : \{1,\ldots,n-1\} \to \{-1,1\}$, let $\sigma_{i,+1}$ (resp. $\sigma_{i,-1}$) stand for the configurations that agrees with σ at all vertices except possibly at vertex i, where $\sigma_{i,+1}(i) = 1$ (resp. $\sigma_{i,-1}(i) = -1$). The transitions of the Markov chain are defined as follows.

(1) Select $i \in \{1,\ldots,n-1\}$ uniformly at random.
(2) Let

$$p = \frac{\exp(-\beta H(\sigma_{i,+1}))}{\exp(-\beta H(\sigma_{i,+1})) + \exp(-\beta H(\sigma_{i,-1}))}.$$

Then with probability p, the new state is $\sigma_{i,+1}$, and with probability $1-p$ the new state is $\sigma_{i,-1}$. For convenience, we imagine fixed boundary conditions at the extra vertices 0 and n.

Henceforth we assume that n is even. Then the segment $\{1,\ldots,n-1\}$ has a middle point $n/2$. We use this fact to form a partition of a set I into two disjoint subsets I_0 and I_1. In other words, represent I as a disjoint union $I_0 \cup I_1$, where I_0 (resp. I_1) is the set of all configuration with $\sigma(n/2) = -1$ (resp. $\sigma(n/2) = 1$). It is useful to consider the restrictions of the Markov chain to I_0 and I_1, and the corresponding projection chain (with two states). See Figure 1.28.

Then the spectral gap of the projection chain is $\overline{\lambda} \geq \dfrac{1}{(\cosh\beta)^2 n}$. The parameter γ satisfies $\gamma \leq \dfrac{1}{(1+e^{-2\beta})n}$. So, denoting by λ_k the spectral gap of the Ising system on $\{0,1,\ldots,k\}$, we have the recurrence

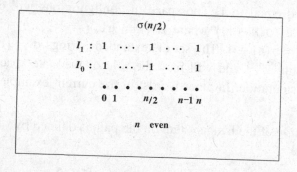

Fig. 1.28

$$\lambda_k \geq \min\left[\frac{1}{3(\cosh\beta)^2 n}, \frac{\lambda_{[k/2]}}{1+3/4(e^{2\beta}+1)}\right]. \tag{1.191}$$

This has the solution

$$\lambda_n = \mathcal{O}(n^{-c}), \quad c = 1 + \log_2\left[1 + \frac{3}{4}(e^{2\beta}+1)\right]$$

In particular, for low temperature T, we obtain that the number of steps for Glauber dynamics to reproduce the unique invariant distribution is

$$N \propto n^{2\log_2 e/T}.$$

Ernst Ising (1900–1998) was a student of W. Lenz in Hamburg. In 1920 Lenz proposed a model of ferromagnetism where particles are attached to sites of a crystal lattice and can have two values of a 'magnetic spin': ± 1. Two particles attached to neighbouring sites are interacting; the interaction depends on the signs of their spins and may favour them to be the same (a 'ferromagnetic system') or to be opposite (an 'anti-ferromagnetic system'). Lenz believed that the model could be 'solved', in the sense that the probability distribution (1.190) (or at least some of its vital characteristics) can be calculated, and suggested that Ising find a solution.

And Ising promptly did find a solution, in the case of a one-dimensional lattice $I \subset \mathbb{Z}$ (one-dimensional magnets). Ising's solution was the main part of his PhD Thesis (1924), and it was based on a straightforward application of the Perron–Frobenius Theorem (see Theorem 1.15.7).

However, for 20 years attempts to reach a solution for multi-dimensional model failed, although the Dutch physicists Kramers and Wannier calculated the numerical value for the so-called 'critical' temperature of the two-dimensional Ising model. The exact and complete solution in two dimensions was first given by the Norwegian-born American chemist L. Onsager in 1944 and turned out to be very complicated, but at the same time inspiring. Further attempts in dimensions three and higher stumbled, but generated a brilliant literature which influenced many areas of mathematics and physics (notably, the theory of *Markov random fields*, where a one-dimensional time is replaced by a multi-dimensional 'argument').

The term 'Ising model' was coined in a 1936 paper "On Ising's model of ferromagnetism" by R. Peierls, a prominent German physicist who moved to Britain in the 1930s. Each year between 500 and 800 papers are published that invoke the Ising model to address problems in such diverse fields as neural networks, machine vision, protein folding, biological membranes and social behaviour.

Meanwhile Ising got married and began a teaching career at high schools in Germany. He was dismissed when the Nazis came into power in 1933, but from 1934

to 1938 managed to keep a job as a teacher and headmaster of a Jewish board-
ing school near Berlin. In this period the Isings found themselves living near to
Albert Einstein's summer house (abandoned by the owners as the Einsteins moved
to America in 1933 and used for the school's needs); Ising enjoyed telling how he
took his daily baths at the Einsteins' because there was no bathtub in his place.

In November 1938, the school was destroyed by the Nazis, and soon after the
Isings were forced to leave Germany. In 1939 they fled to Luxembourg with plans
to emigrate to the US as soon as possible; at the end of that year their only son
Thomas was born. However, the Germans invaded Luxembourg in May 1940, and
the US Consulate was closed when the Isings' visas were about to be granted.
A year later most Jews in Luxembourg were rounded up. Ising and other men
who were married to non-Jews were spared but forced to work on dismantling
the Maginot Line railroad in Lorraine. His wife Johanna worked at menial jobs,
struggling to survive. This continued for the next four years.

The Isings finally got to the US in 1947, and from 1948–1976 Ising taught
physics and mathematics at Bradley University, Peoria, Illinois. He received a number of awards and honorary titles, but never returned to his early research. In fact,
the list of his published papers in physics consists of three titles: his 1924 PhD Thesis, a short 1925 paper (first quoted by W. Heisenberg in 1928 and generating 603
citations between 1975 and 2001), and a beautifully written article "Goethe as a
Physicist", *American Journal of Physics* **18** (1950), 235–236. According to Ising's
own account, it was not until 1949 that he found out from the scientific literature
that his model had become widely known!

1.14 Geometric algebra of Markov chains, III. The Poincaré and Cheeger bounds

> *Manhattan Markov Mystery*
> (From the series 'Movies that never made it to the Big Screen'.)

Proof of Poincaré's inequality. It is convenient to prove a generalisation of bound
(1.151) that holds for an irreducible reversible transition matrix P which is not
necessarily Hermitian. Nevertheless, as was discussed in Section 1.12, due to
reversibility, both P and P^T define Hermitian transformations in $\mathbb{R}^\ell$ and $\mathbb{C}^\ell$,
equipped with the tilted scalar product $\langle \cdot, \cdot \rangle_\pi$; see (1.138). As a result, P and
P^T have orthonormal eigenbases relative to $\langle \cdot, \cdot \rangle_\pi$. Next, P and P^T have the
same real spectrum, and their eigenvalues μ_p, written with their multiplicities and
in the decreasing order, satisfy (1.148). Then the Laplacians $L(P) = \mathbf{I} - P$ and

$L(P^T) = \mathbf{I} - P^T = L(P)^T$, with the same eigenvectors as P and P^T, respectively, have eigenvalues $\beta_p = 1 - \mu_p$; when counted with their multiplicities and arranged in increasing order, the eigenvalues satisfy

$$\beta_0 = 0 < \beta_1 = \delta_1 \le \cdots \le \beta_{\ell-1} \le 2. \tag{1.192}$$

As follows from (1.192), $L(P)$ and $L(P^T)$ define Hermitian, non-negative definite transformations in $\mathbb{R}^\ell$ and $\mathbb{C}^\ell$, relative to the tilted scalar product $\langle\, \cdot\, ,\, \cdot\, \rangle_\pi$. The eigenvector for $L(P)$ with the smallest eigenvalue 0 is π; for $L(P^T)$ it is $\underline{1}^T$.

The general Poincaré bound, for δ_1, is as follows. Set

$$r_{ij} = \pi_i p_{ij}, \quad i, j = 1, \dots, \ell. \tag{1.193}$$

A helpful property here is the symmetry: $r_{ij} = r_{ji}$, valid due to reversibility. Next, for all $i, j = 1, \dots, \ell$, we fix a path $\Gamma_{ij} = (i_0 \to i_1 \to \cdots \to i_m)$, across the diagram, starting at i and ending at j, where every arrow enters not more than once. That is, $i_0 = i$, $i_m = j$, $p_{i_s, i_{s+1}} > 0$ for all $s = 0, \dots, m-1$, and every arrow $\tilde{e} = (\tilde{i} \to \tilde{j})$ appears at most once among the edges $e_s = (i_s \to i_{s+1})$s. Owing to irreducibility, such a path always exists. (In our previous formulation, it was a geodesic from i to j.) Denote the collection of selected paths by $\mathscr{G}$ and set

$$|\Gamma_{ij}| = \sum_{s=0}^{m-1} \frac{1}{r_{i_s i_{s+1}}}, \quad \Gamma_{ij} \in \mathscr{G}. \tag{1.194}$$

In this setting, the Poincaré inequality takes the form

$$\delta_1 \ge \left(\max_{\widehat{e} = (\widehat{i} \to \widehat{j})} \left[\sum_{\Gamma_{ij} \in \mathscr{G}} \mathbf{1}\,(\Gamma_{ij} \ni \widehat{e})\, |\Gamma_{ij}| \pi_i \pi_j \right] \right)^{-1}. \tag{1.195}$$

The RHS in (1.195) is sensitive to the choice of paths Γ_{ij} (as we mentioned on page 118, an appropriate choice was a geodesic). In general, the RHS measures a 'degree of irreducibility' of the matrix P; a shrewd choice of the set $\mathscr{G}$ can prove very helpful. In the case of a uniform equilibrium distribution $\pi = \frac{1}{\ell}\,\underline{1}^T$, the RHS of (1.195) coincides with that of (1.151).

In the course of the proof of (1.195), it will be convenient to work with the Laplacian $L(P)$; for brevity we denote it simply by L. (The convenience of $L(P)$ is related to the fact that the eigenvector with the lowest eigenvalue for $L(P)^T$ is $\underline{1}^T$). The eigenvalue δ_1 can be characterised as the lowest eigenvalue of the matrix L when its action is restricted to the orthocomplement of the vector $\underline{1}$. (The term 'orthogonality' and all related concepts below refer to the tilted scalar product $\langle\, \cdot\, ,\, \cdot\, \rangle_\pi$.) The powerful tool here is the so-called *variational* characterisation (or *minimax* characterisation) of the eigenvalues of a Hermitian matrix, aka the Courant–Fisher Theorem. In fact, it is a generalisation of an earlier statement, often called the Rayleigh–Ritz Theorem, which suffices for our purpose. According to the Rayleigh–Ritz Theorem, the eigenvalue δ_1 forms the solution of the minimisation problem in terms of values $\langle L\mathbf{x}, \mathbf{x}\rangle_\pi$, $||\mathbf{x}||_\pi = \langle \mathbf{x}, \mathbf{x}\rangle_\pi$ and $\langle \mathbf{x}, \underline{1}\rangle_\pi$:

$$
\begin{aligned}
\delta_1 &= \min\left[\langle L\mathbf{x}, \mathbf{x}\rangle_\pi : ||\mathbf{x}||_\pi = 1, \langle \mathbf{x}, \underline{1}\rangle_\pi = 0\right] \\
&= \min\left[\frac{\langle L\mathbf{x}, \mathbf{x}\rangle_\pi}{||\mathbf{x}||_\pi^2} : \mathbf{x} \neq 0, \langle \mathbf{x}, \underline{1}\rangle_\pi = 0\right].
\end{aligned}
\tag{1.196}
$$

In our situation, the handy formula

$$
\langle L\mathbf{x}, \mathbf{x}\rangle_\pi = \frac{1}{2}\sum_{i,j=1}^{\ell} (x_i - x_j)^2 r_{ij}, \quad \mathbf{x} = \begin{pmatrix} x_1 \\ \vdots \\ x_\ell \end{pmatrix} \in \mathbb{R}^\ell,
\tag{1.197}
$$

is deduced from the definitions $L = \mathbf{I} - P^{\mathrm{T}}$ and $r_{ij} = \pi_i p_{ij}$. Observe that the RHS of (1.197) does not change when we transform $\mathbf{x} \mapsto \mathbf{x} + c\underline{1}$, i.e. add a constant c to the entries $x_1, \ldots, x_\ell$. But this operation naturally produces, from a general vector $\mathbf{x}$, a vector orthogonal to $\underline{1}$:

$$
\langle \mathbf{x} + c\underline{1}, \underline{1}\rangle_\pi = 0 \text{ if and only if } c = -\frac{\langle \mathbf{x}, \underline{1}\rangle_\pi}{\langle \underline{1}, \underline{1}\rangle_\pi}.
\tag{1.198}
$$

On the other hand, for a real vector $\mathbf{x}$ with $\langle \mathbf{x}, \underline{1}\rangle_\pi = 0$,

$$
||\mathbf{x}||_\pi^2 = \frac{1}{2}\sum_{i,j=1}^{\ell} (x_i - x_j)^2 \pi_i \pi_j.
\tag{1.199}
$$

Again, this is not affected by adding a constant to the entries. So (1.196) can be written as

$$
\begin{aligned}
\delta_1 &= \min\left[\frac{\langle \phi, L\phi\rangle_\pi}{||\phi||_\pi^2} : \phi = \begin{pmatrix} \phi_1 \\ \vdots \\ \phi_\ell \end{pmatrix} \in \mathbb{R}^\ell, \phi \not\sim \underline{1}\right] \\
&= \min\left[\frac{\mathscr{E}(\phi)}{\mathscr{V}(\phi)} : \phi \text{ real non-constant}\right].
\end{aligned}
\tag{1.200}
$$

Here the vector ϕ is identified by a function $\phi : i \mapsto \phi(i)$, where $\phi(i) = \phi_i$, $i = 1, \ldots, \ell$ (the vector ϕ will play the role of a 'potential' function). Next,

$$
\mathscr{E}(\phi) = \frac{1}{2}\sum_{i,j=1}^{\ell} (\phi(i) - \phi(j))^2 r_{ij},
\tag{1.201}
$$

and

$$
\mathscr{V}(\phi) = \frac{1}{2}\sum_{i,j=1}^{\ell} (\phi(i) - \phi(j))^2 \pi_i \pi_j.
\tag{1.202}
$$

With (1.200) to hand, we can easily prove (1.195). For, for all $i, j = 1, \ldots, \ell$, write the sum along path Γ_{ij}

$$\phi(i) - \phi(j) = \sum_{r=1}^{\ell-1} (\phi(i_{r+1}) - \phi(i_r)) = \sum_{e \in \Gamma_{ij}} \phi(e),$$

where, for arrow $e = (i_s \to i_{s+1})$, $\phi(e)$ is defined as the difference $\phi(i_{s+1}) - \phi(i_s)$. Then re-write (1.202) as

$$\mathcal{V}(\phi) = \frac{1}{2} \sum_{i,j=1}^{\ell} \left[\sum_{e=(i_s \to i_{s+1}) \in \Gamma_{ij}} \left(\frac{r_{i_s i_{s+1}}}{r_{i_s i_{s+1}}} \right)^{1/2} \phi(e) \right]^2 \pi_i \pi_j, \qquad (1.203)$$

and apply the Cauchy–Schwarz inequality to the internal sum

$$\left[\sum_{e=(i_s \to i_{s+1}) \in \Gamma_{ij}} \left(\frac{r_{i_s i_{s+1}}}{r_{i_s i_{s+1}}} \right)^{1/2} \phi(e) \right]^2 \le |\Gamma_{ij}| \sum_{e=(i_s \to i_{s+1}) \in \Gamma_{ij}} r_{i_s i_{s+1}} (\phi(e))^2.$$

So that

$$\begin{aligned}
\mathcal{V}(\phi) &\le \frac{1}{2} \sum_{i,j=1}^{\ell} \pi_i \pi_j |\Gamma_{ij}| \sum_{e=(i_s \to i_{s+1}) \in \Gamma_{ij}} r_{i_s i_{s+1}} \phi(e)^2 \\
&= \frac{1}{2} \sum_{\hat{e}=(\hat{i} \to \hat{j})} r_{\hat{i},\hat{j}} (\phi(\hat{e}))^2 \sum_{\Gamma_{ij}: \Gamma_{ij} \ni \hat{e}} |\Gamma_{ij}| \pi_i \pi_j \\
&\le \mathcal{E}(\phi) \max_{\hat{e}} \sum_{\Gamma_{ij}: \Gamma_{ij} \ni \hat{e}} |\Gamma_{ij}| \pi_i \pi_j. \qquad (1.204)
\end{aligned}$$

After taking maximums over $\hat{e}$, and using (1.196), this implies (1.195).

Proof of bound (1.157). As before, we will prove a more general inequality. Let again P be irreducible and aperiodic. We know that -1 is not an eigenvalue of P (and what we try to do is separate the eigenvalues from -1). In a sense, $\delta_{-1} = 1 + \mu_{\ell-1}$ measures how 'far' our DTMC is from a periodic one, of period 2 (where -1 is an eigenvalue). A periodic DTMC must have a *bipartite* diagram, where the state space is divided into two subsets such that all arrows go from one subset to the other. This means that every loop in the periodic case must have an even number of arrows.

This explains why, for an aperiodic P, we select, for all states $i = 1, \ldots, \ell$, a simple loop $\Sigma_i = (i_0 \to i_1 \to \cdots \to i_{m-1} \to i_m)$, with an odd number m of edges, visiting $i = i_0 = i_m$. As before, let $\mathscr{S}$ denote the set of selected loops. Similarly to (1.194), define the weight of a loop by

$$|\Sigma_i|_Q = \sum_{s=0}^{m-1} \frac{1}{r_{i_s i_{s+1}}}, \quad \Sigma_i \in \mathscr{S}. \qquad (1.205)$$

In this setting, the bound for δ_{-1} becomes

$$\delta_{-1} \geq 2 \left(\max_{\widehat{e}=(\widehat{i}\to\widehat{j})} \left[\sum_{\Sigma_i \in \mathscr{S}} \mathbf{1}\,(\Sigma_i \ni \widehat{e})\, |\Sigma_i|_Q \pi_i \right] \right)^{-1} \tag{1.206}$$

and will again depend on the choice of set $\mathscr{S}$.

To prove (1.206), we begin with the following straightforward identity

$$\begin{aligned}
\frac{1}{2} \sum_{i,j=1}^{\ell} (\phi(i) + \phi(j))^2 r_{ij} &= \sum_{i=1}^{\ell} (\phi(i))^2 \pi_i + \sum_{i,j=1}^{\ell} \phi(i)\phi(j) r_{ij} \\
&= \mathbb{E}_\pi \phi^2 + \langle \phi, P\phi \rangle_\pi, \tag{1.207}
\end{aligned}$$

where $\mathbb{E}_\pi$ stands for the expectation relative to the equilibrium distribution π, and the function $\phi : i \mapsto \phi(i)$ has been identified with the vector $\phi = \begin{pmatrix} \phi_1 \\ \vdots \\ \phi_\ell \end{pmatrix}$, where $\phi_i = \phi(i)$ (a trick we introduced before). Then, if Σ_i is a loop starting and ending at i, we write

$$\phi(i) = \frac{1}{2} \Big[(\phi(i_0) + \phi(i_1)) - (\phi(i_1) + \phi(i_2)) + \cdots + (\phi(i_{m-1}) + \phi(i_m)) \Big];$$

the fact that m is odd helps here. Then, as in the proof of Poincaré's bound, the Cauchy–Schwarz inequality will come into play, with:

$$\begin{aligned}
\mathbb{E}_\pi \phi^2 &= \frac{1}{4} \sum_{i=1}^{\ell} \pi_i \left(\sum_{e=(i_s \to i_{s+1}) \in \Sigma_i} (-1)^s \sqrt{\frac{r_{i_s i_{s+1}}}{r_{i_s i_{s+1}}}} \big[\phi(i_s) + \phi(i_{s+1}) \big] \right)^2 \\
&\leq \frac{1}{4} \sum_{i=1}^{\ell} \pi_i |\Sigma_i|_Q \sum_{e=(i_s \to i_{s+1}) \in \Sigma_i} r_{i_s i_{s+1}} \big[\phi(i_s) + \phi(i_{s+1}) \big]^2 \quad \text{(by CS)} \\
&= \frac{1}{4} \sum_{\widehat{e}=(\widehat{i}\to\widehat{j})} \big[\phi(\widehat{i}) + \phi(\widehat{j}) \big]^2 r_{\widehat{i}\widehat{j}} \sum_{\Sigma_i : \Sigma_i \ni \widehat{e}} |\Sigma_i|_Q \pi_i \\
&= (1/2) \Big(\mathbb{E}_\pi \phi^2 + \langle \phi, P\phi \rangle_\pi \Big) \sum_{\Sigma_i : \Sigma_i \ni \widehat{e}} |\Sigma_i|_Q \pi_i \quad \text{(according to (1.207).}
\end{aligned}$$

Taking maximums yields

$$\mathbb{E}_\pi \phi^2 \leq \frac{1}{2} \Big(\mathbb{E}_\pi \phi^2 + \langle \phi, P\phi \rangle_\pi \Big) \left(\max_{\widehat{e}} \left[\sum_{\Sigma_i : \Sigma_i \ni \widehat{e}} |\Sigma_i|_Q \pi_i : \Sigma_i \in \mathscr{S} \right] \right). \tag{1.208}$$

Next, dividing by $\mathbb{E}_\pi \phi^2 = \|\phi\|_\pi^2$ gives

$$\frac{\langle \phi, P\phi \rangle_\pi}{\|\phi\|_\pi^2} \geq -1 + 2 \left(\max_{\hat{e}} \left[\sum_{\Sigma_i : \Sigma_i \ni \hat{e}} |\Sigma_i| Q \pi_i : \Sigma_i \in \mathscr{S} \right] \right)^{-1},$$

and the Rayleigh–Ritz minimax characterisation (1.148) yields bound (1.206). □

> *When you need to prove you're right, use integration by parts.*
> *If it doesn't work, use the Cauchy–Schwarz inequality.*
> (From the series 'Thus spoke Superviser'.)

Proof of Cheeger's inequality Again, it is instructive to prove the inequality in a more general form than (1.163), assuming that $P = (p_{ij})$ is irreducible and reversible, relative to an arbitrary positive equilibrium distribution $\pi = (\pi_i)$ and considering symmetric weights (1.194). Now, extending (1.163), for a set of states $S \subset \{1, \ldots, \ell\}$, we define *the generalised flow* $Q(S, S^c)$ generated by a reversible (π, P) chain in equilibrium, between S and S^c, by

$$Q(S, S^c) = \sum_{(i,j);\, i \in S, j \in S^c} r_{ij}, \qquad (1.209)$$

and then set

$$h = \min_{S:\, 0 < \pi(S) \leq 1/2} \left[\frac{1}{\pi(S)} Q(S, S^c) \right], \qquad (1.210)$$

where

$$\pi(S) = \sum_{i \in S} \pi_i. \qquad (1.211)$$

In this more general setting, Cheeger's inequality still asserts that

$$\frac{h^2}{2} \leq \delta_1 \leq 2h. \qquad (1.212)$$

The upper bound in (1.212) is straightforward. For, given a set of states $S \subset \{1, \ldots, \ell\}$, with $0 \leq \pi(S) = \sum_{i \in S} \pi_i \leq 1/2$, set

$$\psi_S(i) = \begin{cases} \pi(S^c), & i \in S, \\ -\pi(S), & i \in S^c. \end{cases} \qquad (1.213)$$

Using again (1.200) we obtain

$$\delta_1(P) \leq \frac{\mathscr{E}(\psi_S)}{\mathscr{V}(\psi_S)} = \frac{Q(S, S^c)}{\pi(S)\pi(S^c)} \leq 2\frac{Q(S, S^c)}{\pi(S)}.$$

This implies that $\delta_1 \leq 2h$. The lower bound in (1.212) is more tricky and based on two remarks.

Remark 1.14.1 Given a real function $\psi : i \in \{1,\ldots,\ell\} \mapsto \psi(i)$, let $\psi^+(i) = \max[\psi(i),0] = \psi(i) \vee 0$ and set $S^+(\psi) = \{i \in S : \psi(i) > 0\}$. Assume that $S^+(\psi) \neq \emptyset$.

Owing to (1.197) and (1.201), we can write $\mathscr{E}(\psi) = \langle L\psi, \psi \rangle$ and $\mathscr{E}(\psi^+) = \langle L\psi^+, \psi^+ \rangle$. Here we again use the notation ψ and ψ^+ for ℓ-dimensional vectors $\begin{pmatrix} \psi_1 \\ \vdots \\ \psi_\ell \end{pmatrix}$ and $\begin{pmatrix} \psi_1^+ \\ \vdots \\ \psi_\ell^+ \end{pmatrix}$, where $\psi_i = \psi(i)$ and $\psi_i^+ = \psi(i)^+$, $i = 1,\ldots,\ell$. Now, assume that, for some $\lambda \geq 0$, the following inequality holds true:

$$(L\psi)_i \leq \lambda \psi_i, \ \text{for} \ i \in S^+(\psi).$$

Then the norm $\|\psi^+\|_\pi = \left(\sum_{i=1}^{\ell} (\psi_i \vee 0)^2 \pi_i \right)^{1/2}$ satisfies

$$\lambda \|\psi^+\|_\pi^2 \geq \mathscr{E}(\psi^+). \tag{1.214}$$

Indeed,

$$\lambda \|\psi^+\|_\pi^2 \geq \langle L\psi^+, \psi \rangle_\pi = \frac{1}{2} \sum_{i,j=1}^{\ell} (\psi_i^+ - \psi_j^+)(\psi_i - \psi_j) r_{ij}, \tag{1.215}$$

and the RHS of (1.215) is, rather surprisingly,

$$\geq \langle L\psi^+, \psi^+ \rangle_\pi = \mathscr{E}(\psi^+).$$

In fact, the last inequality follows from an elementary bound

$$(\psi_i^+ - \psi_j^+)(\psi_i - \psi_j) \geq (\psi_i^+ - \psi_j^+)^2.$$

Remark 1.14.2 In the above notation, for all $\psi = \begin{pmatrix} \psi_1 \\ \vdots \\ \psi_\ell \end{pmatrix}$,

$$\mathscr{E}(\psi^+) = \frac{1}{2} \sum_{i,j=1}^{\ell} (\psi_i^+ - \psi_j^+)^2 r_{ij} \geq \frac{1}{2} (h(\psi))^2 \|\psi^+\|_\pi^2. \tag{1.216}$$

Here

$$h(\psi) = \inf \left[\frac{Q(S,S^c)}{\pi(S)} : \emptyset \neq S \subset S^+(\psi) \right]. \tag{1.217}$$

To prove bound (1.216), we assume, without loss of generality, that entries $\psi_i \geq 0$, for all $i = 1,\ldots,\ell$. By Cauchy–Schwarz,

$$\sum_{i,j=1}^{\ell} |\psi_i^2 - \psi_j^2| r_{ij} \leq \sqrt{2} \mathcal{E}(\psi)^{1/2} \left(\sum_{i,j=1}^{\ell} (\psi_i + \psi_j)^2 r_{ij} \right)^{1/2}$$

$$\leq 2^{3/2} \mathcal{E}(\psi)^{1/2} \|\psi\|_{\pi}. \tag{1.218}$$

The LHS of (1.218) can be written as

$$2 \sum_{i,j=1}^{\ell} \mathbf{1}(\psi_j > \psi_i)(\psi(j)^2 - \psi(i)^2) r_{ij} = 4 \sum_{i,j=1}^{\ell} \mathbf{1}(\psi_j > \psi_i) r_{ij} \int_{\psi_i}^{\psi_j} t \, dt$$

$$= 4 \int_0^{\infty} t \sum_{i,j:\psi_i \leq t < \psi_j} r_{ij} \, dt. \tag{1.219}$$

Now, observe that

$$\sum_{i,j=1}^{\ell} \mathbf{1}(\psi_i \leq t < \psi_j) r_{ij} = Q(S^t, S^{tc}), \tag{1.220}$$

for $S^t(= S^t(\psi)) = \{i : \psi_i > t\} \subseteq S^+(\psi)$. Combining (1.214) and (1.216) yields

$$\text{RHS of (1.218)} \geq 4h(\psi) \int_0^{\infty} \pi(S^t) \, t \, dt = 2 \, h(\psi) \|\psi\|_{\pi}^2,$$

i.e. the bound

$$\lambda \geq \frac{1}{2}(h(\psi))^2. \tag{1.221}$$

We may now finish with the lower Cheeger bound $\delta_1 \geq h^2/2$. In view of (1.214), (1.216) and (1.217), if $(L\psi)_i \leq \lambda \psi_i$ for all $i \in S^+(\psi)$, then (1.221) holds. Then take $\lambda = \delta_1$, and let ψ^T be a normalised row eigenvector of L^T with the eigenvalue δ_1: $\psi^T L^T = \delta_1 \psi^T$, i.e. $L\psi = \delta_1 \psi$, where $\|\psi\|_{\pi} = 1$. We know that column vectors ψ and $\underline{1}$ are π-orthogonal, i.e. the π-mean of ψ vanishes: $\sum_{i=1}^{\ell} \psi_i \pi_i = 0$. Hence, we can always arrange that $0 < \pi(S^+(\psi)) \leq 1/2$ and therefore $h(\psi) \geq h$. Then the bound $\delta_1 \geq h^2/2$ follows from (1.221), with this choice of value λ and vector ψ. $\square$

The story we tell at the end of this section has a double moral. One is that a good command of calculus is vital (as has already been manifested in the proofs given in this section). The other (mainly for present or future lecturers) is that delivering dull lectures can be risky. The story comes from the life of the Russian physicist Igor Tamm (1895–1971), Nobel Prize winner in Physics (1958), and a close personal friend of N. Bohr, P. Dirac, R. Peierls and many others. Tamm got to be an extremely popular and respected figure in the Soviet and international physics community; there was a joke that if there is a unit of honesty it is the tamm. Igor Tamm was born in Vladivostock (in the Russian Far East) and begun his undergraduate studies at the University of Edinburgh where he read mathematics. After he returned to spend the summer of 1914 in Russia he was unable to continue his studies because of the outbreak of World War I. He managed to complete his course at Moscow University and for several years taught mathematics in different places. This covered the period of the civil war in Russia (1918–1920) when there were acute shortages of food and clothes. People often resorted to bartering garments for food. Once Tamm travelled to a place near the city of Odessa (now in Southern Ukraine) to get some food for a bag of clothing. The military situation in the region was unstable, the main battles being fought between Reds (followers of the Bolsheviks) and Whites (aiming to restore some form of the old regime), but a mishmash of other forces was also active, including the so-called Greens (not to be confused with modern political and social movements known by the same name). The Greens opposed both the Reds and the Whites (as well as any other other side taking part in the Civil War), and their aim was to establish a 'proper peasant power' (some modern historians consider them as brave fighters for the Ukrainian statehood). One of their slogans was "Beat the reds until they become white, and the whites until they become red". The Greens' tactic was to attack soft spots in the rear of either of the main forces, get a quick bounty and disappear.

Tamm was caught in a sudden attack by the Greens and brought before their commander as a suspect. The picturesquely dressed commander was a bear of a man of approximately the same age as Tamm, wearing, in accordance with the customs of the time, a pair of big Mauser handguns on his belt, his chest crossed with machine-gun bands. His deputy reported that Tamm was arrested as a Bolshevik agitator and should be immediately shot. Tamm protested that he was no political activist, but a professor of mathematics. Unexpectedly, the commander ordered everybody but Tamm to leave. Then he said to Tamm: "Fine. If you're a mathematician, write down the remainder term in the Maclaurin form of the Taylor series." Without blinking, Tamm gave him the answer and commented that the question was rather trivial. The commander was pleased and immediately ordered Tamm to be freed and allowed to go back to Odessa. It turned out that he was a former maths student but found courses too boring.

1.15 Large deviations for discrete-time Markov chains

> ... *do not let it pass, but instantly check them;*
> *you do not know where deviation from truth will end.*
> S. Johnson (1709–1784), English lexicographer and playwright

Large deviation theory describes rare events, with small probabilities. Formally, it is an asymptotical theory, dealing with events A_n such that $\mathbb{P}(A_n) \to 0$ as $n \to \infty$.

A typical example of such an event where a random variable Y_n takes increasingly large values is $\{Y_n \geq n(\mu + a)\}$, $\mathbb{E}Y_n = n\mu$, $a > 0$ (by passing from Y_n to $-Y_n$, we can extend this to increasingly large negative values). The form of the RV Y_n may vary; it is natural to begin with an example where $Y_n = Z_1 + \cdots + Z_n$ is the sum of n IID copies of a RV Z. Assuming, for simplicity, that the RVs Z_i take finitely many values z (more than one), we know that if $\mathbb{E}Z = \mathbb{E}Z_i = \mu$ and var $Z = $ Var$Z_i = \sigma^2 > 0$ then, as $n \to \infty$, the averaged sum Y_n/n converges to μ (the weak and strong laws of large numbers), and the RV $(Y_n - n\mu)/\sigma\sqrt{n}$ has in the limit the normal distribution $N(0,1)$ (the local and integral Central Limit Theorem). However, neither of these statements tells us much about the probability $\mathbb{P}(Y_n > n(\mu + a))$ where $a > 0$. For instance, the Chebyshev inequality

$$\mathbb{P}(Y_n > n(\mu + a)) \leq \left(\text{Var}\,Z_1\right) / na^2$$

guarantees that $\mathbb{P}(Y_n > n(\mu + a))$ converges to 0 in the limit, but how fast? If Z_1, $Z_2, \ldots$ take two values, ± 1 say, we can try to use the precise formula

$$\mathbb{P}(Y_n = m) = \binom{n}{(n+m)/2} p^{(n+m)/2}(1-p)^{(n-m)/2},$$

with $p = \mathbb{P}(Z_1 = 1)$, $m = 0, \pm 1, \ldots, \pm n$, but it is tedious. To cover a decently general case, we need different techniques.

It turns out that the answer is quite simple. Consider the moment-generating function (MGF) of Y_n: $\mathbb{E}e^{\theta Y_n} = (\mathbb{E}e^{\theta Z})^n$. Here, $\mathbb{E}e^{\theta Z} = \sum_k e^{\theta k}\mathbb{P}(Z = k)$, is the MGF of the variable Z: it represents a finite sum of exponentials. Any MGF is a convex function (owing to Jensen's inequality or just by differentiation, viz. $\frac{d^2}{d\theta^2}\mathbb{E}e^{\theta Z} = \mathbb{E}Z^2 e^{\theta Z} > 0$). Under our assumption, $\mathbb{E}e^{\theta Z}$ and $\mathbb{E}e^{\theta Y_n}$ are finite for all real $\theta \in \mathbb{R}$. (They are also obviously positive for all $\theta \in \mathbb{R}$.) Take the logarithm of $\mathbb{E}e^{\theta Y_n}$ and divide by n:

$$\frac{1}{n}\ln \mathbb{E}e^{\theta Y_n} = \ln \mathbb{E}e^{\theta Z} := \Lambda(\theta). \qquad (1.222)$$

This is again a convex function of θ; the shortest way to see it is as follows. Take the derivatives

$$\frac{d}{d\theta}\Lambda(\theta) = \frac{\mathbb{E}Ze^{\theta Z}}{\mathbb{E}e^{\theta Z}}, \quad \frac{d^2}{d\theta^2}\Lambda(\theta) = \frac{\left[\mathbb{E}e^{\theta Z}\mathbb{E}Z^2 e^{\theta Z} - \left(\mathbb{E}Ze^{\theta Z}\right)^2\right]}{(\mathbb{E}e^{\theta Z})^2}.$$

It is convenient to write the second derivative in the form of a variance

$$\frac{\left[\mathbb{E}e^{\theta Z}\mathbb{E}Z^2 e^{\theta Z} - \left(\mathbb{E}Ze^{\theta Z}\right)^2\right]}{(\mathbb{E}e^{\theta Z})^2} = \mathbb{E}\tilde{Z}_1^2 - \left(\mathbb{E}\tilde{Z}_1\right)^2 = \text{Var}\,\tilde{Z}_1 > 0. \qquad (1.223)$$

Here $\tilde{Z}_1$ is a random variable taking the same values as Z, but with 'tilted' probabilities

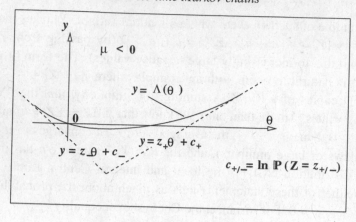

Fig. 1.29

$$\mathbb{P}(\widetilde{Z}_1 = z) = \frac{e^{\theta z}\mathbb{P}(Z = z)}{\mathbb{E}e^{\theta Z}}.$$

(So that $\sum_z \mathbb{P}(\widetilde{Z}_1 = z) = \frac{1}{\mathbb{E}e^{\theta Z}}\sum_z e^{\theta z}\mathbb{P}(Z = z) = 1$, and moments $\mathbb{E}\widetilde{Z}_1^n = \frac{1}{\mathbb{E}e^{\theta Z}}\sum_z z^n e^{\theta z} = \frac{\mathbb{E}Z^n e^{\theta Z}}{\mathbb{E}e^{\theta Z}}, n = 1, 2, \ldots.$) Now, make the other simplifying assumption, that Z takes both positive and negative values, so that the sum $\mathbb{E}e^{\theta Z} = \sum_z e^{\theta z}\mathbb{P}(Z = z)$ includes both positive and negative exponentials. Let $z_+ > 0$ be the maximal and $z_- < 0$ the minimal attained value of Z. The graph of the function $\Lambda(\theta)$ looks like a hyperbola passing through the origin: see Figure 1.29.

Dracula's bloody θs
(From the series *'Movies that never made it to the Big Screen'*.)

Next, consider the so-called *Legendre transform* (also called *Legendre–Fenchel* or, in the context of large deviations, *Legendre–Cramér* transform):

$$\Lambda^*(x) = \max\left[\theta x - \Lambda(\theta), \ \theta \in \mathbb{R}\right] = -\ln\left(\min\left[e^{-\theta x}\mathbb{E}e^{\theta Z} : \theta \in \mathbb{R}\right]\right), \ x \in \mathbb{R}.$$

$$(1.224)$$

The meaning of this operation is that the value $\Lambda^*(x)$ is attained at the point $\theta^* = \theta^*(x)$ where the derivative $\Lambda'(\theta^*) = x$; in our situation, where Z takes finitely many values, both negative and positive, such a point exists and is unique for $z_- \le x \le z_+$ (with the agreement that $\theta^*(z_\pm) = \pm\infty$), and

$$\Lambda^*(x) = \theta^* x - \Lambda(\theta^*).$$

See Figure 1.30.

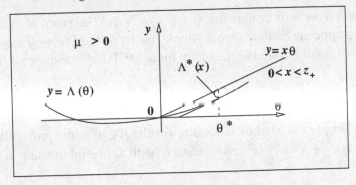

Fig. 1.30

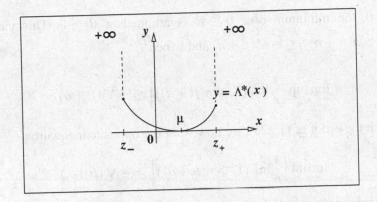

Fig. 1.31

However, for $x < z_-$ and $x > z_+$, no such point exists θ^*, and for such x we set $\Lambda^*(x) = +\infty$. At $x = z_-$ and $x = z_+$, we observe, from a direct calculation, that $\Lambda^*(z_\pm) = -\ln \mathbb{P}(Z = z_\pm)$. See Figure 1.31.

As we will see, for all $a > 0$, the asymptotics of the probability $\mathbb{P}(Y_n > n(\mu + a))$ are exponential, with negative exponent $\Lambda^*(a)$:

$$\lim_{n \to \infty} \frac{1}{n} \ln \mathbb{P}(Y_n > n(\mu + a)) = -\Lambda^*(\mu + a). \tag{1.225}$$

In particular, for $\mu + a \geq z_+$, i.e. $a \geq z_+ - \mu$, the probability $\mathbb{P}(Y_n > n(\mu + a))$ vanishes for all n, and the limit in (1.225) equals $-\infty$, in agreement with the above setting. By the same token, the limit $\lim_{n \to \infty} \frac{1}{n} \ln \mathbb{P}(Y_n < n(\mu - a))$ equals $\Lambda^*(\mu - a)$.

From now on we will assume that $0 < a < z_+ - \mu$. The proof of (1.225) needs to check two opposite bounds. One is simple, based on the *Chernoff inequality*: for all real-valued random variables U with a finite MGF $\mathbb{E}e^{\theta U}$ and for all $\theta \geq 0$,

$$\mathbb{P}(U > b) \leq \frac{1}{e^{\theta b}} \, \mathbb{E}e^{\theta U}. \tag{1.226}$$

In fact, it is simply the Markov inequality for the RV $e^{\theta U}$: the probability $\mathbb{P}(U > b) = \mathbb{P}(e^{\theta U} > e^{\theta b}) \leq \mathbb{E}e^{\theta U}/e^{\theta b}$. Replacing U with Y_n and minimising in $\theta \geq 0$, we obtain

$$\mathbb{P}(Y_n > n\mu + na) \leq \exp\left(n\min\left[\frac{1}{n}\ln\mathbb{E}e^{\theta Y_n} - \theta(\mu + a) : \theta \geq 0\right]\right).$$

For $a > 0$, the minimum over $\theta \in \mathbb{R}$ is attained at $\theta \geq 0$. This yields that $\frac{1}{n}\ln\mathbb{P}(Y_n > n(\mu + a)) \leq -\Lambda^*(\mu + a)$ and hence

$$\limsup_{n \to \infty}\left[\frac{1}{n}\ln\mathbb{P}(Y_n > n(\mu + a))\right] \leq -\Lambda^*(\mu + a). \tag{1.227}$$

To finish the proof of (1.225), we now want the opposite inequality:

$$\liminf_{n \to \infty}\left[\frac{1}{n}\ln\mathbb{P}(Y_n > n(\mu + a))\right] \geq -\Lambda^*(\mu + a) \tag{1.228}$$

This requires some analytical considerations.

> *In Algebra, the equality $A = B$ is usually a triviality.*
> *In Analysis, it is the consequence of two opposite inequalities,*
> *one of which is usually not hard, but the other requires a lot of work.*
> (From the series 'Thus spoke Superviser'.)

First, note that, given $x \in \mathbb{R}$, the Legendre transform $\Lambda^*(x)$ equals $-\ln\mathbb{E}e^{\theta^*(Z-x)} = \theta^* x - \ln\mathbb{E}e^{\theta^* Z}$, for the optimiser $\theta^*(= \theta^*(x))$; see Figure 1.30. Next, pass to the IID copies $\widetilde{Z}_1, \widetilde{Z}_2, \ldots$ of the tilted RVs $\widetilde{Z}$, with $\mathbb{P}(\widetilde{Z}_i = z) = \mathbb{P}(\widetilde{Z} = z) = e^{\theta^* z}\mathbb{P}(Z = z)/\mathbb{E}e^{\theta^* Z}$. Then write

$$\mathbb{P}(Y_n > nx) = \sum_{z_1,\ldots,z_n} \mathbf{1}(z_1 + \cdots + z_n > nx)\mathbb{P}(Z = z_1)\cdots\mathbb{P}(Z_n = z_n)$$

$$= \left(\mathbb{E}e^{\theta^* Z}\right)^n \sum_{z_1,\ldots,z_n} \mathbf{1}(\textstyle\sum z_i > nx)e^{-\theta^*(\Sigma z_i)}\prod_{i=1}^{n}\mathbb{P}(\widetilde{Z}_i = z_i).$$

Now, if both $x, \theta^* > 0$, then, given any $\varepsilon > 0$, the right hand side is

$$\geq \left(\mathbb{E}e^{\theta^* Z}\right)^n \sum_{z_1, \ldots, z_n} \mathbf{1}\left(nx(1+\varepsilon) \geq \sum_{i=1}^{n} z_i > nx\right) e^{-\theta^*(\Sigma z_i)} \prod_{i=1}^{n} \mathbb{P}(\tilde{Z}_i = z_i)$$

$$\geq e^{-n\theta^* x(1+\varepsilon)} \left(\mathbb{E}e^{\theta^* Z}\right)^n \sum_{z_1, \ldots, z_n} \mathbf{1}\left(nx(1+\varepsilon) \geq \sum_{i=1}^{n} z_i > nx\right) \prod_{i=1}^{n} \mathbb{P}(\tilde{Z}_i = z_i)$$

$$= e^{-n\theta^* x(1+\varepsilon)} \left(\mathbb{E}e^{\theta^* Z}\right)^n \mathbb{P}\left(nx(1+\varepsilon) \geq \sum_{i=1}^{n} \tilde{Z}_i > nx\right). \tag{1.229}$$

Observe that the expectation

$$\mathbb{E}(\tilde{Z} - x) = \frac{1}{\mathbb{E}e^{\theta^* Z}} \mathbb{E}(Z - x)e^{\theta^* Z} = 0, \quad \text{i.e. } \mathbb{E}\tilde{Z} = x. \tag{1.230}$$

In fact,

$$\mathbb{E}(\tilde{Z} - x) = \frac{d}{d\theta} \mathbb{E}e^{\theta(Z-x)}\bigg|_{\theta=\theta^*} = 0,$$

as θ^* corresponds to the maximum point of $\ln \mathbb{E}e^{\theta Z} - \theta x$.

Now, the probability on the right hand side of (1.229) is nothing but

$$\mathbb{P}\left(0 < \frac{\tilde{Z}_1 + \cdots + \tilde{Z}_n - nx}{\tilde{\sigma}\sqrt{n}} \leq \frac{\varepsilon x \sqrt{n}}{\tilde{\sigma}}\right),$$

which, by the Central Limit Theorem, tends to

$$\frac{1}{\sqrt{2\pi}} \int_0^\infty e^{-y^2/2} dy = \frac{1}{2}.$$

Hence, if $x, \theta^* > 0$ then for all $\varepsilon > 0$,

$$\liminf_{n \to \infty} \left[\frac{1}{n} \ln \mathbb{P}\left(\frac{1}{n} Y_n > x\right)\right] \geq -\theta^* x(1+\varepsilon) + \ln \mathbb{E}e^{\theta^* Z},$$

or, letting $\varepsilon \to 0$,

$$\liminf_{n \to \infty} \left[\frac{1}{n} \ln \mathbb{P}\left(\frac{1}{n} Y_n > x\right)\right] \geq -\theta^* x + \ln \mathbb{E}e^{\theta^* Z} = -\Lambda^*(x). \tag{1.231}$$

Next, the condition that $x > 0$ and $\theta^* > 0$ implies that $x > \mu = \mathbb{E}Z$. Hence, we can take $x = \mu + a$, with $0 < a < z_+ - \mu$, and (1.228) follows from (1.231).

In general, the large deviation technique is suitable for working with closed as well as open sets. Here we should be careful because the Legendre transform $\Lambda^*(\theta)$, as we saw, is not continuous in $\theta \in \mathbb{R}$ (it may jump to $+\infty$). Still, Λ^* is a convex and lower semi-continuous function. That is, $\Lambda^*(qx_1 + (1-q)x_2) \leq q\Lambda^*(x_1) + (1-q)\Lambda^*(x_2)$ for all $x_1, x_2 \in \mathbb{R}$ and $0 < q < 1$, and if $x_n \to x$ then $\Lambda^*(x) \geq \limsup \Lambda^*(x_n)$. The apparatus of large deviation theory always takes into account this property. The principal theorem, for the sum $Y_n = \sum_{i=1}^{n} Z_i$ of IID summands $Z_1, Z_2, \ldots$, is often called Cramér's Theorem

Theorem 1.15.1 *For all closed sets $F \subset \mathbb{R}$*

$$\limsup_{n \to \infty} \left[\frac{1}{n} \ln \mathbb{P} \left(\frac{Y_n}{n} \in F \right) \right] \leq -\inf \left[\Lambda^*(x) : x \in F \right], \qquad (1.232)$$

while for all open sets $G \subset \mathbb{R}$

$$\liminf_{n \to \infty} \left[\frac{1}{n} \ln \mathbb{P} \left(\frac{Y_n}{n} \in G \right) \right] \geq -\inf \left[\Lambda^*(x) : x \in G \right]. \qquad (1.233)$$

As an example, assume again that the Z_is take finitely many values and consider as the set F the closed semi-infinite interval $[z_+, +\infty)$ where z_+ is the maximal value attained by the RV Z_i. Then the probability $\mathbb{P}(Y_n \in F) = \mathbb{P}(Z_1 = \cdots = Z_n = z_+) = (\mathbb{P}(Z = z_+))^n$, and the limit in the left hand side of (1.232) equals $\ln \mathbb{P}(Z = z_+)$ which coincides with $\Lambda^*(z_+)$; see Figure 1.31.

Example 1.15.2 More precisely, suppose that Z takes value 0 with probability $q = 1 - p$ and value 1 with probability p. Here,

$$\mathbb{E} e^{\theta Z} = 1 - p + p e^{\theta}, \text{ and } \Lambda(\theta) = \ln(1 - p + p e^{\theta}).$$

We know that to calculate the Legendre transform, we have to solve the equation $x = \Lambda'(\theta^*)$ and then to calculate $x\theta^* - \Lambda(\theta^*)$. We saw that this recipe works when $0 \leq x \leq 1$; here,

$$\theta^* = \ln \left(\frac{x(1-p)}{(1-x)p} \right), \text{ and } \Lambda^* = x \ln \left(\frac{x}{p} \right) + (1-x) \ln \left(\frac{1-x}{1-p} \right). \qquad (1.234)$$

For $x < 0$ or $x > 1$, $\Lambda^*(x) = +\infty$.

Expression (1.234) was spotted in Volume 1, p. 83. It is the *relative entropy* of the two-point probability distribution $(1 - p, p)$ on $\{0, 1\}$ with respect to the distribution $(1 - x, x)$; it is clear that $\Lambda^*(p) = 0$. Thus, for the sum $Y_n = \sum_{i=1}^{n} Z_i$ of IID RVs $Z_i \sim Z$, Cramér Theorem's gives:

(a) for all closed sets $F \subset \mathbb{R}$:

$$\lim_{n \to \infty} \frac{1}{n} \ln \mathbb{P}(Y_n \in F) \begin{cases} = -\infty, & \text{if } F \subseteq (-\infty, 0) \cup (1, +\infty), \\ \leq -\Lambda^*(a), & \text{if } a = \min_{x \in F} x \in (p, 1], \\ \leq -\Lambda^*(a), & \text{if } a = \max_{x \in F} x \in [0, p), \\ \leq 0, & \text{if } F \ni p, \end{cases}$$

(b) for all open sets $G \subset \mathbb{R}$:

$$\lim_{n \to \infty} \frac{1}{n} \ln \mathbb{P}(Y_n \in G) \begin{cases} \geq -\infty, & \text{if } G \subseteq (-\infty, 0] \cup [1, +\infty), \\ \geq -\Lambda^*(a), & \text{if } a = \inf_{x \in G} x \in (p, 1], \\ \geq -\Lambda^*(a), & \text{if } a = \sup_{x \in G} x \in [0, p), \\ 0, & \text{if } G \ni p. \end{cases}$$

For $F = [x, \infty)$ and $G = (x, \infty)$, with $p < x \leq 1$, we obtain, by continuity of Λ^* on $[0, 1]$:

$$\mathbb{P}(Y_n > nx) = \left(\frac{x}{p}\right)^{-nx} \left(\frac{1-x}{1-p}\right)^{-n(1-x)} \exp\left(o(n \max[x, 1-x])\right), \quad p < x \leq 1.$$

The aforementioned exact calculation, based on Stirling's formula, gives

$$\mathbb{P}(Y_n > nx) = \frac{1}{\sqrt{2\pi nx(1-x)}} \left(\frac{x}{p}\right)^{-nx} \left(\frac{1-x}{1-p}\right)^{-n(1-x)} \times$$
$$\left[1 + O\left(\frac{1}{n \max[x, 1-x]}\right)\right], \quad p < x \leq 1. \quad (1.235)$$

This argument can be extended to the case of the sum $Y_n = \sum_{i=1}^n Z_i$ where RVs $Z_i \sim Z$, independently, and Z takes a finite number of distinct values, say $z_1, \ldots, z_l$, with probabilities $p_1, \ldots, p_l$. In fact, it turns out to be more convenient to work with l-dimensional vectors $\mathbf{U}_i \sim \mathbf{U}$. Here $\mathbf{U}_i = (U_{i,1}, \ldots, U_{i,l})$ and $\mathbf{U} = (U_1, \ldots, U_l)$, where $U_k = 1$ if $Z = j_k$ and $U_k = 0$ if $Z \neq j_k$, and similarly, $U_{i,k} = 1$ if $Z_i = z_k$ and $U_{i,k} = 0$ if $Z_i \neq z_k$. Then the sum of the vectors $\mathbf{Y}_n = \sum_{i=1}^n \mathbf{U}_i$ counts the numbers of appearances of every value z_k, from which we can reconstruct the original sum: $Y_n = \sum_{i=1}^n \sum_{k=1}^l z_k U_{i,k}$. Then we consider the joint MGF

$$\mathbb{E}e^{\langle \theta, \mathbf{Z} \rangle} = \mathbb{E}\left[\exp\left(\sum_{k=1}^l \theta_k Z_k\right)\right], \quad \theta = (\theta_1, \ldots, \theta_l),$$

and its logarithm $\Lambda(\theta) = \ln \mathbb{E}e^{\langle \theta, \mathbf{Z} \rangle}$. Similarly to the 'scalar' Legendre transform, one introduces the vector version:

$$\Lambda^*(\mathbf{x}) = \sup \left[\langle \theta, \mathbf{x} \rangle - \Lambda(\theta)\right].$$

The analogue of (1.234) identifies the *relative entropy* of the probability distribution (p_j) with respect to a 'test' one (x_j), with $x_j \geq 0$ and $\sum_j x_j = 1$. More precisely, the Legendre transform Λ^* will be a function of a vector $\mathbf{x} \in \mathbb{S}^l$:

$$\Lambda^*(\mathbf{x}) = \begin{cases} \sum_{j=1}^l x_j \ln \frac{x_j}{p_j}, & \mathbf{x} \in \mathbb{S}^l, \\ +\infty, & \mathbf{x} \in \mathbb{R}^l \setminus \mathbb{S}^l. \end{cases} \quad (1.236)$$

Here $\mathbb{S}^l$ stands for the l-dimensional *simplex* of stochastic vectors: $\mathbb{S}^l = \{\mathbf{x} : x_j \geq 0, \, x_1 + \cdots + x_l = 1\}$. The statement of Cramér Theorem's remains valid and this provides vector analogues of the 'scalar' formulas.

Example 1.15.3 Consider the MGF of a Gaussian RV Z:

$$\mathbb{E}e^{\theta Z} = \exp\left(\theta\mu + \frac{1}{2}\theta^2\sigma^2\right), \quad \text{with } \Lambda(\theta) = \ln\left(\mathbb{E}e^{\theta Z}\right) = \theta\mu + \frac{1}{2}\theta^2\sigma^2.$$

To calculate the Legendre transform $\Lambda^*(x)$, equate

$$x = \Lambda'(\theta^*) = \mu + \theta^*\sigma^2, \quad \text{i.e., } \theta^* = \frac{x-\mu}{\sigma^2}.$$

$$\Lambda^*(x) = x\theta^* - \Lambda(\theta^*) = x\frac{x-\mu}{\sigma^2} - \mu\frac{x-\mu}{\sigma^2} - \frac{1}{2}\left(\frac{x-\mu}{\sigma^2}\right)^2\sigma^2 = \frac{1}{2}\frac{(x-\mu)^2}{\sigma^2}.$$

$$(1.237)$$

This is a 'nice' function: infinitely differentiable for all $x \in \mathbb{R}$, strictly increasing for $x > \mu$ and strictly decreasing for $x < \mu$, with the mimimum at $x = \mu$, the mean of the RV Z.

Then, by Cramér Theorem's, for the sum $Y_n = \sum_{i=1}^n Z_i$ of IID Gaussian RVs, $Z_i \sim \mathrm{N}(\mu, \sigma^2)$, for all closed $F \subset (\mu, \infty)$,

$$\limsup_{n\to\infty}\left[\frac{1}{n}\ln\mathbb{P}\left(\frac{1}{n}Y_n \in F\right)\right] \leq -\frac{(x_-^* - \mu)^2}{2\sigma^2},$$

where x_-^* is the left-most point of F: $x_-^* = \min[x : x \in F]$. Similarly, for all open $G \subset (\mu, \infty)$,

$$\liminf_{n\to\infty}\left[\frac{1}{n}\ln\mathbb{P}\left(\frac{1}{n}Y_n \in G\right)\right] \geq -\frac{(y_+^* - \mu)^2}{2\sigma^2},$$

where y_+^* is the right-most point of $\mathbb{R} \setminus G$: $y_+^* = \max[x : x \notin G]$. For $F = [\mu + a, \infty)$ and $G = (\mu + a, \infty)$, $x_-^* = y_+^* = \mu + a$, and we obtain that

$$\mathbb{P}(Y_n > n(\mu+a)) = \mathbb{P}(Y_n \geq n(\mu+a)) \approx e^{-na^2/2\sigma^2}.$$

In fact, a direct calculation, taking into account that $Y_n \sim \mathrm{N}(n\mu, n\sigma^2)$, yields:

$$\mathbb{P}(Y_n > n(\mu+a)) = \frac{1}{\sqrt{2\pi}\sigma}\int_{a\sqrt{n}}^{\infty}e^{-x^2/2\sigma^2}dx$$

$$= \frac{1}{\sqrt{2\pi n}a}\left[1 + O\left(\frac{1}{\sqrt{n}a}\right)\right]e^{-na^2/2\sigma^2}.$$

The last formula follows from the double inequality

$$\frac{1}{a+a^{-1}}\,e^{-a^2/2} \le \int_a^\infty e^{-x^2/2}dx \le \frac{1}{a}\,e^{-a^2/2};\qquad(1.238)$$

which is one of a series of useful bounds for Gaussian integrals.

Example 1.15.4 For a Poisson RV $Z \sim \mathrm{Po}(\lambda)$, the MGF becomes

$$\mathbb{E}e^{\theta Z} = \exp\left[\lambda\left(e^\theta - 1\right)\right], \quad \text{with } \Lambda(\theta) = \ln(\mathbb{E}e^{\theta Z}) = \lambda\left(e^\theta - 1\right).$$

As before, given x, we want to calculate θ^* such that $x = \Lambda'(\theta^*)$; as Λ increases with θ, the value x should be positive. So

$$\theta^* = \ln\left(\frac{x}{\lambda}\right), \quad \text{and } \Lambda^*(x) = \begin{cases} +\infty, & x \le 0, \\ x\left(\ln\left(\frac{x}{\lambda}\right)-1\right)+\lambda, & x > 0. \end{cases}\qquad(1.239)$$

Here $\Lambda^*(0+) = \lambda$, and $\Lambda^*(\lambda) = 0$. Thus, again, Cramér Theorem's yields that the sum $Y_n = \sum_{i=1}^n Z_i$ of IID RVs $Z_i \sim \mathrm{Po}(\lambda)$ has, for $x > \lambda$,

$$\mathbb{P}(Y_n > nx) \approx \left(\frac{x}{\lambda}\right)^{-nx} \exp\left(n(x-\lambda)\right).$$

As before, a more accurate approximation can be produced by invoking the fact that $Y_n \sim \mathrm{Po}(n\lambda)$:

$$\begin{aligned}
\mathbb{P}(Y_n > nx) &= \sum_{j>nx} \frac{(n\lambda)^j}{j!}\,e^{-n\lambda} = \frac{(n\lambda)^{[nx]+1}}{([nx]+1)!}\,e^{-n\lambda+o(nx)} \\
&= \frac{1}{\sqrt{2\pi([nx]+1)}}\left(\frac{x}{\lambda}\right)^{-[nx]-1} e^{n(x-\lambda)}\left(1+O\left(\frac{1}{nx}\right)\right).
\end{aligned}$$

Here, at the last step, we applied Stirling's formula. Note that the exponential term is suppressed by the term x^{-nx}.

Example 1.15.5 For an exponential RV $Z \sim \mathrm{Exp}(\lambda)$:

$$\mathbb{E}e^{\theta Z} = \frac{\lambda}{\lambda - \theta}, \quad \text{with } \Lambda(\theta) = \ln\mathbb{E}(e^{\theta Z}) = \ln\left(\frac{\lambda}{\lambda-\theta}\right), \quad \theta < \lambda.$$

Here, again, for $x > 0$:

$$\theta^* = \frac{x-\lambda}{x}, \quad \text{and } \Lambda^*(x) = \frac{x}{\lambda} - 1 - \ln\frac{x}{\lambda}, \quad \text{with } \Lambda^*(0+) = +\infty,\qquad(1.240)$$

and $\Lambda^*(x) = +\infty$ for $x \leq 0$. So, for $Y_n = \sum_{i=1}^{n} Z_i$, where $Z_i \sim \text{Exp}(\lambda)$, independently, Cramér's Theorem results in

$$\mathbb{P}(Y_n > nx) = \left(\frac{x}{\lambda}\right)^n \exp\left[-n\left(\frac{x}{\lambda} - 1\right) + o(n)\right].$$

> ... *that we should* ... *transform ourselves into beasts.*
> W. Shakespeare (1564–1616), English playwright and poet

The importance of Cramér Theorem's lies in serving as a starting point of a fruitful theory covering a large variety of situations. The mathematical backbone here is provided by the so-called Gärtner–Ellis Theorem. In a non-general form, but convenient for our applications, this theorem states:

Theorem 1.15.6 *Consider an arbitrary sequence of vector-valued RVs* $\mathbf{U}_n = (U_{1n}, \ldots, U_{dn})$, $n = 1, 2, \ldots$. *Assume that there exists a limit*

$$\Lambda(\boldsymbol{\theta}) = \lim_{n \to \infty} \left[\frac{1}{n} \ln\left(\mathbb{E}e^{n\langle \boldsymbol{\theta}, \mathbf{U}_n \rangle}\right)\right], \quad \boldsymbol{\theta} = (\theta_1, \ldots, \theta_d) \in \mathbb{R}^d, \tag{1.241}$$

which is finite for $\boldsymbol{\theta}$ *in a neighbourhood of the origin* $\boldsymbol{\theta} = \mathbf{0}$ *and continuous-differentiable in* $\boldsymbol{\theta}$ *everywhere where it is finite. Then, denoting, as before, by* Λ^* *the Legendre transform,*

$$\Lambda^*(\mathbf{x}) = \sup\left[\langle \boldsymbol{\theta}, \mathbf{x} \rangle - \Lambda(\boldsymbol{\theta}) : \boldsymbol{\theta} \in \mathbb{R}^d, \Lambda(\boldsymbol{\theta}) < \infty\right],$$

equations (1.232) and (1.233) hold true (where F is a closed and G is an open subset of $\mathbb{R}^d$*):*

$$\limsup_{n \to \infty} \left[\frac{1}{n} \ln\left(\mathbb{P}(\mathbf{U}_n \in F)\right)\right] \leq -\inf\left[\Lambda^*(\mathbf{x}) : \mathbf{x} \in F\right], \tag{1.242}$$

while for all open sets $G \subset \mathbb{R}$*,*

$$\liminf_{n \to \infty} \left[\frac{1}{n} \ln\left(\mathbb{P}(\mathbf{U}_n \in G)\right)\right] \geq -\inf\left[\Lambda^*(\mathbf{x}) : \mathbf{x} \in G\right]. \tag{1.243}$$

Condition (1.241) is obviously fulfilled when $\mathbf{U}_n = (\mathbf{Z}_1 + \cdots + \mathbf{Z}_n)/n$ where $\mathbf{Z}_1$, $\mathbf{Z}_2, \ldots$ are IID random vectors. In general, condition (1.241) is non-trivial, and the calculation of the functions Λ and Λ^* is challenging. In popular terminology, one says that the sequence $(\mathbf{U}_n)$ satisfies the *large deviation principle* if relations (1.242) and (1.243) hold true. In this situation Λ^* is called the *large deviation rate function*.

We will analyse the case where a random vector $\mathbf{U}_n$ has

$$U_{jn} = \frac{1}{n} \sum_{i=1}^{n} \mathbf{1}(X_i = j), \quad j = 1, \ldots, \ell, \tag{1.244}$$

where (X_n) is an irreducible and aperiodic DTMC, with finite state space $I = \{1, \ldots, \ell\}$. In other words, U_{jn} represents the portion of time spent in state $j \in I$ between times 1 and n. In general, this RV follows a complicated distribution, but we know that the (weak and strong) Laws of Large Numbers hold: $\frac{1}{n} \sum_{i=0}^{n-1} \mathbf{1}(X_i = j)$ converges to π_j, the equilibrium probability (see Theorem 1.5.2). The original proof appeared in K. Duffy and A.P. Metcalfe. "The large deviations of estimating rate functions". *J. Appl. Prob.*, **42** (2005), 267–274.

A useful statement is the *Perron–Frobenius Theorem* for non-negative matrices. It generalises Theorem 1.12.3 as follows.

Theorem 1.15.7 *Let R be an $\ell \times \ell$ matrix with non-negative entries r_{ij}.*

(a) *Suppose that the following irreducibility condition holds: for all $i, j = 1, \ldots, \ell$, there exists $s(= s(i, j))$ such that $r_{ij}^{(s)}$, the (i, j)th entry of the sth power R^s of R, is positive: $r_{ij}^{(s)} > 0$. Then the norm $||R||$ equals the spectral radius $\rho(R)$ and is always an eigenvalue of R and R^T. Denoting $||R|| = \rho(R) = \mu_0$, the algebraic and geometric dimensions of the eigenvalue μ_0 are equal to 1, and the corresponding eigenspaces of R and R^T are generated by vectors $\psi^{(0)} = \begin{pmatrix} \psi_1^{(0)} \\ \vdots \\ \psi_\ell^{(0)} \end{pmatrix}$*

and $\phi^{(0)} = \begin{pmatrix} \phi_1^{(0)} \\ \vdots \\ \phi_\ell^{(0)} \end{pmatrix}$ with strictly positive components $\psi_i^{(0)}, \phi_i^{(0)} > 0, i = 1, \ldots, \ell$.

(b) *If there exists an s such that $p_{ij}^{(s)} > 0$ for all $i, j = 1, \ldots, \ell$, then all other eigenvalues $\mu_p \neq \mu_0$ of R and R^T satisfy $|\mu_p| \leq \mu_0(1 - \delta)$, i.e. lie within a closed circle of radius $\mu_0(1 - \delta) < \mu_0$ around the origin in the complex plane $\mathbb{C}$, where $\mu_0 \delta > 0$ is the spectral gap.*

Moreover, for all vectors $\mathbf{x} = \begin{pmatrix} x_1 \\ \vdots \\ x_\ell \end{pmatrix} \in \mathbb{R}^\ell$,

$$\mathbf{x}^T R^n = \mu_0^n \left[\langle \mathbf{x}, \phi^{(0)} \rangle \psi^{(0)T} + O((1 - \delta)^n) \right]. \tag{1.245}$$

It is natural to claim that the matrix R is irreducible when it satisfies the condition that for all (i,j) there exists $s = s(i,j)$ such that $r_{ij}^{(s)} > 0$, and irreducible and aperiodic when there exists s such that for all (i,j), $r_{ij}^{(s)} > 0$.

An elegant trick allows converting an irreducible and aperiodic matrix $R = (r_{ij})$ with non-negative entries into a stochastic matrix $\widetilde{P} = (\widetilde{p}_{ij})$ (also irreducible and aperiodic): simply set

$$\widetilde{p}_{ij} = \frac{1}{\mu_0} \left(\phi_i^{(0)} \right)^{-1} r_{ij} \phi_j^{(0)}, \quad i,j = 1, \ldots, \ell. \tag{1.246}$$

Then the (unique) equilibrium distribution $\widetilde{\pi}$ for $\widetilde{P}$ will have probabilities

$$\widetilde{\pi}_i = \frac{1}{\langle \psi^{(0)}, \phi^{(0)} \rangle} \psi_i^{(0)} \phi_i^{(0)}, \quad i = 1, \ldots, \ell. \tag{1.247}$$

In our case of an irreducible and aperiodic Markov chain (X_n), with states $1, \ldots, \ell$ and transition matrix $P = (p_{ij})$, we consider a family of matrices R_θ of the form

$$R_\theta = (p_{ij} e^{\langle \theta, \mathbf{f}(j) \rangle}), \quad i,j = 1, \ldots, \ell. \tag{1.248}$$

Here, for all $j = 1, \ldots, \ell, \mathbf{f}(j) = \begin{pmatrix} f_1(j) \\ \vdots \\ f_\ell(j) \end{pmatrix} \in \mathbb{R}^\ell$ is a real-valued vector, of dimension ℓ, and so is θ. Clearly, the matrix (1.248) exhibits non-negative entries, and it is irreducible and aperiodic. Denote, as before, by $\mu_0(\theta)$ the maximal eigenvalue of R_θ and R_θ^{T}; we know that $\mu_0(\theta) = \|R_\theta\| = \|R_\theta^{\mathrm{T}}\|$, and μ_0 has multiplicity 1. It is also known that $\mu_0(\theta)$ is infinite-differentiable with respect to $\theta \in \mathbb{R}^\ell$. Let again $\psi^{(0)\mathrm{T}} = \psi_\theta^{(0)\mathrm{T}}$ be the corresponding row eigenvector of R_θ and $\phi^{(0)\mathrm{T}} = \phi_\theta^{(0)\mathrm{T}}$ be the corresponding row eigenvector of R_θ^{T}, with positive entries $\psi_j^{(0)}$, $\phi_j^{(0)} > 0$, $j = 1, \ldots, \ell$.

Lemma 1.15.8 *Consider an irreducible and aperiodic DTMC (X_n) with states $1, \ldots, \ell$ and row vector of initial probabilities $\lambda = (\lambda(j))$. Fix a collection of vectors $\mathbf{f}(j) \in \mathbb{R}^\ell$, $j = 1, \ldots, \ell$ and form random vectors $\mathbf{f}(X_n)$, $n = 0, 1, \ldots$. Then the sequence of sums*

$$\mathbf{V}_n = \frac{1}{n} \sum_{i=1}^n \mathbf{f}(X_i), \quad n = 1, 2, \ldots, \tag{1.249}$$

satisfies the large deviation principle. More precisely, for all $\theta = (\theta_1, \ldots, \theta_\ell) \in \mathbb{R}^\ell$, there exists a limit

$$\lim_{n \to \infty} \frac{1}{n} \left[\ln \left(\mathbb{E}_\lambda e^{n \langle \theta, \mathbf{V}_n \rangle} \right) \right] = \ln \left(\mu_0(\theta) \right), \tag{1.250}$$

and the Gärtner–Ellis Theorem implies the large deviation principle for $(\mathbf{V}_n)$. Here $\mathbb{E}_\lambda$ stands for the expectation with respect to the distribution of the DTMC (X_n) with initial probability row vector λ.

Consequently, the large deviation rate function is

$$\Lambda^*(\mathbf{x}) = \sup_{\theta \in \mathbb{R}^\ell} [\langle \mathbf{x}, \theta \rangle - \ln(\mu_0(\theta))], \quad \mathbf{x} \in \mathbb{R}^\ell, \tag{1.251}$$

regardless of the choice of initial distribution λ.

Proof To check (1.250), we write the expectation $\mathbb{E}_\lambda \left[e^{n\langle \theta, V_n \rangle} \right]$ as a sum over the sample values (or more briefly, samples) of $X_0, \ldots, X_n$:

$$\mathbb{E}_\lambda e^{n\langle \theta, V_n \rangle} = \sum_{j_0, \ldots, j_n} \lambda(j_0) p_{j_0 j_1} e^{\langle \theta, \mathbf{f}(j_1) \rangle} \cdots p_{j_{n-1} j_n} e^{\langle \theta, \mathbf{f}(j_n) \rangle} = \sum_{j=1}^{\ell} \left(\lambda (R_\theta)^n \right)_j$$

$$= (\lambda R_\theta^n \mathbf{1}) = \mu_0(\theta)^n \left[\langle \lambda^\mathsf{T}, \phi^{(0)} \rangle \psi^{(0)\mathsf{T}} + O((1-\delta)^n) \right]. \tag{1.252}$$

The last equality in (1.252) holds because of Theorem 1.15.7. Now, it remains to take the logarithm and divide by n, and (1.250) follows. □

In the case where the vector $\mathbf{f}(j)$ has the entry 1 in position j and all other entries 0, the task of calculating $\Lambda^*(\mathbf{x})$ is made easier by the following.

Lemma 1.15.9 *Suppose that the vector* $\mathbf{f}(j) = \begin{pmatrix} 0 \\ \vdots \\ 1 \\ \vdots \\ 0 \end{pmatrix}$, *with entry j equal to 1 and all other entries equal to 0, $j = 1, \ldots, \ell$. Then $\Lambda^*(\mathbf{x})$ equals $+\infty$ unless the vector* $\mathbf{x} = \begin{pmatrix} x_1 \\ \vdots \\ x_\ell \end{pmatrix}$ *has $x_1, \ldots, x_\ell \geq 0$ and $x_1 + \cdots + x_\ell = 1$. For $\mathbf{x}$ satisfying the above conditions,*

$$\Lambda^*(\mathbf{x}) = \sup \left[\sum_{j=1}^{\ell} x_j \ln \left(\frac{u_j}{(\mathbf{u}^\mathsf{T} P)_j} \right) : \mathbf{u} = \begin{pmatrix} u_1 \\ \vdots \\ u_\ell \end{pmatrix}, u_1, \ldots, u_\ell > 0 \right]. \tag{1.253}$$

Here P is the transition matrix of the DTMC (X_n).

Proof First, consider the simplex of stochastic vectors $\mathbb{S}_\ell \subset \mathbb{R}^\ell$ formed by the vectors $\mathbf{x} \in \mathbb{R}^\ell$ with $x_j \geq 0$, and $\sum_{j=1}^{\ell} x_j = 1$. The complement $\mathbb{R}^\ell \setminus \mathbb{S}_\ell$ is an open set.

By Lemma 1.15.8, we can use the large deviation principle for the sequence of random vectors $\mathbf{V}_n = \begin{pmatrix} V_{1n} \\ \vdots \\ V_{\ell n} \end{pmatrix}$, where $V_{jn} = \frac{1}{n} \sum_{i=1}^{n} \mathbf{1}(X_i = j)$,

$$\liminf_{n \to \infty} \frac{1}{n} \left[\ln \left(\mathbb{P}(\mathbf{V}_n \notin \mathbb{S}_\ell) \right) \right] \geq -\inf \left[\Lambda^*(\mathbf{x}) : \mathbf{x} \in \mathbb{R}^\ell \setminus \mathbb{S}_\ell \right]. \tag{1.254}$$

Observe that, for all n, the probability in the LHS of (1.254) equals 0, as $\mathbf{V}_n$ takes values from $\mathbb{S}_\ell$ only. Hence, the logarithm in the LHS is equal to $-\infty$, and so is the RHS. That is,

$$\inf \left[\Lambda^*(\mathbf{x}) : \mathbf{x} \in \mathbb{R}^\ell \setminus \mathbb{S}_\ell \right] = +\infty,$$

i.e. $\Lambda^*(\mathbf{x}) \equiv +\infty$ on $\mathbb{R}^\ell \setminus \mathbb{S}_\ell$.

Henceforth, assume that $\mathbf{x} \in \mathbb{S}_\ell$. To begin with, we check the inequality

$$\Lambda^*(\mathbf{x}) \geq \sup \left[\sum_{j=1}^{\ell} x_j \ln \left(\frac{u_j}{(\mathbf{u}^\mathsf{T} P)_j} \right) : \mathbf{u} = \begin{pmatrix} u_1 \\ \vdots \\ u_\ell \end{pmatrix}, u_1, \ldots, u_\ell > 0 \right]. \tag{1.255}$$

Given the vector $\mathbf{u} \in \mathbb{R}^\ell$ with strictly positive entries $u_1, \ldots, u_\ell$, set:

$$\theta_j^* = \ln \left(\frac{u_j}{(\mathbf{u}^\mathsf{T} P)_j} \right), \ j = 1, \ldots, \ell, \ \text{ with } \ \langle \mathbf{x}, \theta^* \rangle = \sum_{j=1}^{\ell} x_j \ln \left(\frac{u_j}{(\mathbf{u}^\mathsf{T} P)_j} \right), \tag{1.256}$$

and consider the matrix $R \ (= R_{\theta^*})$ with non-negative entries $p_{ij} e^{\theta_j^*}, i, j = 1, \ldots, \ell$. The matrix R is irreducible and aperiodic, hence Theorem 1.15.7 is applicable.

It turns out that $\mathbf{u}^\mathsf{T}$ is the eigenvector of R with eigenvalue 1:

$$\mathbf{u}^\mathsf{T} R = \mathbf{u}^\mathsf{T}, \ \text{ i.e. } \ \mathbf{u}^\mathsf{T} R^n = \mathbf{u} \text{ for all } n \geq 1. \tag{1.257}$$

In fact, for all $j = 1, \ldots, \ell$,

$$\sum_{i=1}^{\ell} u_i p_{ij} e^{\theta_j^*} = \sum_{i=1}^{\ell} u_i p_{ij} \frac{u_j}{(\mathbf{u}^\mathsf{T} P)_j} = (\mathbf{u}^\mathsf{T} P)_j \frac{u_j}{(\mathbf{u}^\mathsf{T} P)_j} = u_j.$$

If the maximal eigenvalue $\mu_0(\theta^*)$ of R is greater than 1, we obtain a contradiction: the corresponding eigenvectors $\psi^{(0)\mathsf{T}}$ and $\phi^{(0)\mathsf{T}}$ have strictly positive components, and we should get the equality

$$\mathbf{u}^\mathsf{T} R^n = \left(\mu_0(\theta^*) \right)^n \left[\langle \mathbf{u}, \phi^{(0)} \rangle \psi^{(0)\mathsf{T}} + O((1-\delta)^n) \right]. \tag{1.258}$$

Since the scalar product $\langle \mathbf{u}, \phi^{(0)} \rangle$ of two positive vectors is > 0, equation (1.258) clashes with (1.257). The only possibility is that $\mu_0(\theta^*) = 1$, in which case $\mathbf{u} = \psi^{(0)}$.

Therefore, $\ln\left(\mu_0(\theta^*)\right) = 0$, and

$$\Lambda^*(\mathbf{x}) = \sup\left[\langle\mathbf{x},\theta\rangle - \ln\mu_0(\theta) : \theta\in\mathbb{R}^\ell\right] \geq \langle\mathbf{x},\theta^*\rangle = \sum_{j=1}^\ell x_j\ln\left(\frac{u_j}{(\mathbf{u}^\mathrm{T}P)_j}\right).$$
(1.259)

Taking the supremum of the RHS of (1.259) over positive vectors $\mathbf{u}$ yields (1.255).

It remains to verify the opposite inequality to (1.255)

$$\Lambda^*(\mathbf{x}) \leq \sup\left[\sum_{j=1}^\ell x_j\ln\left(\frac{u_j}{(\mathbf{u}^\mathrm{T}P)_j}\right) : \mathbf{u} = \begin{pmatrix} u_1 \\ \vdots \\ u_\ell \end{pmatrix}, u_1,\ldots,u_\ell > 0\right].$$
(1.260)

Recall, for all $\mathbf{x}\in\mathbb{R}^\ell$, the value $\Lambda^*(\mathbf{x})$ equals $\sup\left[\langle\mathbf{x},\theta\rangle - \ln\left(\mu_0(\theta)\right) : \theta\in\mathbb{R}^\ell\right]$; see (1.251). Hence, it suffices to check that, for all $\mathbf{x},\theta\in\mathbb{R}^\ell$, there exists a row vector $\mathbf{u}^\mathrm{T}$ with positive entries $u_1,\cdots,u_\ell$ such that

$$\langle\mathbf{x},\theta\rangle - \ln\left(\mu_0(\theta)\right) \leq \sum_{j=1}^\ell x_j\ln\left[\frac{u_j}{(\mathbf{u}^\mathrm{T}P)_j}\right].$$
(1.261)

But such a vector $\mathbf{u}$ is obvious: it is the eigenvector $\psi^{(0)\mathrm{T}}$ of R_θ with the maximal eigenvalue $\mu_0(\theta)$. Indeed, for $\mathbf{u} = \psi^{(0)}$, $\mathbf{u}^\mathrm{T}R_\theta = \mu_0(\theta)\mathbf{u}^\mathrm{T}$, i.e.

$$\sum_{j=1}^\ell x_j\ln\left[\frac{(\mathbf{u}^\mathrm{T}R_\theta)_j}{u_j}\right] = \sum_{j=1}^\ell x_j\ln\left(\mu_0(\theta)\right) = \ln\left(\mu_0(\theta)\right),$$
(1.262)

as the sum $\sum_{j=1}^\ell x_j = 1$.

But the RHS of (1.262) equals

$$\langle\mathbf{x},\theta\rangle + \sum_{j=1}^\ell x_j\ln\left[\frac{(\mathbf{u}^\mathrm{T}P)_j}{u_j}\right] = \langle\mathbf{x},\theta\rangle - \sum_{j=1}^\ell x_j\ln\left[\frac{u_j}{(\mathbf{u}^\mathrm{T}P)_j}\right].$$
(1.263)

Combining (1.263) with (1.262) we achieve the equality in (1.261). This completes the proof of (1.253). $\qquad\square$

Example 1.15.10 Lemma 1.15.9 will enable us to calculate $\Lambda^*(\mathbf{x})$ explicitly in the case of a two-state irreducible aperiodic Markov chain. Here, the transition matrix P takes the form

$$P = \begin{pmatrix} 1-\alpha & \alpha \\ \beta & 1-\beta \end{pmatrix},$$
(1.264)

where $\alpha,\beta\in(0,1)$. The equilibrium distribution $\pi = (\pi_1,\pi_2)$ exhibits the form

$$\pi_1 = \frac{\beta}{\alpha+\beta}, \quad \pi_2 = \frac{\alpha}{\alpha+\beta}.$$
(1.265)

The eigenvalues of the matrix

$$R_\theta = \begin{pmatrix} (1-\alpha)e^{\theta_1} & \alpha e^{\theta_2} \\ \beta e^{\theta_1} & (1-\beta)e^{\theta_2} \end{pmatrix}, \quad \theta = \begin{pmatrix} \theta_1 \\ \theta_2 \end{pmatrix},$$

are easy to find, and the maximal one is

$$\mu_0(\theta) = \frac{1}{2}\Big((1-\alpha)e^{\theta_1} + (1-\beta)e^{\theta_2}$$

$$+\sqrt{4(\alpha+\beta-1)+((1-\alpha)e^{\theta_1}+(1-\beta)e^{\theta_2})^2}\Big). \quad (1.266)$$

To calculate the Legendre transform $\Lambda^*(\mathbf{x})$, where $\mathbf{x} = \begin{pmatrix} x_1 \\ x_2 \end{pmatrix}$, we apply Lemma 1.15.9. We can assume that $x_1, x_2 \geq 0$ and $(x_1 + x_2) = 1$. Set: $x_1 = 1 - c$, $x_2 = c$, $0 \leq c \leq 1$. The terms

$$x_1 \ln\left(\frac{u_1}{(\mathbf{u}^\mathsf{T} P)_1}\right) + x_2 \ln\left(\frac{u_2}{(\mathbf{u}^\mathsf{T} P)_2}\right)$$

from the RHS of (1.253) takes the form

$$-(1-c)\ln(1-\alpha+\beta K) - c\ln\left(1-\beta-\frac{\alpha}{K}\right), \quad \text{where } K = \frac{u_2}{u_1}. \quad (1.267)$$

Maximising in K yields, for $0 < c < 1$:

$$\Lambda^*(\mathbf{x}) = \frac{1}{2\beta(1-\beta)(1-c)} \times \Big(-\alpha\beta(1-2c)$$

$$+\sqrt{(\alpha\beta(1-2c))^2 + 4\alpha\beta c(1-\alpha)(1-\beta)(1-c)}\Big), \quad (1.268)$$

and

$$\Lambda^*(\mathbf{x}) = \begin{cases} -\ln(1-\beta), & c = 1, \\ -\ln(1-\alpha), & c = 0. \end{cases} \quad (1.269)$$

A direct, but somewhat tedious, calculation shows that $\Lambda^*(\mathbf{x}) = 0$ if and only if $x_1 = \pi_1$, $x_2 = \pi_2$.

In the particular case where $\alpha + \beta = 1$, the DTMC $(X_n, n \geq 1)$ becomes a sequence of IID RVs. In this case,

$$\Lambda(\theta) = \beta e^{\theta_1} + \alpha e^{\theta_2}, \quad (1.270)$$

and the Legendre transform $\Lambda^*(\mathbf{x})$, $\mathbf{x} = \begin{pmatrix} x_1 \\ x_2 \end{pmatrix}$, was commented on at the beginning of this section. See Figure 1.32.

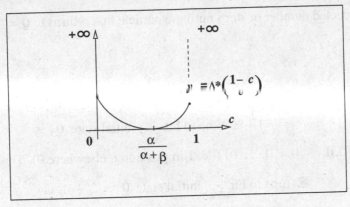

Fig. 1.32

1.16 Examination questions on discrete-time Markov chains

> *O! many a shaft, at random sent,*
> *Find mark the archer little meant,*
> *And many a word, at random spoken,*
> *May soothe or wound a heart that's broken.*
> Walter Scott (1771–1832), Scottish writer and poet

Question 1.16.1 (Markov chains, Part IB, 1991, 108D)
For a finite irreducible Markov chain, what is the relationship between the equilibrium probability distribution and the mean recurrence time of states?

A particle moves on the 2^n vertices of the hypercube $\{0,1\}^n$ in the following way: at each step the particle is equally likely to move to each of n adjacent vertices, independently of its past motion. (Two vertices are *adjacent* if the Euclidean distance between them is 1.) The initial vertex occupied by the particle is $(0,0,\ldots,0)$. Calculate the expected number of steps until the particle

(a) first returns to $(0,0,\ldots,0)$,
(b) first visits $(0,0,\ldots,0,1)$,
(c) first visits $(0,0,\ldots,0,1,1)$.

Solution Use the symmetry of the problem. The invariant probabilities are $\pi(x) = (1/2^n)$, for all $x \in \{0,1\}^n$. For a finite irreducible Markov chain, the mean recurrence time m_x to state x is

$$m_x = \frac{1}{\pi(x)}.$$

Thus the expected number of steps until the particle first returns to $\underline{0} = (0,0,\ldots,0)$ is 2^n.

Also,

$$2^n = m_{\underline{0}} = 1 + \sum_{i=1}^{n} \left(\frac{1}{n} \mathbb{E}\left(\text{time to hit } \underline{0} \mid \text{initial state } \underline{e}_i\right)\right)$$
$$= 1 + \mathbb{E}\left(\text{time to hit } \underline{e}_n \mid \text{initial state } \underline{0}\right),$$

where $\underline{e}_i = (0,0,\ldots,0,1,0,\ldots,0)$ (i.e. 1 in position i, elsewhere 0). Thus

$$\mathbb{E}\left(\text{time to hit } \underline{e}_n \mid \text{initial state } \underline{0}\right) = 2^n - 1.$$

Similarly, considering the first two steps from the initial state

$$2^n = 2 + \frac{1}{n^2}\left[n \cdot 0 + n(n-1)\mathbb{E}\left(\text{time to hit } \underline{0} \mid \text{initial state } (0,\ldots,0,1,1)\right)\right].$$

Next,

$$\mathbb{E}\left(\text{time to hit } \underline{0} \mid \text{initial state } (0,\ldots,0,1,1)\right)$$
$$= \mathbb{E}\left(\text{time to hit } (0,\ldots,0,1,1) \mid \text{initial state } \underline{0}\right).$$

Denoting the last expected value by A, we have

$$2^n = 2 + \left(\frac{n-1}{n}\right)A,$$

whence

$$A = (2^n - 2)\frac{n}{n-1}.$$

$\square$

Question 1.16.2 (Markov chains, Part IB, 1991, 307D)
Three girls A, B and C are playing table tennis. In each game, two of the girls play against each other and the third girl does not play. The winner of any given game n plays again in game $n+1$. The probability that girl x will beat girl y in any game that they play against each other is $s_x/(s_x + s_y)$ for $x, y \in \{A, B, C\}$, $x \neq y$, where s_A, s_B, s_C represent the playing strengths of the three girls.

In what proportion of games does each girl play, in the long run?

Solution Compare with Worked Example 1.1.13. The invariant probability distribution solving $\pi P = \pi$ is found from the detailed balance equations

$$\pi(x) = \frac{1}{2}\left(\frac{s_A + s_B + s_C - s_x}{s_A + s_B + s_C}\right), \quad x = A, B, C.$$

For example,

$$\pi(A)p_{AB} = \frac{1}{2}\left(\frac{s_B + s_C}{s_A + s_B + s_C}\frac{s_C}{s_B + s_C}\right) = \pi(B)p_{BA}.$$

Thus, the girl x plays in a proportion of games $1 - \pi(x)$, i.e.

$$\frac{1}{2}\left(\frac{s_A + s_B + s_C + s_x}{s_A + s_B + s_C}\right).$$

$\square$

Question 1.16.3 (Markov chains, Part IB, 1991, 408D)
Let $(Z_n, n = 0, 1, \ldots)$ be a sequence of discrete random variables. What is meant by saying that $(Z_n, n = 0, 1, \ldots)$ is a Markov chain?

The Markov chain $(X_n, n = 0, 1, \ldots)$ features initial state $X_0 = 0$ and transition probabilities

$$P(i, i+1) = p, \quad P(i, i-1) = q, \quad (i \in \mathbb{Z})$$

where $p + q = 1$. Let $Y_n = |X_n|$. Show that $(Y_n, n = 0, 1, \ldots)$ is a Markov chain and find its transition probabilities.

Solution We see $Y_0 = 0$. Clearly, if $Y_n = i$ then $Y_{n+1} = i \pm 1$, $i \geq 1$, and if $Y_n = 0$, $Y_{n+1} = 1$. Consider the conditional probability

$$\mathbb{P}(Y_{n+1} = i + 1 \mid Y_n = i, Y_{n-1} = y_{n-1}, \ldots, Y_0 = y_0 = 0),$$

with $i > 0$. If $i = 0$ this probability equals 1 and does not depend on $y_1, \ldots, y_{n-1}$. Denote by A the event in the condition

$$A = \{Y_n = i, Y_{n-1} = y_{n-1}, \ldots, Y_0 = y_0 = 0\}$$

and by $A+$ and $A-$ the intersections of A with the events $\{X_n > 0\}$ and $\{X_n < 0\}$:

$$A+ = \{Y_n = i, Y_{n-1} = y_{n-1}, \ldots, Y_0 = y_0 = 0, X_n > 0\},$$
$$A- = \{Y_n = i, Y_{n-1} = y_{n-1}, \ldots, Y_0 = y_0 = 0, X_n < 0\}.$$

Then $\mathbb{P}(Y_{n+1} = i + 1 | A)$ is represented as

$$\mathbb{P}(X_{n+1} = i+1 \mid A+)\mathbb{P}(X_n > 0 | A) + \mathbb{P}(X_{n+1} = -i-1 \mid A-)\mathbb{P}(X_n < 0 | A)$$
$$= p\mathbb{P}(X_n > 0 \mid A) + q\mathbb{P}(X_n < 0 \mid A).$$

Now, each sample path from $X_0 = 0$ to $X_n = i$ has a mirror replica from $X_0 = 0$ to $X_n = -i$, with the same $Y_0, \ldots, Y_n$. For each such pair of paths we have

$$\mathbb{P}_0^X(\text{path}) = \frac{p^i}{q^i}\,\mathbb{P}_0^X(\text{replica}).$$

This is because the probability of 'loops' occurring along the path (between the first and last time the X-chain is in a given state) is the same in both paths, and the only difference in probabilities arises when the chain moves from a state to its neighbour (in the corresponding direction) between loops.

Therefore,

$$\mathbb{P}(A) = \mathbb{P}(\text{path}) + \mathbb{P}(\text{replica}) = \mathbb{P}(\text{path})\left(1 + \frac{q^i}{p^i}\right),$$

and

$$\mathbb{P}(X_n > 0 \mid A) = \frac{\mathbb{P}(\text{path})}{\mathbb{P}(\text{path}) + \mathbb{P}(\text{replica})} = \frac{p^i}{q^i + p^i},$$

$$\mathbb{P}(X_n < 0 \mid A) = \frac{q^i}{q^i + p^i}.$$

Then the probabilities

$$\mathbb{P}(Y_{n+1} = i+1 \mid A) = \left(\frac{p^{i+1} + q^{i+1}}{p^i + q^i}\right)$$

and

$$\mathbb{P}(Y_{n+1} = i-1 \mid A) = 1 - \left(\frac{p^{i+1} + q^{i+1}}{p^i + q^i}\right)$$

do not depend on $y_1, \ldots, y_{n-1}$. Hence, $(Y_n, n = 0, 1, \ldots)$ is a Markov chain, with the above transition probabilities. $\qquad\square$

Question 1.16.4 (Markov chains, Part IB, 1992, 108B)
A frog is trapped in a bank vault, whose floor is divided into N squares numbered from 1 to N. Let X_n denote the number of the square in which the frog is sitting at time n; the sequence $(X_n)_{n \geq 0}$ may be taken to be an irreducible Markov chain with transition probabilities p_{ij} $(1 \leq i, j \leq N)$. A movement-sensitive camera photographs the vault immediately after the frog has jumped from one square to another, but not at any other time. Let Y_n denote the number of the square occupied by the frog in the nth photograph. Show that (Y_n) is also an irreducible Markov chain, and determine its transition probabilities. If $(\pi_1, \ldots, \pi_N)$ is the invariant distribution for the chain (X_n), determine that for the chain (Y_n), and also the average number of photographs taken per unit time.

Now suppose that $N = 9$, the squares are arranged in a 3×3 array

1	2	3
4	5	6
7	8	9

and that in each unit of time the frog is equally likely to remain where it is or jump to any of the adjacent squares (either orthogonally or diagonally; thus, for example, $p_{2j} = 1/6$ for all $j \le 6$). Find the equilibrium probability distribution for the chain (Y_n) and the average number of photographs taken per unit time.

Solution Observe that each $p_{ii} < 1$; otherwise the chain would have been reducible. Suppose that $Y_n = X_m$, i.e. the nth photograph is taken at time m. Then for all $j \ne i$, the conditional probability $\mathbb{P}(Y_{n+1} = j | Y_n = i, Y_{n-1} = i_{n-1}, \dots, Y_0 = i_0)$ is written as

$$\mathbb{P}(X_{m+1} = j | X_m = i) + \mathbb{P}(X_{m+1} = i, X_{m+2} = j | X_m = i) + \cdots$$
$$= p_{ij} + p_{ii}p_{ij} + p_{ii}^2 p_{ij} + \cdots = p_{ij}(1 - p_{ii})^{-1} := q_{ij},$$

and $q_{ii} := \mathbb{P}(Y_{n+1} = i | Y_n = i) = 0$. As this is independent of n and of the values of Y_l for $l < n$, we have that Y_n is a Markov chain with transition matrix

$$\begin{pmatrix} 0 & p_{12}/(1-p_{11}) & p_{13}/(1-p_{11}) & \cdots & p_{1N}/(1-p_{11}) \\ p_{21}/(1-p_{22}) & 0 & p_{23}/(1-p_{22}) & \cdots & p_{2N}/(1-p_{22}) \\ \cdots & \cdots & \cdots & \cdots & \cdots \\ p_{N1}/(1-p_{NN}) & p_{N2}/(1-p_{NN}) & p_{N3}/(1-p_{NN}) & \cdots & 0 \end{pmatrix}.$$

Furthermore, the chain Y_n is irreducible: for $j \ne i$ it is possible to move from state i to j by the same sequence of jumps as in the chain X_n (and from i to i by jumping away from i and then back).

Next,

$$\mathbb{P}(\text{photo at time } n+1 | X_n = i) = 1 - p_{ii}.$$

Hence, in equilibrium, the average number of photographs in the unit time equals

$$\mathbb{P}(\text{photo in a unit time interval}) = \sum_{i=1}^{N} \pi_i (1 - p_{ii}) = 1 - \sum_i \pi_i p_{ii}.$$

The invariant distribution for Y_n has values

$$v_i = \left[\frac{\pi_i(1 - p_{ii})}{1 - \sum_{k=1}^{N} \pi_k p_{kk}} \right], \quad 1 \le i \le N.$$

In fact,

$$\sum_i v_i q_{ij} = \sum_{i: i \ne j} \left[\frac{\pi_i(1 - p_{ii})}{1 - \sum_{k=1}^{N} \pi_k p_{kk}} \right] \frac{p_{ij}}{1 - p_{ii}}$$
$$= \sum_{i \ne j} \left[\frac{\pi_i p_{ij}}{1 - \sum_{k=1}^{N} \pi_k p_{kk}} \right] = \left[\frac{\pi_j(1 - p_{jj})}{1 - \sum_{k=1}^{N} \pi_k p_{kk}} \right] = v_j.$$

In the given example, $\pi_1 = \pi_3 = \pi_7 = \pi_9$ and $\pi_2 = \pi_4 = \pi_6 = \pi_8$ by symmetry. Then the invariance equations become

$$\pi_1 = \tfrac{1}{4}\pi_1 + 2 \times \tfrac{1}{6}\pi_2 + \tfrac{1}{9}\pi_5,$$
$$\pi_2 = 2 \times \tfrac{1}{4}\pi_1 + 3 \times \tfrac{1}{6}\pi_2 + \tfrac{1}{9}\pi_5,$$
$$\pi_5 = 4 \times \tfrac{1}{4}\pi_1 + 4 \times \tfrac{1}{6}\pi_2 + \tfrac{1}{9}\pi_5.$$

Together with $4\pi_1 + 4\pi_2 + \pi_5 = 1$ this yields

$$\pi_1 = \frac{4}{49}, \ \pi_2 = \frac{6}{49}, \ \pi_5 = \frac{9}{49}.$$

The average number of photographs per unit time is then

$$1 - \sum_{i=1}^{9} \pi_i p_{ii} = 1 - 9\frac{1}{49} = \frac{40}{49}.$$

$\square$

Question 1.16.5 (Markov chains, Part IB, 1992, 308B)
A snail crawls on an infinite fence, which may be taken to be a lattice with vertices at the points of $\mathbb{Z} \times \{0, 1, 2\}$. From a vertex of type $(n, 2)$, the snail crawls left or right (that is, to $(n-1, 2)$ or $(n+1, 2)$) with equal probability. From one of type $(n, 1)$, it crawls up to $(n, 2)$ with probability $1/2$, and in any of the other three directions with probability $1/6$. From $(n, 0)$, it necessarily moves to the left if n is even, and to the right if n is odd. Classify the states of the Markov chain corresponding to the sequence of vertices visited by the snail. If it starts at $(0, 1)$, what is the probability that it eventually reaches a positive recurrent state? What is the probability that it eventually visits $(0, 0)$?

Solution The states $(n, 2)$, $n \in \mathbb{Z}$, form a closed communicating class and are all null recurrent. For each n, the pair of states $(2n - 1, 0)$ and $(2n, 0)$ is a closed communicating class, hence all these states are positive recurrent. Finally, every state $(n, 1)$ is transient: $p_{(n,1)(n,2)} > 0$, but there is no return from level 2 to level 1.

Suppose the snail starts on level 1 and consider the probability $\mathbb{P}_{(i,1)}$(eventually reaches level 0). It equals

$$\sum_{n \geq 0} \mathbb{P}_{(i,1)} (\text{stays on level 1 for } n \text{ steps, then moves to level 0})$$

$$= \sum_{n \geq 0} \left(\frac{1}{3}\right)^n \frac{1}{6} = \frac{1}{6}\left(1 - \frac{1}{3}\right)^{-1} = \frac{1}{4}.$$

The snail will eventually visit $(0,0)$ if and only if it eventually crawls down from either $(0,1)$ or $(-1,1)$. Denote

$$h_i = \mathbb{P}(\text{eventually visits } (0,0)|\text{ currently at } (i,1)).$$

Then $h_{-1} = h_0$ (in general, by symmetry $h_{-i} = h_{i-1}$) and

$$h_0 = \tfrac{1}{6} + \tfrac{1}{6}h_0 + \tfrac{1}{6}h_1,$$
$$h_n = \tfrac{1}{6}h_{n-1} + \tfrac{1}{6}h_{n+1}, \quad n \geq 1.$$

Substituting $h_n = At^n$ in the second equation gives $t^2 - 6t + 1 = 0$, i.e. $t = 3 \pm 2\sqrt{2}$. As $h_n \leq 1$, we obtain that $h_n = A(3 - 2\sqrt{2})^n$. From the first equation $A = h_0 = 1/(2(1+\sqrt{2})) = (\sqrt{2}-1)/2$. $\quad\square$

> *Hold infinity in the palm of your hand* . . .
> W. Blake (1757–1827), English poet and artist

Question 1.16.6 (Markov chains, Part IB, 1992, 408D)
Two particles A and B move randomly, at times $t = 1, 2, \ldots$ on the set $M = \{1, \ldots, m\}$, $m \geq 3$, reflecting from the boundary and preserving the order of their positions $X_A(t)$ and $X_B(t)$ (so that $X_A(t) \leq X_B(t)$) according to the following rules.

(a) If the distance between them is greater than 1 then at the next time they proceed independently according to the following criteria.

   (i) When a particle is not in a border site, 1 or m, it jumps to one of its nearest neighbours with probability $1/2$.

   (ii) When a particle is in a border site, it either jumps to its nearest neighbour in M or remains at the same position, again with probability $1/2$.

(b) If they meet each other or are at distance one, they modify their behaviour. They both keep their positions if the jumps attempted would lead to interchanging their order, i.e. to the inequality $X_A > X_B$. (The probabilities of attempts are as before and the attempts are independent.) Otherwise they proceed according to their attempted moves.

Determine the state space of the Markov chain formed by the pair $(X_A(t), X_B(t))$ and find its invariant distribution. Is the chain reversible?

Solution The above description determines a finite irreducible Markov chain (with a single communicating class), on the state space

$$\mathscr{S} = \{(n_A, n_B) : 1 \leq n_A \leq n_B \leq m\},$$

with with total number of states $m(m+1)/2$. Indeed, $p^{(m)}_{(n_A,n_B),(n'_A,n'_B)} > 0$ for all $(n_A,n_B),(n'_A,n'_B) \in \mathscr{S}$. Thus it has a unique equilibrium distribution.

Furthermore, the transition matrix is symmetric: $p_{(n_A,n_B),(n'_A,n'_B)} = p_{(n'_A,n'_B),(n_A,n_B)}$ for all $(n_A,n_B),(n'_A,n'_B) \in \mathscr{S}$. In fact, each of these probabilities is either 0 or $1/4$ or, in the case $(n_A,n_B) = (n'_A,n'_B) = (1,1)$ or $(n_A,n_B) = (n'_A,n'_B) = (m,m)$, $1/2$. Therefore, the equilibrium distribution is equiprobable on $\mathscr{S}$, and the chain is reversible.

□

Question 1.16.7 (Markov chains, Part IB, 1993, 501K)
Consider the 7×7 transition matrix from Worked Example 1.2.8. Find all recurrent states of the associated discrete time Markov chain.

Let p^n_{ij} denote the i,j-entry in P^n. Determine the pairs (i,j) for which the limit $\lim_{n\to\infty} p^n_{ij}$ exists and find for these pairs the limiting values.

Solution See Worked Example 1.2.8. A simpler way is to use the theorem: *If P is irreducible and aperiodic and has invariant distribution π then $p^{(n)}_{ij} \to \pi_j$ for all i,j.*

In fact, for the closed class $\{1,2,6,7\}$, the invariant distribution is $\pi = \left(\dfrac{1}{5},\dfrac{3}{10},\dfrac{1}{5},\dfrac{3}{10}\right)$. This gives the limit $\lim_{n\to\infty} p^{(n)}_{ij}$ for $i,j \in \{1,2,6,7\}$; for $i=3$ it has to be divided by 2.

□

Question 1.16.8 (Markov chains, Part IB, 1993, 502K)
In the setting of Worked Example 1.3.2, suppose the particle starts at a corner A: see Figure 1.8.

(i) Find the average time taken to return to A;

(ii) the expected number of visits to the central vertex C before it returns to A.

Solution See Worked Example 1.3.2 (see also Figure 1.8). For part (i), use the theorem: *For an irreducible Markov chain with equilibrium distribution π, the expected return time to state i equals $1/\pi_i$.*

Here $\pi_A = 3/(18+6) = 1/8$, and the answer is 8.

For part (ii) use the theorem: *For an irreducible Markov chain with invariant distribution π, the expected number of visits to state j before first return to i equals π_j/π_i.*

Here $\pi_C = 1/4$, and the answer is 2.

□

[Poisson processes will be properly introduced in Chapter 2. However, the following problem could be easily solved now if the definition is familiar.]

Question 1.16.9 (Markov chains, Part IB, 1993, 503K)

Suppose that $(X_t)_{t \geq 0}$ and $(Y_t)_{t \geq 0}$ are independent Poisson processes, both of rate λ. Consider the difference $W_t = X_t - Y_t$. Let J_1 denote the first jump time of the process W_t. What is the joint distribution of J_1 and W_{J_1}?

Let M and N be positive integers. Find the probability that $(W_t)_{t \geq 0}$ hits level N before level $-M$.

Show that the average time taken by $(W_t)_{t \geq 0}$ to hit the set $\{-M, N\}$ is $(MN/2\lambda)$.

Solution From the construction, J_1 is an exponential random variable, W_{J_1} takes values ± 1 with probability $1/2$, and J_1 and W_{J_1} are independent. Also, (W_t) is a continuous-time Markov chain on $\mathbb{Z}$, the integer lattice, with holding rate 2λ, and the corresponding jump chain is the simple symmetric random walk.

Hence, if $h_i = \mathbb{P}_i(\text{hit } N \text{ before } -M)$ then

$$h_i = \frac{1}{2} h_{i-1} + \frac{1}{2} h_{i+1}, \quad h_{-M} = 0, \ h_N = 1.$$

The general form $h_i = A + Bi$ yields $h_0 = M/(M+N)$.

Similarly, if $k_i = \mathbb{E}_i(\text{hit } N \text{ or } -M)$ then

$$k_i = \frac{1}{2\lambda} + \frac{1}{2} k_{i-1} + \frac{1}{2} k_{i+1}, \quad k_{-M} = k_N = 0.$$

The general form $k_i = A + Bi + Ci^2$ yields $k_i = (MN/2\lambda) + (N-M)i/2\lambda - (i^2/2\lambda)$, and $k_0 = (MN/2\lambda)$.   □

Question 1.16.10 (Markov chains, Part IB, 1994, 501D)

Let $(X_n)_{n \geq 0}$ and $(Y_n)_{n \geq 0}$ be independent simple random walks on the integers starting from x and y, respectively. At each step $(X_n)_{n \geq 0}$ moves to the right with probability p, to the left with probability $1 - p$. For $(Y_n)_{n \geq 0}$ the corresponding probabilities are q and $1 - q$. Find, for all integers x, y and p, q in $(0, 1)$, the probability $\alpha(x, y, p, q)$ that $X_n = Y_n$ for some $n \geq 0$.

Solution As $X_n = Y_n$ if and only if $X_n - Y_n = 0$, it is convenient to consider $W_n = X_n - Y_n$, which is a random walk on $\mathbb{Z}$ with transition probabilities

$$p_{ij} = \begin{cases} p(1-q), & \text{if } j - i = 2, \\ pq + (1-p)(1-q), & \text{if } j - i = 0, \\ (1-p)q, & \text{if } j - i = -2. \end{cases}$$

Therefore, $\mathbb{P}_{x,y}(X_n = Y_n) = \mathbb{P}_{x-y}(W_n \text{ hits } 0) = 0$ for $|x - y|$ odd. For $x - y = 2k$, denote this probability by h_k. The equations become

$$h_k = (1-p)qh_{k-1} + (pq + (1-p)(1-q))h_k + p(1-q)h_{k+1},$$

or

$$h_k = \left(\frac{(1-p)q}{p+q-2pq}\right) h_{k-1} + \left(\frac{p(1-q)}{p+q-2pq}\right) h_{k+1}, \ k \in \mathbb{Z},$$

with $h_0 = 1$. Of course, we are interested in the minimal non-negative solution. First, consider $k \geq 0$. The recursion

$$(h_k, h_{k+1}) = (h_{k-1}, h_k) \left(\begin{array}{cc} 0 & -(1-p)q/(p(1-q)) \\ 1 & (p+q-2pq)/(p(1-q)) \end{array} \right)$$

has the eigenvalues 1 and $q(1-p)/(p(1-q))$, and hence admits the general solution

$$h_k = A + B \left(\frac{q(1-p)}{p(1-q)}\right)^k, \ \text{if } p \neq q$$

and

$$h_k = A + Bk, \ \text{if } p = q.$$

If $p \leq q$ then $q(1-p)/(p(1-q)) \geq 1$, and the minimal non-negative solution is with $A = 1$, $B = 0$: $h_k \equiv 1$. If $p > q$ then $q(1-p)/(p(1-q)) < 1$, and the minimal non-negative solution is with $A = 0$, $B = 1$:

$$h_k = \left(\frac{q(1-p)}{p(1-q)}\right)^k.$$

For $k \leq 0$ the answer is symmetric: if $q \leq p$ then $h_k \equiv 1$ and if $q > p$ then

$$h_k = \left(\frac{p(1-q)}{q(1-p)}\right)^k.$$

The answer for $\alpha(x, y, p, q)$ is obtained by substitution. $\qquad \square$

Question 1.16.11 (Markov chains, Part IIA, 1994, A101K)
Let $(X_n)_{n \geq 0}$, $(Y_n)_{n \geq 0}$ and $(Z_n)_{n \geq 0}$ be simple symmetric random walks on the integers. Then $V_n = (X_n, Y_n, Z_n)$ is a Markov chain. What are the transition probabilities for this chain?

Show that, with probability 1, $(V_n)_{n \geq 0}$ visits $(0, 0, 0)$ only finitely many times.

Solution The transition probabilities for (V_n) are

$$P_{(i,j,k)(l,m,n)} = \begin{array}{ll} 1/8, & \text{if } |i - l| = |j - m| = |k - n| = 1, \\ 0, & \text{otherwise.} \end{array}$$

and we aim to show that

$$\sum_{n \geq 0} P^{(n)}_{(0,0,0)(0,0,0)} < \infty.$$

As before, $p_{(0,0,0)(0,0,0)}^{(n)} = 0$ when n is odd. Furthermore, $p_{(0,0,0)(0,0,0)}^{(2n)} = \left(p_{00}^{(2n)}\right)^3$ where $p_{00}^{(2n)}$ is as in worked Example 1.6.4. Hence, $\left(p_{00}^{(2n)}\right)^3 \approx (\pi n)^{-3/2}$, and the series converges. So, (V_n) is transient; hence the answer.   □

Question 1.16.12 (Markov chains, Part IB, 1995, 501G)
Consider a discrete time Markov chain with state space $\{0,1,2,3\}$ and transition matrix

$$\begin{pmatrix} 1/3 & 2/3 & 0 & 0 \\ 7/16 & 1/4 & 5/16 & 0 \\ 1/6 & 1/6 & 1/6 & 1/2 \\ 1/2 & 1/6 & 1/6 & 1/6 \end{pmatrix}.$$

Suppose the chain starts in state 0. Determine the expected number of transitions until the chain enters state 3 for the first time.

Determine the probability that the chain enters state 2 for the *first* time on the nth transition.

Solution Set

$$k_i = \mathbb{E}_i(\text{number of steps to hit } 3).$$

The equations are

$$k_0 = 1 + \frac{1}{3}k_0 + \frac{2}{3}k_1,$$

$$k_1 = 1 + \frac{7}{16}k_0 + \frac{1}{4}k_1 + \frac{5}{16}k_2,$$

$$k_2 = 1 + \frac{1}{6}(k_0 + k_1 + k_2),$$

whence

$$k_0 = \frac{77}{6}, \quad k_1 = \frac{34}{3}, \quad k_2 = \frac{181}{30}.$$

Furthermore, from 0 the chain can only enter state 2 via 1. This suggests we consider the reduced chain, with the transition matrix

$$\overline{P} = \begin{pmatrix} 1/3 & 2/3 & 0 \\ 7/16 & 1/4 & 5/16 \\ 0 & 0 & 1 \end{pmatrix}.$$

The eigenvalues are $1, 5/6$ and $-1/4$. Hence,

$$\overline{p}_{02}^{(n)} = \mathbb{P}_0(T_2 \le n) = A + B\left(\frac{5}{6}\right)^n + C\left(\frac{-1}{4}\right)^n,$$

and

$$\mathbb{P}_0(T_2 = n) = \overline{p}_{02}^{(n)} - \overline{p}_{02}^{(n-1)} = \alpha \left(\frac{5}{6}\right)^n + \beta \left(\frac{-1}{4}\right)^n, \quad n \geq 1.$$

with boundary conditions

$$n = 1: \qquad \frac{5}{6}\alpha - \frac{1}{4}\beta = 0,$$

and

$$n = 2: \qquad \frac{25}{36}\alpha + \frac{1}{16}\beta = \frac{5}{24},$$

giving $\alpha = 3/13$, $\beta = 10/13$. The answer is

$$\frac{3}{13}\left(\frac{5}{6}\right)^n + \frac{10}{13}\left(\frac{-1}{4}\right)^n, \quad n \geq 1.$$

$\square$

Question 1.16.13 (Markov chains, Part IB, 1995, 502G)
A sequence of random convex polygons is generated by the following scheme.
At each stage, the current polygon is divided into two polygons by choosing two
distinct edges at random and joining their midpoints; one of these polygons is chosen at random as the current polygon for the next stage. For every choice, each
possible outcome is equally likely and all choices are made independently. For
$n = 0, 1, 2, \ldots$, let $X_n + 3$ be the number of edges of the polygon at the nth stage,
so that X_n takes non-negative integer values. Determine the transition matrix of the
Markov chain $\{X_n, n \geq 0\}$.

Explain why $\lim_{n \to \infty} \mathbb{P}(X_n = i)$ exists and is independent of the number of edges of
the initial polygon and determine this limit for each $i = 0, 1, 2, \ldots$.

Solution If the current polygon is a triangle, the next one will be a triangle or
a quadrilateral with probability $1/2$. If the current polygon is a quadrilateral, the
next one will be a triangle, a quadrilateral or a pentagon, each with probability $1/3$.
Similarly, from a pentagon we can obtain a triangle, a quadrilateral, a pentagon or
a hexagon, each with probability $1/4$. This suggests that the transition matrix, for

$$X_n = (\text{number of edges} - 3) = \begin{cases} 0 \\ 1 \\ 2 \\ \vdots \end{cases} ; \quad n = 1, 2, \ldots,$$

is

$$
\begin{pmatrix}
1/2 & 1/2 & 0 & 0 & 0 & \cdots \\
1/3 & 1/3 & 1/3 & 0 & \ddots & \cdots \\
1/4 & 1/4 & 1/4 & 1/4 & 0 & \cdots \\
& & & & &
\end{pmatrix}.
$$

In fact, if $X_n = m$, i.e. the nth polygon has $m+3$ edges, then the number of choices is $\binom{m+3}{2} = \left(\frac{(m+3)(m+2)}{2}\right)$. If we pick adjacent edges i and $i+1$ mod m ($m+3$ choices) then the next polygon may be a triangle (with $X_{n+1} = 0$) or an $m+4$-gon (with $X_{n+1} = m+1$). Then

$$
\begin{aligned}
\mathbb{P}(X_{n+1} = 0 \mid X_n = m) &= \mathbb{P}(X_{n+1} = m+1 \mid X_n = m) \\
&= \frac{1}{2}(m+3)\frac{2}{(m+3)(m+2)} = \frac{1}{m+2}.
\end{aligned}
$$

In general, if we pick edges i and $i+s$ mod m, with $s-1$ edges between them, (still $m+3$ choices) then the next polygon may be an $s+2$-gon (with $X_{n+1} = s-1$) or an $m+5-s$-gon (with $X_{n+1} = m+2-s$), $1 \le s \le (m+3)/2$. Then

$$
\begin{aligned}
\mathbb{P}(X_{n+1} = s-1 \mid X_n = m) &= \mathbb{P}(X_{n+1} = m+2-s \mid X_n = m) \\
&= \frac{1}{2}(m+3)\frac{2}{(m+3)(m+2)} = \frac{1}{m+2}.
\end{aligned}
$$

[For m odd, the middle value $s = (m+3)/2$ produces the same probability $1/(m+2)$ for $\mathbb{P}(X_{n+1} = s-1 \mid X_n = m)$.] This justifies the claim.

The above matrix is irreducible, as $p_{ij} = 1/(i+2) > 0$ for $j \le i+1$ and $p_{ij}^{(j-i-1)} \ge p_{ii+1}p_{i+1i+2}\cdots p_{j-1j} > 0$ for $j > i+1$. It is aperiodic as $p_{ii} > 0$. Hence, it has at most one equilibrium distribution $\pi = \{\pi_i\}$, with $\pi P = \pi$, and $\pi_j = \lim\limits_{n\to\infty} p_{ij}^{(n)}$ for all $i, j \in I$. To solve $\pi P = \pi$, write

$$
\pi_0 = \frac{1}{2}\pi_0 + \frac{1}{3}\pi_1 + \frac{1}{4}\pi_2 + \cdots = \pi_1,
$$

and

$$
\pi_i = \frac{1}{i+1}\pi_{i-1} + \frac{1}{i+2}\pi_i + \cdots = \frac{1}{i+1}\pi_{i-1} + \pi_{i+1}, \quad i \ge 1,
$$

i.e. $\pi_{i+1} = \pi_i - \pi_{i-1}/(i+1)$. This yields

$$
\pi_2 = \pi_1 - \frac{1}{2}\pi_0 = \frac{1}{2}\pi_0,
$$

and

$$
\pi_3 = \pi_2 - \frac{1}{3}\pi_1 = \frac{1}{3!}\pi_0.
$$

Make the induction hypothesis $\pi_i = \pi_0/i!$. Then

$$\pi_{i+1} = \pi_0 \left(\frac{1}{i!} - \frac{1}{i+1} \frac{1}{(i-1)!} \right) = \frac{\pi_0}{(i+1)!} (i+1-i) = \frac{\pi_0}{(i+1)!}.$$

Hence, $\pi_0 = e^{-1}$ and $\pi_i = 1/(i!e)$, i.e. π is a Poisson distribution of mean 1. $\quad\square$

Leo Tolstoy, in the novel *The Evil*, writes that most implacable conservatives are not some old retired generals but young immature enthusiasts: deprived of their own ideas and personal lifestyle, they hang on someone else's experience and defend it with bureaucratic zeal from any change.

Question 1.16.14 (Markov chains, Part IIA, 1995, A201M)
Suppose that an examiner has to mark a very large number of answers. Each answer, independently of the others, is correct with probability p and incorrect with probability $1-p$. The examiner has two marking strategies. His first strategy (which he uses initially) is to examine every answer, giving correct answers full marks and incorrect answers zero marks. His second strategy is less accurate; for each answer, he either, with probability q and independent of previous choices, marks it as before (giving full marks for a correct answer and zero for an incorrect one); or, with probability $1-q$, gives it full marks without looking at it. He changes from the first strategy to the second when he observes n consecutive correct answers, and from the second strategy to the first when he discovers an incorrect answer. Model this process as a Markov chain. Calculate the long-term proportion of questions that the examiner looks at properly, and the long-term proportion of incorrect answers which obtain full marks.

Solution The states are $0,1,\ldots,n$. In state $i < n$ the examiner observed i subsequent correct answers, and he uses strategy 1. In state n he *either* observed n correct answers and uses strategy 1 *or* he uses strategy 2 and has not observed an incorrect answer so far. This leads to the transition matrix

$$P = \begin{pmatrix} 1-p & p & 0 & 0 & 0 & \cdots & 0 & 0 \\ 1-p & 0 & p & \ddots & \ddots & \cdots & 0 & 0 \\ 1-p & 0 & 0 & p & \ddots & \cdots & 0 & 0 \\ \cdots & \cdots & \cdots & \cdots & \ddots & \ddots & \cdots & \cdots \\ 1-p & 0 & 0 & 0 & 0 & \cdots & 0 & p \\ q(1-p) & 0 & 0 & 0 & 0 & \cdots & 0 & 1-q(1-p) \end{pmatrix}.$$

The invariance equations $\pi P = \pi$ read

$$\pi_0 = (1-p) \sum_{0 \le j \le n-1} \pi_j + q(1-p)\pi_n,$$

$$\pi_i = p\pi_{i-1}, \qquad\qquad 1 \le i \le n-1,$$

$$\pi_n = p\pi_{n-1} + (1-q(1-p))\pi_n,$$

and give

$$\pi = \frac{q(1-p)}{q(1-p^n)+p^n}\left(1, p, \ldots, p^{n-1}, \frac{p^n}{q(1-p)}\right).$$

Hence, the long-term proportion of properly inspected questions equals

$$\sum_{i=0}^{n-1} \pi_i + \pi_n q = \left(\frac{q}{p^n(1-q)+q}\right),$$

and that of incorrectly marked

$$\pi_n(1-q)(1-p) = \left(\frac{p^n(1-q)(1-p)}{p^n(1-q)+q}\right).$$

□

Naturam expellas ... tamen usque recurret.
(You may drive out nature ... yet she will be constantly coming back.)
Horace (65–8 BC), Roman poet

Question 1.16.15 (Markov chains, Part IB, 1996, 501G)
An irreducible Markov chain, with states $0, 1, \ldots, n$, has transition matrix $P = (p(i,j))$. Starting at state a, it is impossible to reach state c without first visiting state b. Further,

$$p(b,c) = p$$
$$p(b,a) = 1-p$$
$$p(b,j) = 0 \quad \text{for} \quad j \ne a, c.$$

Let γ_i^j denote the expected number of steps to reach state j for the first time, starting in state i. Find an expression for γ_a^c in terms of γ_a^b.

A coin with probability p of heads is tossed repeatedly. Calculate:

(i) the expected number of tosses until a run of k consecutive heads occurs;
(ii) the long-run proportion of heads which occur within runs of k or more consecutive heads.

Solution By the strong Markov property

$$\gamma_a^c = p\gamma_a^b + (1-p)(\gamma_a^b + \gamma_a^c) + 1 = \gamma_a^b + (1-p)\gamma_a^c + 1,$$

whence

$$\gamma_a^c = \frac{1}{p}(\gamma_a^b + 1).$$

Next, (i) let the state of the Markov chain be the number of tosses since the last tail. Then

$$\gamma_0^k = \frac{1}{p}(\gamma_0^{k-1} + 1) = \frac{1}{p} + \frac{1}{p^2}(\gamma_0^{k-2} + 1)$$

$$= \frac{1}{p} + \frac{1}{p^2} + \cdots + \frac{\gamma_0^1}{p^{k-1}} = \frac{1}{p} + \frac{1}{p^2} + \cdots + \frac{1}{p^k} = \frac{1-p^k}{p^k(1-p)}.$$

Now, (ii) consider subsequent independent 'blocks' of trials formed by k or more consecutive heads followed by a tail. Within a single block, the expected number of heads equals

$$k + (pq) + 2p^2q + 3p^3q + \cdots = k + \frac{p}{1-p}.$$

The expected length of a block is

$$\gamma_0^k + 1 + \frac{p}{1-p} = 1 + \frac{1-p^k}{p^k(1-p)} + \frac{p}{1-p}.$$

Indeed, the term γ_0^k is the mean number of tosses before k consecutive heads are observed, $p/(1-p)$ is the mean number of additional heads before the first tail appears, and 1 is the contribution of this first tail. Then, by the strong Law of Large Numbers, the long-run proportion of heads occurring within a block equals

$$\left(\frac{k + p/(1-p)}{1 + (1-p^k)/(p^k(1-p)) + p/(1-p)}\right) = (k(1-p) + p)\,p^k.$$

$\square$

Question 1.16.16 (Markov chains, Part IB, 1996, 504G)
The Markov chain $(X_n)_{n\geq 0}$ has initial state $X_0 = 0$ and transition probabilities

$$P(i, i+1) = p, \qquad P(i, i-1) = q \qquad (i \in \mathbb{Z})$$

where $p + q = 1$. Let

$$M_n = \max_{0 \leq r \leq n}\{X_r\} \quad \text{and} \quad Y_n = M_n - X_n.$$

In each of the following cases determine whether or not $(Z_n)_{n\geq 0}$ is a Markov chain, and if it is, find its transition probabilities:

$$\text{(i) } Z_n = M_n; \qquad\qquad \text{(ii) } Z_n = Y_n.$$

Solution (i) $Z_n = M_n$ is not a Markov chain, since, e.g.

$$\mathbb{P}(M_4 = 2|M_3 = 1, M_2 = 0, M_1 = 0, M_0 = 0) = p$$
$$> \mathbb{P}(M_4 = 2|M_3 = 1, M_2 = 1, M_1 = 1, M_0 = 0) = p^2.$$

(ii) But $Z_n = Y_n$ is a Markov chain, since

$$\mathbb{P}(Y_n = y_n|Y_{n-1} = y_{n-1}, \ldots, Y_1 = y_1, Y_0 = 0)$$

$$= \begin{cases} q, & \text{if } y_n = y_{n-1}+1, \\ p, & \text{if } y_n = y_{n-1}-1 \text{ and } y_{n-1} > 0, \\ p, & \text{if } y_n = y_{n-1} = 0, \end{cases}$$

which does not depend on $y_1, \ldots, y_{n-2}$.   □

Question 1.16.17 (Markov chains, Part IIA, 1996, A301E
(i) A random sequence of non-negative integers $(F_n)_{n\geq 0}$ is obtained by setting $F_0 = 0$ and $F_1 = 1$ and, once $F_0, \ldots, F_n$ are known, taking F_{n+1} to be either the sum or the difference of F_n and F_{n-1}, each with probability $1/2$. Is $(F_n)_{n\geq 0}$ a Markov chain?

By considering the Markov chain $X_n = (F_{n-1}, F_n)$, find the probability that $(F_n)_{n\geq 0}$ reaches 3 before first returning to 0.

(ii) Draw enough of the flow diagram for $(X_n)_{n\geq 0}$ to observe a general pattern. Hence, using the strong Markov property, show that the hitting probability for $(1,1)$, starting from $(1,2)$, is $(3 - \sqrt{5})/2$.

Not wrung from speculation and subtleties,
but from common sense and observation.
T. Browne (1605–1682), English author and physician

Solution See Figure 1.33.

We have that (F_n) is not a Markov chain, since F_{n+1} depends on F_n and F_{n-1}; but the pair (F_{n-1}, F_n) is. The initial part of Figure 1.33 shows that the level $F_n = 3$ can be reached from $(F_0, F_1) = (0, 1)$ either at $(2, 3)$ or $(1, 3)$. To hit this level before

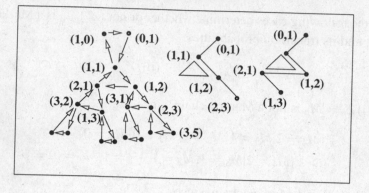

Fig. 1.33

visiting level $F_n = 0$ (i.e. $(1,0)$), we have two straight paths, supplemented with a number of adjacent triangular cycles. The first possibility gives the probability

$$1 \cdot \frac{1}{2} \cdot \frac{1}{2} \left(1 + \frac{1}{8} + \frac{1}{8^2} + \cdots \right) = \frac{2}{7},$$

and the second

$$1 \cdot \frac{1}{2} \cdot \frac{1}{2} \cdot \frac{1}{2} \left(1 + \frac{1}{8} + \frac{1}{8^2} + \cdots \right) = \frac{1}{7},$$

which adds up to $3/7$.

Observe a triangular 'pattern' emerging from Figure 1.33, with tree-like symmetries. In particular,

$$\mathbb{P}_{(1,2)} \big(\mathrm{hit}\ (1,1) \big) = \mathbb{P}_{(2,3)} \big(\mathrm{hit}\ (1,2) \big) = \mathbb{P}_{(1,3)} \big(\mathrm{hit}\ (2,1) \big) := p$$

and

$$\mathbb{P}_{(2,1)} \big(\mathrm{hit}\ (1,1) \big) = \mathbb{P}_{(3,2)} \big(\mathrm{hit}\ (2,1) \big) := p'.$$

Obviously, $0 < p, p' < 1$. Conditioning on the first jump, by the strong Markov property, we can write

$$p = \frac{1}{2} p' + \frac{1}{2} \mathbb{P}_{(2,3)} \big(\mathrm{hit}\ (1,1) \big) = \frac{1}{2} p' + \frac{1}{2} p^2,$$

and

$$p' = \frac{1}{2} + \frac{1}{2} \mathbb{P}_{(1,3)} \big(\mathrm{hit}\ (1,1) \big) = \frac{1}{2} + \frac{1}{2} pp' \ \text{whence} \ p' = \frac{1}{2-p}.$$

This yields

$$p = \frac{1}{2(2-p)} + \frac{1}{2} p^2, \ \text{i.e.} \ p^3 - 4p^2 + 4p - 1 = (p-1)(p^2 - 3p + 1) = 0.$$

The roots are $p = 1$ and $\frac{3 \pm \sqrt{5}}{2}$. We are interested in the minimal non-negative root, i.e. $p = \frac{3 - \sqrt{5}}{2}$. Consequently, $p' = \frac{2}{1 + \sqrt{5}}$.

$\square$

Question 1.16.18 (Markov chains, Part IB, 1997, 503G)
A flea hops at random on the vertices of a triangle; each hop is from the currently occupied vertex to one of the other two vertices each with probability $1/2$. Find the probability that the flea is back where it started after n hops.

A second flea also hops about on the vertices of a triangle, but this flea is twice as likely to jump clockwise as anticlockwise. What is the probability that this second flea is back where it started after n hops?

Solution The transition matrix for the first flea is

$$P = \begin{pmatrix} 0 & 1/2 & 1/2 \\ 1/2 & 0 & 1/2 \\ 1/2 & 1/2 & 0 \end{pmatrix}.$$

By symmetry, all diagonal entries $p_{ii}^{(n)}$ are equal, and all off-diagonal ones $p_{ij}^{(n)}$ are equal. Hence, consider a reduced chain, with two states, 1 and 0 (not 1), and the transition matrix

$$\widetilde{P} = \begin{pmatrix} 0 & 1 \\ 1/2 & 1/2 \end{pmatrix}.$$

Then $p_{11}^{(n)} = \widetilde{p}_{11}^{(n)}$. The eigenvalues of $\widetilde{P}$ are 1 and $-1/2$, and

$$p_{11}^{(n)} = A + B\left(\frac{-1}{2}\right)^n,$$

with $A + B = 1, A - B/2 = 0$. This yields $A = 1/3, B = 2/3$ and

$$p_{ii}^{(n)} = \frac{1}{3} + \frac{2}{3}\left(\frac{-1}{2}\right)^n.$$

For the second flea,

$$P = \begin{pmatrix} 0 & 1/3 & 2/3 \\ 2/3 & 0 & 1/3 \\ 1/3 & 2/3 & 0 \end{pmatrix},$$

with the eigenvalues

$$1, \quad \frac{-1}{2} \pm i \frac{\sqrt{3}}{6} = \frac{-1}{\sqrt{3}} e^{\pm i\pi/6}.$$

Then

$$p_{11}^{(n)} = \alpha + \left(\frac{-1}{\sqrt{3}}\right)^n \left(\beta \cos \frac{\pi n}{6} + \gamma \sin \frac{\pi n}{6}\right).$$

The constant $\alpha = 1/3$, as $p_{11}^{(n)} \to \pi_1$, and $\pi = (1/3, 1/3, 1/3)$ is the (unique) equilibrium distribution. The constants β and γ are then calculated from the equations for $n = 0$ and $n = 1$

$$\alpha + \beta + \gamma = 1, \quad \alpha - \frac{1}{\sqrt{3}}\left(\beta \frac{\sqrt{3}}{2} + \gamma \frac{1}{2}\right) = 0.$$

This gives $\alpha = 1/3, \beta = 2/3$ and $\gamma = 0$, with

$$p_{11}^{(n)} = \frac{1}{3} + \frac{2}{3}\left(\frac{-1}{\sqrt{3}}\right)^n \cos \frac{\pi n}{6}.$$

$\square$

Question 1.16.19 (Markov chains, Part IB, 1997, 504G)
In a simplified computer game, an icon moves at random on an $N \times N$ lattice (where $N \geq 2$), according to the following rules. Let (X_n, Y_n) be the position of the icon after n steps, where X_n and Y_n each has possible values $0, 1, \ldots, N-1$. At step $n+1$, each of the following four possibilities has probability $1/4$:

(a) $X_{n+1} = X_n + 1 \quad \text{(mod) } N, \quad Y_{n+1} = Y_n;$
(b) $X_{n+1} = X_n - 1 \quad \text{(mod) } N, \quad Y_{n+1} = Y_n;$
(c) $X_{n+1} = X_n, \quad Y_{n+1} = Y_n + 1, \quad \text{(mod) } N;$
(d) $X_{n+1} = X_n, \quad Y_{n+1} = Y_n - 1, \quad \text{(mod) } N;$

Calculate:

(i) the expected number of steps to return to $(1, 1)$ starting from $(1, 1)$;
(ii) the expected number of steps to reach $(1, 1)$ starting from $(0, 1)$.

Solution The chain is the random walk on the plane graph, with N^2 vertices that are integer lattice sites (i, j), $i, j = 0, 1, 2, \ldots, N-1$, and edges joining a site with its four neighbours, with the agreement that site $(i, 0)$ neighbours site $(i, N-1)$ and $(0, j)$ neighbours site $(N-1, j)$ (periodic boundary conditions). From the detailed balance equations, the uniform distribution π with

$$\pi_{(i,j)} = \frac{1}{N^2}$$

is invariant. As the graph is connected, the chain is irreducible. Hence, it is positive recurrent, and π is the only equilibrium distribution. Then:

(i) The expected number $m_{(i,j)}$ of steps to return to (i,j) starting from (i,j) is
$1/\pi_{(i,j)}$ N^2 (see Theorem 1.7.8).

ii) By symmetry and the strong Markov property,

$$m_{(i,j)} = 1 + \mathbb{E}_{(i-1,j)}(\text{time to hit } (i,j)),$$

whence

$$\mathbb{E}_{(0,1)}(\text{time to hit } (1,1)) = N^2 - 1.$$

$\square$

Question 1.16.20 (Markov chains, Part IIA, 1997, A201J)
(i) In the Figure 1.34 five numbered rat cages are joined by tunnels. The lines represent tunnels which may be traversed by a rat in either direction, and the arrows represent tunnels having flaps which prevent travel in the opposite direction. A rat runs about the maze, choosing randomly as follows: from each cage the rat chooses a random exit, and at each junction marked • the rat picks a random direction (including possibly the tunnel just traversed, unless prevented by a flap). Let X_n be the number of the nth cage visited. Obtain the transition matrix of the chain $X = (X_n)$.

(ii) Classify (giving reasons) each state of the chain in (i), using, where appropriate, the terms absorbing, aperiodic, non-essential, recurrent, null recurrent, positive recurrent.

Calculate the mean return time of each state.

If the rat begins in cage 3, find the probability that it ends in cage 1.

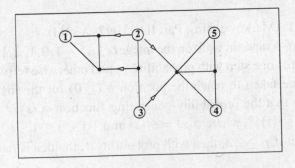

Fig. 1.34

Solution (i) The transition matrix is the 5×5 matrix

$$P = \begin{pmatrix} 1 & 0 & 0 & 0 & 0 \\ 4/9 & 1/9 & 1/9 & 1/6 & 1/6 \\ 1/6 & 1/6 & 1/6 & 1/4 & 1/4 \\ 0 & 0 & 0 & 1/2 & 1/2 \\ 0 & 0 & 0 & 1/2 & 1/2 \end{pmatrix}.$$

(ii) From Figure 1.34 and the matrix $\mathbb{P}$ we conclude that

state 1 is absorbing,
states 2 and 3 are non-essential and aperiodic,
states 4 and 5 are positive recurrent and aperiodic.

The mean return times are

$$m_1 = 1, \quad m_2 = m_3 = \infty, \quad m_4 = m_5 = 2.$$

Finally, for $h_k = \mathbb{P}_k(\text{hit } 1)$

$$h_3 = \frac{1}{6} + \frac{1}{6} h_2 + \frac{1}{6} h_3,$$

and

$$h_2 = \frac{4}{9} + \frac{1}{9} h_2 + \frac{1}{9} h_3,$$

with $h_2 = 7/13$, $h_3 = 4/13$. $\square$

A Countably Many and One Nights
(From the series *'Movies that never made it to the Big Screen'*.)

Question 1.16.21 (Markov chains, Part IIA, 1997, A301J)
(i) A flea performs a random walk on the integers $\{\ldots, -1, 0, 1, \ldots\}$. At each jump, it moves rightwards one step with probability p, and otherwise leftwards two steps. Let T_n be the time taken to reach the position n (≥ 0) for the first time, starting from zero. Show that the probability-generating function $\phi_n(s) = \sum_k s^k \mathbb{P}(T_n = k)$ satisfies $\phi_n(s) = \{\phi(s)\}^n$, where $\phi(s) = \phi_1(s)$ and $-1 < s \leq 1$.

(ii) Deduce that if $p \geq 2/3$, then with probability 1, the flea is bound to visit the integer 1.

Under this condition, find the mean number of steps before the flea visits 1.

Solution (i) One can write

$$T_n = \sum_{k=1}^{n} \tau_k$$

where τ_k is the time between the first hit of $k-1$ and the first hit of k. By the strong Markov property, $\tau_k \sim T_1$, the hitting time of 1 (starting from 0). Then, by using the space-homogeneity of transitions, for all $s \in (0,1]$,

$$\mathbb{E}_0 s^{T_n} = \mathbb{E}_0 \left(s^{\tau_1} \mathbb{E} \left(e^{\tau_2 + \ldots + \tau_n} | \tau_1 \right) \right) = \phi_1(s) \mathbb{E}_0 s^{T_{n-1}} = (\phi_1(s))^n.$$

This calculation is carried regardless of whether $\mathbb{P}_0(T_1 = \infty)$ is 0 or positive. The only difference is that in the second case the limiting value

$$\Phi := \phi_1(s) \big|_{s \to 1-} = \mathbb{P}_0(T_1 < \infty) < 1.$$

(ii) For $\phi(s) = \phi_1(s)$, conditional on the first jump,

$$\phi(s) = ps + (1-p)s\phi_3(s) = ps + (1-p)s(\phi(s))^3.$$

Then the value Φ satisfies

$$(1-p)\Phi^3 - \Phi + p = (\Phi - 1)((1-p)\Phi^2 + (1-p)\Phi - p) = 0.$$

The roots are

$$\Phi = 1, \quad \frac{-(1-p) \pm \sqrt{(1-p)^2 + 4p(1-p)}}{2(1-p)}.$$

For $p \geq 2/3$, the last two roots are ≥ 1 in the absolute value; hence $\Phi = 1$ and so $T_1 < \infty$ with probability 1.

Finally, set

$$M := \phi_1'(s) \big|_{s \to 1-} = \mathbb{E}\, T_1.$$

Then

$$(1-p)\Phi^3 + 3(1-p)\Phi^2 M - M + p = 0.$$

For $p > 2/3$, substituting $\Phi = 1$ yields

$$M = \frac{1}{3p - 2}.$$

In particular, this argument proves that $M = \infty$ for $p = 2/3$. $\qquad\qquad \square$

Question 1.16.22 (Markov chains, Part IIA, 1998, A101E)
Two independent sequences of observations, $Y_1, Y_2, \ldots$ and $Z_1, Z_2, \ldots$ come from some laboratory. They represent independent Bernoulli trials, i.e. random variables taking values 1 and 0, with unknown success probabilities

$$p = \mathbb{P}(Y_i = 1) = 1 - \mathbb{P}(Y_i = 0) \text{ and } q = \mathbb{P}(Z_i = 1) = 1 - \mathbb{P}(Z_i = 0), \ i \geq 1,$$

where $0 < p < 1$ and $0 < q < 1$. To decide whether $p > q$ or $q > p$ we use the following test: choose some positive integer M, and stop taking observations at the first time n at which either

$$Y_1 + \cdots + Y_n - (Z_1 + \cdots + Z_n) = M$$

or

$$Y_1 + \cdots + Y_n - (Z_1 + \cdots + Z_n) = -M;$$

in the former case we decide that $p > q$ and in the latter case that $p < q$. Show that, if $p > q$, the probability of an error (that is, deciding $p < q$) is $(1 + \lambda^M)^{-1}$, where $\lambda = p(1-q)q^{-1}(1-p)^{-1}$.

Solution We introduce a Markov chain $X_n = (Y_1 - Z_1) + \cdots + (Y_n - Z_n)$ on $\{-M, \ldots, M\}$, with transition probabilities

$$p_{ii+1} = p(1-q), \ p_{ii-1} = q(1-p), \ p_{ii} = pq + (1-p)(1-q),$$
$$i = -M+1, \ldots, M-1,$$

and $-M$ and M being absorbing states. We want to compute h_0 where $h_i = \mathbb{P}_i(\text{hit } -M \text{ before } M)$. The equations are

$$
\begin{aligned}
h_{-M} &= 1, \\
h_i &= q(1-p)h_{i-1} + (pq + (1-p)(1-q))h_i + p(1-q)h_{i+1}, \\
&\qquad -M < i < M, \\
h_M &= 0.
\end{aligned}
$$

For $u_i = h_{i+1} - h_i$ the equations become

$$u_i = \frac{1}{\lambda} u_{i-1} = \cdots = \left(\frac{1}{\lambda}\right)^{i+M} u_{-M},$$

where $\lambda = p(1-q)/q(1-p)$. Then

$$h_{i+1} - 1 = u_{-M} \sum_{j=0}^{i+M} \frac{1}{\lambda^j} = \left(\frac{\lambda - 1/\lambda^{i+M}}{\lambda - 1}\right) u_{-M}.$$

At $i = M - 1$:

$$-1 = \frac{1 - 1/\lambda^{2M}}{1 - 1/\lambda} u_{-M}$$

whence

$$u_{-M} = -\frac{1 - 1/\lambda}{1 - 1/\lambda^{2M}}.$$

Then

$$h_0 = 1 - \frac{1 - 1/\lambda^M}{1 - 1/\lambda^{2M}} = \frac{\lambda^M - 1}{\lambda^{2M} - 1} = \frac{1}{1 + \lambda^M}.$$

$\square$

Question 1.16.23 (Markov chains, Part IIA, 1998, A201E)
The vertices in the following graph represent light bulbs (see Figure 1.35). Only one bulb may light at any one time.

Let X_n denote the light bulb that is alight at time $n \geq 0$. The process evolves as follows. If $X_n = i$, the light bulb which will be alight at time $n + 1$ is chosen with equal probabilities among the vertices that are connected to i by an edge. For example, if C is alight at time n, one of R_1, R_4, L_1, L_4 will be alight at time $n + 1$, each with probability $1/4$. Show that $\mathbb{P}(X_n = L_1)$ converges to a limit as $n \to \infty$ and determine the value of this limit.

When $X_0 = C$, find the expected number of times R_1 will light before the first time C lights again.

Now suppose that the amount of time a light bulb lights on each occasion is not 1, as above, but an exponential random variable with mean $2/7$, and the successive lighting times are independent. If C is alight at time $t = 0$, find the expected amount of time the part of the network to the right of C is alight before C lights again.

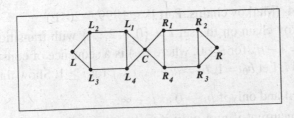

Fig. 1.35

Solution The Markov chain in question is the random walk on a finite graph. It is reversible relative to the equilibrium distribution $\pi = (\pi_i)$ where

$$\pi_i = v_i \Big/ \sum_j v_j$$

and v_i is the valence of vertex i (Theorem 1.10.5). In the example

$$\pi_C = \frac{1}{8}, \ \pi_L = \pi_R = \frac{1}{16}, \ \pi_{L_k} = \pi_{R_k} = \frac{3}{32}, \ 1 \le k \le 4.$$

Because the graph is connected, the chain is irreducible, and the cycles of length 2 and 3 (co-prime), it is aperiodic. Then, as $n \to \infty$, the n-step transition probability $p_{ij}^{(n)}$ converges to π_j and also the probability $\mathbb{P}(X_n = j)$ converges to π_j, for all states i, j and initial distribution λ (Theorem 1.9.1). Hence, $\mathbb{P}(X_n = L_1) \to 3/32$.

Next, the chain is positive recurrent. Then the expected number in question is $\gamma_{R_1}^C = \pi_{R_1}/\pi_C = 3/4$. This follows from Theorem 1.7.7, that *in an irreducible positive recurrent Markov chain with an equilibrium distribution π, $\gamma_j^k = \pi_j/\pi_k$*.

The final part of the question contains a reference to a continuous-time model. But because the holding time rates are all $2/7$, the answer is straightforward:

$$\left(\gamma_{R_1}^C + \gamma_{R_2}^C + \gamma_{R_3}^C + \gamma_{R_4}^C + \gamma_R^C\right) \frac{2}{7} = 1.$$

Alternatively, *the mean return time m_C to state C in an irreducible positive recurrent chain equals $1/\pi_C = 8$* (Theorem 1.7.8). Then the mean return time for a discrete time chain

$$\mathbb{E}_C(\text{time spent out of } C \text{ before returning to } C) = 7.$$

The figure is symmetric; hence

$$\mathbb{E}_C(\text{time spent to the right of } C \text{ before returning to } C) = \frac{7}{2}.$$

which gives the above answer 1 for the continuous-time chain. $\quad\square$

Question 1.16.24 (Markov chains, Part IIA, 1999, A201E)
Consider a Markov chain on the set $I = \{0, 1, 2, \ldots\}$ with transition probabilities $p_{i,i+1} = a_i$, $p_{i,0} = 1 - a_i$, for $i \ge 0$, where (a_i) is a sequence of constants satisfying $0 < a_i < 1$ for all i. Let $b_0 = 1$, $b_i = a_0 a_1 \ldots a_{i-1}$ for $i \ge 1$. Show that the chain is:

(a) recurrent if and only if $b_i \to 0$ as $i \to \infty$;
(b) positive recurrent if and only if $\sum_i b_i < \infty$, and write down the invariant distribution if the latter condition holds.

Solution As the chain is irreducible, we can focus on a fixed state, say 0. Let T be the first passage time for state 0: $T = \inf\{n \geq 1 : X_n = 0\}$. First, we have to analyse $\mathbb{P}_0(T < \infty)$. Write

$$\mathbb{P}_0(T > n) = a_0 a_1 \ldots a_{n-1} = b_n,$$

and

$$\mathbb{P}_0(T < \infty) = \lim_{n \to \infty} \mathbb{P}_0(T \leq n) = 1 - \lim_{n \to \infty} \mathbb{P}_0(T > n) = 1, \text{ iff } b_n \to 0.$$

Hence, 0 is a recurrent state if and only if $b_n \to 0$, otherwise it is transient.

Next,

$$\mathbb{E}_0 T = \sum_{n \geq 0} \mathbb{P}_0(T > n) = \sum_{n \geq 0} b_n,$$

and the chain is positive recurrent if and only if $\sum b_n < \infty$.

The invariance equations are

$$\pi_0 = \sum_{l \geq 0} \pi_k (1 - a_k), \ \pi_i = \pi_{i-1} a_{i-1}, \ i \geq 1,$$

whence

$$\pi_n = \pi_0 b_n \text{ and } \pi_0 = \left(\sum_{i \geq 0} b_n \right)^{-1}.$$

□

Question 1.16.25 (Markov chains, Part IIA, 1999, A401E)
Give some examples of reversible Markov chains.

Solution Not every chain is reversible: a straightforward example is where the matrix P is of finite size ≥ 3 and double-stochastic (with the sums along both the rows and columns equal to 1) and π the uniform distribution, but $P \neq P^{\mathrm{T}}$. Namely

$$P = \begin{pmatrix} 1/4 & 1/4 & 1/2 \\ 0 & 1/2 & 1/2 \\ 3/4 & 1/4 & 0 \end{pmatrix}.$$

On the other hand, there are entire classes of Markov chains which are reversible:

(a) All 2×2 chains.
(b) Finite chains with $P^{\mathrm{T}} = P$: here P is double-stochastic, the uniform distribution π is invariant and π and P are in detailed balance.
(c) Random walks on (finite) non-oriented graphs, with

$$p_{ij} = \frac{1}{v_i} \mathbf{1}(i \text{ and } j \text{ are connected})$$

and

$$\pi_j = v_j \bigg/ \sum_i v_i$$

where v_i is the valency of vertex i (the number of incident edges).

(d) Birth–death processes, with jump probabilities $p_{ii+1} = p_i$, $p_{ii-1} = q_i = 1 - p_i$ and the sum

$$B = 1 + \frac{p_0}{q_1} + \frac{p_0\,p_1}{q_1\,q_2} + \frac{p_0\,p_1\,p_2}{q_1\,q_2\,q_3} + \cdots < \infty.$$

Here $\pi_0 = B^{-1}$,

$$\pi_i = B^{-1} \frac{p_0}{q_1} \cdots \frac{p_i}{q_{i+1}}, \quad i \geq 1,$$

and π is in detailed balance with the transition matrix.

□

Question 1.16.26 (Markov chains, Part IIA, 2000, A101E and Part IIB, 2000, B101E)

A particle performs a random walk on $\{0, 1, 2, \ldots\}$. If it is at position $m \geq 1$ at time n, it moves (at time $n+1$) one step to the left with probability p, one step to the right with probability r, or it remains at the same place with probability $q = 1 - p - r$. If it is at position 0, it moves one step rightwards with probability r, or otherwise remains at 0. Show that the chain is positive recurrent when $p > r$, and derive the invariant distribution in this case. Determine for which values of p, q, r the chain is null recurrent or transient.

Solution The chain is transient when $0 < p < r$, null recurrent when $p = r > 0$, and positive recurrent when $p > r > 0$ (it is trivially positive recurrent when $r = 0$ and transient when $p = 0, r > 0$). For positive recurrence, the detailed balance equations

$$\lambda_i r = \lambda_{i+1}\, p$$

determine the invariant measure $\lambda_i = (r/p)^i$ which is normalizable if and only if $r < p$, with

$$\pi_i = \left(\frac{r}{p}\right)^i \left(1 - \frac{r}{p}\right).$$

To determine null recurrence and transience, consider the hitting probability $h_i = \mathbb{P}(\text{hit } 0 | X_0 = i)$. The equations are

$$h_0 = 1, \quad (p+r)h_i = rh_{i+1} + ph_{i-1}, \quad i \geq 1,$$

and have the minimal non-negative solution

$$h_i = \left(\frac{p}{r}\right)^i, \quad i \geq 0, \quad \text{if } p < r,$$

and

$$h_i \equiv 1, \quad i \geq 0, \quad \text{if } p \geq r > 0.$$

But

$$\mathbb{P}_0(T_0 < \infty) = (1-r)\,\mathbb{P}_0(T_0 < \infty) + rh_1, \quad \text{i.e. } \mathbb{P}_0(T_0 < \infty) = h_1,$$

which gives the answer. $\square$

Question 1.16.27 (Markov chains, Part IIA, 2000, A201E, first half)
A black dog and a white dog are inseparable companions, and are the hosts for N fleas in all. At each epoch of time exactly one flea jumps to the other dog; the jumping flea is chosen uniformly at random from N available, each choice being independent of all earlier choices. Let X_n be the number of fleas on the black dog after n epoch of time.

Show that $X = \{X_n : n \geq 0\}$ is a Markov chain and write down its transition probabilities. Show that the invariant (or 'stationary') distribution is given by

$$\pi_k = \binom{N}{k}\left(\frac{1}{2}\right)^N, \quad k = 0, 1, 2, \ldots, N.$$

Solution We have: $p_{i,i+1} = (N-i)/N$, $p_{i,i-1} = i/N$. This is a Markov chain, owing to independence of choice for the jumping flea of previous events. The detailed balance equations $\pi_i p_{i,i+1} = \pi_{i+1} p_{i+1,i}$ are solved by $\pi_{i+1} = \pi_i (N-i)/(i+1)$, which yields

$$\pi_N = \frac{1 \times 2 \times \cdots \times (N-1) \times N}{N \times (N-1) \times \cdots \times 2 \times 1}\,\pi_0 = \frac{N!}{N!}\,\pi_0 = \pi_0.$$

So for some $0 < M < N$

$$\pi_M = \frac{N-(M-1)}{M}\,\pi_{M-1} = \frac{(N-(M-1)) \times (N-(M-2)) \times \cdots \times N}{M \times (M-1) \times \cdots \times (M-1)}\,\pi_0$$

$$= \frac{N!}{(N-M)!M!}\,\pi_0 = \binom{N}{M}\,\pi_0.$$

The condition $\sum_i \pi_i = \pi_0 \left[\sum_i \binom{N}{i}\right] = 1$ implies $\pi_0 = 2^{-N}$, and

$$\pi_i = 2^{-N} \binom{N}{i}, \quad i = 0, 1, \ldots, N.$$

That is, the invariant distribution is binomial: Bin $(N, 1/2)$. $\square$

A Delicate Detailed Balance
(From the series *'Movies that never made it to the Big Screen'.*)

Question 1.16.28 (Markov chains, Part IIA, 2001, A101D and Part IIB, 2001, B101D)
A finite connected graph G has vertex set V and edge set E, and has neither loops nor multiple edges. A particle performs a random walk on V, moving at each step to a randomly chosen neighbour of the current position, each such neighbour being picked with equal probability, independently of all previous moves. Show that the unique invariant distribution is given by $\pi_v = d_v/(2|E|)$ where d_v is the degree of vertex v.

A rook performs a random walk on a chessboard; at each step, it is equally likely to make any of the moves which are legal for a rook. What is the mean recurrence time of a corner square? (You should give a clear statement of any general theorem used.) [*A chessboard is an* 8×8 *square grid. A legal move is one of any length parallel to the axes.*]

Solution As G is connected, the chain is finite and irreducible, i.e. positive recurrent. Hence, it has a unique invariant (equilibrium) distribution. The distribution $\pi_v = d_v/(2|E|)$ satisfies the detailed balance equations, as $d_v d_{v,v'}/d_v = d_{v'} d_{v'v}/d_{v'}$ for all $v, v' \in V$. Hence, it is invariant, and the chain is reversible. So, π_v is the unique invariant distribution.

In the last part, we use the theorem that *in a positive recurrent chain, with the invariant distribution π, $1/\pi_k$ gives m_k, the mean return time to k.* For the rook at a chessboard corner, or any other of the 64 squares, $d_v = 2 \times 7 = 14, v \in V$. Hence, $|E| = 64 \times 14/2$,

$$\pi_{\text{corner}} = \frac{14}{64 \cdot 14} = \frac{1}{64}, \quad m_{\text{corner}} = 64.$$

$\square$

2

Continuous-time Markov chains

2.1 Q-matrices and transition matrices

> *Markov processes specialists like to do it with chains.*
> (From the series 'How they do it'.)

Definition 2.1.1 A Q-*matrix* on a finite or countable state space I is a real-valued matrix $(q_{ij}, i, j \in I)$ with:

   non-positive diagonal entries $q_{ii} \leq 0, i \in I$,
   non-negative off-diagonal entries $q_{ij} \geq 0, i \neq j, i, j \in I$,
   the row zero-sum condition: $-q_{ii} = \sum_{j \in I: j \neq i} q_{ij}$, i.e. $\sum_j q_{ij} = 0$ for all $i \in I$.

For $i \neq j$, the value q_{ij} represents the *jump*, or *transition* rate from state i to j. The value $-q_{ii} = \sum_{j: j \neq i} q_{ij}$ is denoted by q_i (we will see that it represents the *total jump*, or *exit* rate from state i). A Q-matrix will be denoted by Q (a common abuse of notation). As in Chapter 1, we will denote by $\mathbf{I}$ the unit matrix.

In a general theory of countable continuous-time Markov chains, the row zero-sum condition $\sum_{j \in I: j \neq i} q_{ij} = -q_{ii}$ presumes that the series $\sum_{j: j \neq i} q_{ij} < \infty$. However, a substantial part of the theory can be developed when the equality in this condition is relaxed to the upper bound $\sum_j q_{ij} \leq 0$, i.e. $q_i \geq \sum_{j: j \neq i} q_{ij}$ for all $i \in I$. Then a Q-matrix satisfying the row zero-sum condition is called *conservative*; we will omit this term in the present volume, as we will not consider non-conservative Q-matrices.

As before, a Q-matrix generates a diagram, with an arrow $i \to j$ if and only if $q_{ij} > 0$; see Figure 2.1a.

185

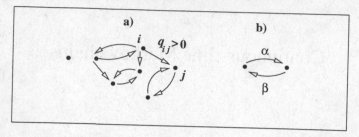

Fig. 2.1

Example 2.1.2 The zero matrix is a Q-matrix:

$$Q = \begin{pmatrix} 0 & \cdots & 0 \\ 0 & \cdots & 0 \\ & \cdots & \\ 0 & \cdots & 0 \end{pmatrix}.$$

The corresponding diagram has no arrows (i.e. consists of isolated points).

Example 2.1.3 A general 2×2 Q-matrix has the form

$$\begin{pmatrix} -\alpha & \alpha \\ \beta & -\beta \end{pmatrix}, \quad \text{with } \alpha, \beta \geq 0;$$

see Figure 2.1b.

In some important examples in this chapter, the Q-matrix will indeed be infinite. However, for a while we focus on finite matrices. An interesting matrix function is the *matrix exponent* $e^{tQ} = \exp(tQ)$:

$$e^{tQ} = \mathbf{I} + \sum_{k \geq 1} \frac{(tQ)^k}{k!} = \sum_{k \geq 0} \frac{(tQ)^k}{k!}, \quad \text{i.e. } \left(e^{tQ} \right)_{ij} = \sum_{k \geq 0} \frac{t^k (Q^k)_{ij}}{k!}. \qquad (2.1)$$

Here, we use a standard agreement that for $k = 0$, $Q^0 = \mathbf{I}$, the unit matrix, and $0! = 1$.

For a finite matrix Q, the parameter t in (2.1) can be any real number, although in applications to continuous-time Markov chains we will assume that $t \geq 0$. Why does series (2.1) converge? This holds because the matrix norm

$$\begin{aligned} \|e^{tQ}\| &= \left\| \sum_{k \geq 0} \frac{(tQ)^k}{k!} \right\| \leq \sum_{k \geq 0} \frac{\|(tQ)^k\|}{k!} \\ &\leq \sum_{k \geq 0} \frac{|t|^k \|Q\|^k}{k!} = \exp\left(|t| \, \|Q\| \right) < \infty. \end{aligned}$$

Basic facts about e^{tQ} which follow immediately from the definition are:

(i) $e^{tQ}e^{sQ} = e^{sQ}e^{tQ} = e^{(t+s)Q}$, $s, t \in \mathbb{R}$ (in particular, $e^{-tQ} = (e^{tQ})^{-1}$).

This property is an extension of the standard multiplicative property of a scalar exponential function $x \mapsto e^{xa}$: $e^{(x+y)a} = e^{xa}e^{ya} = e^{ya}e^{xa}$.

(ii) e^{tQ} depends continuously (in fact, differentiably) on $t \in \mathbb{R}$, with $e^{0 \cdot Q} = \mathbf{I}$. More precisely,

$$\frac{d}{dt}e^{tQ} = Qe^{tQ} = e^{tQ}Q, \ t \in \mathbb{R}$$

i.e.

$$\frac{d}{dt}\left(e^{tQ}\right)_{ij} = \left(Qe^{tQ}\right)_{ij} = \left(e^{tQ}Q\right)_{ij}.$$

Furthermore, for all $n = 0, 1, \ldots,$

$$\frac{d^n}{dt^n}e^{tQ} = Q^n e^{tQ} = e^{tQ}Q^n, \ t \in \mathbb{R}.$$

This is again an extension of the standard equation for a scalar exponential:
$$\frac{d}{dx}e^{xa} = ae^{xa} = e^{xa}a, \ x \in \mathbb{R}.$$

(iii) $\det e^{tQ} = e^{t(\operatorname{tr} Q)}$, $t \in \mathbb{R}$ (this is a bit more tricky: the proof is given after Remark 2.1.12).

We spell out properties (i) and (ii), for $t, s \geq 0$ now.

Theorem 2.1.4 *Let Q be a finite Q-matrix. The family of matrices*

$$P(t) = e^{tQ}, \ t \geq 0, \tag{2.2}$$

satisfies the following properties.

(a) *The semigroup property*

$$P(t+s) = P(s)P(t), \ s, t \geq 0. \tag{2.3}$$

(b) *$P(t)$ is the only solution to*

$$\frac{d}{dt}P(t) = P(t)Q, t \geq 0, \quad \text{the forward equation,}$$

$$\tag{2.4}$$

$$\frac{d}{dt}P(t) = QP(t), t \geq 0, \quad \text{the backward equation,}$$

with $P(0) = \mathbf{I}$.

(c) For all $n = 0, 1, 2, \ldots$,

$$\frac{d^n}{dt^n} P(t) = P(t)Q^n = Q^n P(t); \text{ in particular, } \frac{d^n}{dt^n} P(t)\Big|_{t=0} = Q^n. \quad (2.5)$$

Proof Property (a) follows directly from the definition of the matrix exponent, if we use the binomial expansion

$$((t+s)Q)^k = (t+s)^k Q^k = \sum_{l=0}^{k} \binom{k}{l} t^l s^{k-l} Q^k = \sum_{l=0}^{k} \binom{k}{l} (tQ)^l (sQ)^{k-l}.$$

Property (c) can be proven by iterating (2.4):

$$\frac{d^n}{dt^n} P(t) = \frac{d^{n-1}}{dt^{n-1}} \left(\frac{d}{dt} P(t) \right) = \left(\frac{d^{n-1}}{dt^{n-1}} P(t) \right) Q = \cdots = P(t)Q^n = Q^n P(t).$$

Therefore, we only prove assertion (b). Write

$$\begin{aligned}
\frac{d}{dt} P(t) &= \frac{d}{dt} \sum_{k \geq 0} \frac{t^k Q^k}{k!} = \sum_{k \geq 1} \frac{t^{k-1} Q^k}{(k-1)!} \\
&= Q \sum_{k \geq 1} \frac{(tQ)^{k-1}}{(k-1)!} = QP(t) = P(t)Q. \quad (2.6)
\end{aligned}$$

Note that (2.6) also holds for e^{-tQ}: thus $\frac{d}{dt} e^{-tQ} = -e^{-tQ}Q = -Qe^{-tQ}$.

The initial condition $P(0) = I$ is also verified straight away: for $t = 0$ all terms in (2.1) vanish, except for $k = 0$.

To show uniqueness, let $M(t)$ be any solution to

$$\frac{d}{dt} M(t) = M(t)Q, \text{ with } M(0) = I.$$

Then take

$$\begin{aligned}
\frac{d}{dt} \left(M(t) e^{-tQ} \right) &= \left(\frac{d}{dt} M(t) \right) e^{-tQ} + M(t) \frac{d}{dt} e^{-tQ} \\
&= M(t)Q e^{-tQ} + M(t) \left(-Q e^{-tQ} \right) = 0. \quad (2.7)
\end{aligned}$$

Hence, $M(t)e^{-tQ}$ does not change with t, and it is I at $t = 0$. Therefore, $M(t)e^{-tQ} \equiv I$ and

$$M(t) = e^{tQ}.$$

The same argument works with $\frac{d}{dt} M(t) = QM(t)$. $\square$

Remark 2.1.5 For a finite Q-matrix, (2.3) holds for all $t, s \in \mathbb{R}$ (and is in fact, a group property). Also note that in the proof of uniqueness of the solution to the forward and backward equation, we used e^{-tQ}, i.e. inverting the sign of tQ; see (2.7). However, we need a non-negative t in the following important result.

Theorem 2.1.6 *For a finite matrix* Q, $P(t) = e^{tQ}$ *is a stochastic matrix for all* $t \geq 0$ *if and only if Q is a Q-matrix.*

That is, $P(t)$ contains only strictly positive entries and all rows sum to 1:

$$p_{ij} \geq 0, \quad \text{and} \quad \sum_j p_{ij}(t) = 1. \tag{2.8}$$

Proof Start with the if part of the theorem. That is, suppose that Q is a Q-matrix. First, assume that t is small and positive. Then

$$P(t) = I + tQ + o(t), \quad \text{i.e. } p_{ij}(t) = \delta_{ij} + tq_{ij} + o(t).$$

Hence, for small $t > 0$,

$$p_{ii}(t) > 0, \quad \text{and} \quad p_{ij}(t) > 0 \text{ for } i \neq j \text{ whenever } q_{ij} > 0.$$

Next, if $q_{ij} = 0$ then

$$p_{ij}(t) = \delta_{ij} + \frac{1}{2} t^2 q_{ij}^{(2)} + o(t^2),$$

where $q_{ij}^{(2)}$ is the (i, j)th entry of Q^2. Observe that

$$q_{ij}^{(2)} = \sum_k q_{ik} q_{kj} \geq 0,$$

as the sum $\sum_k q_{ik} q_{kj}$ does not have negative summands (they only could come from $k = i$ or $k = j$, but then $q_{ij} = 0$ cancels them out). So, if $q_{ij} = 0$ then, for small $t > 0$,

$$p_{ij}(t) > 0 \quad \text{whenever } q_{ij}^{(2)} > 0.$$

Continuing in the same fashion, it is not difficult to deduce that, for small $t > 0$,

$$p_{ij}(t) > 0 \quad \text{whenever, for some } n, \text{ entry } q_{ij}^{(n)} \text{ of matrix } Q^n \text{ is} > 0.$$

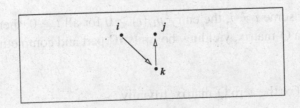

Fig. 2.2

The condition that entry $q_{ij}^{(n)} > 0$ for a given n means that, in the arrow diagram, there exists a directed path $i = k_0 \to k_1 \to \cdots \to k_{n-1} \to k_n = j$ from i to j, of length n. In other words, state i communicates with j.

What if $q_{ij}^{(n)} \equiv 0$ for all n? Then i does not communicate with j, and $p_{ij}(t) \equiv 0$. In any case, we have that, for small $t > 0$,

$$p_{ij}(t) \geq 0, \text{ for all states } i, j.$$

Finally, as Q has 0 row sums, so does the matrix Q^n for each $n \geq 1$:

$$\sum_j q_{ij}^{(n)} = \sum_{j,l} q_{il}^{(n-1)} q_{lj} = \sum_l q_{il}^{(n-1)} \left(\sum_j q_{lj} \right) = 0. \tag{2.9}$$

Then $P(t)$ has row sums 1, for all $t \in \mathbb{R}$:

$$P(t) = \underset{\text{row sum 1}}{\mathbf{I}} + \underset{\text{row sum 0}}{\sum_{k \geq 1} \frac{t^k}{k!} Q^k}.$$

Thus, for $t > 0$ small, we have established that $P(t)$ is a stochastic matrix. This remains true for a general $t > 0$: by the semigroup property we can write

$$P(t) = P\left(\frac{t}{n}\right) \cdots P\left(\frac{t}{n}\right) (n \text{ times}) = \left[P\left(\frac{t}{n}\right) \right]^n.$$

Then we use the fact that a power of stochastic matrix will be a stochastic matrix too. This finishes the proof of the 'if' part of Theorem 2.1.6.

Conversely, if $P(t)$ has row sums 1 for all $t \geq 0$, then

$$0 = \frac{\mathrm{d}}{\mathrm{d}t} \sum_j p_{ij}(t) = \sum_j \frac{\mathrm{d}}{\mathrm{d}t} p_{ij}(t) = \sum_j (QP(t))_{ij}.$$

At $t = 0$, with $P(0) = \mathbf{I}$, this yields

$$\sum_j q_{ij} = 0.$$

Similarly, if, for some $i \neq j$, the entry $p_{ij}(t) \geq 0$ for all $t \geq 0$, then $q_{ij} \geq 0$. This means that Q is a Q-matrix, yielding the 'only if' part and completing the proof of Theorem 2.1.6.

$\square$

Example 2.1.7 For the zero Q-matrix, trivially,

$$P(t) \equiv \mathbf{I}.$$

Example 2.1.8 Now consider a general 2×2 Q-matrix

$$\begin{pmatrix} -\alpha & \alpha \\ \beta & -\beta \end{pmatrix}, \quad \alpha, \beta \geq 0. \tag{2.10}$$

The eigenvalues $\kappa_{1,2}$ of the matrix are the roots of

$$\det \begin{pmatrix} -\kappa - \alpha & \alpha \\ \beta & -\kappa - \beta \end{pmatrix} = (\kappa + \alpha)(\kappa + \beta) - \alpha\beta = \kappa(\alpha + \beta + \kappa) = 0,$$

i.e.

$$\kappa_1 = 0, \quad \kappa_2 = -(\alpha + \beta).$$

The matrix is diagonalisable: $Q = UDU^{-1}$, where

$$D = \begin{pmatrix} 0 & 0 \\ 0 & -\alpha - \beta \end{pmatrix}. \tag{2.11}$$

So,

$$P(t) = \sum_{k \geq 0} \frac{t^k}{k!} Q^k = \sum_{k \geq 0} \frac{t^k}{k!} UD^kU^{-1} = U \sum_{k \geq 0} \frac{t^k}{k!} D^kU^{-1} = Ue^{tD}U^{-1}$$

$$= U \begin{pmatrix} 1 & 0 \\ 0 & e^{-t(\alpha+\beta)} \end{pmatrix} U^{-1}.$$

If $\alpha = \beta = 0$, then $P(t) \equiv I$. Otherwise its each entry follows the form

$$p_{ij}(t) = A + Be^{-(\alpha+\beta)t},$$

with

$$p_{ij}(0) = A + B = \delta_{ij}$$

and

$$\left. \frac{d}{dt} p_{ij}(t) \right|_{t=0} = -(\alpha + \beta)B = q_{ij}.$$

For instance, for the top left entry,

$$A + B = 1, \quad -(\alpha + \beta)B = -\alpha,$$

and

$$A = \frac{\beta}{\alpha + \beta}, \quad B = \frac{\alpha}{\alpha + \beta}.$$

In fact, the whole matrix is given by

$$P(t) = \begin{pmatrix} \dfrac{\beta}{\alpha + \beta} + \dfrac{\alpha}{\alpha + \beta} e^{-(\alpha+\beta)t} & \dfrac{\alpha}{\alpha + \beta} - \dfrac{\alpha}{\alpha + \beta} e^{-(\alpha+\beta)t} \\ \dfrac{\beta}{\alpha + \beta} - \dfrac{\beta}{\alpha + \beta} e^{-(\alpha+\beta)t} & \dfrac{\alpha}{\alpha + \beta} + \dfrac{\beta}{\alpha + \beta} e^{-(\alpha+\beta)t} \end{pmatrix}; \tag{2.12}$$

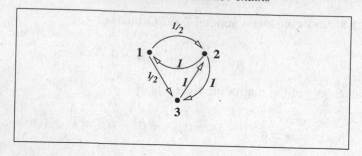

Fig. 2.3

as $t \to \infty$ it converges to

$$\begin{pmatrix} \dfrac{\beta}{\alpha+\beta} & \dfrac{\alpha}{\alpha+\beta} \\[2mm] \dfrac{\beta}{\alpha+\beta} & \dfrac{\alpha}{\alpha+\beta} \end{pmatrix}.$$

Example 2.1.9 Consider a 3×3 Q-matrix

$$Q = \begin{pmatrix} -1 & 1/2 & 1/2 \\ 1 & -2 & 1 \\ 0 & 1 & -1 \end{pmatrix}, \text{ with } Q^2 = \begin{pmatrix} 3/2 & -1 & -1/2 \\ -3 & 11/2 & -5/2 \\ 1 & -3 & 2 \end{pmatrix}.$$

The characteristic equation is

$$\det \begin{pmatrix} -1-\kappa & 1/2 & 1/2 \\ 1 & -2-\kappa & 1 \\ 0 & 1 & -1-\kappa \end{pmatrix}$$

$$= -(\kappa+1)^2(\kappa+2) + \frac{1}{2} + (1+\kappa)\frac{1}{2} + 1 + \kappa$$

$$= (\kappa+1)\left(-(\kappa+1)(\kappa+2) + \frac{1}{2} + 1\right) + \frac{1}{2}$$

$$= (\kappa+1)\left(-\kappa^2 - 3\kappa - 2 + \frac{3}{2}\right) + \frac{1}{2}$$

$$= \kappa\left(-\kappa^2 - 4\kappa - \frac{7}{2}\right) = 0,$$

with the eigenvalues

$$\kappa_1 = 0, \quad \kappa_\pm = -2 \pm \frac{1}{\sqrt{2}} < 0.$$

Then

$$p_{ij}(t) = A + Be^{-(2-1/\sqrt{2})t} + Ce^{-(2+1/\sqrt{2})t}, \quad i,j = 1,2,3, \ t \geq 0,$$

where the constants A, B, C depend on i, j and obey

$$A + B + C = \delta_{ij}, \quad \text{as} \quad P(0) = I,$$

$$-B\left(2 - \frac{1}{\sqrt{2}}\right) - C\left(2 + \frac{1}{\sqrt{2}}\right) = q_{ij}, \quad \text{as} \quad \frac{d}{dt}P(0) = Q,$$

$$B\left(2 - \frac{1}{\sqrt{2}}\right)^2 + C\left(2 + \frac{1}{\sqrt{2}}\right)^2 = q_{ij}^{(2)}, \quad \text{as} \quad \frac{d^2}{dt^2}P(0) = Q^{(2)}.$$

For instance, for $p_{11}(t)$,

$$A = \frac{2}{7}, \quad B = \frac{5 + 3\sqrt{2}}{14}, \quad C = \frac{5 - 3\sqrt{2}}{14},$$

and $p_{11}(t) \to 2/7$ as $t \to \infty$.

Example 2.1.10 It is easy to give an example of a Q-matrix with multiple roots of the characteristic equation $\det(Q - \mu I) = 0$, viz.

$$\begin{pmatrix} -2 & 1 & 1 & 0 \\ 1 & -2 & 1 & 0 \\ 0 & 1 & -2 & 1 \\ 1 & 0 & 1 & -2 \end{pmatrix}.$$

Here the roots are $0, -2, -3$; the last of these has algebraic multiplicity 2, but the geometric multiplicity of each eigenvalue is 1. As a result, the entries $p_{ij}(t)$ of the transition matrix $P(t) = e^{tQ}$ are given by the formula

$$p_{ij}(t) = A_{ij} + B_{ij}e^{-2t} + (C_{ij} + D_{ij}t)e^{-3t}, \quad i,j = 1,2,3,4,$$

where A_{ij}, B_{ij}, C_{ij} and D_{ij} are constants. To find them, we require (2.5) with $k = 0, 1, 2, 3$. The resulting matrix $P(t)$ is

$$\frac{1}{18} \begin{pmatrix} 4 + 6te^{-3t} + 14e^{-3t} & -14e^{-3t} + 9e^{-2t} + 5 - 6te^{-3t} \\ 4 + 6te^{-3t} - 4e^{-3t} & 5 + 4e^{-3t} + 9e^{-2t} - 6te^{-3t} \\ -4e^{-3t} + 4 - 12te^{-3t} & -9e^{-2t} + 5 + 12te^{-3t} + 4e^{-3t} \\ 4 + 6te^{-3t} - 4e^{-3t} & 4e^{-3t} - 9e^{-2t} + 5 - 6te^{-3t} \end{pmatrix}$$

$$\begin{matrix} -6e^{-3t} + 6 & 6e^{-3t} - 9e^{-2t} + 3 \\ -6e^{-3t} + 6 & 6e^{-3t} - 9e^{-2t} + 3 \\ 6 + 12e^{-3t} & -12e^{-3t} + 9e^{-2t} + 3 \\ -6 + 6e^{-3t} & 3 + 6e^{-3t} + 9e^{-2t} \end{matrix} \Bigg)$$

Remark 2.1.11 The square Q^2 of a Q-matrix does not necessarily give a Q-matrix (nor are other powers Q^n). For example, in the 2×2 case:

$$\begin{pmatrix} -\alpha & \alpha \\ \beta & -\beta \end{pmatrix}^2 = \begin{pmatrix} \alpha^2 + \alpha\beta & -\alpha^2 - \alpha\beta \\ -\alpha\beta - \beta^2 & \alpha\beta + \beta^2 \end{pmatrix}.$$

This gives a Q-matrix if and only if $\alpha = \beta = 0$, i.e. $Q = 0$. See also Example 2.1.9 above. However, the sum of the elements along a row in the matrix Q^n is always zero, and we used this property in the proof of Theorem 2.1.6 (see (2.8)). In fact, the structure of the matrix Q^n can be captured from the arrow diagram representing the original Q-matrix Q. So, to calculate the entry $q_{ij}^{(2)}$, we consider all paths of length 2 $i \to k \to j$, multiply q_{ik} and q_{kj} along each of these paths and sum the products. Then for Q^3 take all paths of length 3, and so on.

On the other hand, for all $a \geq 0$, the scalar multiple aQ of a Q-matrix forms a Q-matrix. Also, the sum $Q_1 + Q_2$ of Q-matrices is a Q-matrix (and hence any linear combination $a_1 Q_1 + a_2 Q_2$, or, more generally, $\sum_{j=1}^{n} a_j Q_j$, with non-negative coefficients a_j).

Remark 2.1.12 Given a Q-matrix and any $t > 0$, $P(t) = e^{tQ}$ is a stochastic matrix by Theorem 2.1.6. Therefore, there exists a discrete-time Markov chain for which $P(t)$ is the transition matrix.

However, it is not true that any transition matrix P can be written as e^{tQ} for some Q-matrix and some $t \geq 0$. Here, property (iii) above, $\det e^{tQ} = e^{t(\operatorname{tr} Q)}$, can help. For example, the transition matrices

$$\begin{pmatrix} 0 & 1 \\ 1/2 & 1/2 \end{pmatrix} \quad \text{and} \quad \begin{pmatrix} 1 & 0 & 0 \\ 1 & 0 & 0 \\ 0 & 1 & 0 \end{pmatrix}$$

cannot be written as e^{tQ} since their determinants are zero or negative, while $e^{t(\operatorname{tr} Q)} > 0$.

Now, using the semigroup property to prove that $\det e^{tQ} = e^{t(\operatorname{tr} Q)}$,

$$\det e^{(t+s)Q} = \det \left(e^{sQ} e^{tQ} \right) = \left(\det e^{sQ} \right) \left(\det e^{tQ} \right).$$

Next, $\det e^{tQ}$ is continuous (and even differentiable) in t, with $\det e^{0 \cdot Q} = \det \mathbf{I} = 1$. Hence, $\det e^{tQ} = e^{tq}$ for some $q \in \mathbb{R}$. Finally, to find q, we observe that

$$q = \frac{\mathrm{d}}{\mathrm{d}t} \det e^{tQ} \Big|_{t=0} = \frac{\mathrm{d}}{\mathrm{d}t} \det (\mathbf{I} + tQ) \Big|_{t=0}.$$

That is, q is the first-order coefficient of the polynomial $\det(\mathbf{I} + tQ)$ which is precisely $\operatorname{tr} Q$. Note that unless $Q = 0$, $\det e^{tQ} \to 0$ as $t \to \infty$.

In other cases, a more involved analysis is needed. For instance, take the transition matrix

$$P = \begin{pmatrix} 0 & 1 & 0 \\ 0 & 0 & 1 \\ 1 & 0 & 0 \end{pmatrix}.$$

Suppose it can be written as e^{tQ}. Then consider the transition matrix $P' = e^{tQ/m}$; we will see that $(P')^m = P$. It is easy to find a transition matrix

$$P' = \begin{pmatrix} 0 & 0 & 1 \\ 1 & 0 & 0 \\ 0 & 1 & 0 \end{pmatrix}$$

such that $(P')^2 = P$. But this is impossible for $m = 3$ (the matrix P contains too many 0s). Indeed, the matrix P describes deterministic movements. This implies that P' should be deterministic as well. However, none of the six permutation matrices for 3 states satisfy this condition. Thus, the above matrix P cannot be represented as e^{tQ}.

Remark 2.1.13 A general *multiplicativity* property of matrix exponents is that

$$e^{Q_1 + Q_2} = e^{Q_1} e^{Q_2} = e^{Q_2} e^{Q_1} \tag{2.13}$$

if and only if matrices Q_1 and Q_2 commute, i.e. $Q_1 Q_2 = Q_2 Q_1$. This implies the semigroup property $e^{(s+t)Q} = e^{sQ+tQ} = e^{sQ} e^{tQ}$, as (sQ) and (tQ) commute.

Equation (2.13) is again verified by a direct calculation. Thus, if $Q_1 Q_2 = Q_2 Q_1$ then we can write the binomial matrix expansion:

$$(Q_1 + Q_2)^n = \sum_{k=0}^{n} \frac{n!}{k!(n-k)!} Q_1^k Q_2^{n-k} = \sum_{k=0}^{n} \frac{n!}{k!(n-k)!} Q_2^k Q_1^{n-k}. \tag{2.14}$$

As in the scalar case, the product of matrix series

$$\left(\sum_{k \geq 0} \frac{(Q_1)^k}{k!} \right) \left(\sum_{l \geq 0} \frac{(Q_2)^l}{l!} \right)$$

can then be nicely rearranged into the series $\sum_{n \geq 0} \frac{(Q_1+Q_2)^n}{n!}$, giving $e^{Q_1+Q_2}$. However, without assuming that $Q_1 Q_2 = Q_2 Q_1$, expansion (2.14) will be replaced by a sum of intermittent products of Q_1s and Q_2s, and (2.13) will not hold.

2.2 Continuous-time Markov chains: definitions and basic constructions

The Markov Chain Saw Massacre
(From the series *'Movies that never made it to the Big Screen'*.)

Definition 2.2.1 A *continuous-time Markov chain* (CTMC) with a (finite) Q-matrix Q and initial distribution λ is a family of random variables $(X_t, t \geq 0)$ with values in a set I such that:

(a) $\mathbb{P}(X_0 = i) = \lambda_i$;

(b) for all $0 < t_1 < t_2 < \cdots < t_n$ and states $i_0, \ldots, i_n \in I$, we have

$$\mathbb{P}(X_0 = i_0, X_{t_1} = i_1, \ldots, X_{t_n} = i_n) =$$
$$\lambda_{i_0} p_{i_0 i_1}(t_1) p_{i_1 i_2}(t_2 - t_1) \ldots p_{i_{n-1} i_n}(t_n - t_{n-1}), \qquad (2.15)$$

where $p_{ij}(t)$ are the entries of the matrix $P(t) = e^{tQ}$.

A typical sample path (trajectory) of (X_t) is shown in Figure 2.4.

We will see that a sample path (a) spends a random time $\sim \text{Exp}(q_i)$ in state i and then (b) jumps to $j \neq i$ with probability q_{ij}/q_i. See Property (IX) on page 200.

A standard agreement is that the trajectories of a CTMC are *right-continuous*. This means that with probability 1, for all $t \geq 0$,

$$\lim_{h \searrow 0} X_{t+h} = X_t.$$

However, whenever it bears little importance (which will be often the case), we draw trajectories of a DTMC as 'fully continuous' broken lines.

It is important to understand that the event $\{X_0 = i_0, X_{t_1} = i_1, \ldots, X_{t_n} = i_n\}$ in (2.15) includes all sample trajectories passing through states $i_0, i_1, \ldots, i_n$ at times $t_0 = 0, t_1, \ldots, t_n$; their behaviour between these times can be arbitrary, as illustrated in Figure 2.5.

The matrix Q is often called the *generator (matrix)* of the continuous-time Markov chain (X_t). The chain is called a (λ, Q) Markov chain with continuous time

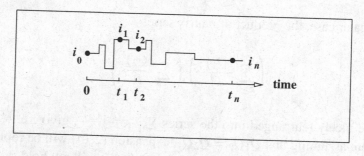

Fig. 2.4

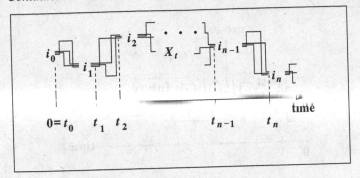

Fig. 2.5

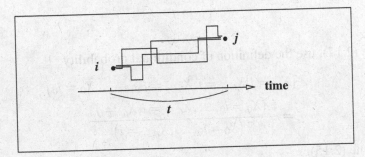

Fig. 2.6

and generator Q. As in the case of discrete-time Markov chains (DTMCs), we will often work with chains starting from a single state, where $\lambda = \delta_i$. Such a chain is called a (δ_i, Q) CTMC.

Definition 2.2.1 implies the following properties:

Property I The matrix $P(t) = (p_{ij}(t))$ is called the transition matrix in time t. It describes the conditional probabilities

$$p_{ij}(t) = \mathbb{P}(X_t = j | X_0 = i) = \mathbb{P}(X_{t+s} = j | X_s = i). \qquad (2.16)$$

In terms of trajectories, $p_{ij}(t)$ gives the total probability of all paths leading from i to j in time t. See Figure 2.6.

Property II The lack of memory in conditional probabilities:

$$\mathbb{P}\left(X_{t_n} = j | X_{t_{n-1}} = i, X_{t_{n-2}} = i_{n-2}, \dots, X_0 = i_0\right)$$
$$= \mathbb{P}(X_{t_n} = j | X_{t_{n-1}} = i) = p_{ij}(t_n - t_{n-1}). \qquad (2.17)$$

That is, the conditional probability

$$\mathbb{P}\left(X_{t_n} = j | X_{t_{n-1}} = i, X_{t_{n-2}} = i_{n-2}, \dots, X_0 = i_0\right)$$

does not depend on $t_1, \dots, t_{n-1}$ and $i_0, i_1, \dots, i_{n-2}$.

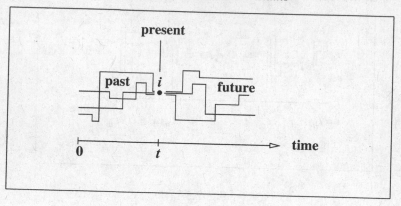

Fig. 2.7

To prove (2.17), use the definition of conditional probability

$$\mathbb{P}\left(X_{t_n} = j \big| X_{t_{n-1}} = i, X_{t_{n-2}} = i_{n-2}, \ldots, X_0 = i_0\right)$$
$$= \frac{\mathbb{P}\left(X_0 = i_0, \ldots, X_{t_{n-1}} = i, X_{t_n} = j\right)}{\mathbb{P}\left(X_0 = i_0, \ldots, X_{t_{n-1}} = i\right)}$$

and substitute (2.15).

Property III The unconditional probability

$$\mathbb{P}(X_t = j) = \sum_i \lambda_i p_{ij}(t) = (\lambda P(t))_j \tag{2.18}$$

describes the total probability of the set of trajectories in state j at time t; see Figure 2.6.

Property IV As in the discrete-time case, the *Markov property* holds, yielding conditional independence of the past and future, given the present state.

Theorem 2.2.2 *Let (X_t) be a (λ, Q)-CTMC. Then, for all given time $t > 0$ and state $i \in I$, conditional on the event $\{X_t = i\}$, the future states $(X_{t+s}, s \geq 0)$ do not depend on past states $(X_s, 0 \leq s < t)$, and (X_{t+s}) is a (δ_i, Q)-CTMC.*

Property V We say that $\lambda = \pi$ is an equilibrium distribution if and only if $\mathbb{P}(X_t = j) = \pi_j$ for all $t \in \mathbb{R}, j \in I$, i.e. $\pi P(t) \equiv \pi$. That is, the row vector π is annihilated by the matrix Q:

$$0 = \frac{\mathrm{d}}{\mathrm{d}t} \pi P(t) = \pi \frac{\mathrm{d}}{\mathrm{d}t} P(t) = \pi P(t) Q,$$
$$0 = \pi P(t) Q \big|_{t=0} = \pi Q. \tag{2.19}$$

In other words, π is a row eigenvector of Q with the eigenvalue zero.

Property VI In fact, every (finite) Q-matrix harbours an equilibrium distribution, i.e. a row eigenvector $\pi = (\pi_i)$, with the eigenvalue zero, which contains non-negative entries and where $\sum_i \pi_i = 1$. In particular, every finite Q-matrix features 0 as an eigenvalue.

Property VII As in the discrete time case, if

$$P(t) \rightarrow \Pi = \begin{array}{c} \pi^{(1)} \\ \pi^{(2)} \\ \cdots \\ \pi^{(m)} \end{array} \left(\begin{array}{ccc} - & - & - \\ - & - & - \\ - & - & - \\ - & - & - \end{array} \right),$$

then each row $\pi^{(i)}$ in $P(t)$ makes up an equilibrium distribution

$$\pi^{(i)} P(t) = \left(\Pi P(t) \right)^{(i)} = \left(\lim_{\tau \to \infty} P(\tau) P(t) \right)^{(i)} = \left(\lim_{\tau \to \infty} P(\tau + t) \right)^{(i)} = \pi^{(i)}. \quad (2.20)$$

See Theorem 2.8.1 below.

For instance, in Example 2.1.3, if $\alpha + \beta > 0$ then

$$\pi = \left(\frac{\beta}{\alpha + \beta}, \frac{\alpha}{\alpha + \beta} \right) \quad (2.21)$$

is the (unique) equilibrium distribution. But for $\alpha = \beta = 0$, $P(t) \equiv I$ and every vector becomes invariant.

Property VIII As in the discrete-time case, we can define the *strong Markov property*. Call a random variable T with values in $[0, \infty]$ a *stopping time* if for all $t > 0$ the event $\{T < t\}$ is determined by $(X_\tau : 0 \leq \tau < t)$. In other words, the indicator $\mathbf{1}(T < t)$ is a function of $(X_\tau : 0 \leq \tau < t)$ only.

Once Upon A Stopping Time

(From the series '*Movies that never made it to the Big Screen*.')

An example of a stopping time is again the *hitting time*, of a subset of states $A \subset I$

$$H^A = \inf[s \geq 0 : X_s \in A]. \quad (2.22)$$

In fact, the indicator

$$\mathbf{1}(H^A < t) = \mathbf{1}(\text{there exists } s \in [0, t) \text{ such that } X_s \in A)$$

depends on $(X_\tau, 0 \leq \tau < t)$.

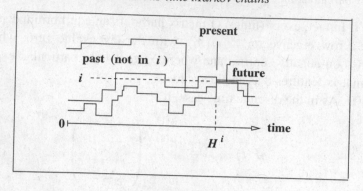

Fig. 2.8

As with DTMCs, we will be particularly interested in the hitting time of a given state i (the *passage time* to i, or, when the initial state in i itself, the *return* time):

$$H_i = \inf\,[s \geq 0 : X_s = i], \qquad\qquad (2.23)$$

Theorem 2.2.3 *Let* (X_t) *be a* (λ, Q)-*CTMC and* T *be a stopping time. Then, for all states* i, *conditional on the event* $\{T < \infty, X_T = i\}$, *future states* $(X_{t+T}, t > 0)$ *do not depend on past states* $(X_t, 0 \leq t < T)$, *and* $(X_{t+T}, t \geq 0)$ *is* (δ_i, Q)-*CTMC.*

Figure 2.8 addresses the case where $T = H_i$, the passage time to state i.

A Passage Time To India
(From the series *'Movies that never made it to the Big Screen.'*)

(IX) A CTMC (X_t) with generator Q illustrates the following property: *given that* $X_t = i$, *with* $q_i = -q_{ii} > 0$, *the residual holding time* R_i *at state* i *follows an exponential distribution, of rate* $q_i = -q_{ii}$:

$$\mathbb{P}(R_i \geq \tau | X_t = i) = \mathbb{P}(X_{t+s} = i \text{ for all } 0 \leq s < \tau | X_t = i) = e^{-q_i \tau}.$$

Further, at time $t + R_i$, *the chain jumps to state* j *with probability* $\widehat{p}_{ij} = q_{ij}/q_i$:

$$\mathbb{P}(X_{t+R_i} = j | X_t = i) = \frac{q_{ij}}{q_i}. \qquad\qquad (2.24)$$

This property will be verified later on. A state i with $q_i = 0$ gets $q_{ij} = 0$ for all j; such a state is called *absorbing*. The corresponding arrow diagram does not contain arrows going out of i.

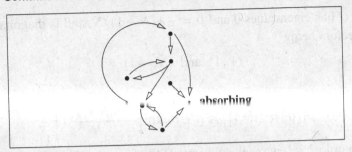

Fig. 2.9

Remark 2.2.4 As in the discrete-time case, Definition 2.2.1 specifies a so-called *time-homogeneous* (or, more briefly *homogeneous*) Markov chain. Here the probabilities $p_{ij}(t)$ depend only on the time t in which a transition from i to j is required. In a more general case of *inhomogeneous* chains, one has to deal with a transition probability $p_{ij}(s, t+s)$ that the state at time $t+s$ is j, given that the state at time s is i, $t, s > 0$. Examples of such chains will appear in Section 2.4.

Worked Example 2.2.5 A virus exists in $N+1$ strains $0, \dots, N$. It keeps its strain for a random time $\sim \mathrm{Exp}(\lambda)$ then mutates to one of the remaining strains, equiprobably. Find the probability that the strain at time t is the same as the initial strain

$$\mathbb{P}(\text{at time } t \text{ the same strain as at time } 0).$$

Solution Due to symmetry,

$$q_i := -q_{ii} = \lambda; \quad q_{ij} = \frac{\lambda}{N}, \quad 1 \le i, j \le N+1, i \ne j,$$

and

$$Q = \begin{pmatrix} -\lambda & \lambda/N & \cdots & \lambda/N \\ \lambda/N & -\lambda & \cdots & \lambda/N \\ & \cdots & \cdots & \\ \lambda/N & \lambda/N & \cdots & -\lambda \end{pmatrix}.$$

We want to compute $p_{ii}(t) = (e^{tQ})_{ii}$. Clearly, $p_{ii}(t) = p_{11}(t)$ for all $i, t \in \mathbb{R}$, again by symmetry.

Consider a reduced (2×2) Q-matrix, over states 0 and 1:

$$\tilde{Q} = \begin{pmatrix} -\lambda & \lambda \\ \lambda/N & -\lambda/N \end{pmatrix}.$$

The matrix $\widetilde{Q}$ has eigenvalues 0 and $\mu = -\lambda(N+1)/N$, and is diagonalisable, its row eigenvectors being

$$(1\,,1) \quad \text{and} \quad (N,\,-1).$$

Hence,

$$p_{11}(t) = \left(e^{t\widetilde{Q}}\right)_{11} = A + B\exp\left[-\frac{\lambda(N+1)}{N}t\right].$$

As in Example 2.1.3, we seek solutions of the form $A + Be^{\mu t}$. We obtain

$$A = \frac{1}{N+1},\ B = \frac{N}{N+1},$$

and

$$p_{11}(t) = \frac{1}{N+1} + \left(\frac{N}{N+1}\right)\exp\left[-\frac{\lambda(N+1)}{N}t\right] = p_{ii}(t).$$

By symmetry,

$$
\begin{aligned}
p_{ij}(t) &= \frac{1}{N}[1 - p_{ii}(t)]\\
&= \frac{1}{N+1} - \left(\frac{1}{N+1}\right)\exp\left[-\frac{\lambda(N+1)}{N}t\right], \quad i \neq j.
\end{aligned}
$$

We conclude that

$$p_{ij}(t) \longrightarrow \frac{1}{N+1} \text{ as } t \to \infty \text{ (equidistribution)}.$$

□

Worked Example 2.2.6 A flea jumps clockwise on the vertices of a triangle ABC; the holding times are independent exponential random variables of rate 1. Find the eigenvalues of the corresponding Q-matrix and express the transition probabilities $p_{xy}(t)$, $t \geq 0$, $x,y = A,B,C$, in terms of these roots. Deduce the formulas for the sums

$$S_0(t) = \sum_{n \geq 0}\frac{t^{3n}}{(3n)!},\ \ S_1(t) = \sum_{n \geq 0}\frac{t^{3n+1}}{(3n+1)!},\ \ S_2(t) = \sum_{n \geq 0}\frac{t^{3n+2}}{(3n+2)!},$$

in terms of the functions $e^t, e^{-t/2}, \cos(\sqrt{3}t/2)$ and $\sin(\sqrt{3}t/2)$.
Find the limits

$$\lim_{t \to \infty} e^{-t}S_j(t), \quad j = 0,1,2.$$

What is the connection between the decompositions

$$e^t = S_0(t) + S_1(t) + S_2(t)$$

and $e^t = (\cosh t + \sinh t)$?

Solution The Q-matrix has the characteristic polynomial

$$\det \begin{pmatrix} -1-\kappa & 1 & 0 \\ 0 & -1-\kappa & 1 \\ 1 & 0 & -1-\kappa \end{pmatrix} = (-\kappa-1)^3 + 1 = -\kappa(\kappa^2+3\kappa+3).$$

The roots (eigenvalues), one real and two complex, are

$$\kappa = 0, \quad -\frac{3}{2} \pm i\frac{\sqrt{3}}{2}.$$

The diagonal transition probabilities are

$$p_{xx}(t) = a + e^{-3t/2} \left(b\cos\left(\frac{\sqrt{3}}{2}t\right) + c\sin\left(\frac{\sqrt{3}}{2}t\right) \right), \quad x = A, B, C,$$

with

$$a = p_{xx}(\infty) = \frac{1}{3} \quad (\text{as } \pi = (1/3, 1/3, 1/3)),$$

$$a+b = p_{xx}(0) = 1, \quad \text{whence } b = 2/3,$$

$$-\frac{3}{2}b + \frac{\sqrt{3}}{2}c = \dot{p}_{xx}(0) = q_{xx} = -1, \quad \text{whence } c = 0.$$

At the same time,

$$p_{xx}(t) = \sum_{n\geq 0} e^{-t} \frac{t^{3n}}{(3n)!}, \quad x = A, B, C.$$

So,

$$S_0(t) = \sum_{n\geq 0} \frac{t^{3n}}{(3n)!} = \frac{1}{3}e^t + \frac{2}{3}e^{-t/2}\cos\left(\frac{\sqrt{3}}{2}t\right).$$

Similarly, the probabilities $p_{AB}(t) = p_{BC}(t) = p_{CA}(t)$ equal

$$\frac{1}{3} - \frac{1}{3}e^{-3t/2}\cos\left(\frac{\sqrt{3}}{2}t\right) + \frac{1}{\sqrt{3}}e^{-3t/2}\sin\left(\frac{\sqrt{3}}{2}t\right),$$

whence $S_1(t) = \sum_{n\geq 0} \frac{t^{3n+1}}{(3n+1)!}$ is equal to

$$\frac{1}{3}e^t - \frac{1}{3}e^{-t/2}\cos\left(\frac{\sqrt{3}}{2}t\right) + \frac{1}{\sqrt{3}}e^{-t/2}\sin\left(\frac{\sqrt{3}}{2}t\right).$$

Finally, the probabilities $p_{AC}(t) = p_{BA}(t) = p_{CB}(t)$ equal

$$\sum_{n\geq 0} e^{-t} \frac{t^{3n+2}}{(3n+2)!} = \frac{1}{3} - \frac{1}{3}e^{-3t/2}\cos\left(\frac{\sqrt{3}}{2}t\right) - \frac{1}{\sqrt{3}}e^{-3t/2}\sin\left(\frac{\sqrt{3}}{2}t\right),$$

and $S_2(t) = \sum_{n \geq 0} \dfrac{t^{3n+2}}{(3n+2)!}$ equals

$$\frac{1}{3} e^t - \frac{1}{3} e^{-t/2} \cos\left(\frac{\sqrt{3}}{2} t\right) - \frac{1}{\sqrt{3}} e^{-t/2} \sin\left(\frac{\sqrt{3}}{2} t\right).$$

The limits

$$\lim_{t \to \infty} e^{-t} S_j(t) = \frac{1}{3}, \quad j = 0, 1, 2,$$

give the equilibrium distribution $\pi = (1/3, 1/3, 1/3)$.

The decomposition $e^t = (\cosh t + \sinh t)$ arises when we consider the Markov chain with the Q-matrix

$$\begin{pmatrix} -1 & 1 \\ 1 & -1 \end{pmatrix},$$

which is a 'reduced' version of the above chain.   □

Example 2.2.7 Consider an $N \times N$ Q-matrix of the form

$$\begin{pmatrix} -\lambda & \lambda & 0 & \cdots & 0 & 0 \\ 0 & -\lambda & \lambda & \cdots & 0 & 0 \\ 0 & 0 & -\lambda & \cdots & 0 & 0 \\ \cdots & \cdots & \cdots & \cdots & \cdots & \cdots \\ 0 & 0 & 0 & \cdots & -\lambda & \lambda \\ 0 & 0 & 0 & \cdots & 0 & 0 \end{pmatrix} = \lambda \begin{pmatrix} -1 & 1 & 0 & \cdots & 0 & 0 \\ 0 & -1 & 1 & \cdots & 0 & 0 \\ 0 & 0 & -1 & \cdots & 0 & 0 \\ \cdots & \cdots & \cdots & \cdots & \cdots & \cdots \\ 0 & 0 & 0 & \cdots & -1 & 1 \\ 0 & 0 & 0 & \cdots & 0 & 0 \end{pmatrix}.$$

$$(2.25)$$

See Figure 2.10. Here the state N is absorbing: $q_{Ni} \equiv 0$, or, equivalently, $q_{NN} = -q_N = 0$. Next, all matrices Q^k are upper triangular (as there is no arrow $j-1 \leftarrow j$). Hence, so is e^{tQ}. The forward equation $\dfrac{d}{dt} P(t) = P(t)Q$ and the initial condition $P(0) = \mathbf{I}$ read

$$\frac{d}{dt} p_{ii} = -\lambda p_{ii}, \qquad p_{ii}(0) = 1, \ 1 \leq i < N,$$

$$\frac{d}{dt} p_{ij} = -\lambda p_{ij} + \lambda p_{ij-1}, \quad p_{ij}(0) = 0, \ 1 \leq i < j < N,$$

$$\frac{d}{dt} p_{iN} = \lambda p_{iN-1}, \qquad p_{iN}(0) = 0, \ 1 \leq i < N,$$

$$\frac{d}{dt} p_{NN} = 0, \qquad p_{NN}(0) = 1,$$

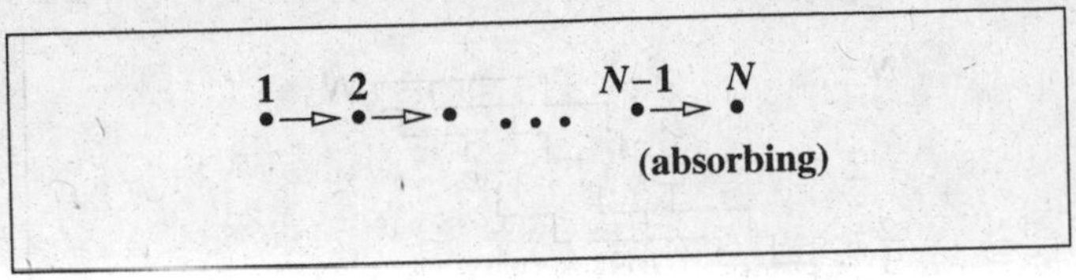

Fig. 2.10

Because, clearly, $p_{ij}(t) = 0$ for all $i > j$, the equations admit the following recursive solution

$$\frac{\mathrm{d}}{\mathrm{d}t} p_{ii} = -\lambda p_{ii} : \ p_{ii}(t) = \mathrm{e}^{-\lambda t}, \ \text{for } 1 \leq i < N,$$

$$\frac{\mathrm{d}}{\mathrm{d}t} p_{ii+1} = -\lambda p_{ii+1} + \lambda \mathrm{e}^{-\lambda t} : \ p_{ii+1}(t) = \lambda t \mathrm{e}^{-\lambda t}, \ \text{for } 1 \leq i < N,$$

$$\frac{\mathrm{d}}{\mathrm{d}t} p_{ii+2} = -\lambda p_{ii+2} + \lambda^2 t \mathrm{e}^{-\lambda t} : \ p_{ii+2}(t) = \frac{(\lambda t)^2}{2} \mathrm{e}^{-\lambda t}, \ \text{for } 0 \leq i < N-1,$$

and so on. In general,

$$p_{ij}(t) = \left(\frac{(\lambda t)^{j-i}}{(j-i)!} \right) \mathrm{e}^{-\lambda t}, \ \text{for } 1 \leq i < j < N-1,$$

and

$$p_{iN}(t) = 1 - \sum_{l=0}^{N-i-1} \frac{(\lambda t)^l}{l!} \mathrm{e}^{-\lambda t}, \ \text{for } 0 \leq i < N;$$

finally,

$$p_{NN} \equiv 1.$$

In the matrix form

$$P(t) = \mathrm{e}^{tQ} = \begin{pmatrix} \mathrm{e}^{-\lambda t} & (\lambda t)\,\mathrm{e}^{-\lambda t} & (\lambda t)^2 \mathrm{e}^{-\lambda t}/2! & \cdots & \displaystyle\sum_{l \geq N-1} (\lambda t)^l \mathrm{e}^{-\lambda t}/l! \\ 0 & \mathrm{e}^{-\lambda t} & (\lambda t)\mathrm{e}^{-\lambda t} & \cdots & \displaystyle\sum_{l \geq N-2} (\lambda t)^l \mathrm{e}^{-\lambda t}/l! \\ \cdots & \cdots & \cdots & \cdots & \cdots \\ 0 & 0 & 0 & 0 & 1 \end{pmatrix}. \tag{2.26}$$

We see that, as $t \to \infty$,

$$p_{ij}(t) \to 0 \text{ for all } 0 \leq i < j < N \text{ and } p_{iN}(t) \to 1 \text{ for all } 0 \leq i < N.$$

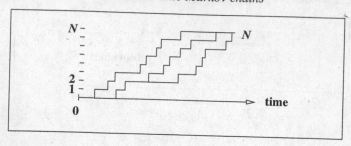

Fig. 2.11

That is,

$$\lim_{t\to\infty} p_{ij}(t) = \delta_{jN} \text{ i.e. } P(t) \to \begin{pmatrix} 0 & 0 & 0 & \cdots & 1 \\ 0 & 0 & 0 & \cdots & 1 \\ \cdots & \cdots & \cdots & \cdots & \cdots \\ 0 & 0 & 0 & \cdots & 1 \end{pmatrix} \begin{matrix} \sim \delta_N \\ \sim \delta_N \\ \vdots \\ \sim \delta_N \end{matrix},$$

where δ_N forms the probability distribution concentrated at state N. The chain eventually ends up in state N.

A useful observation is that an absorbing state, with $q_i = 0$, has $p_{ii}(t) \equiv 1$ for all $t \in \mathbb{R}^+$ (and vice versa).

Typical trajectories of the $(\delta^{(0)}, Q)$ continuous-time Markov chain are shown in Figure 2.11. They all jump upwards by one unit and, as was noted, eventually reach level N where they stay forever.

Example 2.2.8 If we modify the Q-matrix to

$$\begin{pmatrix} -\lambda & \lambda & 0 & \cdots & 0 & 0 \\ 0 & -\lambda & \lambda & \cdots & 0 & 0 \\ 0 & 0 & -\lambda & \cdots & 0 & 0 \\ \cdots & \cdots & \cdots & \cdots & \cdots & \cdots \\ 0 & 0 & 0 & \cdots & -\lambda & \lambda \\ \lambda & 0 & 0 & \cdots & 0 & -\lambda \end{pmatrix} = \lambda \begin{pmatrix} -1 & 1 & 0 & \cdots & 0 & 0 \\ 0 & -1 & 1 & \cdots & 0 & 0 \\ 0 & 0 & -1 & \cdots & 0 & 0 \\ \cdots & \cdots & \cdots & \cdots & \cdots & \cdots \\ 0 & 0 & 0 & \cdots & -1 & 1 \\ 1 & 0 & 0 & \cdots & 0 & -1 \end{pmatrix},$$

$$(2.27)$$

it will generate a cycle, as shown in Figure 2.12.

Here, the matrix $P(t) = e^{tQ}$ will exhibit a cyclic structure: for all i, j, l,

$$p_{ij}(t) = p_{i+l,j+l} \text{ (addition mod } N).$$

In fact, summing up the transition probabilities for all the states which are 'projected' into j when the circle is produced from the real line, one obtains the answer

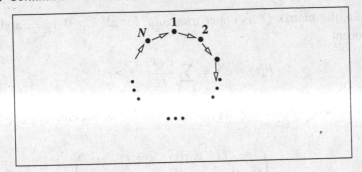

Fig. 2.12

$$p_{ij}(t) = p_{1,j+1-i}(t) = \sum_{l \geq 0} \frac{(\lambda t)^{j-i+lN}}{(j-i+lN)!} \, e^{-\lambda t}, \ 1 \leq i \leq j \leq N,$$

and

$$p_{ij}(t) = p_{1,N+j-i+1}(t), \ 1 \leq j < i \leq N.$$

As $t \to \infty$, $p_{ij}(t) \to 1/N$, giving the uniform distribution. This makes up the unique solution of $\pi Q = 0$ with $\pi_i > 0$, and $\sum_{i=1}^{N} \pi_i = 1$.

Example 2.2.9 Many properties observed so far can be extended to infinite matrices without much difficulty. For example, consider an infinite Q-matrix

$$Q = \begin{pmatrix} -\lambda & \lambda & 0 & 0 & \cdots \\ 0 & -\lambda & \lambda & 0 & \ddots \\ 0 & 0 & -\lambda & \lambda & \ddots \\ \cdots & \ddots & \ddots & \ddots & \ddots \end{pmatrix} = \lambda \begin{pmatrix} -1 & 1 & 0 & 0 & \cdots \\ 0 & -1 & 1 & 0 & \ddots \\ 0 & 0 & -1 & 1 & \ddots \\ \cdots & \ddots & \ddots & \ddots & \ddots \end{pmatrix}, \ (2.28)$$

with the diagram in Figure 2.13.

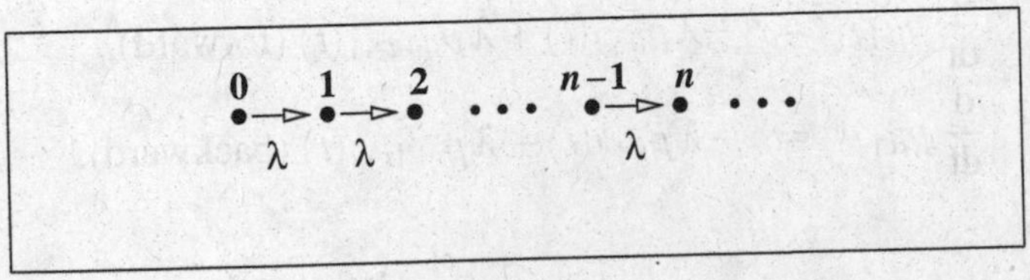

Fig. 2.13

Here again, the matrix Q^k is upper triangular for all $k = 0, 1, \ldots,$ and so is the matrix exponent

$$P(t) = e^{tQ} = \sum_{k \geq 0} \frac{t^k Q^k}{k!}, \quad t \geq 0. \tag{2.29}$$

We may not be sure about elements $p_{ii+l}(t)$ with $l \geq 0$, but entries $p_{ii-l}(t)$ clearly equal 0:

$$P(t) = \begin{pmatrix} p_{00}(t) & p_{01}(t) & p_{02}(t) & \cdots \\ 0 & p_{11}(t) & p_{12}(t) & \cdots \\ 0 & 0 & p_{22}(t) & \cdots \\ & \mathbf{0} & & \ddots \end{pmatrix}.$$

To find the entries $p_{ii+l}(t)$, we can use the forward or backward equation

$$\frac{d}{dt} P(t) = P(t)Q \text{ or } \frac{d}{dt} P(t) = QP(t), \tag{2.30}$$

with the initial condition $P(0) = \mathbf{I}$. For $l = 0$ (diagonal entries), the equations become

$$\frac{d}{dt} p_{ii} = -\lambda p_{ii}(t), \quad p_{ii}(0) = 1,$$

whence

$$p_{ii}(t) = e^{-\lambda t} \text{ for all } i = 0, 1, \ldots \text{ and } t \geq 0. \tag{2.31}$$

For $l = 1$ (one step above the main diagonal), we see

$$\frac{d}{dt} p_{ii+1} = -\lambda p_{ii+1}(t) + \lambda p_{ii}(t) \text{ (forward)},$$

$$\frac{d}{dt} p_{ii+1} = -\lambda p_{ii+1}(t) + \lambda p_{i+1i+1}(t) \text{ (backward)},$$

whence

$$p_{ii+1}(t) = \lambda t\, e^{-\lambda t} \text{ for all } i = 0, 1, \ldots, \text{ and } t \geq 0. \tag{2.32}$$

In general, for all $l = 0, 1, \ldots$:

$$\frac{d}{dt} p_{ii+l} = -\lambda p_{ii+l}(t) + \lambda p_{ii+l-1}(t) \text{ (forward)},$$

$$\frac{d}{dt} p_{ii+l} = -\lambda p_{ii+l}(t) + \lambda p_{i+1i+l}(t) \text{ (backward)},$$

yielding

$$p_{ii+l}(t) = \frac{(\lambda t)^l}{l!} e^{-\lambda t} \text{ for all } i = 0, 1, \ldots, \text{ and } t \geq 0. \tag{2.33}$$

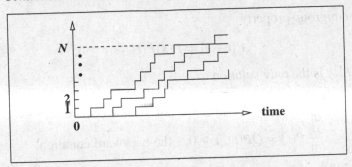

Fig. 2.14

Finally, as transitions $i \to j$ for $j < i$ are impossible:

$$p_{ii-l}(t) = 0, \text{ for all } i = 0,1,\ldots, \, l = 1,2,\ldots \text{ and } t \geq 0. \tag{2.34}$$

In matrix form

$$P(t) = \begin{pmatrix} e^{-\lambda t} & \dfrac{\lambda t}{1} e^{-\lambda t} & \dfrac{(\lambda t)^2}{2!} e^{-\lambda t} & \cdots \\ 0 & e^{-\lambda t} & \dfrac{\lambda t}{1} e^{-\lambda t} & \ddots \\ 0 & 0 & e^{-\lambda t} & \ddots \\ \cdots & & \ddots & \ddots \end{pmatrix} = \begin{pmatrix} \text{Poisson } (\lambda t) & & \\ 0 \; \text{Poisson } (\lambda t) & \\ 0 \; 0 \; \text{Poisson } (\lambda t) \\ \cdots \qquad \cdots \end{pmatrix}. \tag{2.35}$$

Equivalently, we could have arrived at the same result by taking the limit

$$P(t) = e^{tQ} = \lim_{N \to \infty} e^{tQ^{(N)}}, \; t \geq 0, \tag{2.36}$$

where the matrices $Q^{(N)}$ and $p^{(N)}(t) = e^{tQ^{(N)}}$ were seen in (2.25) and (2.27).

Matrices $P(t)$, $t \geq 0$, defined by (2.35) satisfy the properties listed in Theorem 2.1.4 above. Obviously, each $P(t)$ constitutes a stochastic matrix defining a collection of transition probabilities on $\mathbb{Z}_+$. We could repeat Definition 2.2.1, with $I = \mathbb{Z}_+ = \{0,1,\ldots\}$ and an (infinite) transition probability matrix $P(t)$ as specified in (2.35). Typical trajectories of the $\left(\delta^{(0)}, Q\right)$ Markov chain (starting at state 0) will appear as in Figure 2.14. The paths jump upwards by one and increase indefinitely over time.

So, in summary:

Theorem 2.2.10 *Let Q be of the form (2.28). The family of matrices P(t) from (2.35) satisfies (2.29) and (2.36) and features the following properties:*

(a) *the semigroup property*

$$P(t+s) = P(s)P(t), \quad s,t \geq 0; \tag{2.37}$$

(b) *that $P(t)$ is the only solution to*

$$\frac{\mathrm{d}}{\mathrm{d}t} P(t) = P(t)Q, \, t \geq 0, \quad \text{the forward equation,}$$
$$\frac{\mathrm{d}}{\mathrm{d}t} P(t) = QP(t), \, t \geq 0, \quad \text{the backward equation,} \tag{2.38}$$

with $P(0) = \mathbf{I}$;

(c) *for all $k = 1, 2, \ldots,$*

$$\frac{\mathrm{d}^k}{\mathrm{d}t^k} P(t) \Big|_{t=0} = Q^k. \tag{2.39}$$

The backward and forward equations are often called the *Kolmogorov equations*, after Andrey Nikolaievich Kolmogorov (1903–1987), the great Russian mathematician who made important contributions to many areas of theoretical and applied mathematics. Kolmogorov is credited with providing a rigorous foundation for the whole of probability theory, and for more than 50 years was the recognised leader of the Soviet mathematics community. Unlike other Soviet mathematicians and physicists of the period, he never held particularly high administrative positions and did not participate directly in nuclear or space programmes. However, he had an unquestionable moral authority on many issues beyond mathematics and was greatly admired as an intellectual icon nationwide as well as internationally.

Another name mentioned in this context is William Feller (1906–1970), a famous Yugoslav-born American mathematician who greatly clarified the rôle of the forward and backward equations and helped to build a unified view of probability theory and its numerous applications. He wrote a classic book in two volumes (Wiley, 1968, 1971) which remains recommended reading for students in probability.

2.3 The Poisson process

The Poisson Adventure
(From the series *'Movies that never made it to the Big Screen'*.)

The Poisson process is precisely that introduced in Example 2.2.9 above. In view of its importance, we introduce some special notation.

Definition 2.3.1 Fix $\lambda > 0$. A family of random variables $(N_t, \, t \geq 0)$ with values in $\mathbb{Z}_+ = \{0, 1, \ldots\}$ is called a *Poisson process of rate* (or intensity) λ if

(i) $N_0 = 0$,

(ii) for all $0 < t_1 < t_2 < \cdots < t_n$ and non-negative integers $i_1, \ldots, i_n \in I$

$$\mathbb{P}(N_{t_1} = i_1, \ldots, N_{t_n} = i_n) = p_{0i_1}(t_1) p_{i_1 i_2}(t_2 - t_1) \cdots p_{i_{n-1} i_n}(t_n - t_{n-1}), \quad (2.40)$$

the $p_{ij}(t)$ being the entries of the matrix $P(t) = e^{tQ}$ specified in (2.35).

For brevity, we refer to the Poisson process of rate λ as PP(λ); its distribution will be denoted simply by $\mathbb{P}$ (rather than $\mathbb{P}_0$). We see that PP(λ) is a process with piecewise constant, non-decreasing (right-continuous) sample trajectories $N_t, t \geq 0$, starting at 0 and jumping by 1. Various aspects of the sample behaviour of the process are featured in Figures 2.15–2.29.

Equation (2.40) implies that a Poisson process has *independent incre-ments* $N_{t_{j+1}} - N_{t_j}$ over disjoint intervals (t_j, t_{j+1}) which are distributed as Po $(\lambda(t_{j+1} - t_j))$. In fact, setting $t_0 = 0$ and $i_0 = 0$, the probability $\mathbb{P}(N_{t_1} = i_1, \ldots, N_{t_n} = i_n)$ coincides with

$$\mathbb{P}(N_{t_1} - N_{t_0} = i_1 - i_0, \ldots, N_{t_n} - N_{t_{n-1}} = i_n - i_{n-1}),$$

the probability of observing a prescribed succession of increments. Then, repeating Definition 2.3.1, we see that for $0 = t_0 < t_1 < \cdots < t_n$ and $0 = i_0, i_1, \cdots, i_n \in \mathbb{Z}_+$

$$\mathbb{P}(N_{t_1} - N_{t_0} = i_1 - i_0, \ldots, N_{t_n} - N_{t_{n-1}} = i_n - i_{n-1})$$

$$= \begin{cases} \displaystyle\prod_{k=0}^{n-1} \frac{(\lambda(t_{k+1} - t_k))^{i_{k+1} - i_k}}{(i_{k+1} - i_k)!} e^{-\lambda(t_{k+1} - t_k)}, & \text{if } 0 \leq i_1 \leq \cdots \leq i_n, \\ 0, & \text{otherwise.} \end{cases} \quad (2.41)$$

Conversely, property (2.41) implies (2.40). This fact will prove important for our understanding of Poisson processes.

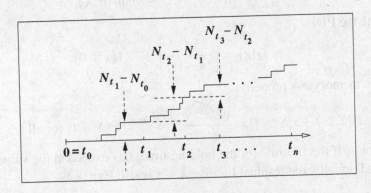

Fig. 2.15

Before we move further, recall some basic properties of the exponential distributions $\text{Exp}(\lambda)$.

(a) The PDF

$$f(x) = \lambda e^{-\lambda x}\mathbf{1}(x > 0).$$

(b) The CDF

$$F(x) := \mathbb{P}(X < x) = \int_{-\infty}^{x} f(y)\mathrm{d}y = \left(1 - e^{-\lambda x}\right) \times \mathbf{1}(x > 0).$$

(c) The tail probability

$$1 - F(x) = \mathbb{P}(X \geq x) = \begin{cases} e^{-\lambda x}, & x > 0, \\ 1, & x \leq 0. \end{cases}$$

(d) The mean value

$$\mathbb{E}X = \int_{0}^{\infty} x f(x)\mathrm{d}x = \lambda^{-1}.$$

(e) The variance

$$\mathbb{E}(X - \mathbb{E}X)^2 = \int_{0}^{\infty} \mathrm{d}x\,(x - \lambda^{-1})^2 f(x) = \lambda^{-2}.$$

(f) If $X_1 \sim \text{Exp}(\lambda_1), \ldots, X_n \sim \text{Exp}(\lambda_n)$, independently. Then

$$W = \min[X_1, \ldots, X_n] \sim \text{Exp}\left(\sum_{i=1}^{n} \lambda_i\right).$$

In fact,

$$\mathbb{P}(W > x) = \mathbb{P}(X_i > x,\ 1 \leq i \leq n) = \prod_{i=1}^{n}\mathbb{P}(X_i > x) = \prod_{i=1}^{n} e^{-\lambda_i x}.$$

(g) If $X_1 \sim \text{Exp}(\lambda_1), \ldots, X_n \sim \text{Exp}(\lambda_n)$, independently, then

$$Z = X_1 + \cdots + X_n \sim \text{Gam}(n, \lambda),$$

with the PDF

$$f_Z(x) = \left[\frac{\lambda^n x^{n-1}}{(n-1)!}\right] e^{-\lambda x}\mathbf{1}(x > 0).$$

(h) The memoryless property:

$$\mathbb{P}(X > t + s \mid X > s) = \frac{e^{-\lambda(t+s)}}{e^{-\lambda s}} = e^{-\lambda t} = \mathbb{P}(X > t) \text{ for all } t, s > 0.$$

That is, if the lifetime (or the holding time) has exceeded the value s then, conditionally, the residual lifetime $X - s$ is still $\text{Exp}(\lambda)$.

From now on we write variously N_t and $N(t)$ whichever notation for the Poisson process $\text{PP}(\lambda)$ would be more suitable. The main result of this section is

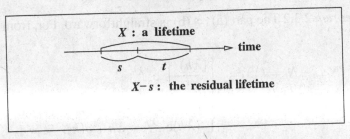

Fig. 2.16

Theorem 2.3.2 *The Poisson process* $\mathrm{PP}(\lambda)$ *can be characterised in three equivalent ways: a process taking values in* $\mathbb{Z}_+ = \{0, 1, \ldots\}$, *with* $N_0 = 0$ *and:*

(a) *satisfying* (2.40), *where* $P(t) = (p_{ij}(t))$, $t \geq 0$, *is the stochastic matrix given in* (2.35); *equivalently, as a process with independent Poisson distributed increments:*

$$N(t_k) - N(t_{k-1}) \sim \mathrm{Po}\left(\lambda(t_k - t_{k-1})\right), \quad \text{for all } 0 = t_0 < t_1 < \cdots < t_n; \quad (2.42)$$

or

(b) *with independent increments* $N(t_1) - N(t_0), \ldots, N(t_n) - N(t_{n-1})$, *for all* $0 = t_0 < t_1 < \cdots < t_n$, *and the following infinitesimal probabilities: for all* $t \geq 0$, *as* $h \searrow 0$

$$\left.\begin{array}{l} \mathbb{P}\left(N(t+h) - N(t) = 0\right) = 1 - \lambda h + o(h), \\ \mathbb{P}\left(N(t+h) - N(t) = 1\right) = \lambda h + o(h), \\ \mathbb{P}\left(N(t+h) - N(t) \geq 2\right) = o(h), \end{array}\right\} \quad (2.43)$$

where terms $o(h)$ *do not depend on* t;

or

(c) *spending a random time* $S_i \sim \mathrm{Exp}(\lambda)$ *in each state* i *independently, and then jumping to* $i+1$, $i = 0, 1, \ldots$.

We say that (a) forms the definition (or characterisation) via independent Poissonian increments, (b) via infinitesimal probabilities and (c) via exponential *holding times*.

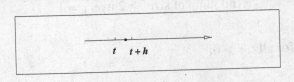

Fig. 2.17

Proof of Theorem 2.3.2 The part (a) ⇒ (b) is straightforward. For, from (a)

$$\mathbb{P}\left(N_{t+h} - N_t = \ell\right) = \left[\frac{(\lambda h)^{\ell}}{\ell!}\right] \mathrm{e}^{-\lambda h}$$

$$= \begin{cases} \mathrm{e}^{-\lambda h} = 1 - \lambda h + o(h), & \ell = 0, \\ (\lambda h)\mathrm{e}^{-\lambda h} = \lambda h + o(h), & \ell = 1, \end{cases}$$

and

$$\mathbb{P}\left(N_{t+h} - N_t \geq 2\right) = 1 - \mathbb{P}\left(N_{t+h} - N_t = 0 \text{ or } 1\right)$$
$$= 1 - (1 - \lambda h + \lambda h + o(h)) = o(h).$$

This yields (b).

(b) ⇒ (c) proves more involved. First check that no double jump exists, i.e.

$$\mathbb{P}\left(\text{no jump of size} \geq 2 \text{ occurs in } (0,t]\right)$$

$$= \mathbb{P}\left(\text{no such jump occurs in } \left(\frac{k-1}{m}t, \frac{k}{m}t\right] \text{ for all } k = 1,\ldots,m\right)$$

$$= \prod_{k=1}^{m} \mathbb{P}\left(\text{no such jump occurs in } \left(\frac{k-1}{m}t, \frac{k}{m}t\right]\right), \quad \text{by (b)},$$

$$\geq \prod_{k} \mathbb{P}\left(\text{no jump at all or a single jump of size 1 in } \left(\frac{k-1}{m}t, \frac{k}{m}t\right]\right)$$

$$= \left(1 - \lambda \frac{t}{m} + \lambda \frac{t}{m} + o\left(\frac{t}{m}\right)\right)^m$$

$$= \left(1 + o\left(\frac{t}{m}\right)\right)^m \to 1 \quad \text{as } m \to \infty.$$

This is true for all $t > 0$, so

$$\mathbb{P}\left(\text{no jump of size} \geq 2 \text{ ever}\right) = 1.$$

Next, check that for all $t, s > 0$:

$$\mathbb{P}\left(N_{t+s} - N_s = 0\right) = \mathrm{e}^{-\lambda t}.$$

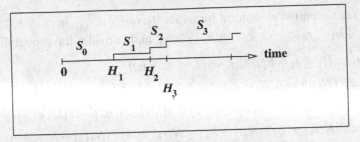

Fig. 2.18

In fact, as before

$$\mathbb{P}(N_{t+s} - N_s = 0)$$
$$= \mathbb{P}(\text{no jump in } (s, t+s])$$
$$= \mathbb{P}\left(\text{no jump in } \left(s + \frac{k-1}{m}t, \; s + \frac{k}{m}t\right] \text{ for all } k = 1, \ldots, m\right)$$
$$= \prod_{1}^{m} \mathbb{P}\left(\text{no jump in } \left(s + \frac{k-1}{m}t, \; s + \frac{k}{m}t\right]\right) \quad \text{by (b)}$$
$$= \left(1 - \lambda \frac{t}{m} + o\left(\frac{t}{m}\right)\right)^{m} \rightarrow e^{-\lambda t} \text{ as } m \rightarrow \infty.$$

Now introduce holding times

$$S_0 = \sup[t \geq 0 : N(t) = 0],$$
$$S_1 = \sup[t \geq 0 : N(S_0 + t) = 1]$$

and so on. Then the times of jump can be written

$$S_0 \quad (= H_1, \text{ the hitting time of state 1})$$
$$S_0 + S_1 \quad (= H_2, \text{ the hitting time of state 2})$$
$$\vdots$$
$$S_0 + \cdots + S_{k-1} \quad (= H_k, \text{ the hitting time of state } k)$$
$$\vdots$$

Further, take n pairwise disjoint intervals $[t_1, t_1 + h_1), \ldots, [t_n, t_n + h_n)$, with $0 = t_0 < t_1 < t_1 + h_1 < t_2 < \cdots < t_{n-1} + h_{n-1} < t_n$, and consider the probability

$$\mathbb{P}(t_1 < H_1 \leq t_1 + h_1, \ldots, t_n < H_n \leq t_n + h_n)$$
$$= \mathbb{P}(N(t_1) = 0, N(t_1 + h_1) - N(t_1) = 1, N(t_2) - N(t_1 + h_1) = 0,$$
$$\cdots, N(t_n) - N(t_{n-1} + h_{n-1}) = 0, N(t_n + h_n) - N(t_n) = 1)$$
$$= \mathbb{P}(N(t_1) = 0)\mathbb{P}(N(t_1 + h_1) - N(t_1) = 1)\mathbb{P}(N(t_2) - N(t_1 + h_1) = 0)$$
$$\times \cdots \times \mathbb{P}(N(t_n) - N(t_{n-1} + h_{n-1}) = 0)\mathbb{P}(N(t_n + h_n) - N(t_n) = 1)$$
$$= e^{-\lambda t_1}(\lambda h_1 + o(h_1))e^{-\lambda(t_2 - t_1 - h_1)}$$
$$\times \cdots \times e^{-\lambda(t_n - t_{n-1} - h_{n-1})}(\lambda h_n + o(h_n))$$

Divide by $h_1 \times \cdots \times h_n$ and let $h_k \to 0$. Then

$$\text{the LHS} \quad \to \quad \text{the joint PDF } f_{H_1, \ldots, H_n}(t_1, \cdots, t_n),$$

and

$$\text{the RHS} \quad \to \quad (e^{-\lambda t_1}\lambda)[e^{-\lambda(t_2 - t_1)}\lambda] \cdots [e^{-\lambda(t_n - t_{n-1})}\lambda].$$

Thus,

$$f_{H_1, \ldots, H_n}(t_1, \cdots, t_n) = \prod_{k=1}^{n}(\lambda \exp[-\lambda(t_k - t_{k-1})])\,\mathbf{1}(0 < t_1 < \cdots < t_n)$$
$$= \lambda^n e^{-\lambda t_n}\mathbf{1}(0 < t_1 < \cdots < t_n).$$

Now,

$$S_0 = H_1, \; S_2 = H_2 - H_1, \; \ldots, \; S_{n-1} = H_n - H_{n-1}.$$

The change of variables

$$s_0 = t_1, \; s_1 = t_2 - t_1, \; \ldots, \; s_{n-1} = t_n - t_{n-1}$$

gives

$$\text{the Jacobian} \left(\frac{\partial(s_0, \ldots, s_{n-1})}{\partial(t_1, \ldots, t_n)}\right) = 1$$
$$= \left(\frac{\partial(t_1, \ldots, t_n)}{\partial(s_0, \ldots, s_{n-1})}\right), \text{ the inverse Jacobian}.$$

Thus the joint PDF

$$f_{S_0, \ldots, S_{n-1}}(s_0, \ldots, s_{n-1}) = f_{H_1, \ldots, H_n}(s_0, s_0 + s_1, \ldots, s_0 + \cdots + s_{n-1})$$
$$= \prod_{k=0}^{n-1}[\lambda e^{-\lambda s_k}\mathbf{1}(s_k > 0)] = \prod_{0}^{n-1}f_{S_k}(s_k).$$

So, $S_0, S_1, \ldots$ are $\text{Exp}(\lambda)$, independently, which gives (c).

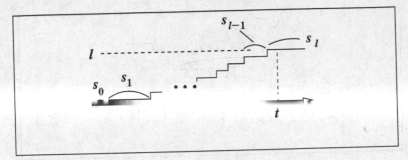

Fig. 2.19

Finally, (c) $\Rightarrow$ (a). To check the following equality,

$$\mathbb{P}\big(N(t_1) - N(0) = \ell_1,\ N(t_2) - N(t_1) = \ell_2, \ldots, N(t_n) - N(t_{n-1}) = \ell_n\big)$$
$$= \prod_1^n \frac{(\lambda(t_k - t_{k-1}))^{\ell_k}}{\ell_k!}\, e^{-\lambda t_n} \quad \text{(by independent increments)},$$

we use induction on n. The first step is $n = 1$, and we set $t_1 = t$, $\ell_1 = \ell$. With $H_\ell = S_0 + \cdots + S_{\ell-1}$, write

$$\mathbb{P}\big(N(t) = \ell\big) = \mathbb{P}\big(H_\ell < t < H_\ell + S_\ell\big) \quad \text{by (c)},$$

and use the fact that $H_\ell \sim \text{Gam}(\ell, \lambda)$, with $f_{H_\ell}(x) = \lambda^\ell x^{\ell-1} e^{-\lambda x}/(\ell-1)!,\ x > 0$:

$$\mathbb{P}\big(N(t) = \ell\big) = \int_0^t f_{H_\ell}(x) \mathbb{P}(S_\ell > t - x)\, dx = \int_0^t \frac{\lambda^\ell x^{\ell-1}}{(\ell-1)!} e^{-\lambda x} e^{-\lambda(t-x)} dx$$

$$= \frac{\lambda^\ell e^{-\lambda t}}{(\ell-1)!} \int_0^t x^{\ell-1} dx = \frac{(\lambda t)^\ell}{\ell!} e^{-\lambda t}. \tag{2.44}$$

Finally, to make the induction step from $n-1$ to n, it suffices to prove that the conditional probability

$$\mathbb{P}\big(N(t_n) - N(t_{n-1}) = \ell_n \mid N(t_k) - N(t_{k-1}) = \ell_k,\ 1 \le k \le n-1\big)$$
$$= \mathbb{P}\big(N(t_n - t_{n-1}) = \ell_n\big) = \left[\frac{(\lambda(t_n - t_{n-1}))^{\ell_n}}{\ell_n!}\right] e^{-\lambda(t_n - t_{n-1})}. \tag{2.45}$$

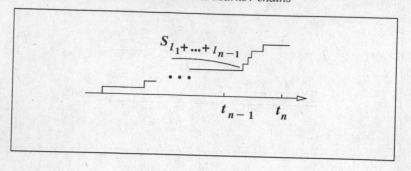

Fig. 2.20

But thanks to the memoryless property of the holding time

$$S_{\ell_1+\cdots+\ell_{n-1}} \sim \text{Exp}(\lambda),$$

this process thus yields (a).

$\square$

Comparing the above definitions (a)–(c) leads to the following insightful representation of the Poisson probability as an integral over subsequent jump times: for all $t, s > 0$ and $n, i = 0, 1, \ldots$

$$
\begin{aligned}
\frac{(\lambda t)^n}{n!} e^{-\lambda t} &= p_{0n}(t) = \mathbb{P}(N_t = n) \\
&= p_{i\,i+n}(t) = \mathbb{P}(N_{t+s} - N_s = n \mid N_s = i) \\
&= \mathbb{P}(N_{t+s} - N_s = n) \\
&= e^{-\lambda t} \lambda^n \int_0^t \int_0^t \cdots \int_0^t \mathbf{1}(t_1 < \cdots < t_n)\, dt_n\, dt_{n-1} \cdots dt_2\, dt_1 \\
&= \int_0^t \int_0^{t_n} \cdots \int_0^{t_2} \underbrace{\lambda \exp(-\lambda t_1)}_{\text{first jump between 0 and } t \text{ at } t_1}
\end{aligned}
$$

$$
\times \underbrace{(\lambda \exp[-\lambda(t_2 - t_1)])}_{\text{second jump between 0 and } t \text{ at } t_2} \times \cdots
$$

$$
\times \underbrace{(\lambda \exp[-\lambda(t_n - t_{n-1})])}_{n\text{th jump between 0 and } t \text{ at } t_n}
$$

$$
\times \underbrace{\exp[-\lambda(t - t_n)]}_{\text{no jump between } t_n \text{ and } t} \quad dt_1\, dt_2 \cdots dt_{n-1}\, dt_n. \quad (2.46)
$$

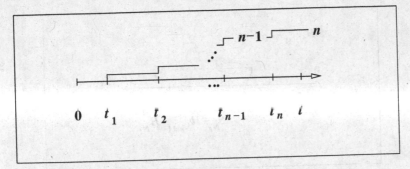

Fig. 2.21

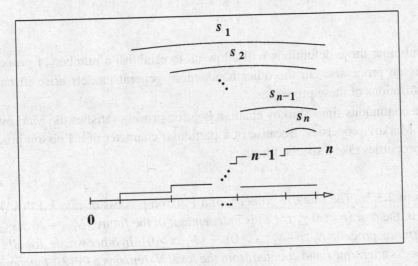

Fig. 2.22

An alternative representation comes in terms of the times $s_k = t - t_k$ between t_k, the point of the kth jump, and t:

$$
\frac{(\lambda t)^n}{n!} e^{-\lambda t} = \int_0^t \int_0^{s_1} \cdots \int_0^{s_{n-1}} \lambda \exp\left[-\lambda(t-s_1)\right]
$$
$$
\times \left(\lambda \exp\left[-\lambda(s_1-s_2)\right]\right) \times \cdots
$$
$$
\times \left(\lambda \exp\left[-\lambda(s_{n-1}-s_n)\right]\right)
$$
$$
\times \exp\left(-\lambda s_n\right) ds_n\, ds_{n-1} \cdots ds_2\, ds_1. \qquad (2.47)
$$

It is also useful to know that the Poisson process $PP(\lambda)$ $\left(N_t^{(\lambda)}\right)$ is obtained from $PP(1)$ $\left(N_t^{(1)}\right)$ by the *time change*

$$
\left(N_{\lambda t}^{(1)}\right) \sim \left(N_t^{(\lambda)}\right). \qquad (2.48)
$$

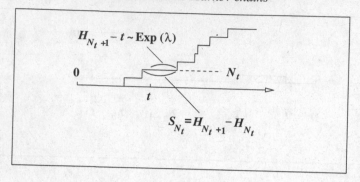

Fig. 2.23

Combining these definitions will allow us to establish a number of properties of Poisson processes. On the other hand, more general models arise in natural generalisations of these properties.

As a continuous-time Markov chain, a Poisson process satisfies the Markov and strong Markov properties. Because of a particular character of a Poisson process, these properties take a special form:

Theorem 2.3.3 *(The Markov property of a Poisson process of rate λ PP(λ)) For all $t > 0$, the past $(N_\tau : 0 \le \tau < t)$ is independent of the future $(N_{t+s} - N_t : s \ge 0)$. Furthermore, process $(N_{t+s} - N_t : s \ge 0) \sim (N_s, s \ge 0)$. In other words, for all $t > 0$ the process after time t and counted from the level N_t remains a PP(λ) independent of its past history $(N_\tau : 0 \le \tau < t)$.*

Observe that the RV $S_{N(t)} = H_{N(t)+1} - H_{N(t)}$ (the holding time covering point t) is *not* exponentialy distributed (the so-called inspection paradox, see below).

Theorem 2.3.4 *(The strong Markov property, with the stopping time H_k (the time of the kth jump)) The past $(N_s : 0 \le s < H_k)$ is independent of the future $(N_{H_k+s} - k : s \ge 0)$. Furthermore, the process $(N_{H_k+s} - k : s \ge 0) \sim (N_s, s \ge 0)$. In other words, the process observed from time H_k and counted from the level k is a PP(λ) independent of its past $(N_s : 0 \le s < H_k)$.*

We skip the proof of Theorem 2.3.4; notice only that the memoryless property of the exponential distribution plays an important role here.

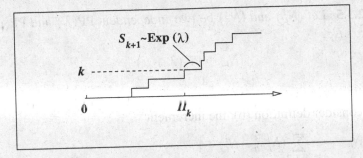

Fig. 2.24

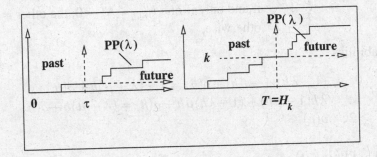

Fig. 2.25

It is instructive to compare graphically the Markov and the strong Markov properties for the Poisson process $PP(\lambda)$ (N_t):

Markov: for all $\tau > 0$

$(N_t, 0 \leq t \leq \tau)$ (past)

is independent
of the future

$(N_{\tau+t} - N_\tau, t \geq 0)$
$\sim PP(\lambda)$

strong Markov:
for all $k = 1, 2, \ldots$

$(N_t, 0 \leq t \leq H_k)$ (past)

$(N_{H_k+t} - k, t \geq 0)$
$\sim PP(\lambda)$

Markov and Strong Markov Private Property
(From the series '*Movies that never made it to the Big Screen*'.)

Our next result concerns the *sum* of independent Poisson processes.

Theorem 2.3.5 Let (N_t^1) and (N_t^2) be two independent $\mathrm{PP}(\lambda)$ and $\mathrm{PP}(\mu)$. Then, for $N_t = N_t^1 + N_t^2$,

$$(N_t) \sim \mathrm{PP}\,(\lambda + \mu). \tag{2.49}$$

Proof Both Poisson processes (N_t^i) feature independent increments. Hence, so does (N_t). Then consider definition (b): the increment

$$
\begin{aligned}
N_{t+h} - N_t &= \sum_{i=1,2} (N_{t+h}^i - N_t^i) \\
&= \begin{cases} 0 & \text{if and only if } N_{t+h}^i - N_t^i = 0, i = 1,2 \\ 1 & \text{if and only if one } N_{t+h}^i - N_t^i = 0, \text{ the other} = 1, \\ \geq 2 & \text{otherwise,} \end{cases}
\end{aligned}
$$

with probabilities

$$
\begin{aligned}
0: & \quad (1-\lambda h)(1-\mu h) + o(h) = 1 - (\lambda + \mu)h + o(h), \\
1: & \quad \lambda h(1-\mu h) + (1-\lambda h)\mu h + o(h) = (\lambda + \mu)h + o(h) \\
\geq 2: & \quad o(h).
\end{aligned}
$$

So, $(N_t) \sim \mathrm{PP}(\lambda + \mu)$. $\qquad\square$

This result should not be too surprising as the sum of independent Poisson random variables forms another Poisson variable. Adding Poisson processes can also be described as an operation of *superposition*: we count, in the appropriate order, all jumps in several processes. An 'inverse' operation can be described as 'thinning'.

Theorem 2.3.6 Let $(N_t) \sim \mathrm{PP}(\lambda)$ and $0 < p < 1$. Let (M_t) be the process where each jump in (N_t) is allowed with probability p, otherwise discarded (a thinned or slowed process). Then

$$(M_t) \sim \mathrm{PP}(p\lambda). \tag{2.50}$$

Proof Use definition (c): consider the holding times $S_0^M, S_1^M, \ldots$ in the process (M_t). The MGF equals

$$
\begin{aligned}
\mathbb{E}\!\left[e^{sS_0^M}\right] &= \mathbb{E}\left[\mathbb{E}\!\left(e^{sS_0^M} \middle| \text{number of discards}\right)\right] \\
&= \sum_{k \geq 0} (1-p)^k p\, \mathbb{E}\!\left(e^{sS_0^M} \middle| \text{number of discards} = k\right) \\
&= \sum_{k \geq 0} (1-p)^k p\left(\mathbb{E}\!\left[e^{sS_0^N}\right]\right)^{k+1} = \frac{p}{1-p} \sum_{k \geq 0} \left((1-p)\frac{\lambda}{\lambda - s}\right)^{k+1} \\
&= \frac{p}{1-p}\left(\frac{(1-p)\lambda/(\lambda - s)}{1 - (1-p)\lambda/(\lambda - s)}\right) = \frac{p\lambda}{p\lambda - s}.
\end{aligned}
$$

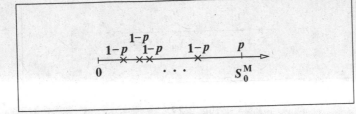

Fig. 2.26

We immediately deduce that $S_0^M \sim \mathrm{Exp}(p\lambda)$. Similarly, $S_1^M, S_2^M, \cdots \sim \mathrm{Exp}(p\lambda)$. The independence follows obviously. So, $(M_t) \sim \mathrm{PP}(\lambda p)$. $\qquad\square$

Summarising, both operations (adding independent Poisson processes and thinning one Poisson process) still leaves us in the class of Poisson processes; this plays an important rôle in a number of applications; see below.

Theorem 2.3.7 *Let $(N_t) \sim \mathrm{PP}(\lambda)$. Then for all $s, t > 0$ and $m = 1, 2, \ldots$, conditional on $N_{t+s} - N_s = m$, the jump points $J_1 = J_1(s,t), \ldots, J_m = J_m(s,t)$ in $(s, s+t)$ exhibit the joint PDF*

$$f_{J_1,\ldots,J_m}(x_1,\ldots,x_m \mid m \text{ jumps in total in } (s,s+t))$$
$$= \left(\frac{m!}{t^m}\right)\mathbf{1}(s < x_1 < \cdots < x_m < t+s). \qquad (2.51)$$

That is, 'conditional' RVs

$$(J_1, \cdots, J_m \mid m \text{ jumps in interval } (s, s+t))$$

are obtained by throwing m uniform IID points on the interval $(s, s+t)$ and listing them in the increasing order.

In particular, conditional on $m = 1$ (a single jump), the point J_1 is distributed $U(s, s+t)$.

Proof Use definition (b): for all $s < x_1 < \cdots < x_m < (t+s)$ and small h_i

$$\mathbb{P}(x_i < J_i < x_i + h_i,\ 1 \le i \le m;\ m \text{ points in total in } (s, s+t))$$
$$= \mathbb{P}(N_{x_k} - N_{x_{k-1}+h_{k-1}} = 0, N_{x_k+h_k} - N_{x_k} = 1, 1 \le k \le m$$
$$\text{(with } x_0 + h_0 = s\text{)},\ N_{t+s} - N_{x_m+h_m} = 0)$$
$$= e^{-\lambda(x_1-s)}(\lambda h_1)e^{-\lambda(x_2-x_1-h_1)}(\lambda h_2) \times \cdots \times (\lambda h_m)e^{-\lambda(t+s-x_m-h_m)}$$
$$= \left(\prod_{i=1}^m e^{-\lambda(x_i-x_{i-1}-h_{i-1})}(\lambda h_i)\right) \times e^{-\lambda(t+s-x_m-h_m)}.$$

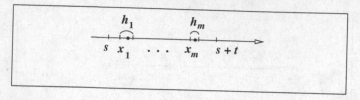

Fig. 2.27

Next,

$$\mathbb{P}(m \text{ jump points in total in } (s, s+t)) = \mathbb{P}(N_{t+s} - N_s = m) = \left(\frac{\lambda^m t^m}{m!}\right) e^{-\lambda t}.$$

Divide by $h_1 h_2 \ldots h_m$, and let $h_i \to 0$:

$$f_{J_1,\ldots,J_m}(x_1,\ldots,x_m \mid m \text{ in all}) = \frac{m!}{t^m}.$$

And if $\mathbf{1}(s < x_1 < \cdots < x_m < s+t) = 0$, then $f_{J_1,\ldots,J_m}(x_1,\ldots,x_m \mid m \text{ in all}) = 0$.

$\square$

Worked Example 2.3.8 Guillemots and puffins arrive to make their nests at a cliff. Arrivals of the two species are independent, as are arrivals of each species in disjoint intervals. It is observed that in any small interval, of length h say, the probability that no guillemots arrive is $1 - \lambda h + o(h)$; and that one guillemot arrives with probability $\lambda h + o(h)$. The corresponding probabilities for puffins are $1 - \mu h + o(h)$ and $\mu h + o(h)$. Determine from first principles the distribution of the total number of birds arriving in an interval of length t.

What is the probability that the first three birds to arrive are all puffins?

Suppose that by time t, just one guillemot and one puffin have arrived. What is the probability that both arrived by time s, for $s \leq t$.

What is the probability that the puffin arrived first?

Solution By independence,

$$\mathbb{P}(\text{no birds in } (t, t+h)) = (1 - \lambda h + o(h))(1 - \mu h + o(h)) = 1 - (\lambda + \mu)h + o(h).$$

Also,

$$\begin{aligned}
\mathbb{P}(\text{one bird in } (t, t+h))) \\
= (1 - \lambda h + o(h))(\mu h + o(h)) + (\lambda h + o(h))(1 - \mu h + o(h)) \\
= (\lambda + \mu)h + o(h),
\end{aligned}$$

and

$$\mathbb{P}(\text{two or more birds in } (t, t+h))) = o(h).$$

Let (Z_t) be the superposition of the two processes and $p_{ij}(t)$ stand for $\mathbb{P}_i(Z_t = j)$. Set $v = \lambda + \mu$. Still by independence,

$$p_{00}(t+h) = p_{00}(t)(1 - vh + o(h)), \text{ implying that } \frac{d}{dt} p_{00}(t) = -v p_{00}(t),$$

whence $p_{00}(t) = e^{-vt}$ (as $p_{00}(0+) - 1$). Similarly,

$$p_{0i}(t+h) = p_{0i-1}(t)(vh + o(h)) + p_{0i}(t)(1 - vh + o(h)),$$

implying that

$$\frac{d}{dt} p_{0i}(t) = v(p_{0i-1}(t) - p_{0i}(t)),$$

whence $p_{0i}(t) = e^{-vt}(vt)^i/i!$. That is, (Z_t) is Poisson, of rate $\lambda + \mu$.

Next,

$$\mathbb{P}\big(\text{a puffin in } (t, t+h) \,\big|\, \text{one bird in } (t, t+h)\big) = \frac{\mu}{\lambda + \mu} + o(1).$$

Thus,

$$\mathbb{P}\big(\text{1st three birds are puffins}\big) = \left(\frac{\mu}{\lambda + \mu}\right)^3,$$

as the arrival times are irrelevant.

Further,

$$\mathbb{P}\big(\text{both birds arrive before } s \,\big|\, \text{one bird of each type lands before } t\big)$$

$$= \frac{e^{-v(t-s)} e^{-vs}(vs)^2 2\lambda\mu/(v^2)}{e^{-vt}(vt)^2 2\lambda\mu/(v^2)} = \frac{s^2}{t^2}.$$

Finally,

$$\mathbb{P}\big(\text{puffin arrives first} \,\big|\, \text{one bird of each type lands before } t\big)$$

$$= \frac{e^{-vt}(vt)^2\lambda\mu/(v^2)}{e^{-vt}(vt)^2 2\lambda\mu/(v^2)} = \frac{1}{2}.$$

$\square$

We now pass to the analysis of the asymptotic properties of Poisson processes. This will lead us to the important concepts of explosion in a CTMC.

Theorem 2.3.9 *In the Poisson process* $\mathrm{PP}(\lambda)$ (N_t):

$$N_t \nearrow \infty \text{ and } H_n = \inf\,[t \geq 0 : N_t = n] \nearrow \infty \text{ a.s.}$$

That is,

$$\lim_{t \to \infty} N_t = \lim_{n \to \infty} H_n = \infty \text{ with probability 1.}$$

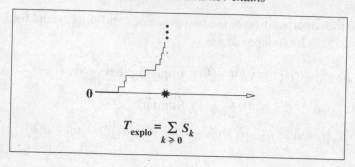

Fig. 2.28

Proof Begin with the hitting times $H_n = S_0 + \cdots + S_{n-1}$; recall, $S_i \sim \text{Exp}(\lambda)$, independently. So, $H_{n+1} > H_n$, i.e. H_n monotonically increases with n. Thus, either $H_n \nearrow \infty$ or H_n remains bounded. Our goal is to check that

$$\mathbb{P}\big(H_n \text{ remains bounded}\big) = \mathbb{P}\left(\sum_{k\geq 0} S_k < \infty\right) = 0.$$

The event $\{\sum_{k\geq 0} S_k < \infty\}$ means 'explosion'. See Figure 2.28.

A comprehensible argument runs as follow. Set

$$T_{\text{explo}} = \sum_{k\geq 0} S_k = \lim_{K\to\infty} \sum_{k=0}^{K} S_k = \lim_{K\to\infty} H_{K+1}$$

and use the MGF $\mathbb{E}\big[e^{\theta T_{\text{explo}}}\big]$, for $\theta = -1$. Formally, the random variable $e^{-T_{\text{explo}}}$ is defined as the limit

$$\lim_{K\to\infty} \exp\left[-\sum_{k=0}^{K} S_k\right] = \lim_{K\to\infty} \prod_{k=0}^{K} e^{-S_k}.$$

The convergence is monotone, and the variable $e^{-T_{\text{explo}}}$ is confined between 0 and 1 and equals 0 whenever $T_{\text{explo}} = \infty$. Consequently, the expectation $\mathbb{E}\big[e^{-T_{\text{explo}}}\big]$ should be understood as $\mathbb{E}\big[e^{-T_{\text{explo}}}\mathbf{1}(T_{\text{explo}} < \infty)\big]$. On the other hand,

$$\mathbb{E}\big[e^{-T_{\text{explo}}}\big] = \lim_{K\to\infty} \mathbb{E}\left(\prod_{k=0}^{K} e^{-S_k}\right),$$

by the monotone convergence theorem (one can also invoke the bounded convergence theorem, otherwise known as Lebesgue's dominated convergence theorem).

Because of the independence of $S_0, S_1, \ldots,$

$$\mathbb{E}\left(\prod_{k=0}^{K} e^{-S_k}\right) = \prod_{k=0}^{K} \mathbb{E}\left[e^{-S_k}\right] = \left(\frac{\lambda}{\lambda+1}\right)^{K},$$

which tends to 0 as $K \to \infty$. Thus,

$$\mathbb{E}\left[e^{-T_{\text{explo}}}\right] = 0.$$

We conclude that, with probability 1, the variable $e^{-T_{\text{explo}}}$ equals 0 and hence $T_{\text{explo}} = \infty$, meaning that the series $\sum_k S_k$ diverges and $N_n \nearrow \infty$:

$$\mathbb{P}\left(\sum_{k \geq 0} S_k = \infty\right) = \mathbb{P}\left(\lim_{n \to \infty} H_n = \infty\right) = 1.$$

Next, similarly, either $N_t \nearrow \infty$ or N_t remains bounded (actually not changing after a certain (random) time). In other words, the event $\{N_t \nrightarrow \infty\}$ coincides with the events

$$\{\text{for some } t > 0: N_{t+s} \equiv N_t \text{ for all } s > 0\} = \left\{\sum_k L_k < \infty\right\}.$$

Here $L_k = N_{k+1} - N_k$ represents the increment over a unit time interval $[k, k+1)$; we know that $L_k \sim \text{Po}(1)$, independently, $k = 0, 1, \ldots$. The MGF $\mathbb{E}\left[e^{\theta L_k}\right] = \exp\left(e^{\theta} - 1\right)$.

Mimicking the above argument, set $U = \sum_k L_k$ and use the MGF $\mathbb{E}\left[e^{-U}\right]$. Arguing as before,

$$\mathbb{E}\left[e^{-U}\right] = \lim_{K \to \infty} \left[\exp\left(e^{-1} - 1\right)\right]^{K} = 0,$$

which implies that, with probability 1, the variable e^{-U} equals 0 and that $U = \infty$, meaning that the series $\sum_k L_k$ diverges and $N_t \nearrow \infty$. $\qquad \square$

We conclude this section with a brief discussion of the so-called *inspection paradox* for a Poisson process.

Consider the length $S_{N(t)}$ of the holding interval containing the time-point t. It follows this distribution function:

$$\mathbb{P}(S_{N(t)} \leq x) = 1 - (1 + \lambda \min[t, x]) e^{-\lambda x}, \quad x \geq 0, \tag{2.52}$$

with the expected value

$$\mathbb{E}[S_{N(t)}] = \int_0^{\infty} (1 + \lambda \min[t, x]) e^{-\lambda x} dx = (2 - e^{-\lambda t})/\lambda. \tag{2.53}$$

The key remark here lies in that $S_{N(t)} \sim \left(\min[t, S^-] + S^+\right)$ where $S^{\pm} \sim \text{Exp}(\lambda)$, independently.

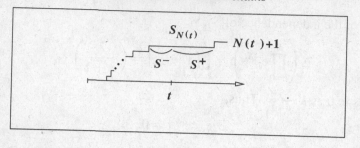

Fig. 2.29

In particular, the tail probability forms a cut-off exponential: for all $y > 0$,

$$\mathbb{P}\big(\min[t, S^-] > y\big) = \mathbf{1}(0 < y < t)e^{-\lambda y}, \tag{2.54}$$

which yields

$$\mathbb{E}\big(\min[t, S^-]\big) = \int_0^\infty \mathbb{P}\big(\min[t, S^-] > y\big)\, dy = \int_0^t e^{-\lambda y}dy = \frac{1}{\lambda}\big(1 - e^{-\lambda t}\big).$$

It also gives the convolution formula

$$
\begin{aligned}
F_{S_{N(t)}}(x) &= \mathbb{P}(S_{N(t)} \le x) = \int_0^x \mathbb{P}(\min[t, S^-] \le x - s)\big(\lambda e^{-\lambda s}\big)\, ds \\
&= \lambda \int_0^{(x-t)_+} e^{-\lambda s}ds + \lambda \int_{(x-t)_+}^x \big(1 - e^{-\lambda(x-s)}\big)e^{-\lambda s}ds \\
&= \lambda \int_0^x e^{-\lambda s}ds - \lambda e^{-\lambda x}\int_{(x-t)_+}^x ds \\
&= 1 - e^{-\lambda x} - \lambda e^{-\lambda x}\big[x - (x-t)_+\big]
\end{aligned}
$$

where $(x-t)_+ = \max[x - t, 0]$. As $x - (x-t)_+ = \min[x, t]$, we obtain (2.52).

For the expected value:

$$\mathbb{E}\big[S_{N(t)}\big] = \mathbb{E}\big[\min[t, S^-]\big] + \mathbb{E}\big[S^+\big] = \frac{1}{\lambda}\big(1 - e^{-\lambda t}\big) + \frac{1}{\lambda} = \frac{2 - e^{-\lambda t}}{\lambda},$$

which yields (2.53).

It is worth noting that, as $t \to \infty$, the tail probability

$$\mathbb{P}(S_{N(t)} > x) = \big(1 + \lambda \min[t, x]\big)e^{-\lambda x} \to (1 + \lambda x)e^{-\lambda x}, \quad x > 0. \tag{2.55}$$

The RHS of (2.55) is the tail probability of the $\text{Gam}(2, \lambda)$ distribution. The corresponding PDF is

$$f_{\text{Gam}(2,\lambda)}(x) = \lambda^2 x e^{-\lambda x}\mathbf{1}(x > 0).$$

In other words, the random variable $S_{N(t)}$ converges in distribution to the sum of two IID $\mathrm{Exp}(\lambda)$ random varaibles. Clearly, one of them can be identified with S^+, the other with S^-. Also observe that for $0 < t < \infty$,

$$e^{-\lambda x} < (1 + \lambda \min[t, x])e^{-\lambda x} < (1 + \lambda x)e^{-\lambda x},$$

i.e. the tail probabilities satisfy

$$\mathbb{P}(S^+ \geq x) < \mathbb{P}(S_{N(t)} > x) < \mathbb{P}(S^+ + S^- > x), \; x > 0.$$

In this situation, one says that $S_{N(t)}$ is *stochastically larger* than S^+, but *stochastically smaller* than $(S^- + S^+)$. It is a specific example of *stochastic order* between random variables where $X \prec Y$ means that the cumulative distribution functions $F_X(t)$ and $F_Y(t)$ and tail probabilities $\overline{F}_X(t)$ and $\overline{F}_Y(t)$ satisfy

$$F_X(t) = \mathbb{P}(X \leq t) \geq \mathbb{P}(Y \leq x) = F_Y(t), \; t \in \mathbb{R},$$

or

$$\overline{F}_X(t) = \mathbb{P}(X > t) \leq \mathbb{P}(Y > t) = \overline{F}_Y(t), \; t \in \mathbb{R}.$$

The inspection paradox has attracted a lot of attention in the literature. Perhaps the most spectacular attempts to exploit it are related to the search for UFOs and extra-terrestrial civilisations.

Worked Example 2.3.10 A Poisson process of rate λ is observed by someone who believes that the first holding time is longer than all subsequent times. How long on average will it take before the observer is proved wrong?

Solution The observer detects the first holding time J_1; to prove that he is wrong, one must wait for a holding time which is at least of the same length. Given that $J_1 = s$, the conditional expected time till such an event becomes $s + \mathbb{E}T(s)$ where

$$T(s) = \inf \{t \geq s : N_t = N_{t-s}\}.$$

For $\mathbb{E}T(s)$ we have the equation

$$\mathbb{E}T(s) = se^{-\lambda s} + \int_0^s da \, (a + \mathbb{E}T(s))\lambda e^{-\lambda a}$$

obtained by conditioning on the first jump. Hence,

$$\mathbb{E}T(s) = (e^{\lambda s} - 1)/\lambda.$$

Then the mean time until one sees the holding time greater or equal to J_1 is given by

$$\int_0^\infty \lambda e^{-\lambda s}\left(s + \frac{e^{\lambda s} - 1}{\lambda}\right) ds = \infty.$$

□

Worked Example 2.3.11 **(i)** In each of the following cases, the state space I and non-zero transition rates q_{ij} ($i \neq j$) of a continuous-time Markov chain are given. Determine in which cases the chain is explosive.

(a) $I = \{1, 2, 3, \ldots\}$, $q_{i,i+1} = i^2$, $i \in I$,

(b) $I = \mathbb{Z}$, $q_{i,i+1} = q_{i,i-1} = 2^i$, $i \in I$.

(ii) Children arrive at a see-saw according to a Poisson process of rate 1. Initially there are no children. The first child to arrive waits at the see-saw. When the second child arrives, they play on the see-saw. When the third child arrives, they all decide to go and play on the merry-go-round. The cycle then repeats. Show that the number of children at the see-saw evolves as a Markov chain and determine its generator matrix. Find the probability that there are no children at the see-saw at time t.

Hence obtain the identity

$$\sum_{n=0}^\infty e^{-t} \frac{t^{3n}}{(3n)!} = \frac{1}{3} + \frac{2}{3} e^{-3t/2} \cos\left(\frac{\sqrt{3}}{2} t\right).$$

Solution **(i)** (a) The worst case surfaces when $X_0 = 1$. The nth holding time $J_n \sim \text{Exp}(n^2)$. So, the expected time till explosion is

$$\mathbb{E}\left(\sum_n J_n\right) = \sum_n \mathbb{E}J_n = \sum_n \frac{1}{n^2} < \infty,$$

and the chain is explosive.

(b) The jump chain is the simple symmetric random walk on $\mathbb{Z}$ which is recurrent and visits 0 infinitely often. Denote the holding times on successive visits to 0 by $T_1, T_2, \ldots$. Then $T_i \sim \text{Exp}(2)$, independently, and $\mathbb{P}(\sum_i T_i = \infty) = 1$. The explosion time exceeds $\sum_i T_i$, hence the chain is non-explosive.

(ii) Owing to Poisson arrival, the number of children at the see-saw is a three-state Markov chain with generator (Q-matrix)

$$Q = \begin{pmatrix} -1 & 1 & 0 \\ 0 & -1 & 1 \\ 1 & 0 & -1 \end{pmatrix}.$$

To find $p_{00}(t)$, compute

$$\det(\kappa \mathbf{I} - Q) = \det \begin{pmatrix} \kappa+1 & -1 & 0 \\ 0 & \kappa+1 & -1 \\ -1 & 0 & \kappa+1 \end{pmatrix} = (\kappa+1)^3 - 1 = \kappa(\kappa^2 + 3\kappa + 3).$$

So, the eigenvalues of Q are 0 and $-3/2 \pm \sqrt{3}i/2$. By a standard diagonalization argument

$$p_{00}(t) = A + e^{-\frac{3}{2}t}\left(B\cos\left(\frac{\sqrt{3}}{2}t\right) + C\sin\left(\frac{\sqrt{3}}{2}t\right)\right)$$

for some constants A, B, C. Observe that

$$\frac{1}{3} = p_{00}(\infty) = A, 1 = p_{00}(0) = A + B,$$

$$-1 = q_{00} = \frac{d}{dt}p_{00}(0) = -\frac{3}{2}B + \frac{\sqrt{3}}{2}C,$$

implying $B = 2/3$ and $C = 0$. We conclude that

$$p_{00}(t) = \frac{1}{3} + \frac{2}{3}e^{-\frac{3}{2}t}\cos\left(\frac{\sqrt{3}}{2}t\right).$$

Alternatively, since the see-saw gets vacant exactly when the total number of arrivals is a multiple of 3,

$$p_{00}(t) = \sum_{n=0}^{\infty} e^{-t}\frac{t^{3n}}{(3n)!}.$$

This yields the identity. □

2.4 Inhomogeneous Poisson process

> ... *and since a woman must wear chains,*
> *I would have the pleasure of hearing 'em rattle a bit.*
> G. Farquhar (1678–1707), Irish dramatist

The concept of a Poisson process turns out to be extremely fruitful and leads to numerous generalisations.

We begin with *inhomogeneous* Poisson processes (IPP). Here we generalise definitions (a) (independent Poisson distributed increments) and (b) (infinitesimal

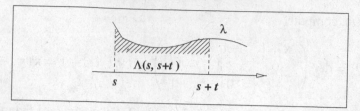

Fig. 2.30

probabilities) established in Theorem 2.3.2. It is convenient to start with characterisation (b): here we simply replace the constant rate λ by $\lambda(t)$, the rate varying in
time.

$$\mathbb{P}\big(\text{no jump in } (t,t+h)\big) = 1 - \lambda(t)h + o(h)$$
$$\mathbb{P}\big(\text{one jump in } (t,t+h)\big) = \lambda(t)h + o(h) \tag{2.56}$$
$$\mathbb{P}\big(\text{two or more jumps in } (t,t+h)\big) = o(h).$$

We keep the assumption of the increments over non-overlapping time intervals
being independent.

Characterisation (a) can also be re-phrased in a straightforward manner: for all
$s, t > 0$,

$$\text{number of jumps in } (s, t+s) \sim \text{Po}\left(\Lambda(s, t+s)\right), \tag{2.57}$$

independently for non-overlapping intervals. Here

$$\Lambda(s, t+s) = \int_s^{s+t} \lambda(u)\, du \quad (\text{previously } \lambda t\, \mathbf{1}(s > 0))$$
$$= \mathbb{E}\big(\text{number of jumps in } (s, s+t)\big).$$

We assumed here that $\Lambda(s, t)$ is finite for all $0 < s < t < \infty$.

We call this process an inhomogeneous Poisson process of rate $\lambda(t)$ (an
IPP$(\lambda(t))$ for short). Formally,

Definition 2.4.1 An IPP of rate $\lambda(t), t > 0$, is a non-decreasing process $(N_t, t \geq 0)$
with values in $\mathbb{Z}_+$ and $N_0 = 0$ satisfying the following equivalent properties (a) and
(b).

(a) The process has independent increments $N(t_{j+1}) - N(t_j) \sim \text{Po}\left(\Lambda\left(t_j, t_{j+1}\right)\right)$
over disjoint time intervals. That is, for all $0 = t_0 < t_1 < \cdots < t_n$ and $0 = i_0, i_1, \ldots, i_n \in \mathbb{Z}_+$:

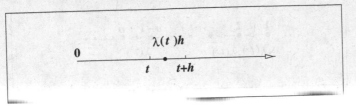

Fig. 2.31

$$\mathbb{P}\left(N_{t_1} - N_{t_0} = i_1 - i_0, \ldots, N(t_n) - N(t_{n-1}) = i_n - i_{n-1}\right)$$

$$= \begin{cases} \displaystyle\prod_{k=0}^{n-1} \frac{\left(\Lambda(t_k, t_{k+1})\right)^{i_{k+1}-i_k}}{(i_{k+1} - i_k)!} \exp\left(-\int_{t_0}^{t_n} \lambda(u)\, du\right), & \text{if } 0 \leq i_1 \leq \cdots \leq i_n, \\ & \hspace{3.5cm} (2.58) \\ 0, & \text{otherwise.} \end{cases}$$

(b) The process has independent increments over disjoint time intervals, and obeys the following asymptotics as $h \to 0+$:

$$\left.\begin{array}{c} \mathbb{P}\left(N_{t+h} - N_t = 0\right) = 1 - \lambda(t)h + o(h), \\ \mathbb{P}\left(N_{t+h} - N_t = 1\right) = \lambda(t)h + o(h), \\ \mathbb{P}\left(N_{t+h} - N_t > 1\right) = o(h). \end{array}\right\} \qquad (2.59)$$

As in the case of a homogeneous Poisson process, $\mathbb{P}$ stands for the distribution of an IPP $(\lambda(t))$.

The characterisation of an inhomogeneous Poisson process IPP$(\lambda(t))$ via part (c) from Theorem 2.3.2 does not feel so natural as it needs more substantial changes. For example, inhomogeneous Poisson processes exhibit non-independent and non-exponential holding times. In fact, for all $s > 0$ and $n \geq 1$, the conditional probability satisfies

$$\mathbb{P}\left(S_n = \infty \big| H_n = S_0 + \cdots + S_{n-1} = s\right) = e^{-\Lambda(s,\infty)}, \qquad (2.60)$$

where

$$\Lambda(s, \infty) = \int_s^\infty \lambda(u)\, du.$$

That is, if $\Lambda(s, \infty) < \infty$, the process will 'stagnate' with the positive probability $\exp\left[-\Lambda(s, \infty)\right]$. This means that, with positive probability, only finitely many jumps will occur on the whole time interval $(0, \infty)$. (This does not take place for a homogeneous Poisson process.)

Next, the conditional PDF

$$f_{S_n|H_{n-1}}\left(x \big| H_{n-1} = s,\, S_n < \infty\right) = \lambda(s+x)\left[\frac{e^{-\Lambda(s, s+x)}}{1 - e^{-\Lambda(s, \infty)}}\right], \quad x > 0; \qquad (2.61)$$

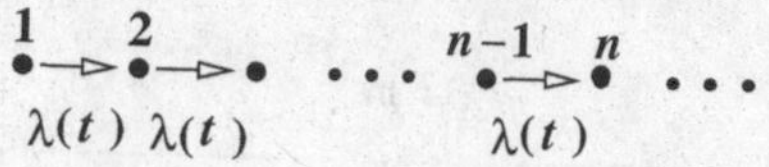

Fig. 2.32

when $\Lambda(s,\infty) = \infty$, the probability $\mathbb{P}(S_n < \infty) = 1$ and the denominator $1 - e^{-\Lambda(s,\infty)} \equiv 1$. In this case the inequality $S_n < \infty$ can be omitted from the condition. It is easy to see that, assuming the local rate $\lambda(t)$ to be differentiable in t, the PDF $f_{S_n|H_{n-1}}$ does not depend on s (e.g. satisfies $\dfrac{\partial}{\partial s} f_{S_n|H_{n-1}}(x|s) = 0$) if and only if $\lambda(u) \equiv \lambda$ (that is, the inhomogeneous Poisson process turns out to be homogeneous).

An inhomogeneous Poisson process provides an example of an inhomogeneous CTMC where the transition rates, and hence transition probabilities, vary with the time. More precisely, the transition probability

$$\mathbb{P}(N_{t+s} = i + k | N_t = i) = p_{0k}(t,t+s) = \left(\frac{(\Lambda(t,t+s))^k}{k!} \right) \exp\left(-\Lambda(t,t+s)\right).$$

$$(2.62)$$

Conventionally, the arrow diagram of an inhomogeneous Poisson process $\text{IPP}(\lambda(t))$ is drawn as in Figure 2.32.

If $\Lambda(s,s+t) = \infty$, we see an accumulation of jump points in interval $(s,s+t)$ (explosion). We will explore these 'pathologies' in more detail for other classes of CTMCs.

Worked Example 2.4.2 The fire alarm in a University building on Martingale Close is set off at random times:

$$\mathbb{P}\left(\text{alarm in } (u,u+h)\right) = \lambda(u)h + o(h).$$

The rate $\lambda(u)$ may vary with u. Let N_t be the number of alarms by time t. Show, by introducing reasonable additional assumptions, that

$$p_i(t) = \mathbb{P}(N_t = i)$$

obeys

$$\begin{aligned}
\dot{p}_0(t) &= -\lambda(t)p_0(t), \\
\dot{p}_i(t) &= \lambda(t)(p_{i-1}(t) - p_i(t)), \ i \geq 1,
\end{aligned}$$

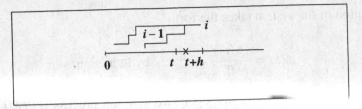

Fig. 2.33

and check that $N_t \sim \mathrm{Po}(\Lambda(t))$, with

$$\Lambda(t) = \int_0^t \lambda(u)\,\mathrm{d}u.$$

Solution Assume $N_0 = 0$, λ is continuous, and for all $t \geq 0$, $i = 1$ and $h \searrow 0$:

$$
\begin{aligned}
\mathbb{P}(N_{t+h} - N_t = 0 \mid N_t = i) &= 1 - \lambda(t)h + o(h), \\
\mathbb{P}(N_{t+h} - N_t = 1 \mid N_t = i) &= \lambda(t)h + o(h),
\end{aligned}
$$

and

$$\mathbb{P}(N_{t+h} - N_t > 1 \mid N_t = i) = o(h).$$

Then

$$
\begin{aligned}
p_i(t+h) &= \mathbb{P}(N_{t+h} = i) = \mathbb{P}(N_t = N_{t+h} = i) \\
&\quad + \mathbb{P}(N_t = N_{t+h} - 1 = i - 1) + o(h) \\
&= p_i(t)\mathbb{P}(N_{t+h} - N_t = 0 \mid N_t = i) \\
&\quad + p_{i-1}(t)\mathbb{P}(N_{t+h} - N_t = 1 \mid N_t = i - 1) + o(h) \\
&= p_i(t)(1 - \lambda(t)h) + p_{i-1}(t)\lambda(t)h + o(h)
\end{aligned}
$$

and

$$\frac{1}{h}(p_i(t+h) - p_i(t)) = -\lambda(t)(p_i(t) - p_{i-1}(t)) + o(1).$$

Letting $h \to 0$ yields

$$\frac{\mathrm{d}}{\mathrm{d}t}p_i(t) = -\lambda(t)(p_i(t) - p_{i-1}(t)),$$

Also, $p_i(0) = \delta_{i0}$.

$\square$

The solution of the system takes the form

$$p_i(t) = \frac{(\Lambda(t))^i}{i!} e^{-\Lambda(t)}, \quad \text{by induction.}$$

In fact, (N_t) constitutes an inhomogeneous Poisson process $\text{IPP}(\lambda(t))$, as the increments $N_{t+s} - N_s \sim \text{Po}(\Lambda(s,t))$, independently over non-overlapping time intervals. Here

$$\Lambda(s,t) = \int_s^t \lambda(u)\, du.$$

Example 2.4.3 It is instructive to understand that an inhomogeneous Poisson process $\text{IPP}(\lambda(t))$ can be produced by a *time change*

$$t \mapsto \Lambda(0,t) = \int_0^t \lambda(u)\, du \tag{2.63}$$

from a Poisson process of constant rate 1. This means the following. Assume that $\lambda(t)$ is a continuous positive function such that $\int_0^\infty \lambda(t)\, dt = \infty$. Let (N_t) be a Poisson process $\text{PP}(1)$. Set

$$N_t^{\text{IH}} = N_{\Lambda(0,t)}, \quad t > 0.$$

Then (N_t^{IH}) is an inhomogeneous Poisson process $\text{IPP}(\lambda(t))$. In particular, a homogeneous process $\text{PP}(\lambda)$ results from the time change $t \mapsto \lambda t$.

The simplest way to check this fact is by using the infinitesimal definition (b): the image of the Poisson process $\text{PP}(1)$ under the suggested time change feartures independent increments over disjoint intervals, and probabilities $1 - \lambda(t)h + o(h)$ and $\lambda(t)h + o(h)$ of observing no jump or a single jump over the increment $[t, t+h)$ when $h \searrow 0$.

Example 2.4.4 (Record processes) Inhomogeneous Poisson processes play an important rôle in a variety of applications (physics, technology, biology), where the lifetime distribution does depend on the current position on the time axis. We show here one particular application of inhomogeneous Poisson processes related to *record values*, or *records*. Let $X_1, X_2, \ldots$ be a sequence of independent random variables with continuous and strictly increasing distribution function F. We say that X_n is a record value, if

$$X_n > \max\{X_1, \ldots, X_{n-1}\}.$$

Then the sequence of record values forms an inhomogeneous Poisson process; we want to determine its intensity.

First, consider the distribution function of the RV $F(X_1)$: for all $0 < x < 1$

$$
\begin{aligned}
F_{F(X_1)}(x) &= \mathbb{P}(F(X_1) \le x) = \mathbb{P}(X_1 \le F^{-1}(x)) \\
&= F(F^{-1}(x)) = x,
\end{aligned}
\tag{2.64}
$$

where F^{-1} stands for the inverse of F (which exists under the assumptions on F). Then $-\ln(1 - F(X_1)) \sim \text{Exp}(1)$.

Now let V_n^R be the subsequent record values:

$$
V_0^R = 0, \quad V_1^R = X_1, \quad V_2^R = \sum_{k>1} X_k \mathbf{1}(X_2, \ldots, X_{k-1} < X_1 < X_k), \text{ etc.}
\tag{2.65}
$$

Consider the *record process* (R_t)

$$
R_t = \text{number of } \{n \ge 1 : V_n^R \le t\}, \quad t > 0.
\tag{2.66}
$$

The pattern of producing a new record arises as follows. Suppose we look at a sequence of n previous record values $x_1 < \cdots < x_n$ achieved by $X_1, \ldots, X_{k_n}$. Then we either find $X_{k_n+1} > x_n$ or we see a number $m \ge 1$ of unsuccessful attempts on $X_{k_n+1}, \ldots, X_{k_n+m}$ and then obtain $X_{k_n+m+1} > x_n$. Hence,

$$
\mathbb{P}(\text{a new record} > x_n + y \mid \text{previous records } x_1 < \cdots < x_n)
$$

$$
= \sum_{m \ge 0} F(x_n)^m (1 - F(x_n + y)) = \frac{1 - F(x_n + y)}{1 - F(x_n)},
$$

regardless of n and $x_1, \ldots, x_{n-1}$.

In the particular case where $F(x) = 1 - e^{-x}$, we obtain, for the nth holding time $S_n^R = V_n^R - V_{n-1}^R$:

$$
\mathbb{P}(S_n^R > y \mid V_{n-1}^R = x, \ldots, V_1^R = x_1) = \frac{e^{-(x+y)}}{e^{-x}} = e^{-y}.
$$

That is, $S_n^R \sim \text{Exp}(1)$, independently of $S_0^R, \ldots, S_{n-1}^R$. Hence, (R_t) is Poisson process $PP(1)$.

For a general F, the above probability becomes

$$
\frac{1 - F(x + y)}{1 - F(x)} = e^{-(\Lambda(x+y) - \Lambda(x))} = \exp\left(-\int_x^{x+y} \lambda(s)\, ds\right).
$$

Here

$$\Lambda(t) = \int_0^t \lambda(s)\,\mathrm{d}s = -\ln[1 - F(t)], \quad \text{with } \lambda(t) = \frac{f(t)}{1 - F(t)},$$

where $f(t) = F'(t)$ is the probability density function of X_1. Hence, (R_t) forms an inhomogeneous Poisson process, with rate $\lambda(t)$. It is mapped to the Poisson process of rate 1 by replacing $X_i \sim F$ with $-\ln(1 - F(X_i)) \sim \mathrm{Exp}(1)$.

It was already mentioned that important contributions to the theory of Markov chains (and more general random processes) were made by Joseph Doob (1910–2004) and William Feller (1906–1970), two famous American mathematicians of the 1930–1960s. Both Doob and Feller were of Eastern European origin and both were more than ordinary personalities, with a strong sense of humour and leadership qualities. Below, we present a folklore account of the origin of the term 'random variable' that is repeatedly used in this volume (as well as in Vol. 1). The term acquired popularity in the Western literature in the late 1940s, when both Doob and Feller were working on their respective monographs, which shaped the theory of random processes as we know it nowadays. (In the English language literature, the term 'random variable' can be traced to a paper by A. Winter "On analytic convolutions of Bernoulli distributions", *American Journ. Math.*, **56** (1934), 659–663, but Cantelli has already used it in Italian in 1916.) According to an account by Doob, he and Feller had an argument. Feller "asserted that everyone said 'random variable' while Doob asserted that everyone said 'chance variable'. The issue was decided by a stochastic procedure: they tossed a coin, and Feller won." However, Doob entitled his book *Stochastic Processes*, not 'Random Processes' (apparently their gentleman's agreement did not go beyond the concept of a single variable). It must be said that the Russian (Soviet) school of probability used, painlessly, the commonly accepted term 'sluchainaya velichina' from the mid-1920s. This term somehow combines both versions of its English counterpart; it could also be the case that Kolmogorov was a unique leader of the Russian probability community, and his verdict was final. (However, in our opinion, a drawback of the Russian terminology is the use of the term 'dispersion' for the variance. This creates a confusion as in physics this term is used widely and has a different meaning.)

On a more serious tone, Doob is credited by some specialists with introducing the term 'Markov chain', in "Topics in the theory of Markov chains", *Transactions of the AMS*, **52** (1942), 37–64. However, Kolmogorov introduced the term in German in 1936 and in Russian in 1938, both times in the titles of his papers, whereas W. Döblin did it in French in 1937. In the opinion of Russian historians of mathematics, the term 'Markov chain' was proposed by J. Hadamard, the famous French mathematician, in the 1920s, although the word 'chain' can be found in Markov's original works.

Markov himself was the first to use the term 'probability density' (obviously, in Russian).

Doob's name is a modification of 'Dub' which in Czech (and other Slavic languages) means 'oak'. His father changed it when the family emigrated to the USA, to avoid jokes and being teased. Amongst probabilists, Doob is widely remembered, inter alia, for his perhaps unique quality of making no mistakes in his papers or books. For example, in the above-mentioned volume, not a single mistake has been found, not even a typo, to the amazement of the Russian translators and many other assiduous readers.

This did not happen with Feller's book. His highly acclaimed and hugely influential two-volume monograph *An Introduction to Probability Theory and its Applications* (Wiley 1968, 1971) contained a number of errors (most of which were correctable and many of which were corrected by the author in later editions). In the 1960s it attracted the attention of writers to, and readers of, a mural newspaper at the Department of Mechanics and Mathematics, Moscow State University. The newspaper was called, in Soviet traditions, 'For an Advanced Department'; it appeared periodically and was officially recognised as an opinion channel approved by the Departmental Chairmanship, the local Communist Party Bureau and the local Trade Union Committee (of these three bodies, only the first survives at the present time; the newspaper publication was stopped in 1990). Articles were typed on a typewriter and sometimes handwritten and glued to a broadsheet of thick white paper (the original destination of such large sheets of paper was technical drawings, and they were produced in large quantities for the needs of the Soviet industry). At periods of political 'thaws', newspaper editors might be able to allow authors to adopt a somewhat flirtatious attitude presented as a 'socialist satire and humour' and aiming, officially, to 'remove existing deficiencies hindering our progress', and, unofficially, to amuse (and attract) readers. Indeed, many mathematicians and other academics rarely missed an opportunity to see a fresh issue of the mural newspaper; many travelled for hours, from regions around Moscow (we suppose it was an equivalent of a society website nowadays). Some articles caused considerable professional or social controversy, and exerted a protracted impact on mathematical life in Moscow and beyond. One series of (anonymous) articles appeared in the paper a dozen or so times; its running title was clearly mocking the Cold War rhetoric dominating the Soviet press of the time. In a loose translation from Russian, the title read 'A fresh error of the American author', and it always started with the sentence: "Our readers discovered yet another error by the American author Feller." In the Russian volume of memoirs *Kolmogorov in Recollections of his Students* (A.N. Shiryaiev, N.G. Khimchenko, Eds). Moscow: MCCME, 2006, this episode is described slightly differently, although the two accounts presented also differ from each other.

2.5 Birth-and-death process. Explosion

> *And such a yell was there,*
> *Of sudden and portentous birth,*
> *As if men fought upon the earth,*
> *And fiends in upper air.*
> W. Scott (1771–1832), Scottish writer and poet

Another useful generalisation of Poisson processes is when we take independent holding times

$$S_k \sim \text{Exp}(\lambda_k), \tag{2.67}$$

with $\lambda_k \geq 0$ depending on state k. (If $\lambda_k = 0$, the state becomes absorbing, with $S_k = \infty$. The process stops developing further after entering such a state.) This generalisation yields a *birth process* of rates (λ_k) (BP(λ_k) for short). See Figure 2.34.

As a CTMC, a birth process BP(λ_i) operates over the state space $\mathbb{Z}_+$ and with the Q-matrix

$$Q = \begin{pmatrix} -\lambda_0 & \lambda_0 & 0 & 0 & \cdots \\ 0 & -\lambda_1 & \lambda_1 & 0 & \ddots \\ 0 & 0 & -\lambda_2 & \lambda_2 & \ddots \\ \cdots & \ddots & \ddots & \ddots & \ddots \end{pmatrix}. \tag{2.68}$$

We can hope that, as in the case of a Poisson process PP(λ), the matrix exponent

$$P(t) = e^{tQ} = \sum_{k \geq 0} \frac{(tQ)^k}{k!}$$

will give the transition probabilities of this process. However, it proves convenient to start with a definition of the last two characterisations, which extends that of the corresponding features of a Poisson process PP(λ).

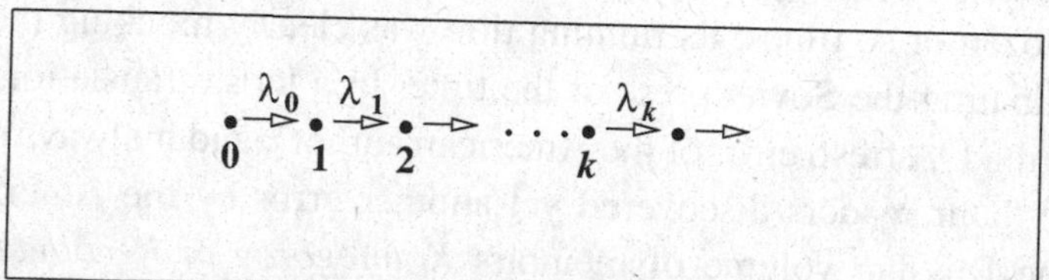

Fig. 2.34

Definition 2.5.1 Let $\lambda_0, \lambda_1, \ldots$ be a collection of non-negative numbers. A birth process of rates $\lambda_0, \lambda_1, \ldots$, starting from state $i_0 \in \mathbb{Z}_+$ is a non-decreasing process $(N_t^B, t \geq 0)$ with $N_0^B = i_0$ and with values $i \in \mathbb{Z}_+$, $i \geq i_0$, characterised in the following two ways:

(b) as a process such that, for all $t > 0$ and $i \in \mathbb{Z}_+$, conditional on $N_t^B = k$, the increment $(N_{t+h}^B - N_t^B)$ over a future time interval $[t, t+h)$ is conditionally independent of the past $(N_s^B, 0 \leq s < t)$, and with the following infinitesimal probabilities:

$$\left. \begin{array}{l} \mathbb{P}\left(N_{t+h}^B = k \mid N_t^B = k\right) = 1 - \lambda_k h + o(h), \\ \mathbb{P}\left(N_{t+h}^B = k+1 \mid N_t^B = k\right) = \lambda_k h + o(h), \\ \mathbb{P}\left(N_{t+h}^B > k+1 \mid N_t^B = k\right) = o(h), \end{array} \right\} \tag{2.69}$$

where the remainders may depend on k but not on t;

(c) as a process that

  (i) spends a random time $\sim \mathrm{Exp}(\lambda_k)$ in a state $k \geq i_0$ with rate $\lambda_k > 0$, independently of the previous history, and then jumps to $k+1$, or

  (ii) stays forever in this state if $\lambda_k = 0$, independently of the previous history.

Clearly, these characterisations follow properties (b) and (c) of a Poisson process $\mathrm{PP}(\lambda)$. In future, we will assume that all $\lambda_k > 0$ (no absorbing states).

In applications, a birth process $\mathrm{BP}(\lambda_k)$ often represents the size of a growing population (e.g. of living organisms or physical particles), growing at a rate depending on the state the process is in at a given point t in time.

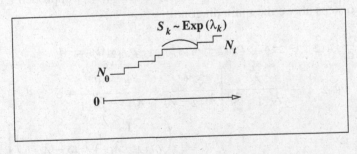

Fig. 2.35

We now turn to the first characterisation (a) of a birth process. It is natural to assume that the matrix Q would lead to an upper-triangular matrix exponent $P(t) = e^{tQ}$:

$$
P(t) = e^{tQ} = \begin{pmatrix}
p_{00}(t) & p_{01}(t) & p_{02}(t) & \cdots \\
0 & p_{11}(t) & p_{12}(t) & \cdots \\
0 & 0 & p_{22}(t) & \cdots \\
& & \mathbf{0} & \ddots
\end{pmatrix},
\tag{2.70}
$$

with $P(0) = \mathbf{I}$. To find the matrix $P(t)$, we can again use the forward and backward equations

$$
\frac{d}{dt}P = PQ = QP, \quad \text{with } P(0) = \mathbf{I}
$$

(the argument $t \geq 0$ will often be omitted). The equations for a given entry are

$$
\frac{d}{dt}\,p_{ij} = -\lambda_j p_{ij} + \lambda_{j-1} p_{ij-1} \quad \text{(forward)}
$$
$$
= -\lambda_i p_{ij} + \lambda_i p_{i+1\,j} \quad \text{(backward)}, \qquad p_{ij}(0) = \delta_{ij},\ i \leq j. \tag{2.71}
$$

The assumption of upper triangularity of $P(t)$ allows us to solve these equations. For example, consider the forward equations. Here, on the main diagonal we find

$$
\frac{d}{dt}\,p_{ii} = -\lambda_i p_{ii}, \ p_{ii}(0) = 1 \ \Rightarrow \ p_{ii}(t) = e^{-\lambda_i t}, \ t \geq 0.
$$

Then, one step above the main diagonal:

$$
\frac{d}{dt}p_{ii+1} = -\lambda_{i+1} p_{ii+1} + \lambda_i p_{ii}, \ p_{ii+1}(0) = 0
$$
$$
\Rightarrow \quad p_{ii+1} = \frac{\lambda_i}{\lambda_{i+1} - \lambda_i}\left(e^{-\lambda_i t} - e^{-\lambda_{i+1} t}\right)
$$
$$
= \frac{\lambda_i}{\lambda_{i+1} - \lambda_i}\, e^{-\lambda_i t}\left(1 - e^{-(\lambda_{i+1} - \lambda_i)t}\right), \quad \lambda_i \neq \lambda_{i+1}, \ t \geq 0.
$$

We see that $p_{ii+1}(t) > 0$ and becomes $\lambda t e^{-\lambda t}$ when λ_i, λ_{i+1} approach λ. Next, two steps above the main diagonal:

$$
\frac{d}{dt}p_{ii+2} = -\lambda_{i+2} p_{ii+2} + \lambda_{i+1} p_{ii+1}, \ p_{ii+2}(0) = 0
$$
$$
\Rightarrow \quad p_{ii+2} = \frac{\lambda_i}{\lambda_{i+1} - \lambda_i}\left[\frac{e^{-\lambda_i t}}{\lambda_{i+2} - \lambda_i} - \frac{e^{-\lambda_{i+1} t}}{\lambda_{i+2} - \lambda_{i+1}} + e^{-\lambda_{i+2} t}\right.
$$
$$
\left. \times \left(\frac{-1}{\lambda_{i+2} - \lambda_i} + \frac{1}{\lambda_{i+2} - \lambda_{i+1}}\right)\right],
$$
$$
\lambda_i \neq \lambda_{i+1} \neq \lambda_{i+2} \neq \lambda_i, \ t \geq 0.
$$

Again: $p_{ii+2}(t) > 0$ and tends to $\dfrac{(\lambda t)^2}{2} e^{-\lambda t}$ as $\lambda_i, \lambda_{i+1}, \lambda_{i+2}$ approach λ, and so on.

An elegant way of solving both forward and backward equations is to use characterisation (c) from Definition 2.5.1 above. That is, we write

$$p_{ii}(t) = e^{-\lambda_i t} \quad \text{(no jump up to time } t), \tag{2.72}$$

$$p_{ii+1}(t) = \int_0^t \lambda_i\, e^{-\lambda_i t_1}\, e^{-\lambda_{i+1}(t-t_1)} \,dt_1$$

$$= \int_0^t \lambda_i\, e^{-\lambda_i(t-s_1)}\, e^{-\lambda_{i+1}s_1} \,ds_1 \text{ (a single jump up to time } t), \tag{2.73}$$

$$p_{ii+2}(t) = \int_0^t \int_0^t \lambda_i\, e^{-\lambda_i t_1}\, \lambda_{i+1}\, e^{-\lambda_{i+1}(t_2-t_1)}\, e^{-\lambda_{i+2}(t-t_2)} \mathbf{1}(t_1 < t_2)\, dt_2\, dt_1$$

$$= \int_0^t \int_0^{t_2} \lambda_i e^{-\lambda_i t_1} \lambda_{i+1}\, e^{-\lambda_{i+1}(t_2-t_1)}\, e^{-\lambda_{i+2}(t-t_2)} dt_1\, dt_2$$

$$= \int_0^t \int_0^{s_1} \lambda_i e^{-\lambda_i(t-s_1)} \lambda_{i+1}\, e^{-\lambda_{i+1}(s_1-s_2)}\, e^{-\lambda_{i+2}s_2} ds_2\, ds_1$$

$$\text{(precisely two jumps up to time } t), \tag{2.74}$$

and so on. The solution for p_{ii+n} becomes

$$p_{ii+n}(t)$$

$$= \int_0^t \cdots \int_0^t \lambda_i\, e^{-\lambda_i t_1}\, \lambda_{i+1}\, e^{-\lambda_{i+1}(t_2-t_1)} \cdots \lambda_{i+n-1}\, e^{-\lambda_{i+n-1}(t_n-t_{n-1})}$$

$$\times e^{-\lambda_{i+n}(t-t_n)} \mathbf{1}(t_1 < \cdots < t_n) dt_n \cdots dt_1 \tag{2.75}$$

$$= \int_0^t \cdots \int_0^{t_2} \lambda_i\, e^{-\lambda_i t_1} \cdots \lambda_{i+n-1}\, e^{-\lambda_{i+n-1}(t_n-t_{n-1})}\, e^{-\lambda_{i+n}(t-t_n)} dt_1 \cdots dt_n$$

$$= \int_0^t \cdots \int_0^{s_{n-1}} \lambda_i\, e^{-\lambda_i(t-s_1)} \cdots \lambda_{i+n-1}\, e^{-\lambda_{i+n-1}(s_{n-1}-s_n)}\, e^{-\lambda_{i+n}s_n} ds_n \cdots ds_1,$$

$$\text{(precisely } n \text{ jumps up to time } t), \tag{2.76}$$

and so on. Here, the version of the integrand in the variables $0 < t_1 < \cdots < t_n < t$ (times of jumps) is suitable for checking the forward equations and the version in

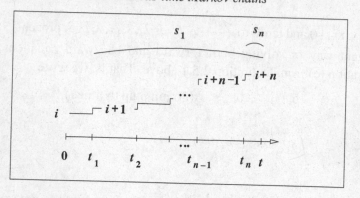

Fig. 2.36

variables $s_l = t - t_l$ (times from the jumps till t, the transition time) for checking the backward equations. Observe that formulas (2.72)–(2.75) do not require that $\lambda_i = \lambda_{i+1}$. But, when all the λ_is coincide, we obtain (2.46)–(2.47). The development of the sample path of a birth process $BP(\lambda_k)$ is outlined in Figure 2.36.

Worked Example 2.5.2   (Continued from Worked Example 2.4.2) A different alarm is set in a University building on Quantum Road: for all $t \geq 0$, $i = 0, 1, \ldots$, and $h \searrow 0$

$$
\begin{aligned}
\mathbb{P}(M_{t+h} - M_t = 0 | M_t = i) &= 1 - \lambda_i h + o(h), \\
\mathbb{P}(M_{t+h} - M_t = 1 | M_t = i) &= \lambda_i h + o(h), \\
\mathbb{P}(M_{t+h} - M_t > 1 | M_t = i) &= o(h).
\end{aligned}
$$

Find the equations for $P_i(t) = \mathbb{P}(M_t = i)$. Check that for $\lambda_i = \alpha i + \beta$,

$$
m(t) = \mathbb{E}M_t = \frac{\beta}{\alpha} \left(e^{\alpha t} - 1 \right)
$$

and find $\operatorname{Var} M_t$.

Solution With $M_0 = 0$, (M_t) represents a birth process $BP(\lambda_i)$ (with birth rates λ_i). The equations are

$$
\begin{aligned}
\frac{\mathrm{d}}{\mathrm{d}t} p_0 &= -\lambda_0 p_0 \\
\frac{\mathrm{d}}{\mathrm{d}t} p_i &= \lambda_{i-1} p_{i-1} - \lambda_i p_i, \ i \geq 1.
\end{aligned}
$$

For $\lambda_i = \alpha i + \beta$ the process is non-explosive. Consider the probability generating function

$$
G(s,t) = \sum_{i \geq 0} s^i p_i(t),
$$

with

$$G(1,t) = 1 \quad \text{(non-explosive), and} \quad \frac{\partial}{\partial s} G(s,t) = \sum_{i \geq 1} is^{i-1} p_i(t).$$

Then

$$\frac{\partial}{\partial t} G(s,t) = (s-1)\beta G(s,t) + s(s-1)\alpha \frac{\partial}{\partial s} G(s,t).$$

and for $m(t) = \frac{\partial}{\partial s} G(s,t) \Big|_{s=1}$,

$$\frac{dm}{dt} = \alpha m(t) + \beta, \quad m(0) = 0.$$

This implies $m(t) = \frac{\beta}{\alpha}(e^{\alpha t} - 1)$.

Finally, consider

$$v(t) = \mathbb{E}[M_t(M_t - 1)] = \frac{\partial^2}{\partial s^2} G(s,t) \Big|_{s=1},$$

with

$$\operatorname{Var} M_t = v(t) + m(t) - m(t)^2.$$

Then

$$\dot{v}(t) = 2\alpha v(t) + (2\alpha + 2\beta)m(t), v(0) = 0.$$

This implies

$$v(t) = \frac{(\alpha + \beta)\beta}{\alpha^2} (e^{\alpha t} - 1)^2$$

and

$$\operatorname{Var} M_t = \frac{\beta}{\alpha} e^{\alpha t}(e^{\alpha t} - 1).$$

$\square$

A birth process with 'linear' rates $\lambda_k = \alpha k + \beta$ is called the *Yule–Furry* process; it finds several practical applications.

Million Dollar β

(From the series *'Movies that never made it to the Big Screen'*.)

Next, we briefly discuss the possibility of *explosions*, that is, infinitely many jumps (i.e. indefinitely growing paths) in a finite time interval; see Figure 2.28. We have seen that when $\lambda_k \equiv \lambda > 0$ (i.e. for a Poisson process (N_t^B)), then $\mathbb{P}\left(\sum_{k \geq 0} S_k < \infty\right)$, which is the probability of an explosion in a finite time, equals 0. Now, a more general result.

Theorem 2.5.3 *For the birth process* $BP(\lambda_k)$, *with rates* $\lambda_k > 0$, *the following dichotomy holds:*

(i)   *if* $\sum_k \left(\dfrac{1}{\lambda_k}\right) = \infty$, *no explosion:* $\mathbb{P}\left(\sum_k S_k < \infty\right) = 0$;   (2.77)

(ii)   *if* $\sum_k \left(\dfrac{1}{\lambda_k}\right) < \infty$, *explosion:* $\mathbb{P}\left(\sum_k S_k < \infty\right) = 1$.   (2.78)

Proof (i) Following the proof of Theorem 2.3.9, set $T_{\text{explo}} = \sum_k S_k$ and consider the MGF $\mathbb{E}(e^{-T_{\text{explo}}})$. Using again monotone convergence of the partial sums $\sum_{k=0}^K S_k \nearrow T_{\text{explo}}$ and independence of the holding times $S_0, S_1, \ldots$, write:

$$\mathbb{E}(e^{-T_{\text{explo}}}) = \lim_{K \to \infty} \prod_{k=0}^K \mathbb{E}e^{-S_k} = \lim_{K \to \infty} \prod_{k=0}^K \frac{\lambda_k}{\lambda_k + 1} = \lim_{K \to \infty} \left[\prod_{k=0}^K (1 + 1/\lambda_k)\right]^{-1}.$$

Observe that the last product $\prod_{k=0}^K (1 + 1/\lambda_k)$ increases with K. By using an elementary bound

$$\prod_{k=0}^K \left(1 + \frac{1}{\lambda_k}\right) = 1 + \sum_{k=0}^K \frac{1}{\lambda_k} + \sum_{k_1, k_2 = 0}^K \frac{1}{\lambda_{k_1}} \frac{1}{\lambda_{k_2}} + \cdots \geq \sum_{k=0}^K \frac{1}{\lambda_k},$$

we conclude that

$$\lim_{K \to \infty} \left[\prod_{k=0}^K (1 + 1/\lambda_k)\right]^{-1} \leq \left(\sum_k 1/\lambda_k\right)^{-1} = 0,$$

that is,

$$\mathbb{E}(e^{-T_{\text{explo}}}) = 0.$$

Again as in the proof of Theorem 2.3.9, we deduce that $e^{-T_{\text{explo}}} = 0$ and $T_{\text{explo}} = \infty$ with probability 1, meaning (2.77).

(ii) The random variable T_{explo} takes values in $[0, \infty)$ and, possibly the value $+\infty$. However, the expectation

$$\mathbb{E}(T_{\text{explo}}) = \lim_{K \to \infty} \mathbb{E} \sum_{k=0}^K S_k = \sum_k \lambda_k^{-1} < \infty,$$

by the monotone convergence theorem. We conclude that value $+\infty$ is taken with probability 0, that is, (2.78) holds. $\square$

Remark 2.5.4 Theorem 2.5.3 considers explosions from the initial state 0. However, the conclusion holds if we start from an arbitrary state i, as it simply means that the event $\{T_{\text{explo}} < \infty\}$ coincides with $\{\sum_{k \geq i} S_k\}$.

Therefore, the result remains true for any initial distribution.

Definition 2.5.5 In case (i) in Theorem 2.5.3 (i.e. under condition (2.77)), we call the birth process BP(λ_k) (or the matrix Q from (2.68)) *non-explosive*. In case (ii) (i.e. under condition (2.78)), we call it *explosive*.

For a non-explosive birth process BP(λ_k), where $\sum_k \lambda_k^{-1} = \infty$: the forward/backward equations (2.71) possess a unique solution $P(t) = (p_{ij}(t))$, for all $t \geq 0$, given by (2.72)–(2.75), and such that

$$0 < p_{ij}(t) < 1, \ i \leq j, \ \text{and} \ \sum_{j \geq i} p_{ij}(t) = 1, \ i = 0, 1, \cdots, \ \text{for all } t \geq 0. \qquad (2.79)$$

In this case the matrix $P(t) = (p_{ij}(t))$ forms a 'genuine' transition matrix for all $t \geq 0$. (Some authors use the term an 'honest' transition matrix). In addition, the family of matrices $P(t)$ exhibits the semigroup property $P(t+s) = P(t)P(s)$, for all $t, s \geq 0$. We see a 'nice' situation, where the process is determined by its Q-matrix (and the initial distribution).

> *[This is] what I understand by 'philosopher': a terrible explosive in the presence of which everything is in danger.*
>
> F. Nietzsche (1844–1900), German philosopher

But for an explosive birth process BP(λ_k), the situation turns more complicated. The solution $P(t) = (p_{ij}(t))$ to problems (2.71) specified in (2.72)–(2.75) does not yield $\sum_{j \geq i} p_{ij}(t) = 1$; on the contrary, $\sum_{j \geq i} p_{ij}(t) < 1$ for all $t > 0$ and $i = 0, 1, \ldots$. Still, this solution remains rather special: it is *minimal*, in the sense that for any family of matrices $R(t) = (r_{ij}(t))$ satisfying $\dfrac{d}{dt}R = RQ = QR$, with $R(0) = \mathbf{I}$, the entries $r_{ij}(t)$ obey $r_{ij}(t) \geq p_{ij}(t)$, for all $t > 0$ and $i, j = 0, 1, \ldots$. The minimality property follows from our assumption that the matrix $P(t)$ is upper-triangular. The minimal solution still offers some nice properties: for instance, it can be written as

$$P(t) = e^{tQ} = \sum_{k \geq 0} \frac{(tQ)^k}{k!} \ \text{and} \ P(t) = \lim_{n \to \infty} e^{tQ(n)}, \qquad (2.80)$$

with $Q(n)$ an $n \times n$ modification (a 'truncation') of the matrix Q applied in (2.69). It also presents the semigroup property $P(t+s) = P(t)P(s)$, for all $t, s \geq 0$.

In the non-explosive case, the minimal solution is formed by stochastic matrices $P(t)$, where (2.79) hold true, and there is no place for other solutions. However, in an explosive case, when the minimal solution does not yield stochastic matrices, it opens a Pandora's box of surprises, perhaps interesting, but rather unsettling. In this volume, we will not discuss this issue in detail: see, for example, the following

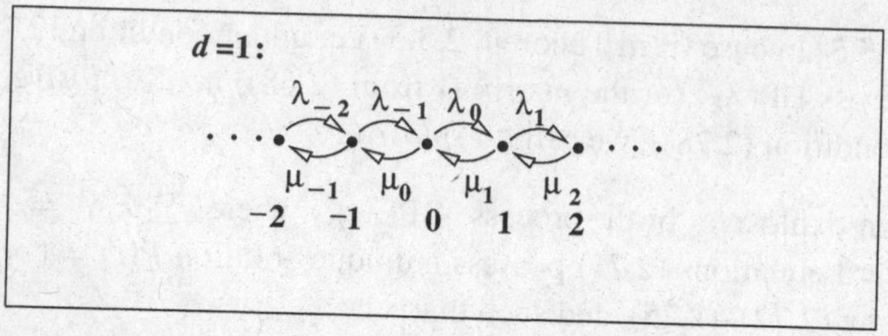

Fig. 2.37

papers: Karlin, S. & McGregor, J. "The classification of birth and death processes." *Trans. Amer. Math. Soc.*, **86** (1957), 366–400; Ledermann, W. & Reuter, G.E.H. "Spectral theory for the differential equations of simple birth and death processes." *Philos. Trans. Roy. Soc. London.* Ser. A. **246** (1954), 321–369; Reuter, G.E.H. & Ledermann, W. "On the differential equations for the transition probabilities of Markov processes with enumerably many states." *Proc. Cambridge Philos. Soc.* **49** (1953), 247–262.

For a non-explosive birth-and-death process we observe $\mathbb{P}_i(T_{\text{explo}} = +\infty) = 1$, for all states $i \in \mathbb{Z}_+$.

> *Anyone can stop a man's life, but no one*
> *his death: a thousand doors open to it.*
> Seneca (4B.C.–65A.D.), Roman philosopher

The next class of processes to mention are called *continuous-time random walks* (CTRWs) on $\mathbb{Z}^d$. Consider first the one-dimensional case, where $d = 1$. The corresponding diagram is given in Figure 2.37.

If $\lambda_i = \mu_i$, the RW is called *symmetric*; if $\lambda_i \equiv \lambda$ and $\mu_i \equiv \mu$, the RW is called *homogeneous* (more precisely, *space-homogeneous*).

The Q-matrix here becomes double-infinite in both directions

$$
Q = \begin{pmatrix}
\ddots & \ddots & \ddots & & \ddots & & \ddots & \ddots \\
\ddots & \mu_{i-1} & -(\lambda_{i-1}+\mu_{i-1}) & \lambda_{i-1} & & 0 & \cdots & \ddots \\
\ddots & 0 & \mu_i & -(\lambda_i+\mu_i) & \lambda_i & & 0 & \ddots \\
\ddots & \ddots & 0 & \mu_{i+1} & -(\lambda_{i+1}+\mu_{i+1}) & \lambda_{i+1} & & \ddots \\
& \ddots & \ddots & & \ddots & & \ddots & \ddots
\end{pmatrix}.
$$

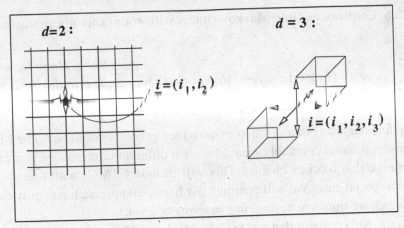

Fig. 2.38

In the two-dimensional case ($d = 2$), the states become sites $\underline{i} = (i_1, i_2)$, $i_1, i_2 \in \mathbb{Z}$: see Figure 2.38. Here we mainly concentrate on the homogeneous and symmetric case, where

$$q_{\underline{i}} := -q_{\underline{ii}} \equiv q > 0, \quad \underline{i} = (i_1, i_2) \in \mathbb{Z}^2,$$

and

$$q_{\underline{ii}'} = \frac{q}{4} \text{ if the site } \underline{i}' \text{ sits next to } \underline{i}, \text{ i.e. } \underline{i}' = (i_1 \pm 1, i_2) \text{ or } (i_1, i_2 \pm 1).$$

For a homogeneous and symmetric continuous time random walk in $d = 3$ there exist six possibilities:

$$q_{\underline{ii}'} = \frac{q}{6} \quad \text{for} \quad \underline{i}' \quad \text{next to} \quad \underline{i} = (i_1, i_2, i_3).$$

The general case of $\mathbb{Z}^d$, $d \geq 1$, should be considered similarly. Here, the entries of the Q-matrix are labelled by sites $\underline{i} \in \mathbb{Z}^d$:

$$q_{\underline{i}} = -q_{\underline{ii}} = 1, \ q_{\underline{ij}} = \left(\frac{1}{2d}\right) \mathbf{1}(\|\underline{i} - \underline{j}\| = 1) \text{ for } \underline{i} \neq \underline{j}, \ \underline{i}, \underline{j} \in \mathbb{Z}^d.$$

Again, in a more general model the exit rate $q_{\underline{i}}$ may depend on $\underline{i} \in \mathbb{Z}^d$.

2.6 Continuous-time Markov chains with countably many states

Countably Many Dalmatians
(From the series 'Movies that never made it to the Big Screen'.)

We now find ourselves in a position to introduce general continuous-time Markov chains with (at most) countably many states (in different terminology, continuous-time denumerable Markov chains). This will include CTMCs with finitely many states as a special case. We will continue our hands-on approach and provide only proofs which we think are appropriate in terms of level.

The state space remains denoted by I; we deal with Q-*matrices* $Q = (q_{ij})$ whose entries q_{ij} are labelled by pairs of states $i, j \in I$. The conditions stay the same (Q contains non-negative off-diagonal terms and is conservative):

$$\text{the off-diagonal entry } q_{ij} \geq 0 \text{ for all pair } i \neq j \in I, \tag{2.81}$$

and

$$-\infty < q_{ii} \leq 0, \ q_i := -q_{ii} = \sum_{j \in I: j \neq i} q_{ij} \text{ for all } i \in I. \tag{2.82}$$

Recall, for all $j \neq i$, the entry q_{ij} represents the transition rate from i to j, and for all $i \in I$, the value q_i gives a total exit rate from state i.

In addition, we will assume that convergence of the series $\sum_{j: j \neq i} q_{ij}$ happens uniformly (although, admittedly, the general statements below do not require such a restriction). The uniform convergence means that states $i \in I$ can be enumerated as, say $j_0, j_1, j_2, \ldots$, in such a way that the tail of the series, formed by summands $q_{j_l j_k}$ with large labels k, gets uniformly small:

$$\limsup_{n \to \infty} \left[\sum_{k: |j_k - j_l| > n} q_{j_l j_k} : j_l \in I \right] = 0. \tag{2.83}$$

Most of the time, the particular order of summation in (2.83) will not be necessary, and the series $\sum_{j \in I}$ may be understood in any order of summation.

Our strategy basically has not changed: we treat Q as a *generator*, and want to construct a semigroup $(P(t), t \geq 0)$ of transition matrices $P(t) = (p_{ij}(t))$, with $P(0) = \mathbf{I}$, associated with Q, and study properties of the corresponding continuous time Markov chain. When possible, we used the matrix exponent; and the helpful tool of a pair of forward/backward equations

$$\frac{d}{dt} P = PQ = QP, \text{ with } P(0) = \mathbf{I}, \tag{2.84}$$

or, for individual entries,

$$\frac{d}{dt} p_{ij} = \sum_k p_{ik} q_{kj} \quad \text{(forward)},$$
$$= \sum_k q_{ik} p_{kj} \quad \text{(backward)}, \qquad p_{ij}(0) = \delta_{ij}.$$

We saw that for finite matrices, (2.84) exhibits a unique solution,

$$P(t) = e^{tQ} = \sum_{k \geq 0} \frac{(tQ)^k}{k!}, \quad t \geq 0, \tag{2.85}$$

and it was composed of stochastic matrices. We were thinking of possible general-isations of this fact to an infinite case. In the countable case, more precisely, in the case of a birth process $BP(\lambda_k)$, we noted in Section 2.5 that the solution can give a sub-stochastic matrix, which leads to explosion; a way out was to extend the state space, by adding an absorbing state ∞. We then discussed the more general model of a birth-and-death process $BDP(\lambda_k, \mu_k)$.

In this section, we continue with general theorems (Theorems 2.6.1–2.6.11) given without proof (see, e.g., Bharucha-Reid, 1960, Kemeny, Snell and Knapp, 1966 and original papers quoted therein). In these statements, Q is assumed to be a Q-matrix satisfying (2.81)–(2.83).

Theorem 2.6.1 *The forward and backward equations* (2.84) *always give a solution* $P(t), t \geq 0$, *satisfying the semigroup property*

$$P(t+s) = P(t)P(s), \quad t, s \geq 0. \tag{2.86}$$

In general, matrices $P(t) = (p_{ij}(t))$ *will only be sub-stochastic: for all* $t > 0$,

$$p_{ij}(t) \geq 0, \text{ for all } i, j \in I, \text{ and } \sum_{j \in I} p_{ij}(t) \leq 1, \text{ for all } i \in I. \tag{2.87}$$

Such a solution is in general not unique, and the forward and backward equations will generally differ in their sets of solutions.

However, there always exists a unique minimal sub-stochastic solution $P^{\min}(t) = \left(p_{ij}^{\min}(t)\right)$ to (2.84), satisfying (2.87). Minimality means that for all non-negative solutions $R(t) = (r_{ij}(t)), t \geq 0$,

$$p_{ij}^{\min}(t) \leq r_{ij}(t). \tag{2.88}$$

The minimal sub-stochastic solution is given by

$$P^{\min}(t) = \lim_{n \to \infty} e^{tQ(n)} := e^{tQ},$$

where the Q-matrix $Q(n) = (q_{j_l j_k}(n))$ is an $n \times n$ truncation of Q, viz.

$$q_{j_l j_k}(n) = q_{j_l j_k}, \quad \text{for } l, k = 0, \ldots, n-1, \text{ with } j_l \neq j_k;$$

$$q_{j_l j_l}(n) = - \sum_{0 \leq k < n: \, j_k \neq j_l} q_{j_l j_k} \quad \text{for } l, k = 0, \ldots, n-1.$$

Here we use the same enumeration of states $i \in I$ as in (2.83).

Finally, the minimal solution posseses the semigroup property (2.86).

Thus if we face a case where the minimal solution is formed by stochastic matrices $P^{\min}(t) = (p_{ij}^{\min}(t))$, $t > 0$, with $\sum_{j \in I} p_{ij}^{\min}(t) = 1$ for all $t > 0$ and $i \in I$, then the issue of *sub*-stochasticity goes, as $P^{\min}(t)$ will be the only solution to (2.84). (More precisely, any sub-stochastic solution $P(t)$ equals $P^{\min}(t)$ and hence will be stochastic.)

A useful specification of the minimal solution turns out to be that $p_{ij}^{\min}(t)$ can be written as a series over the number of jumps performed by the chain in the time interval $(0, t)$.

Theorem 2.6.2 *For the minimal solution* $P^{\min}(t) = (p_{ij}^{\min}(t))$, *the following representation holds true: for all $t > 0$ and $i, j \in I$,*

$$p_{ij}^{\min}(t) = e^{-t q_i} \mathbf{1}(i = j) \qquad \qquad \text{(no jump)}$$

$$+ \mathbf{1}(i \neq j) \mathbf{1}(q_i > 0) \int_0^t e^{-t_1 q_i} q_{ij} e^{-(t-t_1)q_j} dt_1 \qquad \text{(one jump)}$$

$$+ \mathbf{1}(q_i > 0) \sum_{k \in I} \mathbf{1}(k \neq i, j) \mathbf{1}(q_k > 0) \int_0^t \int_0^t e^{-t_1 q_i}$$

$$\times q_{ik} e^{-(t_2 - t_1)q_k} q_{kj} e^{-(t-t_2)q_j} \mathbf{1}(t_1 < t_2) dt_2 \, dt_1 \quad \text{(two jumps)}$$

$$+ \cdots .$$

$$(2.89)$$

A general term in the RHS of (2.89) equates to a sum over sequences of jumps through states $i = i_0, i_1, \ldots, i_n = j$, where $i_l \neq i_{l-1}$, and takes the form

$$\sum_{i=i_0,\ldots,i_n=j} \prod_{l=0}^{n-1} \mathbf{1}(q_l > 0, i_{l+1} \neq i_l) \int_0^t \cdots \int_{t_{n-1}}^t \exp[-(t-t_n)q_{i_n}]$$

$$\times \prod_{k: n \to 1} \left(q_{i_{k-1} i_k} \exp\left[-q_{i_{k-1}}(t_k - t_{k-1}) \right] \right) dt_1 \cdots dt_n$$

$$= \sum_{i=i_0,\ldots,i_n=j} \prod_{l=0}^{n-1} \mathbf{1}(q_l > 0, i_{l+1} \neq i_l) \int_0^t \cdots \int_0^{s_{n-1}} \exp[-(t-s_1)q_{i_0}]$$

$$\times \prod_{k: 1 \to n} \left(q_{i_{k-1} i_k} \exp\left[-(s_k - s_{k+1})q_{i_k} \right] \right) ds_n \cdots ds_1$$

$$(2.90)$$

with $t_0 = 0$ and $s_{n+1} = 0$, and after changing variables with $t_1 = t - s_1, t_2 = t - s_2$, etc.

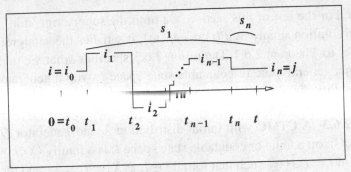

Fig. 2.39

As before, (2.89) and (2.90) yield an important 'step-by-step description' of the entries $p_{ij}^{\min}(t)$, which helps represent the 'contribution' to a given entry coming from various trajectories of the chain; see Figure 2.39. This representation will form the basis for our definitions and constructions. It is worth remarking that (2.89) and (2.90) provide generalisations of (2.46)–(2.47); this emphasizes the role played by the Poisson process in various constructions below.

We will bear in mind two situations. A 'nice' situation like the one mentioned before, where the minimal solution $P^{\min}(t) = \left(p_{ij}^{\min}(t)\right), t > 0$, described in (2.87) is formed by stochastic matrices. Then the entry $p_{ij}^{\min}(t)$ can be taken as the transition probability from i to j in time t. We call this case 'non-explosive', as the CTMC (X_t) defined by transition matrices $P^{\min}(t)$ 'lives' forever in the original state space I. A complication arises when the matrices $P^{\min}(t)$ stop being stochastic. Then we add an absorbing state at ∞ and consider a 'minimal' CTMC with state space $\bar{I} = I \cup \{\infty\}$. This will equip us with a 'first' definition of a CTMC in terms of transition probabilities. See Definition 2.6.3 below.

We see that the minimal solution $P^{\min}(t)$ leads to a *minimal* CTMC with generator matrix Q and an added absorbing state at ∞; in the 'nice' situation where the minimal solution was stochastic, the added state was not needed, and we considered a CTMC on the original state space I.

It is worth mentioning that if the situation ceases to be 'nice', i.e. with a non-stochastic minimal solution, then a sub-stochastic solution to one or both equations in (2.84) is not unique. Other solutions $P(t)$ may be stochastic. However, in general, another complication arises: the forward and the backward equations may yield different sets of solutions.

So, the variety of cases turns out to be quite wide. There exist general conditions on Q guaranteeing that the minimal solution $P^{\min}(t)$ to (2.84) is formed by stochastic matrices. These conditions prove rather complicated (and sometimes lack a clear 'physical sense'); one such condition is quoted in Question 2.10.17 at the end of this chapter. For the rest of this section, we omit the superscript 'min' and denote the minimal solution simply by $P(t) = (p_{ij}(t))$; it satisfies the semigroup property (2.86) thanks to Theorem 2.6.1. Definition 2.6.3 specifies what we understand by a CTMC on a general finite or countable state space I, with a generator Q and an initial probability distribution λ.

Definition 2.6.3 A CTMC with initial distribution λ and generator Q (briefly, a (λ, Q)-CTMC), on a finite or countable state space I, is a family $(X_t, t \geq 0)$ of RVs with values in $\bar{I} = I \cup \{\infty\}$ such that for all $i \in I$, $\mathbb{P}(X_0 = i) = \lambda_i$, and the following property holds.

For all $n = 1, 2, \ldots$, time points $0 = t_0 < t_1 < \cdots < t_n$ and states $i_0, \ldots, i_n \in I$,

$$\mathbb{P}\left(X_0 = i_0, X_{t_1} = i_1, \ldots, X_{t_n} = i_n\right) = \lambda_{i_0} \prod_{k=1}^{n} p_{i_{k-1} i_k}(t_k - t_{k-1}). \tag{2.91}$$

Moreover, for all extended sequences of time points $t_n < t_{n+1} < \cdots < t_{n+l}$,

$$\mathbb{P}\left(X_0 = i_0, X_{t_1} = i_1, \ldots, X_{t_n} = i_n, X_{t_{n+1}} = \cdots = X_{t_{n+l}} = \infty\right)$$
$$= \lambda_{i_0} \prod_{k=1}^{n} p_{i_{k-1} i_k}(t_k - t_{k-1}) \left(1 - \sum_{j \in I} p_{i_n j}(t_{n+1} - t_n)\right). \tag{2.92}$$

Here $p_{ij}(t)$ is defined in (2.89)–(2.90). Remember that matrices $P(t) = (p_{ij}(t))$, $t \geq 0$, give the minimal sub-stochastic solution to (2.84) described in Theorems 2.6.1 and 2.6.2.

As in the discrete-time case, we deduce from Definition 2.6.3 that $p_{ij}(t)$ gives in fact the transition probability in (X_t) from state i to j in time t, i.e. the conditional probability $\mathbb{P}(X_{s+t} = j | X_s = i)$. Furthermore, the difference $1 - \sum_j p_{ij}(t)$ provides the transition probability from i to ∞ in time t; i.e., the conditional probability $\mathbb{P}(X_{s+t} = \infty | X_s = i)$; this represents the probability of explosion in time t from state i. Finally, by definition, $\mathbb{P}(X_{s+t} = \infty | X_s = \infty) \equiv 1$.

By analogy with the case of an explosive birth or birth-and-death process, one would suggest that Definition 2.6.3 specifies a *minimal* chain with given λ and Q.

As in similar definitions above, we have defined what is often called a *homogeneous* Markov chain, where transition probabilities $p_{ij}(t)$ depend only on the lapsed time, not on the position of the pair s, $t + s$ on the time half-axis. A more general class includes *inhomogeneous* chains, one such example being an inhomogeneous Poisson process IPP$(\lambda(t))$ as discussed in Section 2.4.

In what follows the sums $\sum_i$ and $\sum_j$ are understood as $\sum_{i \in I}$ and $\sum_{j \in I}$, without reference to a particular order of summation. Also, expressions like 'for all states i', or briefly, 'for all i' mean for all states $i \in I$.

Definition 2.6.4 We call a (λ, Q) CTMC (X_t) (or generator matrix Q) *non-explosive* if for all i and $t \geq 0$

$$\sum_j p_{ij}(t) = 1, \qquad (2.93)$$

i.e. matrices $P(t)$ are stochastic. Otherwise (X_t) is called explosive.

For non-explosive chains all probabilities (2.92) are zeros. That is, we do not need the state ∞: the chain started from any state i remains in I for all times. Thus we can use

Definition 2.6.5 Assume that Q is non-explosive, in the sense of Definition 2.6.3. We define the following properties.

(b) For all states $i \neq j$, time point $t \geq 0$, and $h \searrow 0+$, the conditional probabilities have the following asymptotics:

$$\left. \begin{array}{l} \mathbb{P}(X_{t+s} \text{ has no jump for } 0 < s < h \\ \quad | X_t = i, \text{ plus a past history prior time } t) \\ \quad\quad = 1 - q_i h + o(h) = 1 + q_{ii} h + o(h), \end{array} \right\} \qquad (2.94)$$

which specifies $q_i = -q_{ii}$ as the rate of leaving state i, and

$$\left. \begin{array}{l} \mathbb{P}(X_{t+s} \text{ has a single jump } i \to j \text{ for } 0 < s < h \\ \quad | X_t = i, \text{ plus a past history prior time } t) = q_{ij} h + o(h), \end{array} \right\} \qquad (2.95)$$

which specifies q_{ij} as the rate of leaving state i for j. Essentially, (2.94) and (2.95) mean that

$$\mathbb{P}(X_{t+s} = j \mid X_t = i, \text{ plus a past history prior time } t)$$
$$= \begin{cases} 1 - q_i h + o(h), & j = i, \\ q_{ij} h + o(h), & j \neq i, \end{cases}$$

regardless of the past history.

(c) For all i with $q_i > 0$, conditional on $X_0 = i$, process (X_t) spends at state i a random time $\sim \mathrm{Exp}(q_i)$ then jumps to state $j \neq i$ with probability

$$\widehat{p}_{ij} = \frac{q_{ij}}{q_i}. \qquad (2.96)$$

Then if J_1 is the time of jump, process X_{J_1+t} behaves as (X_t) conditional on $X_0 = j$, and so on. If $q_i = 0$, then, conditional on $X_0 = i$, process (X_t) stays at state i forever.

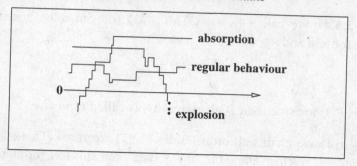

Fig. 2.40

In (b), the euphemism 'past history' means any event generated by random variables X_s when s varies within a specified range of values: $0 \le s < t$ in property (b), and $0 \le s < H_k^i$ in property (c), where H_k^i is the kth hitting time of state i. There is also a subtlety as to whether the remainder terms $o(h)$ in (b) depend on i, j and t; we will not go into further detail here. However, under appropriate specific conditions on terms $o(h)$ in property (b) one can prove:

Theorem 2.6.6 *For a non-explosive chain, the above characterisations from Definitions 2.6.3 and 2.6.5 are equivalent.*

Finally, it is convenient to characterise the case of explosion in terms of jump times $J_0, J_1, J_2, \cdots$. These are defined by

$$J_0 = 0, \; J_1 = \inf[t > 0 : X_t \ne X_0], \; J_2 = \inf[t > J_1 : X_t \ne X_{J_1}], \; \dots. \tag{2.97}$$

Definition 2.6.7 We say that (λ, Q) CTMC (X_t) is *non-explosive*, if, for all states i, conditional on $X_0 = i$, with probability 1, the jump times J_n increase to $+\infty$:

$$\mathbb{P}_i \left(\lim_{n \to \infty} J_n = \infty \right) = 1.$$

Otherwise, when $\mathbb{P}_i \left(\lim_{n \to \infty} J_n = \infty \right) < 1$, i.e., $\mathbb{P}_i \left(\lim_{n \to \infty} J_n < \infty \right) > 0$, the state i is called *explosive* (in (X_t)). The types of sample paths of a CTMC are outlined in Figure 2.40.

Theorem 2.6.8 *The definitions of explosiveness in Definitions 2.6.4 and 2.6.7 are equivalent.*

In sum, we could envisage three types of behaviour of a trajectory of a CTMC: regular, absorbing, and explosive. As was said above, the latter is also converted into absorbing, by adding state ∞.

As in the discrete-time case, single-time probabilities $\mathbb{P}(X_t = i)$ form a vector obtained from $\lambda = (\lambda_i)$ by the action of a transition matrix $P(t)$: for all $t \geq 0$ and states j

$$\mathbb{P}(X_t = j) = \left(\lambda P(t) \right)_j = \sum_i \lambda_i p_{ij}(t), \tag{2.98}$$

In future, we set

$$\widehat{P} = (\widehat{p}_{ij}), \quad \text{where} \quad \widehat{p}_{ij} = \begin{cases} q_{ij}/q_i, & j \neq i, \\ 0, & j = i. \end{cases} \tag{2.99}$$

Assuming that $q_i > 0$ for all i, we obtain a correctly defined stochastic matrix $\widehat{P}$ with diagonal entries zero. The matrix $\widehat{P}$ defines the *jump chain* for CTMC (X_t). More precisely, it is the $(\lambda, \widehat{P})$ DTMC (Y_n) (some authors call it an embedded jump chain, as it follows the jumps in the original CTMC (X_t)). The term 'embedded' is significant here: chain (Y_n) is coupled to (X_t), i.e. its trajectory is constructed from that of (X_t).

Physically speaking, (Y_n) is an observation of jumps in (X_t). Formally,

$$Y_n = X_{J_n}, \quad \text{where } 0 = J_0 < J_1 < \cdots \text{ are jump times in } (X_t). \tag{2.100}$$

Observe that the sample trajectory of the jump chain (Y_n) can always be continued indefinitely in the discrete time $n = 0, 1, \ldots$; it is insensitive to the question of explosiveness.

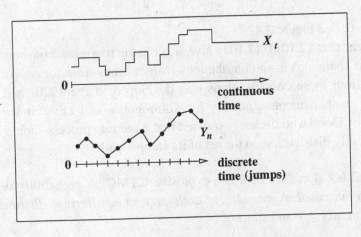

Fig. 2.41

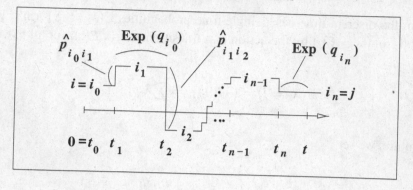

Fig. 2.42

We would like to rewrite (2.89) in terms of jump probabilities $\widehat{p}_{ij}$:

$$p_{ij}(t) = e^{-tq_i} \mathbf{1}(i = j) + \mathbf{1}(i \neq j) \int_0^t q_i e^{-t_1 q_i} \widehat{p}_{ij} \, e^{-(t-t_1)q_j} \, dt_1$$

$$+ \sum_k \mathbf{1}(q_i > 0, k \neq i, j) \int_0^t \int_0^{t-1} q_i e^{-t_1 q_i} \widehat{p}_{ik} q_k e^{-(t_2-t_1)q_k} \widehat{p}_{kj} e^{-(t-t_2)q_j} dt_2 dt_1 + \cdots,$$

$$(2.101)$$

with a general term

$$\int_0^t \int_{t_1}^t \cdots \int_{t_{n-1}}^t \prod_{k=1}^n \left(q_{i_{k-1}} \exp\left[-(t_k - t_{k-1})q_{i_{k-1}}\right] \right) \widehat{p}_{i_{k-1}i_k}$$

$$\times \exp\left[-(t - t_n)q_{i_n}\right] dt_n \cdots dt_1,$$

$$(2.102)$$

where $t_0 = 0$. See Figure 2.42.

We repeat that (2.101), (2.102) give a 'constructive view' on how probability $p_{ij}(t)$ is 'built up' from contributions from various trajectories. An approach emerging from these considerations was developed in the 1920s and 30s by a number of mathematicians, notably by Kolmogorov and Lévy. It was strongly advocated by Doob who tirelessly stressed that a random process should be treated as a probability distribution on the set of its sample paths.

Definition 2.6.9 Let (X_t) be a non-explosive CTMC. A probability distribution $\lambda = (\lambda_i)$ on I is called an *invariant*, or *stationary*, or *equilibrium*, *distribution* (ED) for (X_t) if for all $t \geq 0$ and states j,

$$\mathbb{P}(X_t = j) = \lambda_j, \quad \text{i.e.} \quad \lambda P(t) = \lambda. \tag{2.103}$$

If we only have $\lambda_i \geq 0$ but not $\sum_i \lambda_i = 1$, we say λ is an *invariant measure* (IM) for (X_t). If $\sum_i \lambda_i < \infty$ then

$$\pi_i = \lambda_i \left(\sum_j \lambda_j \right)^{-1}$$

yields an ED $\pi = (\pi_i)$.

Remark 2.6.10 For a minimal explosive chain, the definition only makes sense if distribution π is concentrated at the absorbing state ∞, although one can still think of a measure λ, with $\sum_i \lambda_i = \infty$, satisfying (2.103). We avoid this avenue by always assuming that the chain is non-explosive when speaking of invariant measures and equilibrium distributions.

Theorem 2.6.11 *Assume that (X_t) is a non-explosive CTMC with generator Q. Then $\lambda = (\lambda_i)$ is an IM for (X_t) if and only if for all j,*

$$\sum_i \lambda_i q_{ij} = 0, \quad \text{i.e. the vector } \lambda Q = 0. \tag{2.104}$$

Remark 2.6.12 The statement of Theorem 2.6.11 for finite continuous time Markov chains was proved in (2.19), but we have extended it now to a general non-explosive case. The argument in (2.19) would not work for an explosive chain. More precisely, in the case of an explosive chain one cannot guarantee that $\frac{d}{dt}(\lambda P(t)) = 0$, although it is true that $\frac{d}{dt} P(t) = QP(t) = P(t)Q$.

Theorem 2.6.13 *Assume that (X_t) is a non-explosive continuous time Markov chain with generator Q and that $q_i > 0$ for all i. Let $\lambda = (\lambda_i)$ be an invariant measure for (X_t). Then $\mu = (\mu_i)$ is an invariant measure for the jump chain (Y_n), where*

$$\mu_i = \lambda_i q_i = -\lambda_i q_{ii}. \tag{2.105}$$

Conversely, if μ is an IM for (Y_n) then λ determined from (2.105) is an invariant measure for (X_t).

Proof Write the equation $\mu \widehat{P} = \mu$ as $\mu \widehat{P} - \mu = 0$, or

$$\left(\mu \widehat{P} - \mu \right)_j = \sum_{i: i \neq j} \mu_i \frac{q_{ij}}{q_i} - \mu_j = \sum_i \mu_i \left[(1 - \delta_{ij}) \frac{q_{ij}}{q_i} - \delta_{ij} \right]$$

$$= \sum_i \mu_i \left[\frac{q_{ij}}{q_i} - \underbrace{\delta_{ij} \left(1 + \frac{q_{ij}}{q_i} \right)}_{0} \right] = \sum_i \lambda_i q_{ij} = (\lambda Q)_j.$$

The LHS is zero if and only if the RHS is. $\qquad \square$

So, if λ is an invariant measure for a non-explosive CTMC (X_t), with $\sum \lambda_i q_i < \infty$, and $\mu_i = \lambda_i q_i$, then $\widehat{\pi} = (\widehat{\pi}_i)$ is an equilibrium distribution for (Y_n) where

$$\widehat{\pi}_i = \frac{\mu_i}{\sum_j \mu_j} = \frac{\lambda_i q_i}{\sum_j \lambda_j q_j}. \tag{2.106}$$

Definition 2.6.14 Let (X_t) be a (λ, Q) CTMC (possibly explosive). We say that states i, j *communicate* in (X_t) if and only if i, j communicate in the jump chain (Y_n). Thus, the class division in both (X_t) and (Y_n) is the same. In particular, we say that (λ, Q) CTMC (X_t) (or its generator Q) is *irreducible* if it has a unique communicating class, coinciding with the whole space I.

> *Members of all communicating classes, unite!*
> (From the series 'When they go political'.)

Theorem 2.6.15 Let (X_t) be a (λ, Q) CTMC (possibly explosive). Then states i, j communicate in (X_t) if and only if one of the following equivalent conditions holds:

(a) $q_{i_0 i_1} \cdots q_{i_{n-1} i_n} > 0$ for some states $i = i_0, i_1, \ldots, i_n = j$;   (2.107)

(b) $p_{ij}(t) > 0$ for all $t > 0$.   (2.108)

Proof (b) $\Longrightarrow$ (a). If (b) holds then i and j communicate in (Y_n), hence i and j communicate in (X_t). Then $p_{i_0 i_1}(t) \cdots p_{i_{n-1} i_n}(t) > 0$ for all $t > 0$ and some path $i = i_0, i_1, \ldots, i_n = j$. Hence, $q_{i_0 i_1} \cdots q_{i_{n-1} i_n} > 0$. This yields (a).

(a) $\Longrightarrow$ (b). If (a) holds, then for all pairs $i_{k-1} i_k$ and any $t > 0$:

$$p_{i_{k-1} i_k}(t) > \mathbb{P}\big(\text{a single jump in } (0,t); X_t = i_k | X_0 = i_{k-1}\big)$$

$$= \int_0^t \underbrace{q_{i_{k-1}} \exp\left(-q_{i_{k-1}} s\right)}_{\text{1st jump at } s} \underbrace{\frac{q_{i_{k-1} i_k}}{q_{i_{k-1}}}}_{\text{jump } i_{k-1} \to i_k} \underbrace{\exp\left(-q_{i_k}(t-s)\right)}_{\text{no jump in } (s,t)} ds > 0.$$

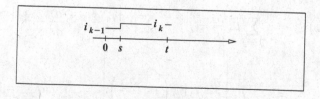

Fig. 2.43

Then

$$p_{ij}(t) > p_{i_0 i_1}\left(\frac{t}{n}\right) \cdots p_{i_{n-1} i_n}\left(\frac{t}{n}\right) > 0,$$

which implies (b).   □

States of all classes, communicate!
(From the series 'When they go political.')

A general countable CTMC (explosive or not) possesses the Markov and strong Markov properties in exactly the same form as stated in Theorems 2.2.2 and 2.2.3 for a finite CTMC. However, manipulations with states can destroy the Markov property. For example, glueing (merging) several states of a chain (DTMC or CTMC) into a single state may or may not produce a Markov chain.

Worked Example 2.6.16 (i) Consider the continuous-time Markov chain $(X_t)_{t \geq 0}$ with state space $\{1, 2, 3, 4\}$ and Q-matrix

$$Q = \begin{pmatrix} -2 & 0 & 0 & 2 \\ 1 & -3 & 2 & 0 \\ 0 & 2 & -2 & 0 \\ 1 & 5 & 2 & -8 \end{pmatrix}.$$

Set

$$Y_t = \begin{cases} X_t & \text{if } X_t \in \{1, 2, 3\}, \\ 2 & \text{if } X_t = 4 \end{cases}$$

and

$$Z_t = \begin{cases} X_t & \text{if } X_t \in \{1, 2, 3\}, \\ 1 & \text{if } X_t = 4 \end{cases}.$$

Determine which, if any, of the processes $(Y_t)_{t \geq 0}$ and $(Z_t)_{t \geq 0}$ are Markov chains.

(ii) Find an invariant distribution for the chain $(X_t)_{t \geq 0}$ given in Part (i). Suppose $X_0 = 1$. Find, for all $t \geq 0$, the probability that $X_t = 1$.

Solution (i) The passage from (X_t) to (Y_t) consists in merging states 2 and 4 into a single state, say $2 * 4$; see Figure 2.44. It is clear that in order to prove that (Y_t) is a Markov chain, we only need to check that the holding time $J^Y(2 * 4)$ at state $2 * 4$ is exponential (an educated guess is that it is $\mathrm{Exp}(3)$). In chain (X_t) the holding time $J^X(2)$ at state 2 is $\mathrm{Exp}(3)$ and the holding time $J^X(4)$ at state 4 is $\mathrm{Exp}(8)$.

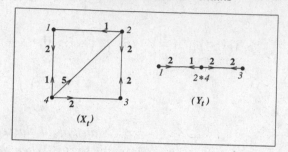

Fig. 2.44

Let $H_n^Y(2*4)$ be the *n*th hitting time of merged state $2*4$ in (Y_t), then $X_{H_n^Y(2*4)}$, the state of chain (X_t) at time $H_n^Y(2*4)$, will be either 2 or 4. Thus,

$$\mathbb{P}(J_n^Y(2*4) > x) = \mathbb{P}\left(J_n^Y(2*4) > x, X_{H_n^Y(2*4)} = 2\right)$$
$$+ \mathbb{P}\left(J_n^Y(2*4) > x, X_{H_n^Y(2*4)} = 4\right).$$

The fact that $J_n^Y(2*4) > x$ means that (Y_t) didn't jump in the time interval $\left(H_n^Y(2*4), H_n^Y(2*4)+x\right)$. But if $X_{H_n^Y(2*4)} = 2$, this means that X_t didn't jump in $\left(H_n^Y(2*4), H_n^Y(2*4)+x\right)$. Correspondingly,

$$\mathbb{P}\left(J_n^Y(2*4) > x, X_{H_n^Y(2*4)} = 2\right)$$
$$= \mathbb{P}\left(X_{H_n^Y(2*4)} = 2\right) \mathbb{P}\left(J_n^Y(2*4) > x \middle| X_{H_n^Y(2*4)} = 2\right)$$
$$= \mathbb{P}\left(X_{H_n^Y(2*4)} = 2\right)$$
$$\times \mathbb{P}\left(X_t \text{ didn't jump in } \left(H_n^Y(2*4), H_n^Y(2*4)+x\right) \middle| X_{H_n^Y(2*4)} = 2\right)$$
$$= \mathbb{P}\left(X_{H_n^Y(2*4)} = 2\right) e^{-3x}.$$

Similarly,

$$\mathbb{P}\left(J_n^Y(2*4) > x, X_{H_n^Y(2*4)} = 4\right)$$
$$= \mathbb{P}\left(X_{H_n^Y(2*4)} = 4\right)$$
$$\times \left[\mathbb{P}\left(X_t \text{ didn't jump in } \left(H_n^Y(2*4), H_n^Y(2*4)+x\right) \middle| X_{H_n^Y(2*4)} = 4\right)\right.$$
$$+ \mathbb{P}\left(X_t \text{ had a single jump } 4 \to 2\right.$$
$$\left.\left. \text{ in } \left(H_n^Y(2*4), H_n^Y(2*4)+x\right) \middle| X_{H_n^Y(2*4)} = 4\right)\right].$$

The sum in the square brackets equals

$$e^{-8x} + \int_0^x ds\,\left(8e^{-8s}\right)\frac{5}{8}e^{-3(x-s)} = e^{-8x} + 5e^{-3x}\int_0^x ds\,e^{-5x}$$

$$= e^{-8x} + e^{-3x}\left(1 - e^{-5x}\right) = e^{-3x}.$$

Hence,

$$\mathbb{P}(J_n^Y(2*4) > x) = e^{-3x}\left[\mathbb{P}\left(X_{H_n^Y(2*4)} = 2\right) + \mathbb{P}\left(X_{H_n^Y(2*4)} = 4\right)\right] = e^{-3x},$$

as the sum $\mathbb{P}\left(X_{H_n^Y(2*4)} = 2\right) + \mathbb{P}\left(X_{H_n^Y(2*4)} = 4\right)$ equals 1.

So, (Y_t) is a Markov chain; its Q-matrix is

$$Q^Y = \begin{pmatrix} -2 & 2 & 0 \\ 1 & -3 & 2 \\ 0 & 2 & -2 \end{pmatrix} \begin{matrix} 1 \\ 2*4 \\ 3 \end{matrix} \;.$$

Analysing the above argument, this fact holds because in chain (X_t), the jump rates from state 2 to states 1 and 3 are the same as from state 4 to states 1 and 3 (the jump from 4 to 2 is discarded when we pass from (X_t) to (Y_t)).

In contrast, (Z_t) is not a Markov chain, since the above property does not hold for states 1 and 4. In fact, given $s,t > 0$, start both processes (Z_t) and (X_t) from state 2. We can write

$$\mathbb{P}(Z_{t+s} = 3 \mid Z_s = 1*4, Z_0 = 2)$$

$$= \frac{\mathbb{P}(Z_{t+s} = 3, Z_s = 1*4 \mid Z_0 = 2)}{\mathbb{P}(Z_s = 1*4 \mid Z_0 = 2)}$$

$$= \frac{\mathbb{P}(X_{t+s} = 3, X_s = 1 \mid X_0 = 2) + \mathbb{P}(X_{t+s} = 3, X_s = 4 \mid X_0 = 2)}{\mathbb{P}(X_s = 1 \mid X_0 = 2) + \mathbb{P}(X_s = 4 \mid X_0 = 2)}$$

$$= \frac{\mathbb{P}(X_s = 1 \mid X_0 = 2)\mathbb{P}(X_t = 3 \mid X_0 = 1)}{\mathbb{P}(X_s = 1 \mid X_0 = 2) + \mathbb{P}(X_s = 4 \mid X_0 = 2)}$$

$$+ \frac{\mathbb{P}(X_s = 4 \mid X_0 = 2)\mathbb{P}(X_t = 3 \mid X_0 = 4)}{\mathbb{P}(X_s = 1 \mid X_0 = 2) + \mathbb{P}(X_s = 4 \mid X_0 = 2)}$$

$$= \frac{\mathbb{P}(X_t = 3 \mid X_0 = 1)}{1 + q} + \frac{\mathbb{P}(X_t = 3 \mid X_0 = 4)}{1 + q^{-1}}.$$

Here q stands for the ratio:

$$q = \frac{\mathbb{P}(X_s = 4 \mid X_0 = 2)}{\mathbb{P}(X_s = 1 \mid X_0 = 2)} = \frac{\left(e^{sQ}\right)_{24}}{\left(e^{sQ}\right)_{21}}.$$

Similarly, starting both (Z_t) and (X_t) from state 3,

$$\mathbb{P}\big(Z_{t+s} = 3 \,|\, Z_s = 1*4, Z_0 = 3\big)$$

$$= \frac{\mathbb{P}\big(Z_{t+s} = 3, Z_s = 1*4 \,|\, Z_0 = 3\big)}{\mathbb{P}\big(Z_s = 1*4 \,|\, Z_0 = 3\big)}$$

$$= \frac{\mathbb{P}\big(X_t = 3 \,|\, X_0 = 1\big)}{1+r} + \frac{\mathbb{P}\big(X_t = 3 \,|\, X_0 = 4\big)}{1+r^{-1}},$$

where

$$r = \frac{\mathbb{P}\big(X_s = 4 \,|\, X_0 = 3\big)}{\mathbb{P}\big(X_s = 1 \,|\, X_0 = 3\big)} = \frac{\big(e^{sQ}\big)_{34}}{\big(e^{sQ}\big)_{31}}.$$

But the ratios q and r are not identically equal:

$$\frac{\big(e^{sQ}\big)_{24}}{\big(e^{sQ}\big)_{21}} \neq \frac{\big(e^{sQ}\big)_{34}}{\big(e^{sQ}\big)_{31}},$$

which implies that the conditional probability $\mathbb{P}\big(Z_{t+s} = 3 \,|\, Z_s = 1*4, Z_0 = i\big)$ still depends on i, i.e. condition $Z_0 = i$ cannot be dropped.

To show that $q \not\equiv r$, consider the case of small s: $s \to 0+$. It is useful to calculate

$$Q^2 = \begin{pmatrix} 6 & 10 & 4 & -20 \\ -5 & 13 & -10 & 2 \\ 2 & -10 & 8 & 0 \\ -5 & -51 & -10 & 66 \end{pmatrix} \text{ and } Q^3 = \begin{pmatrix} -22 & -122 & -28 & 172 \\ 25 & -49 & 50 & -26 \\ -14 & 46 & -36 & 4 \\ 25 & 463 & 50 & -538 \end{pmatrix},$$

(in reality, we will only need entries Q_{21}, $(Q^2)_{31}$, $(Q^2)_{31}$ and $(Q^3)_{34}$ which can be computed straightaway). In fact, the entries

$$\big(e^{sQ}\big)_{ij} = \sum_{k \geq 0} \frac{s^k}{k!} \big(Q^k\big)_{ij} \to \delta_{ij}, \quad s \to 0+.$$

More precisely, $\big(e^{sQ}\big)_{34}$, $\big(e^{sQ}\big)_{31}$, $\big(e^{sQ}\big)_{24}$ and $\big(e^{sQ}\big)_{21}$ have the following leading terms:

$$\big(e^{sQ}\big)_{34} = \frac{s^3}{3!}\big(Q^3\big)_{34} + O(s^4) = \frac{s^3}{3!} \times 4 + O(s^4)$$

$$\big(e^{sQ}\big)_{31} = \frac{s^2}{2!}\big(Q^2\big)_{31} + O(s^3) = \frac{s^2}{2!} \times 2 + O(s^3),$$

$$\big(e^{sQ}\big)_{24} = \frac{s^2}{2!}\big(Q^2\big)_{24} + O(s^3) = \frac{s^2}{2!} \times 2 + O(s^3),$$

$$\big(e^{sQ}\big)_{21} = \frac{s}{1!}Q_{21} + O(s^2) = \frac{s}{1!} \times 1 + O(s^2).$$

This yields

$$q \approx s, \ r \approx \frac{2}{3}s,$$

verifying the above claim.

(ii) The detailed balance equations for chain (Y_t) are

$$2\lambda_1 = \lambda_{2*4}, \ \lambda_{2*4} = \lambda_3.$$

The (unique) normalised solution is $\lambda = (1/5, 2/5, 2/5)$ which is the equilibrium distribution for (Y_t). Then the ED π for (X_t) has $\pi_1 = \lambda_1$, $\pi_3 = \lambda_3$, and $\pi_2 + \pi_4 = \lambda_{2*4}$. In fact, it is easy to see that

$$\pi = \left(\frac{1}{5}, \frac{7}{20}, \frac{2}{5}, \frac{1}{20} \right).$$

Finally, observe that $\mathbb{P}(X_t = 1 | X_0 = 1) = \mathbb{P}_1(Y_t = 1 | Y_0 = 1)$, as state 1 is identical in both chains (X_t) and (Y_t). The Q^Y-matrix eigenvalues for chain (Y_t) are solutions to

$$\det \begin{pmatrix} -2-\mu & 2 & 0 \\ 1 & -3-\mu & 2 \\ 0 & 2 & -2-\mu \end{pmatrix} = 0$$

and are 0, -2 and -5. So, by a standard diagonalization argument, entry $p_{11}^Y(t)$ of the transition matrix $P^Y(t) = \exp(tQ^Y)$ is of the form

$$p_{11}^Y(t) = A + Be^{-2t} + Ce^{-5t}, \ t \geq 0.$$

Constants A, B and C satisfy

$$A + B + C = p_{11}^Y(0) = 1, \ -2B - 5C = \frac{d}{dt} p_{11}^Y(0) = -2, \ A = p_{11}^Y(\infty) = \frac{1}{5},$$

whence $B = 2/3$, $C = 2/15$. The final answer is

$$\mathbb{P}(X_t = 1 | X_0 = 1) = \mathbb{P}(Y_t = 1 | Y_0 = 1) = \frac{1}{5} + \frac{2}{3} e^{-2t} + \frac{2}{15} e^{-5t}.$$

$\square$

2.7 Hitting times and probabilities. Recurrence and transience

> *The workers have nothing to lose*
> *... but their chains.*
> K. Marx (1818–1883), German philosopher

We begin this section with the following

Definition 2.7.1 Let (X_t) be a (λ, Q) CTMC (possibly explosive). Given a set of states $A \subset I$, we define the *hitting time* H^A (of set A in chain (X_t)) by

$$H^A = \begin{cases} \inf\,[t \geq 0: X_t \in A], & \text{if } X_t \in A \text{ for some } t \geq 0, \\ \infty & \text{if } X_t \notin A \text{ for all } t \geq 0. \end{cases} \tag{2.109}$$

To stress connections with (X_t) and (Y_n), we often use the notation H_X^A and H_Y^A.

Next, we repeat the definition of jump times:

Definition 2.7.2 The times of subsequent jumps in (X_t) are defined by

$$J_0 = 0, \; J_1 = \inf\,[t > 0: X_t \neq X_0], \; J_2 = \inf\,[t > J_1: X_t \neq X_{J_1}], \ldots. \tag{2.110}$$

To stress their origin we often write $J_1^X, J_2^X, \ldots$.

For (X_t) non-explosive, obviously,

$$H_X^A < \infty \text{ if and only if } H_Y^A < \infty; \quad \text{in fact, } H_X^A = J_{H_Y^A}^X. \tag{2.111}$$

Then the *hitting probabilities* h_i^A (of set A from state i in chain (X_t)) are defined in the same way as for the discrete time case:

$$h_i^A = \mathbb{P}_i(H_X^A < \infty) = \mathbb{P}(H_X^A < \infty | X_0 = i) = \mathbb{P}(H_Y^A < \infty | Y_0 = i). \tag{2.112}$$

As in preceding sections, $\mathbb{P}_i$ stands for the probability distribution of the CTMC with initial distribution δ_i, i.e. starting from state i. Similarly, $\mathbb{E}_i$ denotes the expectation relative to $\mathbb{P}_i$.

Consider the column vector $h^A = (h_i^A)$ formed by the hitting probabilities of the set A from various initial states.

Theorem 2.7.3 Let (X_t) be a (λ, Q) CTMC (possibly explosive). Assume that $q_i > 0$ for all states i. The vector h^A gives the minimal non-negative solution to the following equations:

$$\begin{cases} h_i^A = 1, & i \in A, \\ (Qh^A)_i = \sum_j q_{ij} h_j^A = 0, & i \notin A. \end{cases} \tag{2.113}$$

Proof The case $i \in A$ is obvious: $h_i^A = \mathbb{P}_i(\text{hit } A) = 1$. So, let $i \notin A$. Then in the $(\delta_i, \widehat{P})$ jump chain (Y_n), by conditioning on the first jump,

$$h_i^A = \sum_{j: j \neq i} \left(\frac{q_{ij}}{q_i} \right) h_j^A, \quad \text{i.e.} \quad -q_{ii} h_i^A = \sum_{i \neq i} q_{ij} h_j^A, \quad \text{or} \quad (Q h^A)_i = 0.$$

Thus, h^A always yields a non-negative solution. Minimality is proven exactly as in the discrete-time case. $\quad\square$

Remark 2.7.4 Observe that $h_i \equiv 1$, i.e. the vector $h = \mathbf{1}$ is always a non-negative solution, as for all i,

$$(Q\mathbf{1})_i = \sum_j q_{ij} = 0$$

(the sum along row i of Q). But it is not always minimal.

Definition 2.7.5 Next, we define the *mean* hitting times:

$$k_i^A = \mathbb{E}_i H_X^A = \mathbb{E}\left(H_X^A | X_0 = i \right). \tag{2.114}$$

Note that k_i^A can be infinite.

Theorem 2.7.6 *Let* (X_t) *be a* (λ, Q) *continuous time Markov chain (possibly explosive). Assume that* $q_i > 0$ *for all states* i. *The column vector* k^A *gives the minimal non-negative solution (possibly with some entries* $k_i^A = +\infty$*) to*

$$\begin{cases} k_i^A = 0, & i \in A, \\ k_i^A = \dfrac{1}{q_i} + \sum_{j: j \neq i} \left(\dfrac{q_{ij}}{q_i} \right) k_j^A = \dfrac{1}{q_i} + (\widehat{P}k^A)_i, & i \notin A. \end{cases} \tag{2.115}$$

If $k_i^A < +\infty$ *for all* i, *then* $k^A = (k_i^A)$ *solves*

$$\begin{cases} k_i^A = 0, & i \in A, \\ \sum_j q_{ij} k_j^A = (Q k^A)_i = -1, & i \notin A. \end{cases} \tag{2.116}$$

Proof The equality $k_i^A = 0$ for $i \in A$ holds trivially. If $X_0 = i \notin A$ then the hitting time $H_X^A \geq J_1^X$, the time of the first jump in (X_t). Conditional on the first jump

$$
\begin{aligned}
k_i^A &= \mathbb{E}\big[\mathbb{E}_i\big(H_X^A\big|\text{position after } J_1^X\big)\big] \\
&= \mathbb{E}\big[\mathbb{E}_i\big(J_1^X + (H_X^A - J_1^X)\big|\text{position after } J_1^X\big)\big] \\
&= q_i^{-1} + \sum_{j:j\neq i} q_{ij}q_i^{-1}\mathbb{E}_j H_X^A \quad \text{(by the strong Markov property)} \\
&= q_i^{-1} + \sum_{j:j\neq i} \widehat{p}_{ij} k_j^A .
\end{aligned}
$$

This yields (2.115). If we know that all entries $k_i^A < \infty$, we can transfer terms from the LHS to the RHS and vice versa. Multiplying by q_i this leads to (2.116).

To prove minimality, let $g = (g_i)$ be any solution. Then $g_i = k_i^A = 0$ for $i \in A$. For $i \neq A$, let $J_0 = 0$ and $J_1, J_2, \dots$ be the times of subsequent jumps. (The index X referring to (X_t) will be now omitted.) Divide by q_i, rearrange and iterate the equation. We obtain that

$$
\begin{aligned}
g_i &= q_i^{-1} + \sum_{j\notin A} \widehat{p}_{ij} g_j \\[4pt]
&= \mathbb{E}_i(J_1 - J_0) + \sum_{j\notin A} \widehat{p}_{ij}\left(q_j^{-1} + \sum_{k\notin A} \widehat{p}_{jk} g_k \right) \\[4pt]
&= \mathbb{E}_i\big[(J_1 - J_0)\mathbf{1}(H_Y^A \geq 1)\big] + \mathbb{E}_i\big[(J_2 - J_1)\mathbf{1}(H_Y^A \geq 2)\big] \\
&\quad + \sum_{j,k\notin A} \widehat{p}_{ij}\widehat{p}_{jk} g_k \\
&= \cdots \\
&= \mathbb{E}_i\big[(J_1 - J_0)\mathbf{1}(H_Y^A \geq 1)\big] + \mathbb{E}_i\big[(J_2 - J_1)\mathbf{1}(H_Y^A \geq 2)\big] + \cdots \\
&\quad + \mathbb{E}_i\big[(J_n - J_{n-1})\mathbf{1}(H_Y^A \geq n)\big] + \sum_{j_1,\dots,j_n\notin A} \widehat{p}_{ij_1} \prod_{1\leq l<n} \widehat{p}_{j_l j_{l+1}} g_{j_n} \\
&= \sum_{k=1}^{n} \mathbb{E}_i\big[(J_k - J_{k-1})\mathbf{1}(H_Y^A \geq k)\big] + \sum_{j_1,\dots,j_n\notin A} \widehat{p}_{ij_1} \prod_{1\leq l<n} \widehat{p}_{j_l j_{l+1}} g_{j_n} .
\end{aligned}
$$

If $g \geq 0$ then, for all n, the last sum is ≥ 0. Then, with $H_Y^A \wedge n = \min\big[n, H_Y^A\big]$,

$$
\begin{aligned}
g_i &\geq \sum_{k=1}^{n} \mathbb{E}_i\big[(J_k - J_{k-1})\mathbf{1}(H_Y^A \geq k)\big] = \mathbb{E}_i\left[\sum_{k=1}^{H_Y^A \wedge n}(J_k - J_{k-1})\right] \\
&= \mathbb{E}_i J_{H_Y^A \wedge n} \nearrow \mathbb{E}_i J_{H_Y^A} = \mathbb{E}_i H_X^A = k_i^A, \quad \text{as } n \to \infty.
\end{aligned}
$$

$\square$

Remark 2.7.7 In some chains, the mean times come out as

$$
k_i^A = \begin{cases} 0, & i \in A, \\ +\infty, & i \notin A \end{cases} . \tag{2.117}
$$

Compare with Example 1.3.5. Then (2.117) will satisfy (2.115); formally, it will give a non-negative solution. Conversely, if any non-negative solution to (2.115) is of the form (2.117) then the mean times $\mathbb{E}_i H_X^A \equiv +\infty$, $i \notin A$.

Definition 2.7.8 Let (X_t) be a (λ, Q) CTMC (possibly explosive). We describe state i as $\begin{cases} \text{recurrent} & \text{(R)} \\ \text{transient} & \text{(T)} \end{cases}$ in chain (X_t) if

$$\mathbb{P}_i\big(\sup[t \geq 0 : X_t = i] = \infty\big)$$

$$= \mathbb{P}_i(i \text{ visited in } (X_t) \text{ at arbitrarily large times}) = \begin{cases} 1 \\ 0 \end{cases} . \quad (2.118)$$

Remark 2.7.9 If (X_t) explodes from state i then i is transient.

Theorem 2.7.10 Let (X_t) be a (λ, Q) CTMC (possibly explosive). Assume that $q_i > 0$ for all states i. Then:

  (i)   *each state i is recurrent or transient in (X_t) and (Y_n) at the same time;*
  (ii) *each state must be either recurrent or transient in (X_t);*
  (iii) *recurrence and transience represent class properties in (X_t).*

Proof (i) Let state i be recurrent in (Y_n). Then, starting from i, $Y_n = X_{J_n} = i$ infinitely often. Then (X_t) does not explode from i (the explosion time would contain infinitely many holding times at i):

$$\mathbb{P}_i(T_{\text{explo}} < \infty) \leq \mathbb{P}_i\left(\sum_{k \geq 1} S_k^{(i)} < \infty\right) = 0$$

as

$$S_1^{(i)}, S_2^{(i)}, \ldots \sim \text{Exp}(q_i), \text{ independently.}$$

We deduce that $\mathbb{P}_i(J_n \nearrow \infty) = 1$. Also, $Y_n = X_{J_n} = i$ infinitely often. It then follows that $X_t = i$ for indefinitely large t. Hence, i is recurrent in (X_t), and state i will be revisited.

Now let state i be transient in (Y_n). Then, with $X_0 = Y_0 = i$:

$$\mathbb{P}_i(\bar{n} = \sup[n : Y_n = i] < \infty) = 1.$$

But this can be written as

$$\mathbb{P}_i(\bar{t} = \sup[t : X_t = i] = J_{\bar{n}+1} < \infty).$$

Thus, i is transient in (X_t), and (i) is proved.

  (ii) This statement follows from (i), as it holds for (Y_n).
  (iii) The same argument works, as (ii) also applies. $\square$

Theorem 2.7.11 *Let (X_t) be a (λ, Q) CTMC (possibly explosive). Given state i, the following dichotomy holds. Either*

(i) $q_i = 0$ *(absorption) or* $\mathbb{P}_i(T_i^X < \infty) = 1$. *In the latter case i is* R, *and*

$$\int_0^\infty p_{ii}(t)\, dt = \infty,$$

where

$$T_i^X = \inf [t \geq J_1 : X_t = i], \text{ the return time to } i \text{ in } (X_t), \qquad (2.119)$$

(ii) *or* $q_i > 0$ *and* $\mathbb{P}_i(T_i^X < \infty) < 1$; *then i is* T, *and*

$$\int_0^\infty p_{ii}(t)\, dt < \infty,$$

where T_i^X is defined in (2.119).

Proof The case $q_i = 0$ should be obvious, so suppose that $q_i > 0$. Then, for T_i^Y the return time to i in (Y_n), events $\{T_i^X < \infty\} = \{T_i^Y < \infty\}$, and

$$\mathbb{P}_i(T_i^X < \infty) = \mathbb{P}_i(T_i^Y < \infty).$$

By Theorem 2.7.10(i), state i is recurrent if and only if $\mathbb{P}_i(T_i^X < \infty) = 1$ (and hence transient if and only if $\mathbb{P}_i(T_i^X < \infty) < 1$).

Finally,

$$
\begin{aligned}
\int_0^\infty p_{ii}(t)\, dt &= \int_0^\infty \mathbb{E}_i\left[\mathbf{1}(X_t = i)\right] dt = \mathbb{E}_i\left[\int_0^\infty \mathbf{1}(X_t = i) dt\right] \\
&= \mathbb{E}_i\left[\sum_n (J_{n+1} - J_n)\, \mathbf{1}(Y_n = i)\right] \\
&= \sum_n \mathbb{E}\left(S_n^{(i)}\big|Y_n = i\right) \mathbb{P}_i(Y_n = i) \\
&= \frac{1}{q_i}\sum_n \mathbb{P}(Y_n = i) \\
&= \frac{1}{q_i}\sum_n \hat{p}_{ii}^{(n)}.
\end{aligned}
$$

We deduce that $\int_0^\infty p_{ii}(t)\, dt = \infty$ if and only if $\sum_n \hat{p}_{ii}^{(n)} = \infty$. This proves the theorem. $\square$

Remark 2.7.12 As in the discrete time case, if the hitting probability $h_j^{\{i\}} = \mathbb{P}_j$ (hit i) $= 1$ for all $j \neq i$, state i will be recurrent. Otherwise, state i may be recurrent or transient (it is transient if $h_j^{(i)} < 1$ for some $j \neq i$ and the chain is irreducible). More precisely,

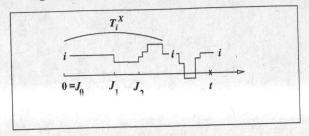

Fig. 2.45

$$\mathbb{P}_i\big(T_i < \infty\big) = \sum_{j:j\neq i} \widehat{p}_{ij} h_j^{(i)} \begin{cases} = 1, & \text{if } h_j^{(i)} \equiv 1, \text{ for all } j \neq i, \\ < 1, & \text{if } \widehat{p}_{ij} h_j^{(i)} < \widehat{p}_{ij} \text{ for some } j \neq i. \end{cases}$$

In fact, if the chain is irreducible and $h_j^{\{i\}} < 1$ for some $j \neq i$, then we can take n such that $\widehat{p}_{ij}^{(n)} > 0$. Writing

$$\mathbb{P}_i\big(i \text{ visited at indefinitely large times}\big) \;\leq\; \sum_l \widehat{p}_{il}^{(n)} h_l^{\{i\}}$$
$$< \sum_l \widehat{p}_{il}^{(n)} = 1 \ (\text{as } h_j^{\{i\}} < 1),$$

we see that in this case i is transient.

Recurrence and transience may be detected from a *discrete-time sampling* of (X_t) at the 'spacing' h:

$$Z_n = X_{nh}, \ n = 0, 1, \dots; \quad (Z_n) \text{ forms a } (\lambda, P(h)) \text{ DTMC.} \tag{2.120}$$

We call (Z_n) the *embedded h-spacing* DTMC for (X_t).

Theorem 2.7.13 *For all $h > 0$, with $Z_n = X_{nh}$, a state i is R in (X_t) if and only if i is R in (Z_n). Hence, i is T in (X_t) if and only if it is T in (Z_n).*

Proof If i is transient for (X_t), it is transient for (Z_n), trivially. Conversely, let i be recurrent for (X_t). Then, for $nh < t < (n+1)h$:

$$p_{ii}\big((n+1)h\big) \;\geq\; p_{ii}(t)e^{-q_ih}$$
$$= \mathbb{P}_i\big(X_t = i \text{ and no jump in } (t, t+h)\big).$$

Consequently,

$$\int_0^\infty p_{ii}(t)\, \mathrm{d}t \leq he^{q_ih}\sum_{n\geq1} p_{ii}(nh),$$

and i is recurrent for (Z_n). $\qquad\square$

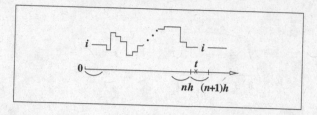

Fig. 2.46

Definition 2.7.14 Suppose that the (λ, Q) CTMC (X_t) is irreducible. We call (X_t) (or its Q-matrix Q) *recurrent* (R) if every state i is recurrent and *transient* (T) if every state i is transient this property holds for every state i.

Theorem 2.7.15 *Let Q be irreducible and recurrent. Then the invariant measure λ_i, with $\lambda Q = 0$, is unique up to scalar factors.*

Proof Excluding the trivial case where the chain contains a single absorbing state and assuming $q_i > 0$ for all i, the jump chain transition matrix $\widehat{P} = (q_{ij}/q_i, \ i \neq j)$ is recurrent.

Then, from Theorem 1.7.5, all invariant measures $\mu = (\mu_i)$ for (Y_n) become proportional. Thus, all IMs $\lambda = (\lambda_i)$ for (X_t), with $\lambda_i = \mu_i/q_i$, will be proportional as well. □

Theorem 2.7.16 *Let CTMC (X_t) be irreducible. Then if (X_t) is R it is non-explosive, i.e. if $J_1 < J_2 < \cdots$ make up the jump times, then for all states i,*

$$\mathbb{P}_i\left(\lim_{n\to\infty} J_n = \infty\right) = 1.$$

Proof For any given state i, the jump chain (Y_n) keeps returning to i. Let $S_0^{(i)}$, $S_1^{(i)}, \ldots$ be the successive holding times at i. Then

$$\lim_{n\to\infty} J_n \geq \lim_{n\to\infty}\left(S_1^{(i)} + \cdots + S_n^{(i)}\right) = \infty \quad \text{a.s.}$$

as the $(S_k^{(i)})$ are IID $\sim \text{Exp}(q_i)$. □

From now on, until the end of Section 2.7, we assume that (X_t) is irreducible, with all $q_i > 0$. So, if (X_t) is recurrent, it will be non-explosive.

Recall that a state i is recurrent if and only if $\mathbb{P}_i(T_i < \infty) = 1$, so it is revisited for indefinitely large times.

Definition 2.7.17 We say a state i is *positive recurrent* (PR) in a CTMC (X_t), if

$$m_i = \mathbb{E}_i T_i < \infty, \tag{2.121}$$

and *null-recurrent* (NR), if

$$\mathbb{P}_i(T_i < \infty) = 1, \text{ but } m_i = \infty. \tag{2.122}$$

Recall that $\mathbb{E}_i$ stands for the expectation under the probability distribution $\mathbb{P}_i$ of the CTMC starting from state i. As we will see from Theorem 2.7.18 below, if a chain is irreducible and recurrent, then we have a dichotomy: either all states are positive recurrent or all states are null recurrent. In the first case we say that the chain (or its generator matrix) is *positive recurrent* and in the second case that it is *null recurrent*.

Theorem 2.7.18 *Let* (X_t) *be an irreducible and recurrent* (λ, Q) *CTMC. Then:*

(i) *either every state i is PR or every state i is NR;*

(ii) *or Q is PR if and only if it has a (unique) equilibrium distribution $\pi = (\pi_i)$, in which case*

$$\pi_i > 0 \text{ and } m_i = \frac{1}{\pi_i q_i} \quad \text{for all } i. \tag{2.123}$$

Proof Given state i, split the mean value m_i according to the outcome of the first jump:

$$
\begin{aligned}
m_i &= \text{ mean return time to } i \\
&= \text{ mean holding time at } i \\
&\quad + \sum_{j:j\neq i} (\text{mean time spent at } j \text{ before returning to } i).
\end{aligned}
$$

The first summand equals q_i^{-1}. So, set $\gamma_i = 1/q_i$ and for $j \neq i$ write:

$$
\begin{aligned}
\gamma_j &= \mathbb{E}_i (\text{time spent at } j \text{ before returning to } i) \\
&= \mathbb{E}_i \left[\int_{J_1}^{T_i} \mathbf{1}(X_t = j)\, dt \right] \\
&= \int_0^\infty \mathbb{E}_i \mathbf{1}(X_t = j, J_1 < t < T_i)\, dt. \tag{2.124}
\end{aligned}
$$

Then

$$m_i = \sum_j \gamma_j = \frac{1}{q_i} + \sum_{j:j\neq i} \gamma_j \begin{cases} < \infty, & \text{if state } i \text{ is positive recurrent,} \\ = \infty, & \text{if state } i \text{ is null recurrent.} \end{cases}$$

This determines a vector $\gamma = (\gamma_j)$; to stress that it depends on the choice of reference point i, we will sometimes write $\gamma^i = (\gamma_j^i)$.

Next, if T_i^Y is the return time to i in the jump chain (Y_n) then

$$
\begin{aligned}
\gamma_j &= \mathbb{E}_i \left[\sum_{n\geq 0} (J_{n+1} - J_n) \mathbf{1}(Y_n = j, n < T_i^Y) \right] \\
&= \sum_{n\geq 0} \underbrace{\mathbb{E}_i \left[(J_{n+1} - J_n) | Y_n = j \right]}_{q_j^{-1}} \underbrace{\mathbb{P}_i(Y_n = j, 1 \leq n < T_i^Y)}_{\mathbb{E}_i \mathbf{1}(Y_n = j, 1 \leq n < T_i^Y)} \\
&= \left(\frac{1}{q_j} \right) \mathbb{E}_i \left[\sum_{n\geq 1} \mathbf{1}(Y_n = j, 1 \leq n < T_i^Y) \right] \\
&= \left(\frac{1}{q_j} \right) \mathbb{E}_i \left[\sum_{n=1}^{T_i^Y - 1} \mathbf{1}(Y_n = j) \right] := \frac{\widehat{\gamma}_j}{q_j}.
\end{aligned}
$$

Here we have set $\widehat{\gamma}_i = 1$, and for $j \neq i$

$$
\begin{aligned}
\widehat{\gamma}_j &= \mathbb{E}_i \left[\sum_{n=1}^{T_i^Y - 1} \mathbf{1}(Y_n = j) \right] \\
&= \mathbb{E}_i (\text{time spent at } j \text{ in } (Y_n) \text{ before returning to } i) \\
&= \mathbb{E}_i (\text{number of visits to } j \text{ before returning to } i), \qquad (2.125)
\end{aligned}
$$

cf. equation (1.52). This defines the vector $\widehat{\gamma} = (\widehat{\gamma}_j, j \in I)$; to stress its dependence on the choice of reference point i, we may write $\widehat{\gamma}^i = (\widehat{\gamma}_j^i)$.

Now, if the chain (X_t) is recurrent, then so is (Y_n). Then, from Theorem 1.7.4, for all states i, the vector $\widehat{\gamma}^i = (\widehat{\gamma}_j^i)$ gives an invariant measure for the chain (Y_n) with $0 < \widehat{\gamma}_j^i < \infty$ for all j. Next, all IMs for (Y_n) are proportional to $\widehat{\gamma}^i$. Then $\gamma^i = (\gamma_j^i)$, with $\gamma_j^i = \widehat{\gamma}_j^i / q_j$, gives an IM for the chain (X_t); all such IMs being again proportional to γ^i. (In particular, the vector $\gamma^i = \widehat{\gamma}_k^i \times \gamma^k$, for all states i, k.)

Furthermore, if the state i is positive recurrent, then

$$
m_i = \sum_j \gamma_j^i < \infty.
$$

But then

$$
m_k = \sum_j \gamma_j^k < \infty, \text{ for all } k,
$$

i.e. all states become positive recurrent. Similarly, if i is null recurrent, then that applies to all states k. We deduce that positive recurrence and null recurrence form class properties. Hence (i).

Also, if Q is positive recurrent, then

$$
\pi_i = \frac{\gamma_i}{\sum_j \gamma_j} = \frac{1}{q_i m_i} \quad \text{yields a (unique) equilibrium distribution } \pi.
$$

Obviously, $\pi_i > 0$, for all states i. Moreover, $\gamma^k = m_k \pi$, i.e.

$$\mathbb{E}_k (\text{time spent at } j \text{ before returning to } k) = \frac{\pi_j}{\pi_k q_k}, \qquad (2.126)$$

II. Vyullion (1 60)

Conversely, if (X_t) features an ED π then all invariant measures $\lambda = (\lambda_i)$ have $\sum_j \lambda_j < \infty$. So, for all states i, the vector $\gamma^i = (\gamma^i_j)$ yields $m_i = \sum_j \gamma^i_j < \infty$. So, i is positive recurrent. This proves (ii). □

It is now time to give a concise summary of recurrence and transience properties of CTMCs.

(I) Irreducible CTMCs with more than one state show rates $q_i > 0$ for all states i (no absorption).

(II) Non-explosive irreducible CTMCs can be transient or recurrent:

(i) transience: $\mathbb{P}_i(\text{return time } T_i < \infty) < 1$, i.e. $\mathbb{P}_i(T_i = \infty) > 0$, for all i. Equivalently: $\mathbb{P}_i(i \text{ is not visited in } (X_t) \text{ after some finite time}) = 1$ and $\int_0^\infty p_{ii}(t) \, dt < \infty$, for all i. Equivalently: $h_j^{\{i\}} = \mathbb{P}_j (\text{hit } i) < 1$, for some j and i.

(ii) recurrence: $\mathbb{P}_i(\text{return time } T_i < \infty) = 1$, i.e. $\mathbb{P}_i(T_i = \infty) = 0$, for all i. Equivalently: $\mathbb{P}_i(i \text{ visited in } (X_t) \text{ at arbitrarily large times}) = 1$ and $\int_0^\infty p_{ii}(t) \, dt = \infty$, for all i. Equivalently: $h_j^{\{i\}} = \mathbb{P}_j (\text{hit } i) = 1$, for all j and i. In this case, for all i, the vector $\gamma^i = (\gamma^i_j)$ from (2.123) satisfies $0 < \gamma^i_j < \infty$ and gives an invariant measure for (X_t), all such invariant measures being of the form $\alpha \gamma^i$. In particular, $\gamma^k = (\gamma^i_k)^{-1} \times \gamma^i$, for all i, k.

(III) Next, an irreducible recurrent CTMC can be:

(i) null recurrent: $m_i = \mathbb{E}_i(\text{return time } T_i) = \infty$, for all i; in this case no invariant measure $\lambda = (\lambda_i)$ with $\sum_j \lambda_j < \infty$ exists. Hence, there is no equilibrium distribution.

(ii) positive recurrent: $m_i < \infty$, for all i; in this case any invariant measure $\lambda = (\lambda_i)$ presents $\sum_j \lambda_j < \infty$, and a unique equilibrium distribution $\pi = (\pi_i)$ exists, where $\pi_i = \dfrac{\lambda_i}{\sum_j \lambda_j} > 0$. In this case, the vector $\gamma^k = m_k \pi$. Furthermore,

$$\mathbb{E}_i T_i = \frac{1}{\pi_i q_i}, \quad \text{and} \quad \mathbb{E}_i(\text{time at } j \text{ before } T_i) = \frac{\pi_k}{\pi_i q_i}, \quad \text{for all } i, k.$$

Finite irreducible CTMCs are always positive recurrent.

(IV) Explosive irreducible CTMCs are always transient.

Before we discuss some examples, let us present without proof a useful result about the long-run, or long-term, proportion for CTMCs. Here, we use the notation $\overset{\text{a.s.}}{\to}$ for convergence with probability 1 (relative to the probability distribution $\mathbb{P}$ of the (λ, Q) Markov chain; cf. Theorem 2.8.6 below).

Theorem 2.7.19 *Let* (X_t) *be a* (λ, Q) *irreducible positive recurrent CTMC with an equilibrium distribution* $\pi = (\pi_i)$. *Then, for all states* i, *as* $t \to \infty$,

$$\frac{1}{t} \int_0^t \mathbf{1}(X_s = i) \, ds = \text{fraction of time at } i \text{ in } (0, t)$$

$$\overset{\text{a.s.}}{\to} \pi_i = \frac{1}{m_i q_i} = \frac{\text{mean holding time at } i}{\text{mean return time to } i}.$$

(2.127)

Also, the expected value

$$\left(\frac{1}{t}\right) \mathbb{E} \int_0^t \mathbf{1}(X_s = i) \, ds = \frac{1}{t} \int_0^t \mathbb{P}(X_s = i) \, ds \to \pi_i.$$

(2.128)

In particular, for an (δ_i, Q) *irreducible positive recurrent CTMC:*

$$\left(\frac{1}{t}\right) \mathbb{E}_i \int_0^t \mathbf{1}(X_s = i) \, ds = \frac{1}{t} \int_0^t p_{ii}(s) \, ds \to \pi_i,$$

(2.129)

emphasizing the divergent character of the integral $\int_0^\infty p_{ii}(s) \, ds$.

Remark 2.7.20 Equation (2.128) can be derived from the statement of Theorem 2.8.1 below; see (2.134).

> *Remember this: if something possesses a frequency,*
> *then it will eventually occur with that frequency.*
> (From the series *'Thus spoke Superviser'*.)

Worked Example 2.7.21 (i) Consider the continuous-time Markov chain $(X_t)_{t \geq 0}$ on $\{1, 2, 3, 4, 5, 6, 7\}$ with generator matrix

$$Q = \begin{pmatrix} -6 & 2 & 0 & 0 & 0 & 4 & 0 \\ 2 & -3 & 0 & 0 & 0 & 1 & 0 \\ 0 & 1 & -5 & 1 & 2 & 0 & 1 \\ 0 & 0 & 0 & 0 & 0 & 0 & 0 \\ 0 & 2 & 2 & 0 & -6 & 0 & 2 \\ 1 & 2 & 0 & 0 & 0 & -3 & 0 \\ 0 & 0 & 1 & 0 & 1 & 0 & -2 \end{pmatrix}.$$

Compute the probability that X_t, starting from state 3, hits state 2 eventually.

Deduce that

$$\lim_{t \to \infty} \mathbb{P}(X_t = 2 \mid X_0 = 3) = \frac{4}{15}.$$

(ii) A colony of cells contains immature and mature cells. Each immature cell after an exponential time of parameter 2, becomes a mature cell. Each mature cell after, an exponential time of parameter 3, divides into two immature cells. Suppose we begin with one immature cell and let $n(t)$ denote the expected number of immature cells at time t. Show that

$$n(t) = (4e^t + 3e^{-6t})/7.$$

Solution (i) States $1,2,6$ form a closed communicating class: once (X_t) enters it, X_t stays there forever. Another closed communicating class consists of state 4; states $3,5,7$ form an open communicating class. From 3 one can enter $\{1,2,6\}$ only via state 2. See Figure 2.47.

Set $h_i = \mathbb{P}_i(\text{hit } 2)$. Then

$$5h_3 = 1 + 2h_5 + h_7, \quad 2h_7 = h_3 + h_5, \quad \text{and} \quad 6h_5 = 2 + 2h_3 + 2h_7,$$

so

$$10h_3 = 2 + 4h_5 + h_3 + h_5, \quad 6h_5 = 2 + 2h_3 + h_3 + h_5,$$

and

$$9h_3 = 2 + 5h_5, \quad 5h_5 = 2 + 3h_3,$$

or

$$9h_3 = 4 + 3h_3, \quad 6h_3 = 4, \quad \text{and} \quad h_3 = \frac{2}{3}.$$

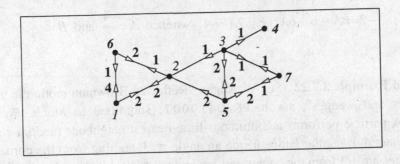

Fig. 2.47

By symmetry, the jump chain on $\{1, 2, 6\}$ takes the invariant measure $(1, 1, 1)$, so $(X_t)_{t \geq 0}$ follows the invariant distribution $(1/5, 2/5, 2/5)$. Hence, by standard arguments,

$$\lim_{t \to \infty} \mathbb{P}(X_t = 2 \mid X_0 = 3) = h_3 \pi_2 = \frac{4}{15}.$$

(ii)   Let $m(t)$ denote the expected number of immature cells at the time t when we start with one mature cell at time 0. By conditioning on the first event, with $n(t)$ being the number of immature cells at time t,

$$n(t) = e^{-2t} + \int_0^t 2e^{-2s} m(t - s) \, ds, \text{ and } m(t) = \int_0^t 3e^{-3s} 2n(t - s) \, ds.$$

So

$$e^{2t} n(t) = 1 + \int_0^t 2e^{2u} m(u) \, du, \text{ and } e^{3t} m(t) = \int_0^t 6e^{3u} n(u) \, du.$$

Differentiating yields

$$\frac{dn}{dt} + 2n = 2m, \text{ and } \frac{dm}{dt} + 3m = 6n,$$

whence

$$\frac{d^2 n}{dt^2} + 2\frac{dn}{dt} = 2\frac{dm}{dt} = 12n - 6m = 12n - 3\frac{dn}{dt} - 6n.$$

Thus,

$$\frac{d^2 n}{dt^2} + 5\frac{dn}{dt} - 6n = 0, \text{ with } n(0) = 1, \frac{dn}{dt}(0) = -2.$$

Hence

$$n(t) = Ae^t + Be^{-6t}, \text{ where } 1 = A + B, \ -2 = A - 6B.$$

Then

$$-2 = A - 6 + 6A \text{ i.e., } 7A = 4, \text{ whence } A = \frac{4}{7} \text{ and } B = \frac{3}{7}.$$

□

Worked Example 2.7.22 (See W. Kager. "Reflected Brownian motion in generic triangles and wedges", math-PR/0410007; submitted to *Stoch. Processes Appl.*) A particle performs a continuous-time nearest neighbour random walk on an equilateral triangular lattice inside an angle $\pi/3$, starting from the corner. The jump rates are 1/3 from the corner and 1/6 in each of the six directions if the particle sits inside the angle. However, if the particle is located on the edge of the angle,

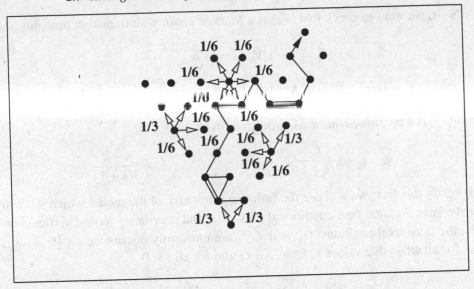

Fig. 2.48

the rate is 1/3 along the edge away from the corner and 1/6 to each of the three other neighbouring sites in the angle. See Figure 2.48 where a typical trajectory is also shown.

The particle position at time $t \geq 0$ is determined by its vertical level V_t and its horizontal position G_t; if $V_t = k$ then $G_t = 0, \ldots, k$. Here $1, \ldots, k-1$ are positions inside, and 0 and k positions on the edge of the angle, at vertical level k.

Let $J_1^V, J_2^V, \ldots$ be the times of subsequent jumps of the process (V_t) and consider embedded discrete-time processes

$$Y_n^{\text{in}} = \left(\widehat{G}_n^{\text{in}}, \widehat{V}_n\right) \text{ and } Y_n^{\text{out}} = \left(\widehat{G}_n^{\text{out}}, \widehat{V}_n\right)$$

where (a) $\widehat{V}_n$ is the vertical level immediately after time J_n^V, (b) $\widehat{G}_n^{\text{in}}$ is the horizontal position immediately after time J_n^V, (c) $\widehat{G}_n^{\text{out}}$ is the horizontal position immediately before time J_{n+1}^V.

We first remark that (Y_n^{in}) and (Y_n^{out}) are Markov chains. Indeed, (Y_n^{in}) is a Markov chain because the probability of transition from $Y_{n-1}^{\text{in}} = (i_{n-1}, k_{n-1})$ to $Y_n^{\text{in}} = (i_n, k_n)$, is completely determined by the pair (i_{n-1}, k_{n-1}) and does not depend on the previous values $Y_{n-1}^{\text{in}}, Y_{n-2}^{\text{in}}, \ldots$ Similarly for (Y_n^{out}). Observe that $\widehat{V}_n = \widehat{V}_{n-1} \pm 1$, i.e. we always jump to nearest neigbour vertical levels.

Next, we want to check that $(\widehat{V}_n)$ is a Markov chain with transition probabilities

$$\mathbb{P}\big(\widehat{V}_n = k+1 \big| \widehat{V}_{n-1} = k\big) = \frac{k+2}{2(k+1)},$$

$$\mathbb{P}\big(\widehat{V}_n = k-1 \big| \widehat{V}_{n-1} = k\big) = \frac{k}{2(k+1)},$$

and (V_t) is a continuous-time Markov chain with rates

$$q_{kk-1} = \frac{k}{3(k+1)}, \quad q_{kk} = -\frac{2}{3}, \quad q_{kk+1} = \frac{k+2}{3(k+1)}.$$

To verify this fact, we will use the following property of the model which we will prove later. Assume that, conditional on $\widehat{V}_n = k$ and previously passed vertical levels, the horisontal positions $\widehat{G}_n^{\text{in}}$ and $\widehat{G}_n^{\text{out}}$ are uniformly distributed on $\{0, \dots, k\}$, i.e. for all attainable values $k, k_{n-1}, \dots, k_1$ and for all $i = 0, \dots, k$

$$\mathbb{P}\big(\widehat{G}_n^{\text{in}} = i \big| \widehat{V}_n = k, \widehat{V}_{n-1} = k_{n-1}, \dots, \widehat{V}_1 = k_1, \widehat{V}_0 = k_0\big)$$
$$= \mathbb{P}\big(\widehat{G}_n^{\text{out}} = i \big| \widehat{V}_n = k, \widehat{V}_{n-1} = k_{n-1}, \dots, \widehat{V}_1 = k_1, \widehat{V}_0 = k_0\big)$$
$$= \frac{1}{k+1}. \qquad (2.130)$$

In fact, owing to (2.130), we can write

$$\mathbb{P}\big(\widehat{V}_n = i \big| \widehat{V}_{n-1} = k_{n-1}, \dots, \widehat{V}_0 = 0\big)$$

$$= \sum_{j=0}^{k_{n-1}} \mathbb{P}\big(\widehat{V}_n = k, \widehat{G}_{n-1}^{\text{out}} = j \big| \widehat{V}_{n-1} = k_{n-1}, \dots, \widehat{V}_0 = 0\big)$$

$$= \frac{1}{k_{n-1}+1} \times \sum_{j=0}^{k_{n-1}} \mathbb{P}\big(\widehat{V}_n = k \big| \widehat{G}_{n-1}^{\text{out}} = j, \widehat{V}_{n-1} = k_{n-1}, \dots, \widehat{V}_0 = 0\big).$$

Next,

$$\mathbb{P}\big(\widehat{V}_n = k \big| \widehat{G}_{n-1}^{\text{out}} = j, \widehat{V}_{n-1} = k_{n-1}, \dots, \widehat{V}_0 = 0\big)$$

$$= \begin{cases} 3/4, & \text{if } j = 0 \text{ or } k_{n-1} \text{ and } k = k_{n-1}+1, \\ 1/4, & \text{if } j = 0 \text{ or } k_{n-1} \text{ and } k = k_{n-1}-1, \\ 1/2, & \text{if } j = 1, \dots, k_{n-1}-1. \end{cases}$$

Here we used the jump probabilities $\{1/4, 1/4, 1/4, 1/4\}$ and $\{1/2, 1/4, 1/4\}$ emerging from rates $\{1/6, 1/6, 1/6, 1/6, 1/6, 1/6\}$ and $\{1/3, 1/6, 1/6, 1/6\}$ after discarding the rates in the horizontal directions. See Figure 2.49.

This yields

$$\mathbb{P}\big(\widehat{V}_n = k+1 \big| \widehat{V}_{n-1} = k, \dots, \widehat{V}_0 = 0\big) = \frac{1}{2} \times \frac{k+2}{k+1},$$

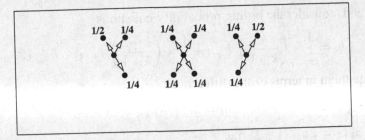

Fig. 2.49

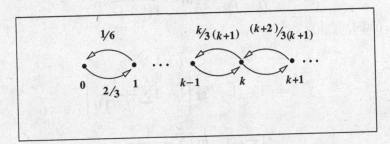

Fig. 2.50

and

$$\mathbb{P}\big(\widehat{V}_n = k-1 \,\big|\, \widehat{V}_{n-1} = k, \ldots, \widehat{V}_0 = 0\big) = \frac{1}{2} \times \frac{k}{k+1},$$

i.e. $(\widehat{V}_n)$ is indeed a Markov chain as suggested.

Consequently, (V_t) is a continuous-time Markov chain. The holding times are $\mathrm{Exp}(2/3)$, and rates

$$q_{kl} = \begin{cases} \dfrac{1}{3} \times \dfrac{k+2}{k+1}, & \text{if } l = k+1, \\[2mm] \dfrac{1}{3} \times \dfrac{k}{k+1}, & \text{if } l = k-1, \quad k \geq 1, \\[2mm] -\dfrac{2}{3}, & \text{if } l = k, \end{cases}$$

and $q_{01} = -q_{00} = 2/3$. See Figure 2.50.

Further, introduce the hitting probabilities

$$h_i = \mathbb{P}\big(\widehat{V}_n \text{ hits } 0 \,\big|\, \widehat{V}_0 = i\big), \; i = 0, 1, \ldots,$$

with $h_0 = 1$. If we show that $h_1 < 1$, it will obviously imply that

$$\mathbb{P}\big(\widehat{V}_n \text{ returns } 0 \,\big|\, \widehat{V}_0 = 0\big) < 1,$$

i.e. that 0 is a transient state. As $(\widehat{V}_n)$ is irreducible, it will be transient.

To this end, consider the hitting probability equations

$$h_k = \frac{1}{2} \times \left(\frac{k}{k+1}\right) h_{k-1} + \frac{1}{2} \times \left(\frac{k+2}{k+1}\right) h_{k+1}, \quad k \geq 1,$$

and re-write them in terms of the differences

$$u_k = h_{k-1} - h_k.$$

We obtain $u_{k+1} = k\, u_k/(k+2)$, i.e.

$$u_k = \frac{(k-1)\cdots 1}{(k+1)k\cdots 3} u_1 = \frac{2}{k(k+1)} u_1.$$

Then, as usual,

$$
\begin{aligned}
h_k &= -u_k - \cdots - u_1 + 1 \\
&= 1 - (1-h_1)\left[1 + 2\sum_{l=2}^{k}\frac{1}{(l+1)l}\right] \\
&= 1 - 2(1-h_1)\sum_{l=1}^{k}\frac{1}{l(l+1)},
\end{aligned}
$$

and the minimal solution is given by

$$h_k = 1 - \sum_{l=1}^{k}\frac{1}{l(l+1)} \Bigg/ \sum_{m=1}^{\infty}\frac{1}{m(m+1)}, \quad k \geq 1.$$

Hence,

$$h_1 = 1 - \left(\frac{1}{2} \Bigg/ \sum_{m \geq 1}\frac{1}{m(m+1)}\right) < 1.$$

Thus $(\widehat{V}_n)$ is transient. As it is the jump chain for (V_t), we deduce that (V_t) is also transient.

Finally, we have to prove property (2.130) for chains (Y_n^{in}) and (Y_n^{out}). To prove this, we use induction in n: for $n = 1$ the assertion holds for $\widehat{G}_n^{\text{out}}$ trivially and $\widehat{G}_n^{\text{in}}$ because the transition probabilities from the corner equal 1/2. Next, write

$$
\begin{aligned}
&\mathbb{P}\big(\widehat{G}_n^{\text{in}} = i \,\big|\, \widehat{V}_n = k, \widehat{V}_{n-1} = k_{n-1}, \ldots, \widehat{V}_1 = k_1, \widehat{V}_0 = 0\big) \\
&= \frac{\mathbb{P}\big(Y_n^{\text{in}} = (i,k) \,\big|\, \widehat{V}_{n-1} = k_{n-1}, \ldots, \widehat{V}_1 = k_1, \widehat{V}_0 = 0\big)}{\mathbb{P}\big(\widehat{V}_n = k \,\big|\, \widehat{V}_{n-1} = k_{n-1}, \ldots, \widehat{V}_0 = 0\big)}.
\end{aligned}
\tag{2.131}
$$

To complete the induction for $\widehat{G}_n^{\text{in}}$ it suffices to check that the numerator

$$\mathbb{P}\big(Y_n^{\text{in}} = (i,k) \,\big|\, \widehat{V}_{n-1} = k_{n-1}, \ldots, \widehat{V}_0 = 0\big) \tag{2.132}$$

does not depend on $i = 0, \ldots, k_{n-1}$.

As was observed above, the value k may arise from $k_{n-1}+1$ or $k_{n-1}-1$. Write (2.131) as the sum

$$\sum_{j=0}^{k_{n-1}} \mathbb{P}\big(Y_n = (i,k), \widehat{G}_{n-1}^{\text{out}} = j \big| \widehat{V}_{n-1} = k_{n-1}, \ldots, \widehat{V}_0 = 0 \big). \qquad (2.133)$$

By the induction hypothesis, the conditional distribution of $\widehat{G}_{n-1}^{\text{out}}$ is uniform. Next, observe that the number of non-zero summands in (2.133) will be one or two, depending on i. However, the coefficients will adjust it to make the sum independent of i. More precisely, arguing as above, the sum (2.133) equals

$$\mathbb{P}\big(\widehat{V}_{n-1} = k_{n-1}, \ldots, \widehat{V}_0 = 0\big)$$
$$\times \begin{cases} \dfrac{1}{2} \times \dfrac{1}{k_{n-1}+1}, & \text{if } k = k_{n-1}+1 \text{ and } i = 0 \text{ or } k, \\[2mm] \left(\dfrac{1}{4}+\dfrac{1}{4}\right) \times \dfrac{1}{k_{n-1}+1}, & \text{if } k = k_{n-1}+1 \text{ and } i = 1, \ldots, k-1, \\[2mm] \left(\dfrac{1}{4}+\dfrac{1}{4}\right) \times \dfrac{1}{k_{n-1}+1}, & \text{if } k = k_{n-1}-1 \text{ and } i = 0, \ldots, k. \end{cases}$$

We see that indeed probability (2.131) does not depend on i which proves that

$$\mathbb{P}\big(\widehat{G}_n^{\text{in}} = i \big| \widehat{V}_n = k, \widehat{V}_{n-1} = k_{n-1}, \ldots, \widehat{V}_1 = k_1, \widehat{V}_0 = k_0 \big) = \frac{1}{k+1}, \quad i = 0, \ldots, k.$$

To finish the proof, we need to reproduce a similar equality for $\widehat{G}_n^{\text{out}}$. But between times J_n^V and J_{n+1}^V the random walk jumps in the horizontal direction only, keeping the vertical level $k_n = k$ intact. These jumps happen at the same rate $1/6$ and preserve the uniform distribution. (The horizontal random walk is reversible, and the uniform distribution is in detailed balance with the family of constant rates). Hence, the conditional distribution of $\widehat{G}_n^{\text{out}}$ also proves uniform.

2.8 Convergence to an equilibrium distribution. Reversibility

> *Action is transitory,– a step, a blow,*
> *The motion of a muscle, this way or that–...*
> *Suffering is permanent, obscure and dark,*
> *And shares the nature of infinity.*
> W. Wordsworth (1770–1850), English poet

Throughout this section we assume that a CTMC (X_t) under consideration is irreducible and non-explosive. We begin with a continuous-time analogue of Theorem 1.9.2.

Theorem 2.8.1 *An irreducible positive recurrent* CTMC (X_t) *has*

$$p_{ij}(t) \to \pi_j \text{ for all states } i, j \quad \text{as } t \to \infty. \tag{2.134}$$

In other words, the transition matrix $P(t)$ converges, as $t \to \infty$, to a matrix with constant entries along the columns and whose rows repeat the vector π:

$$P(t) \to \Pi = \begin{pmatrix} -- & -- & -- \\ & \vdots & \\ -- & -- & -- \\ & \vdots & \end{pmatrix} \begin{matrix} \pi \\ \\ \pi \\ \end{matrix} .$$

Here $\pi = (\pi_i)$ forms the (unique) equilibrium distribution of the chain.

Proof Omitted: it is essentially similar to that of Theorem 1.9.2. □

Remark 2.8.2 Comparing with the corresponding DTMC result (Theorem 1.9.1), observe that no aperiodicity condition is mentioned here. In fact, any CTMC (X_t) will be aperiodic. Moreover, for all $h > 0$, the h-spacing imbedded DTMC (Z_n) will always be aperiodic, where $Z_n = X_{nh}$. Consequently, if π makes up an ED for the h-spacing chain (Z_n) then it also does for CTMC (X_t), and if we see convergence $P(nh) \to \Pi$ as $n \to \infty$ then convergence $P(t) \to \Pi$ as $t \to \infty$ follows.

On the other hand, the jump chain (Y_n) could be periodic, viz.

$$\widehat{P} = \begin{pmatrix} 0 & 1 \\ 1 & 0 \end{pmatrix}, \quad \text{with } \widehat{P}^{2n} = \mathbf{I}, \ \widehat{P}^{2n+1} = \widehat{P}.$$

Clearly, the power $\widehat{P}^n$ does not converge as $n \to \infty$. Still, in continuous time, the chain has the generator $Q = \begin{pmatrix} -\alpha & \alpha \\ \beta & -\beta \end{pmatrix}$, and the transition matrix

$$P(t) = \begin{pmatrix} \dfrac{\beta}{\alpha+\beta} + \dfrac{\alpha}{\alpha+\beta} e^{-(\alpha+\beta)t} & \dfrac{\alpha}{\alpha+\beta} - \dfrac{\alpha}{\alpha+\beta} e^{-(\alpha+\beta)t} \\ \dfrac{\beta}{\alpha+\beta} - \dfrac{\beta}{\alpha+\beta} e^{-(\alpha+\beta)t} & \dfrac{\alpha}{\alpha+\beta} + \dfrac{\beta}{\alpha+\beta} e^{-(\alpha+\beta)t} \end{pmatrix}$$

$$\to \Pi = \begin{pmatrix} \dfrac{\beta}{\alpha+\beta} & \dfrac{\alpha}{\alpha+\beta} \\ \dfrac{\beta}{\alpha+\beta} & \dfrac{\alpha}{\alpha+\beta} \end{pmatrix}, \quad \text{as } t \to \infty.$$

See Figure 2.51.

Remark 2.8.3 An easy observation arises as follows. Assume that, for an irreducible positive recurrent CTMC (X_t), a limiting matrix $\Pi = (\pi_{ij})$ exists and is

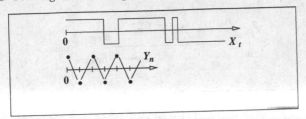

Fig. 2.51

stochastic (this is a part of the statement of Theorem 2.8.1). Then the rows of Π must give exactly the equilibrium distribution for chain (X_t). In fact, for all $t \geq 0$,

$$\Pi P(t) = \lim_{s \to \infty} P(s)P(t) = \lim_{s \to \infty} P(s+t) = \Pi,$$

and

$$\Pi \frac{\mathrm{d}}{\mathrm{d}t} P(t) = \Pi Q P(t) = 0.$$

For $t = 0$ this gives $\Pi Q = 0$. So, every row of the limiting matrix Π is an equilibrium distribution for Q. For an irreducible positive recurrent chain there exists a unique equilibrium distribution π, so all rows equal π. Conversely, suppose that, for a CTMC (X_t), the transition matrix $P(t)$ converges, as $t \to \infty$, to a limiting stochastic matrix $P(\infty)$ whose rows are repetitions of a stochastic vector π. Then π forms the unique ED, and the chain exhibits a unique closed communicating class where it achieves positive recurrence.

Remark 2.8.4 For a transient or null recurrent irreducible CTMC, $\lim_{t \to \infty} P(t) = \mathbf{0}$, the zero matrix. Compare with Remark 1.9.5.

Definition 2.8.5 A non-explosive (λ, Q) CTMC (X_t) is called *reversible* if, for all $T > 0$, $n = 1, 2, \ldots$, and time points $0 = t_0 < t_1 < \cdots < t_n = T$, the joint distribution of random variables $X_0 = X_{t_0}, X_{t_1}, \ldots, X_{t_n} = X_T$ matches that of $X_T = X_{T-t_0}, X_{T-t_1}, \ldots, X_{T-t_n} = X_0$. That is, for all states $i_0, \ldots, i_n$,

$$\mathbb{P}\big(X_0 = i_0, X_{t_1} = i_1, \ldots, X_T = i_n\big)$$
$$= \mathbb{P}\big(X_0 = i_n, \ldots, X_{T-t_1} = i_1, X_T = i_0\big). \tag{2.135}$$

In short,

$$(X_t, 0 \leq t \leq T) \sim (X_{T-t}, 0 \leq t \leq T), \tag{2.136}$$

In other words, on any interval $[0, T]$, the process stays stochastically the same, regardless of the direction of time.

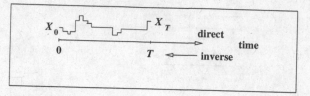

Fig. 2.52

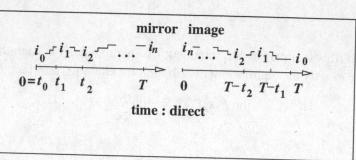

Fig. 2.53

If we prefer to work in the original 'direct' time, then (2.135) and (2.136) mean that the 'mirror image' of a sample path carries the same weight as the original path: see Figure 2.53.

In general, the RHS of (2.136) defines the distribution of a *time reversal* process $(X_t^{\mathrm{TR}}, 0 \leq t \leq T)$ (about point T):

$$
\begin{aligned}
\mathbb{P}\big(X_0^{\mathrm{TR}} &= i_0, X_{t_1}^{\mathrm{TR}} = i_1, \ldots, X_T^{\mathrm{TR}} = i_n\big) \\
&= \mathbb{P}\big(X_0 = i_n, \ldots, X_{T-t_1} = i_1, X_T = i_0\big),
\end{aligned} \tag{2.137}
$$

and reversibility means that $(X_t, 0 \leq t \leq T) \sim (X_{T-t}, 0 \leq t \leq T)$, for all $T > 0$. As we will see shortly, reversibility implies that the vector $\lambda Q = 0$, i.e. $\lambda = \pi$, the equilibrium distribution of the chain (as with DTMCs). But (again as in the discrete-time case), it means the stronger property, that λ is in a *detailed balance* with Q.

Theorem 2.8.6 *A non-explosive* (λ, Q) *CTMC* (X_t) *is reversible if and only if the following detailed balance equations hold*

$$
\lambda_i q_{ij} = \lambda_j q_{ji}, \text{ for all states } i, j \text{ where } i \neq j. \tag{2.138}
$$

Proof   (a) The if part: suppose the detailed balance equations (DBEs) hold (trivially, (2.138) always holds for $i = j$). Then λ forms an ED: for all states j,

$$
(\lambda Q)_j = \sum_i \lambda_i q_{ij} = \lambda_j \sum_i q_{ji} = 0 \quad \text{(the sum along row } i\text{)} \tag{2.139}
$$

Also, by induction, the DBEs hold for all powers of Q

$$\lambda_i q_{ij}^{(k)} = \lambda_i \sum_l q_{il} q_{lj}^{(k-1)} = \sum_l q_{li} \lambda_l q_{lj}^{(k-1)} = \sum_l q_{jl}^{(k-1)} q_{li} \lambda_j = \lambda_j q_{ji}^{(k)}.$$

This fact immediately implies that in the case of a finite CTMC, the DBEs hold for transition probability matrices $P(t) = e^{tQ} = \sum_{k \geq 0} \dfrac{(tQ)^k}{k!}$, for all $t \geq 0$,

$$\lambda_i p_{ij}(t) = \lambda_j p_{ji}(t), \text{ for all states } i, j. \tag{2.140}$$

For a general non-explosive chain, we go back to (2.89) and use (2.139) to check that the equality holds for every summand (2.90):

$$\pi_i \sum_{i=i_0,\ldots,i_n=j} \prod_{l=0}^{n-1} \mathbf{1}(q_l > 0, i_{l+1} \neq i_l) \int_0^t \cdots \int_0^{s_{n-1}} \exp\left[-(t-s_1)q_{i_0}\right]$$

$$\times \prod_{k:\, 1 \to n} \left(q_{i_{k-1}i_k} \exp\left[-(s_k - s_{k+1})q_{i_k}\right]\right) \mathrm{d}s_n \cdots \mathrm{d}s_1$$

$$= \sum_{j=j_0,\ldots,j_n=i} \prod_{l=0}^{n-1} \mathbf{1}(q_l > 0, j_{l+1} \neq j_l) \int_0^t \cdots \int_0^{t_2} \exp\left[-(t-t_n)q_{j_n}\right]$$

$$\times \prod_{k:\, n \to 1} \left(q_{j_{k-1}j_k} \exp\left[-q_{j_k}(t_k - t_{k-1})\right]\right) \mathrm{d}t_1 \cdots \mathrm{d}t_n \, \pi_j$$

The argument is then extended to the whole sum (2.89) again leading to (2.140).

Now we want to check (2.135): for all $0 = t_0 < t_1 < \cdots < t_n < t_{n+1} = T$ and states $i_0, i_1, \ldots, i_{n+1}$,

$$\mathbb{P}\left(X_{t_k} = i_k, \, 0 \leq k \leq n\right) = \mathbb{P}\left(X_{T-t_k} = i_k, \, n \geq k \geq 0\right). \tag{2.141}$$

By using (2.140),

$$\begin{aligned}
\text{the LHS of (2.141)} &= \lambda_{i_0} p_{i_0 i_1}(t_1 - t_0) p_{i_1 i_2}(t_2 - t_1) \cdots p_{i_{n-1} i_n}(t_n - t_{n-1}) \\
&= p_{i_1 i_0}(t_1 - t_0) \lambda_{i_1} p_{i_1 i_2}(t_2 - t_1) \cdots p_{i_{n-1} i_n}(t_n - t_{n-1}) \\
&= \cdots \\
&= p_{i_1 i_0}(t_1 - t_0) p_{i_2 i_1}(t_2 - t_1) \cdots p_{i_n i_{n-1}}(t_n - t_{n-1}) \lambda_{i_n}.
\end{aligned}$$

Re-arrange the RHS:

$$\lambda_{i_n} p_{i_n i_{n-1}}(t_n - t_{n-1}) \cdots p_{i_1 i_0}(t_1 - t_0) = \mathbb{P}\left(X_{T-t_k} = i_k, \, n \geq k \geq 0\right).$$

(b) The only if part: suppose (2.135) holds. For $n = 1$ and $i_0 = i$, $j_0 = j$, this yields (2.140):

$$\lambda_i p_{ij}(T) = \lambda_j p_{ji}(T).$$

Differentiate in T and set $T = 0$, with $\dfrac{\mathrm{d}}{\mathrm{d}t} p_{ij}(0) = q_{ij}$, to get (2.138). $\qquad\square$

Remark 2.8.7 As in the discrete-time setting, the detailed balance equations make a convenient tool to find an equilibrium distribution (or more generally, an invariant measure). Hence, if you cannot find an ED (or an invariant measure) from $\lambda Q = 0$, try the DBEs. If you succeed, you achieve two goals: finding an invariant measure (and getting an ED by normalising when possible), and proving reversibility.

Example 2.8.8 What if the DBEs fail to hold for a given CTMC (X_t)? Then the time reversal process $(X_t^{TR}, 0 \le t \le T)$ differs from the original chain $(X_t, 0 \le t \le T)$. Still, the process $(X_t^{TR}, 0 \le t \le T)$ remains a CTMC, albeit *inhomogeneous*, with transition probabilities $p_{ij}^{TR}(t, t+s)$ depending on 'initial' and 'terminal' times $t, t+s$. In other words, these probabilities require knowledge of initial time t and lapsed time s (as in an inhomogeneous Poisson process $IPP(\lambda(t))$ from Section 2.4; see (2.62)). More precisely, for all $0 = t_0 < t_1 < \cdots < t_n < t < t+s < T$ and states $i_0, \ldots, i_n, i, j$, the conditional probability

$$\mathbb{P}\big(X_{t+s}^{TR} = j \,|\, X_t^{TR} = i \text{ and } X_{t_k} = i+k, \ 0 \le k \le n\big)$$
$$= \mathbb{P}\big(X_{t+s}^{TR} = j \,|\, X_t^{TR} = i\big)$$
$$= \frac{\mathbb{P}(X_{T-t-s} = j)}{\mathbb{P}(X_{T-t} = i)} \, p_{ji}(s) := p_{ij}^{TR}(t, t+s). \tag{2.142}$$

Here, $p_{ij}(s)$ is the transition probability for the original CTMC (X_t).

Further, if the chain (X_t) is in equilibrium, then $\mathbb{P}(X_{T-t-s} = j) = \pi_j$, $\mathbb{P}(X_{T-t} = i) = \pi_i$, and $p_{ij}^{TR}(t, t+s) = \pi_j p_{ji}(s)/\pi_i$ loses its dependence on t. This means that (X_t^{TR}) would be a homogeneous CTMC in equilibrium, with transition matrix $P^{TR}(t) = (\pi_j p_{ji}(t)/\pi_i)$ (note that the sum over j equals 1). Obviously, the generator $Q^{TR} = (\pi_j q_{ji}/\pi_i)$ (with the sum over j equal to 0). See Worked Example 1.10.7. However, to guarantee that chain (X_t^{TR}) remains the same as (X_t), we need a stronger property of detailed balance, which will result in $Q^{TR} = Q$ and $P^{TR}(t) = P(t)$ for all $t > 0$.

It is also instructive to see that the DBEs (2.138) are equivalent to the fact that Q is self-adjoint relative to the scalar product $\langle \cdot, \cdot \rangle_\pi$.

Example 2.8.9 A birth-and-death process (BDP), when it is positive recurrent, is reversible. In fact, the DBEs for such processes can be easily solved recursively. Indeed, in a $BDP(\lambda_j, \mu_k)$:

$$q_{jj+1} = \lambda_j, \ q_{j+1j} = \mu_{j+1}, \ j = 0, 1, \ldots. \tag{2.143}$$

Here, the equations become

$$\pi_0 \lambda_0 = \pi_1 \mu_1, \quad \pi_1 \lambda_1 = \pi_2 \mu_2, \quad \ldots, \tag{2.144}$$

whence

$$\pi_1 = \frac{\lambda_0}{\mu_1} \pi_0, \quad \pi_2 = \frac{\lambda_0 \lambda_1}{\mu_1 \mu_2} \pi_0, \quad \ldots, \quad \pi_i = \frac{\lambda_0 \cdots \lambda_{i-1}}{\mu_1 \cdots \mu_i} \pi_0, \quad \ldots; \tag{2.145}$$

We see that if the series emerging from (2.145) converges, i.e.

$$\sum_{n \geq 1} \prod_{1 \leq j \leq n} \frac{\lambda_{j-1}}{\mu_j} < \infty, \tag{2.146}$$

then the solution π_i is given by

$$\left.\begin{aligned}
\pi_0 &= \left(1 + \sum_{n \geq 1} \prod_{1 \leq j \leq n} \frac{\lambda_{j-1}}{\mu_j} \right)^{-1}, \\[2mm]
\pi_i &= \frac{\lambda_0 \cdots \lambda_{i-1}}{\mu_1 \cdots \mu_i} \left(1 + \sum_{n \geq 1} \prod_{1 \leq j \leq n} \frac{\lambda_{j-1}}{\mu_j} \right)^{-1}, \quad i \geq 1.
\end{aligned}\right\} \tag{2.147}$$

Remark 2.8.10 Solving the detailed balance equations for a birth-and-death process is only a half of the job. We also must guarantee that $\mathrm{BDP}(\lambda_k, \mu_k)$ is non-explosive: only then solution (2.147) will give a genuine equilibrium distribution and hence define a CTMC which is reversible. An elegant necessary and sufficient condition for $\mathrm{BDP}(\lambda_k, \mu_k)$ to be positive recurrent and hence reversible is that (2.147) is complemented by a certain divergence requirement:

$$\sum_{n \geq 1} \prod_{1 \leq j \leq n} \frac{\lambda_{j-1}}{\mu_j} < \infty, \quad \text{and} \quad \sum_{n \geq 1} \frac{1}{\lambda_n} \prod_{1 \leq j \leq n} \frac{\mu_j}{\lambda_{j-1}} = \infty. \tag{2.148}$$

So, it is under condition (2.148) that measure the $\pi = (\pi_i)$ from (2.147) is an ED for $\mathrm{BDP}(\lambda_k, \mu_k)$. We are not going to prove this, but note that

$$\sum_{n \geq 1} \prod_{1 \leq j \leq n} \frac{\lambda_{j-1}}{\mu_j} = \infty, \quad \text{and} \quad \sum_{n \geq 1} \frac{1}{\lambda_n} \prod_{1 \leq j \leq n} \frac{\mu_j}{\lambda_{j-1}} = \infty \tag{2.149}$$

constitutes a necessary and sufficient condition for $\mathrm{BDP}(\lambda_k, \mu_k)$ to be null recurrent, and

$$\sum_{n \geq 1} \frac{1}{\lambda_n} \prod_{1 \leq j \leq n} \frac{\lambda_{j-1}}{\mu_j} = \infty, \quad \text{and} \quad \sum_{n \geq 1} \prod_{1 \leq j \leq n} \frac{\mu_j}{\lambda_{j-1}} < \infty \tag{2.150}$$

forms a necessary and sufficient condition for $\mathrm{BDP}(\lambda_k, \mu_k)$ to be transient (which includes the possibility to be explosive). See S. Karlin, 1968.

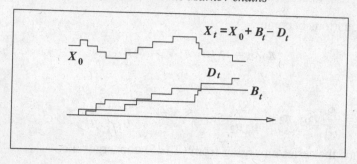

Fig. 2.54

If $\lambda_i \equiv \lambda$ and $\mu_i \equiv \mu$, the birth-and-death process is homogeneous; in this case it will always be non-explosive. Then condition (2.146) is equivalent to $\lambda < \mu$; under this condition the process will be positive recurrent and reversible. If $\lambda = \mu$, the process will be null recurrent, and if $\lambda > \mu$, transient.

Birth-and-death processes display the following surprising result.

Theorem 2.8.11 *Assume that condition (2.148) holds and let (X_t) be the non-explosive, positive recurrent and reversible* $BDP(\lambda_i, \mu_i)$, *with equilibrium distribution π given by (2.147). Consider the process in equilibrium (i.e. with $X_0 \sim \pi$) and write*

$$X_t = X_0 + B_t - D_t, \quad t > 0, \tag{2.151}$$

where processes (B_t) and (D_t) give the birth and death accounts of (X_t). (That is, the process (B_t) increases by one every time a jump up in (X_t) occurs while (D_t) increases by one every time there is a jump down in (X_t).) Then

$$(B_t) \sim (D_t). \tag{2.152}$$

This appears surprising because (B_t) is associated with rates λ_i and (D_t) with rates μ_i, which can be completely different.

Proof We know that (X_t) is reversible, i.e. $(X_t) \sim (X_t^{\mathrm{TR}})$, with (X_t^{TR}) being the original process (X_t) viewed in reversed time. But, in reversed time, the jumps up become jumps down. In other words,

$$(B_t) \sim (D_t^{\mathrm{TR}}), \text{ the death account in } (X_t^{\mathrm{TR}}),$$

and

$$(D_t) \sim (B_t^{\mathrm{TR}}), \text{ the birth account in } (X_t^{\mathrm{TR}}).$$

But $\left(D_t^{\mathrm{TR}}\right) \sim (D_t)$ and $\left(B_t^{\mathrm{TR}}\right) \sim (B_t)$ by reversibility. Combine all equivalences:

$$(B_t) \sim \left(D_t^{\mathrm{TR}}\right) \sim (D_t) \sim \left(B_t^{\mathrm{TR}}\right) \sim (B_t).$$

This yields (2.152). □

The point here lies in that processes (B_t) and (D_t) are not independent. In general, neither of them need even be Markov (although when $\lambda_i \equiv \lambda$, process (B_t) becomes Poisson PP(λ)). However, the equilibrium distribution π provides a delicate link between the two which results in the identity of their distributions.

Remark 2.8.12 Theorem 2.8.11 plays a big rôle in queueing theory. Here a jump in (B_t) is interpreted as the arrival of a new task (alternatively, a client or a customer) in a Markovian queue, while a jump in (D_t) as the departure (of a fully served task, client or customer). Then X_t represents the queue size (or length) at time t, i.e. the number of customers in the system (including the one(s) currently being served). In this context, the assertion of Theorem 2.8.11 says that the arrival process in the queue is *stochastically equivalent* to the departure process. See Section 2.9.

2.9 Applications to queueing theory. Markovian queues

> *A Room With A Queue*
>
> (From the series *'Movies that never made it to the Big Screen'*.)

In this section we focus on various popular Markovian queueing models. All of these models, except for one, will be 'genuine' birth-and-death processes, where the state space I is the set of non-negative integers $\mathbb{Z}_+ = \{0, 1, \ldots\}$ and where the jump rate to the right will remain a constant, $q_{ii+1} \equiv \lambda$, while the jump rate q_{ii-1} to the left will depend on i. The remaining model forms a simplification of this where the state space is reduced to a finite set $\{0, 1, \ldots, c\}$ for some positive integer c. As a result, the rates $q_{ii+1} = \lambda$ for $i = 0, 1, \ldots, c - 1$, but the rate q_{cc+1} and subsequent jump rates to the right vanish (in some aspects, it makes things more complicated). We can say that the former models operate with an infinite buffer while the latter work with a finite buffer.

First, consider models with an infinite buffer: see Figure 2.55.

As was mentioned, the jump rate to the left $q_{ii-1} = \mu_i$, $i \geq 1$, may vary; popular examples being:

(a) $\mu_i \equiv \mu$, a constant;
(b) $\mu_i = i\mu$, μ_i proportional to i;
(c) $\mu_i = \mu \min[i, r]$ with μ and r both constants.

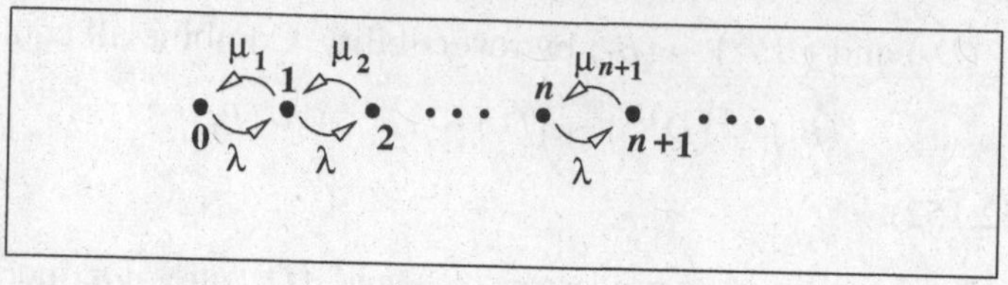

Fig. 2.55

To start with, let us assume that the chain begins at time 0 from state 0 (in most cases it will not matter).

Model (a) corresponds to the so-called M/M/1/∞ queue: *Markov* arrival, *Markov* service, a single server, an infinite-size buffer. When it does not create a confusion, the last symbol ∞ is often omitted, and one uses the simplified notation M/M/1.

In this model, customers join the queue in a process $(A(t)) \sim \mathrm{PP}(\lambda)$ and are served one by one (you may think of a village barber shop with an unlimited waiting space; the shop opens at time 0 with no customers). We call λ the arrival rate. The service times are IID $\mathrm{Exp}(\mu)$. After service, the customer leaves the system (and never comes back) while the server immediately starts dealing with the next customer (if any are in the queue) or stays idle until the next customer arrives. The order in which customers are served is usually supposed to be FCFS (First Come First Served), relative to customers' arrival times; an alternative notation is FIFO (First In First Out). However, for a number of important questions (although not always) it does not matter.

We find interest in the process $(Q(t), t \geq 0)$ representing the queue length (queue size); that is, the number of customers $N(t)$ in the queue at time $t \geq 0$ (including the customer currently being served). A convenient formula for expressing this is:

$$Q(t) = A(t) - D(t), \quad t \geq 0. \tag{2.153}$$

Here $A(t) \sim \mathrm{Po}(\lambda t)$ gives the number of customers arrived by time t, and $D(t)$ the number of customers served by then.

We see that $Q(t)$ jumps up when a new customer arrives and down when the served customer leaves. Then $(Q(t))$ constitutes a birth-and-death process, with the generator

$$Q = \begin{pmatrix} -\lambda & \lambda & 0 & 0 & \cdots \\ \mu & -(\lambda+\mu) & \lambda & 0 & \ddots \\ 0 & \mu & -(\lambda+\mu) & \lambda & \ddots \\ \cdots & & \ddots & \ddots & \ddots \end{pmatrix}. \tag{2.154}$$

Model (b) corresponds to the so-called M/M/∞ queue (*Markov* arrival, *Markov* service, infinitely many servers). Customers again arrive in a process $(A(t)) \sim$ PP(λ), but upon arrival each of them gets a 'personal' server to be taken care of immediately. As before, the service times of customers are IID Exp(μ). Once again $(Q(t))$ is a birth-and-death process. Here, for i customers with remaining service times $S^{(1)}, \ldots, S^{(i)}$ in the system, the jump $i \to i-1$ occurs at rate $i\mu$, as the time of a potential jump is

$$\min\left[S^{(1)}, \ldots S^{(i)}\right] \sim \text{Exp}(i\mu). \tag{2.155}$$

If no new customer arrives during this time, we replace the rate $i\mu$ by $(i-1)\mu$, otherwise (i.e. if the new customer comes before jump $i \to i-1$), we replace i by $i+1$, by virtue of the memoryless property of exponential distributions.

The corresponding generator is

$$Q = \begin{pmatrix} -\lambda & \lambda & 0 & 0 & \cdots \\ \mu & -(\lambda+\mu) & \lambda & 0 & \ddots \\ 0 & 2\mu & -(\lambda+2\mu) & \lambda & \ddots \\ \cdots & \ddots & \ddots & \ddots & \ddots \end{pmatrix}. \tag{2.156}$$

Model (c) describes an r-server queue M/M/r/∞ (or, in the simplified notation, M/M/r). Here, the barber shop employs r hairdressers: all are busy if $Q(t) \geq r$, but otherwise (i.e. when $0 \leq Q(t) < r$), $(r-Q(t))$ of them sit idle.

In all models, because of (2.153), the sample trajectory $Q(t)$ stays below $A(t)$.

We begin our analysis with (and spend most of the time on) model (a). The corresponding CTMC is called the *M/M/1 chain*. Consider the probabilities of hitting state 0 from state i:

$$h_i = \mathbb{P}_i(\text{hit } 0), \quad i \geq 0. \tag{2.157}$$

Then $h_0 = 1$ and $(hQ)_j = 0$ for all $j \geq 1$. That is

$$h_0 = 1, \quad (\lambda+\mu)h_i = \lambda h_{i+1} + \mu h_{i-1}, \quad i \geq 1. \tag{2.158}$$

The general solution to (2.158) is given as

$$h_i = \begin{cases} A + B\left(\dfrac{\mu}{\lambda}\right)^i, & \text{if } \lambda \neq \mu, \\ A + Bi, & \text{if } \lambda = \mu. \end{cases}$$

We see that if the arrival rate does not exceed the service rate: $\lambda \leq \mu$, then the minimal non-negative solution to (2.158) must have $B = 0, A = 1$:

$$h_i \equiv 1, \quad i \geq 0,$$

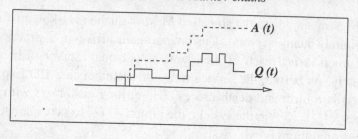

Fig. 2.56

and the M/M/1 chain is recurrent. But if $\lambda > \mu$ then $A = 0$ and $B = 1$:

$$h_i = \left(\frac{\mu}{\lambda}\right)^i, \quad i \geq 0, \tag{2.159}$$

and the chain is transient. It means that the process drifts towards $+\infty$ (an indefinitely growing queue)

$$\mathbb{P}_i\left(\lim_{t\to\infty} Q(t) = +\infty\right) = 1.$$

To find the equilibrium distribution, attempt the DBEs; that is $\pi_i\lambda = \pi_{i+1}\mu$, $i \geq 0$, i.e.

$$\pi_{i+1} = \left(\frac{\lambda}{\mu}\right)\pi_i = \cdots = \left(\frac{\lambda}{\mu}\right)^i \pi_0.$$

The normalising condition

$$1 = \sum_{i\geq 0} \pi_i = \pi_0 \sum_{i\geq 0} \left(\frac{\lambda}{\mu}\right)^i = \pi_0 \left(1 - \frac{\lambda}{\mu}\right)^{-1} = 1$$

yields that for all $i = 0, 1, \cdots$

$$\pi_i = (1-\rho)\rho^i, \quad \text{where } \rho = \frac{\lambda}{\mu}. \tag{2.160}$$

Let us summarise.

Theorem 2.9.1 *For $\lambda < \mu$, i.e., $\rho < 1$, the M/M/1 chain is positive recurrent and reversible, with a geometric equilibrium distribution $\pi = (\pi_i)$, as in equation (2.160). Hence, it converges to the equilibrium:*

$$\lim_{t\to\infty} \mathbb{P}\big(Q(t) = j\big) = (1-\rho)\rho^j, \tag{2.161}$$

regardless of the initial distribution.

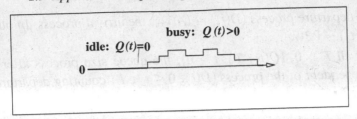

idle: $Q(t)=0$ busy: $Q(t)>0$

Fig. 2.57

Also, the mean return time to 0

$$m_0 = \frac{1}{\pi_0 q_0} = \frac{\mu}{\lambda(\mu - \lambda)} \qquad (2.162)$$

gives the mean time of the server's cycle (idle plus busy period). See Figure 2.57.

Then the mean busy period

$$m_0 - \frac{1}{\lambda} = \frac{\mu}{\lambda(\mu - \lambda)} - \frac{1}{\lambda} = \frac{1}{\mu - \lambda}. \qquad (2.163)$$

Finally, the mean waiting time of a customer in equilibrium

$$\mathbb{E} W = \mathbb{E}[\mathbb{E}(W|Q)] = \sum_{i \geq 0} \mathbb{E}(W|Q = i)\pi_i$$

$$= \sum_{i \geq 0} \frac{i}{\mu}(1 - \rho)\rho^i = \frac{\lambda}{\mu(\mu - \lambda)}, \qquad (2.164)$$

and the mean sojourn time (waiting plus service)

$$\mathbb{E} W + \frac{1}{\mu} = \frac{1}{\mu - \lambda} \qquad (2.165)$$

(which equals the mean length of the busy period).

The condition $\lambda < \mu$, or $\rho < 1$, offers a transparent meaning: the arrival at the queue is overpowered by the service. In other words, if the expected inter-arrival time $1/\lambda$, amounts to less than the expected length of the service time $1/\mu$ then the M/M/1/∞ queue becomes 'stable' and reaches an equilibrium from any initial distribution. Or, to put it differently, stability in mean implies stability almost surely.

Now, the most surprising fact arises as a corollary of Theorem 2.8.11.

Theorem 2.9.2 *Suppose that $\rho < 1$ and consider the M/M/1 chain $(Q(t))$ in equilibrium. Then:*

(i) *In the representation*

$$Q(t) = Q(0) + A(t) - D(t), \qquad (2.166)$$

the departure process $(D(t)) \sim (A(t))$, the arrival process. In other words, $(D(t)) \sim \mathrm{PP}(\lambda)$.

(ii) For all $T > 0$, $(Q(t+T),\ t \geq 0)$, the queue size process after time T is independent of the process $(D(t),\ 0 \leq t < T)$ counting departures prior to time T.

Again, the surprise lies in that $(D(t))$ is associated with the rate λ, not μ, and, in addition, that the queue after a given time evolves independently of the departure process prior to this time. (So, for instance, if you previously observed a number of rare departures, it tells you nothing about how low the queue level drops at present, let alone how high it will be in the future. The explanation proves the same: in equilibrium, processes (A_t) and (D_t) are dependent in a specific way, which creates delicate phenomena (e.g. $Q_t = Q_0 + A_t - D_t$ is always non-negative).

Proof The statement (i) follows formally from Theorem 2.8.11. Recall, the key point in the proof of that theorem was twofold: (i) that process $(Q(t))$ under the condition $\lambda < \mu$ is reversible; and (ii) the fact that the jumps up of trajectory $Q(t)$ become jumps down when we reverse the time. In other words, (i) if $(\widetilde{Q}_t)$ is the reverse process of (Q_t), relative to time T, with

$$\widetilde{Q}_t = Q_{T-t}, \text{ and } \widetilde{Q}_t = \widetilde{Q}_0 + \widetilde{A}_t - D_t^{\mathrm{TR}}, \ t \geq 0,$$

then $(\widetilde{Q}_t) \sim (Q_t)$. Next, (ii), the number of jumps, in $(0,T)$, in the processes $(\widetilde{A}_t)$ and (D_t) will be the same (as well as the number of jumps, in $(0,T)$, in processes (D_t^{TR}) and (A_t)). Finally, given that the number of jumps, in $(0,T)$, in $(\widetilde{A}_t)$ and (D_t) equals n, the jump times $H_1^{\widetilde{A}}, H_2^{\widetilde{A}}, \ldots$ (arrivals in $(\widetilde{A}_t,\ 0 \leq t < T)$) are related to the jump times $H_1^D, H_2^D, \ldots$ (departures in $(D_t,\ 0 \leq t < T)$) by the 'conditional' equation

$$\left(H_1^D, \ldots, H_n^D \,\middle|\, n \text{ jumps in } (D_t) \text{ on } [0,T) \right)$$
$$= \left(T - H_n^{\widetilde{A}}, \ldots, T - H_1^{\widetilde{A}} \,\middle|\, n \text{ jumps in } (\widetilde{A}_t) \text{ on } [0,T) \right).$$

Also, the sample trajectory $(Q_{t+T},\ t \geq 0)$ is mirrored by a trajectory of $(\widetilde{Q}_t,\ t \leq 0)$. See Figure 2.58.

But in $(\widetilde{Q}_t)$, the future arrivals $(\widetilde{A}_t,\ t \geq 0)$ are independent of the past queue size $(\widetilde{Q}_t,\ t \leq 0)$ (as usual, we set here $\widetilde{A}_0 = 0$). Hence, in the original M/M/1 chain (Q_t), the present and future queue size $(Q_{t+T},\ t \geq 0)$ are independent of the past departures $(D(t),\ 0 \leq t < T)$. $\qquad \square$

Theorem 2.9.2 is known as Burke's Theorem. It plays an important role in the theory of queueing networks where customers travel from one node (station) to the

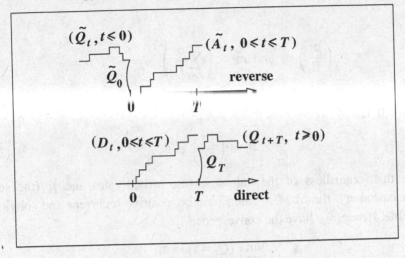

Fig. 2.58

other and join various queues along their paths. In turn, the theory of queueing networks underpins a number of applications, notably telecommunications, computer networks and transport network research.

For $\lambda = \mu$, i.e., $\rho = 1$, the M/M/1 chain will be null recurrent. In fact, if (π_i) is an invariant measure, then

$$\pi_0 = \pi_1, \ 2\pi_i = \pi_{i-1} + \pi_{i+1}, \ i \geq 1.$$

Hence $\pi_i = \pi_{i+1}$, $i \geq 0$ (i.e. the measure (π_i) satisfies the DBEs). So, $\sum_i \pi_i = \infty$ unless $\pi_i \equiv 0$. In this case, $\mathbb{P}_i \left(\lim_{t \to \infty} N(t) = +\infty \right) = 0$ but the process lacks an equilibrium distribution, and the queue size oscillates between 0 and ∞. For $\lambda > \mu$, i.e., $\rho > 1$, the M/M/1 chain turns transient and grows indefinitely:

$$\mathbb{P} \left(\lim_{t \to \infty} Q_t = \infty \right) = 1, \text{ i.e. } Q_t \overset{\text{a.s.}}{\to} \infty, \text{ as } t \to \infty. \tag{2.167}$$

ρ and His Brothers

(From the series *'Movies that never made it to the Big Screen'*.)

The analysis of models (b) and (c) moves along similar lines. The M/M/∞ model, with infinitely many servers, is relatively straightforward. Here, the DBEs are given by

$$\lambda \pi_{i-1} = i\mu \pi_i, \ i = 1, 2, \ldots,$$

implying

$$\pi_i = \left(\frac{\rho^i}{i!}\right)\pi_0 \text{ and } \pi_0 = \left(\sum_{i\geq 0}\frac{\rho^i}{i!}\right)^{-1} = e^{-\rho}, \ \rho = \frac{\lambda}{\mu}.$$

In other words, the equilibrium distribution $\pi = (\pi_i)$ is Poisson, with parameter ρ: for all $i = 0,1,\ldots$

$$\pi_i = \left(\frac{\rho^i}{i!}\right)e^{-\rho}. \tag{2.168}$$

We see that, regardless of the values λ (the arrival rate) and μ (the service rate per customer), the M/M/∞ chain remains positive recurrent and (obviously) irreducible. Hence, we have the convergence

$$\lim_{t\to\infty}\mathbb{P}(Q_t = i) = \pi_i,$$

regardless of the initial distribution.

So, the M/M/∞ queue always stays stable, for all $\lambda, \mu > 0$.

Burke's theorem is extended to this model in a straightforward way. As a result, we obtain that, in equilibrium,

(i) the arrival process (A_t) and the departure process (D_t) forming the representation

$$Q_t = Q_0 + A_t - D_t$$

are stochastically equivalent: $(A_t) \sim (D_t) \sim PP(\lambda)$, and

(ii) for all $T > 0$, $(Q_{t+T}, \ t \geq 0)$, the queue size at and after time T, is independent of the departure process $(D_t, \ 0 \leq t < T)$ prior to time T.

Finally, the M/M/r/∞ chain features the DBEs

$$\lambda\pi_{i-1} = i\mu\pi_i, \ i = 1,\ldots,r,$$
$$\lambda\pi_i = r\mu\pi_{i+1}, \ i = r,r+1,\ldots, \tag{2.169}$$

with the sole solution

$$\pi_i = \begin{cases} \left(\dfrac{\rho^i}{i!}\right)\pi_0, & i = 1,\ldots,r, \\[2ex] \left(\dfrac{\rho^r}{r!}\right)\left(\dfrac{\rho^{i-r}}{r^{i-r}}\right)\pi_0, & i = r+1,\ldots, \end{cases} \qquad \rho = \frac{\lambda}{\mu}. \tag{2.170}$$

We see that when $\rho < r$, the series $\sum_{j\geq 1}(\rho/r)^j < \infty$, and we have a correctly defined ED $\pi = (\pi_i)$. Namely,

$$\pi_0 = \left(1 + \sum_{i=1}^{r}\frac{\rho^i}{i!} + \frac{\rho^r}{r!}\sum_{j\geq 1}\frac{\rho^j}{r^j}\right)^{-1} = \left[\sum_{i=0}^{r-1}\frac{\rho^i}{i!} + \left(\frac{\rho^r}{r!}\right)\left(\frac{1}{1-\rho/r}\right)\right]^{-1},$$

and π_i, $i \geq 1$, are given by (2.170). Alternatively, for all $i = 0, 1, \ldots,$

$$\pi_i = \left[\sum_{k=0}^{r-1} \frac{\rho^k}{k!} + \frac{\rho^r}{r!} \left(1 - \frac{\rho}{r} \right)^{-1} \right]^{-1} \left(\frac{\rho^{i \wedge r}}{(i \wedge r)!} \right) \left(\frac{\rho}{r} \right)^{(i-r) \wedge 0}. \tag{2.171}$$

Here, as before, $a \wedge b$ stands for the minimum $\min[a,b]$.

Conditions $\rho < r$, or $\lambda < r\mu$, mean the same: the system, on average, is able to cope with the arriving customers.

Consequently, for $\rho < r$, the M/M/r/∞ chain is positive recurrent and reversible. Again, it is obviously irreducible. In addition, Burke's theorem still holds. Therefore,

Theorem 2.9.3 *Suppose that $\lambda < r\mu$, i.e., $\rho < r$. Then the M/M/r/∞ chain $(Q(t))$ is positive recurrent and reversible, with equilibrium distribution π given by (2.170), (2.171). Therefore, for all initial distributions,*

$$\lim_{t \to \infty} \mathbb{P}(Q(t) = i) = \pi_i, \ \ i = 0, 1, \ldots.$$

Next, in the representation

$$Q(t) = Q(0) + A(t) - D(t),$$

with $Q(0) \sim \pi$, the departure process $(D(t))$ is stochastically equivalent to $(A(t))$, the arrival process. In other words, in equilibrium, $(D(t)) \sim \mathrm{PP}(\lambda)$. Finally, for all $T > 0$, $(Q(t+T), t \geq 0)$, the queue process after time T is independent of the process $(D(t), 0 \leq t < T)$, counting departures before time T.

Formulas analogous to (2.161)–(2.164) can be established for both M/M/∞ and M/M/r/∞ models; we omit the details.

So far we have considered models with infinite buffers: here all customers are admitted to the system (but in general will have to wait for service, sometimes in a long queue). In models with a finite buffer, the customers are not allowed in (or considered 'lost') when the buffer gets full. These models are also called *loss models*. The notation here is M/M/r/c: Markov arrival, Markov service, r servers and a buffer of size $c \geq r$ (in other words, the number of waiting places is $c - r$). So, we speak here about *M/M/r/c chains*. Thus the queue size $Q(t)$, i.e. the number of customers in the system, at time t, satisfies $0 \leq Q(t) \leq c$; if $0 \leq Q(t) \leq r$ then all customers are being served, while if $Q(t) \geq r$ then r customers are served and $Q(t) - r$ are waiting for service.

Consider the arrival process of admitted customers which is denoted, for simplicity, by the same symbol (A_t). It constitutes a process 'embedded' in the 'nominal'

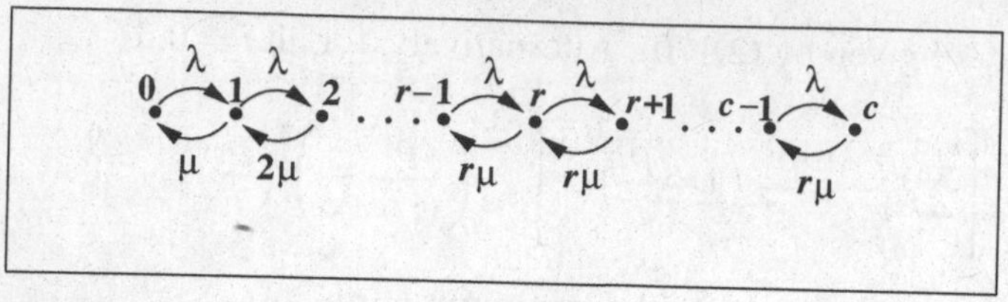

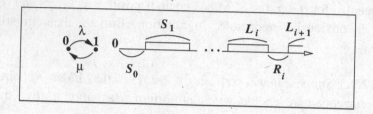

Fig. 2.59

Fig. 2.60

arrival Poisson process $(N(t))$ (which is also called an exogeneous arrival process). More precisely, a customer arriving in $(N(t))$ is admitted if and only if, at the time of arrival, $Q(t) < c$.

The diagram of a M/M/r/c chain is shown in Figure 2.59.

A simple case is where $r = c = 1$ (a single server, no waiting place). Here, $Q(t)$ takes two values: 0 (an idle server) and 1 (a busy server). The chain starting from, say, the empty state, with $Q(0) = 0$, spends a random holding time $S_0 \sim \text{Exp}(\lambda)$ in this state then jumps to state 1. After the holding time $S_1 \sim \text{Exp}(\mu)$, the chain jumps back to state 0, and so on.

Here, the DBEs are

$$\lambda \pi_0 = \mu \pi_1, \text{ implying that}$$
$$\pi_0 = \frac{\mu}{\lambda + \mu}, \text{ and } \pi_1 = \frac{\lambda}{\lambda + \mu}. \qquad (2.172)$$

The jump up process $(A(t))$, counting admitted arrivals in the M/M/1/1 chain, spends in each state $i = 1, 2 \ldots$ a holding time S_i^A set as the sum $L_i + R_i$ of independent summands, $L_i \sim \text{Exp}(\mu)$ (the service length of the ith admitted customer) and $R_i \sim \text{Exp}(\lambda)$ (the time till the next arrival); after that $A(t)$ jumps to $i+1$. The PDF $f_{S_i^A}$ of the random variable S_i^A, $i \geq 1$, is found by the convolution formula

$$f_{S_i^A}(x) = \int_0^x f_{L_i}(y) f_{R_i}(x-y)\, dy$$

$$= \int_0^x \mu e^{-\mu y} \lambda e^{-\lambda(x-y)} dy = \lambda \mu e^{-\lambda x} \int_0^x e^{(\lambda-\mu)y} dy$$

$$= \begin{cases} \dfrac{\lambda\mu}{\lambda-\mu}\left(e^{-\mu x} - e^{-\lambda x}\right), & \lambda \neq \mu, \\ \lambda^2 x e^{-\lambda x}, & \lambda = \mu. \end{cases} \tag{2.173}$$

The initial holding time is an exception: $S_0^A \sim \text{Exp}(\lambda)$ (the time till the first arrival in the empty queue). Holding times $S_0^A, S_1^A, \ldots$ are, obviously, independent. The process $(A(t))$ is not Markov, but belongs to the class of *renewal processes* which possess many properties similar to CTMCs; we will investigate these processes in a subsequent volume.

Similarly, the jump down process $(D(t))$ counting departures in the loss chain M/M/1/1 spends in each state $i = 0, 1, 2, \ldots$ a holding time S_i^D which is the sum $R_i + L_{i+1}$; the PDF of the RV S_i^D coincides with that of S_i^A and is given by (2.173). (Here, S_0^D is no exception.) Again, $(D(t))$ forms a renewal process. Note though that $(A(t))$ and $(D(t))$ are dependent (e.g. the term R_i contributes to both holding times S_i^A and S_i^D).

The key argument from the proof of Burke's theorem, obviously, remains applicable here: you can repeat, without any notable change, the proof of Theorem 2.9.2(i) for an M/M/1/1 chain. Thus, in equilibrium, the admitted arrival process $(A(t))$ and the departure process $(D(t))$ are stochastically equivalent. In fact, each of these processes in equilibrium represents a stationary version of the same renewal process, determined by the holding time PDF of the form (2.173), common for both of them. However, they remain dependent as the above correlation between S_i^A and S_i^D also appears in equilibrium.

Formulas similar to (2.168) can be derived for a general M/M/r/c chain. In fact, the DBEs, with $r \leq c$, generalise to

$$\lambda \pi_0 = \mu \pi_1, \ \ldots, \quad \lambda \pi_{r-1} = r\mu \pi_{r+1}, \ \ldots,$$
$$\lambda \pi_r = r\mu \pi_r, \ldots \quad \lambda \pi_{c-1} = r\mu \pi_c.$$

These equations can be easily solved. For instance, for $r = c$

$$\pi_i = \left(\frac{\rho^i}{i!}\right) \pi_0, \ i = 0, \ldots, r, \quad \text{whence} \quad \pi_0 = \left(\sum_{i=0}^r \frac{\rho^i}{i!}\right)^{-1},$$

where, as before, $\rho = \lambda/\mu$. Therefore, for all $i = 0, 1, \ldots, r$,

$$\pi_i = \left(\sum_{i=0}^r \frac{\rho^i}{i!}\right)^{-1} \left(\frac{\rho^i}{i!}\right). \tag{2.174}$$

These expressions are known as *Erlang's formulas*. For $r < c$

$$\pi_i = \left(\sum_{i=0}^{r} \frac{\rho^i}{i!} + \frac{\rho^r}{r!} \sum_{i=1}^{c-r} \frac{\rho^i}{r^i} \right)^{-1} \times \begin{cases} \left(\dfrac{\rho^i}{i!} \right), & i = 0, \dots, r, \\[3mm] \left(\dfrac{\rho^i}{r! r^{i-r}} \right), & i = r+1, \dots, c. \end{cases} \tag{2.175}$$

Again, the first half of Burke's theorem holds true. Summarising, we obtain

Theorem 2.9.4 *Let $(Q(t))$ be an M/M/r/c chain, with $Q(t) = Q(0) + A(t) - D(t)$ where $(A(t))$ is the admitted arrival process and $(D(t))$ is the departure process. Then:*

(i) *$(Q(t))$ is positive recurrent and reversible, with equilibrium distribution $\pi = (\pi_i)$ of the form (2.175), and the distribution of $Q(t)$ converges to π: for all $i = 0, \dots, c$ and initial distributions,*

$$\lim_{t \to \infty} \mathbb{P}(Q(t) = i) = \pi_i;$$

(ii) *in equilibrium, $(A(t)) \sim (D(t))$.*

However, the independence property manifested, for M/M/1/∞ chains, Theorem 2.9.2 (ii), is absent here. This is because the future admitted arrival process $(A(t+T) - A(T), \, t \geq 0)$ is correlated with $Q(T)$, the queue size at time T.

The final theorem of this section, Theorem 2.9.5, describes the long-term proportion formulas for Markovian queues (stated without proof). The assertion of Theorem 2.9.5 is, of course, a corollary of Theorem 2.7.19.

Theorem 2.9.5 *Set $\rho = \lambda/\mu$. In the M/M/1/∞ chain with $\rho = \lambda/\mu < 1$ and in the M/M/r/∞ chain with $\rho = \lambda/\mu < r$, for all $i = 0, 1, \dots,$*

$$\frac{1}{t} \int_0^t \mathbf{1}(Q_s = i) \, \mathrm{d}s \overset{\text{a.s.}}{\longrightarrow} \pi_i, \tag{2.176}$$

where the equilibrium probability π_i is determined in (2.160) and (2.171), respectively. In the M/M/∞ system and in the M/M/r/c chain, with $r \leq c$, for all $\lambda, \mu > 0$ and $i = 0, 1, \dots, c$, relation (2.176) holds for the equilibrium probability π_i determined in (2.168), (2.174) and (2.175), respectively.

Worked Example 2.9.6 Customers arrive in a barber's shop according to a Poisson process of rate $\lambda > 0$. The shop has s barbers and N waiting places; each barber works (on a single customer) provided that there is a customer to serve, and any customer arriving when the shop is full (i.e. the numbers of customers present is $N + s$) is not admitted and never returns. Every admitted customer waits in the queue and is then served, on a first-come-first-served order (say), the service taking

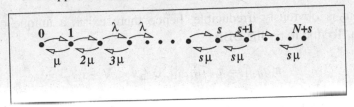

Fig. 2.61

an exponential time of rate $\mu > 0$; the service times of admitted customers are independent. After completing the hair cut, the customer leaves the shop and never returns.

Set up a Markov chain model for the number X_t of customers in the shop at time $t \geq 0$. Calculate the equilibrium distribution π of this chain and explain why it is unique. Show that (X_t) in equilibrium is reversible, i.e. for all $T > 0$, $(X_t, 0 \leq t \leq T)$ has the same distribution as $(Y_t, 0 \leq t \leq T)$ where $Y_t = X_{T-t}$, and $X_0 \sim \pi$.

Solution The chain (X_t) represents a birth-and-death process on the state space $\{0, 1, \ldots, N+s\}$, with the rates

$$q_{ii+1} = \lambda, \quad q_{ii-1} = \begin{cases} \mu i & \text{for } i = 1, \ldots, s, \\ \mu s & \text{for } i = s+1, \ldots, s+N. \end{cases}$$

See Figure 2.61.

We have used here the fact that if $X_k \sim \text{Exp}(\theta_k)$ are IID variables, then

$$\min(X_1, \ldots, X_l) \sim \text{Exp}\left(\sum_{k=1}^{l} \theta_k\right).$$

So, the generator is an $(N+s+1) \times (N+s+1)$ matrix given by

$$
\begin{array}{ccccccccc}
0 & 1 & 2 & \cdots & s & & \cdots & & N+s
\end{array}
$$
$$
\begin{pmatrix}
-\lambda & \lambda & 0 & \cdots & 0 & 0 & \cdots & 0 & 0 \\
\mu & -(\lambda+\mu) & \lambda & \cdots & 0 & 0 & \cdots & 0 & 0 \\
0 & 2\mu & -(\lambda+2\mu) & \cdots & 0 & 0 & \cdots & 0 & 0 \\
\cdots & \cdots & \cdots & \cdots & \cdots & \cdots & \cdots & \cdots & \cdots \\
0 & 0 & 0 & \cdots & -(\lambda+s\mu) & \lambda & \cdots & 0 & 0 \\
\cdots & \cdots & \cdots & \cdots & \cdots & \cdots & \cdots & \cdots & \cdots \\
0 & 0 & 0 & \cdots & 0 & 0 & \cdots & s\mu & -s\mu
\end{pmatrix}.
$$

The chain is obviously irreducible, hence there exists a unique equilibrium distribution. To find it, try the DBEs

$$\pi_i q_{i i+1} = \pi_{i+1} q_{i+1 i}, \ 0 \le i < N+s.$$

More precisely, we have

$$\pi_0 \lambda = \pi_1 \mu, \ \ldots, \ \ \pi_{s-1} \lambda = s \pi_s \mu,$$
$$\pi_s \lambda = s \pi_{s+1} \mu, \ldots, \ \ \pi_{N+s-1} \lambda = s \pi_{N+s} \mu.$$

These equations have a unique normalised solution

$$\pi_n = \pi_0 \times \begin{cases} \dfrac{\lambda^n}{n! \mu^n}, & \text{for } n = 1, \ldots, s, \\[2mm] \dfrac{\lambda^n}{s! s^{n-s} \mu^n}, & \text{for } n = s+1, \ldots, s+N, \end{cases}$$

where

$$\pi_0 = \left(\sum_{l=0}^{s} \frac{\lambda^l}{l! \mu^l} + \frac{\lambda^s}{s! \mu^s} \sum_{l=1}^{N} \frac{\lambda^l}{\mu^l} \right)^{-1}.$$

The fact that $(X_t, 0 \le t \le T)$ has the same distribution as $(Y_t, 0 \le t \le T)$ where $Y_t = X_{T-t}$, is now checked in the standard way. So, chain (X_t) is reversible if and only if it is in its (unique) equilibrium regime. $\qquad \square$

Worked Example 2.9.7 Consider a queue with a Poisson arrival of rate λ and IID exponential service times, of rate μ. The queue is modified in such a way that after being served, a customer goes away with probability $(1-p)$ and joins the queue with probability p, $0 < p < 1$. Argue that the modified queue is equivalent to an M/M/1 queue and determine the condition for positive recurrence. Calculate the probability that the queue is empty in equilibrium.

Solution (sketch) Let Q_t be the number of customers in the queue at time t. Then $Q_t \sim \tilde{Q}_t$ where $\tilde{Q}_t$ stands for the number of customers in the M/M/1 queue with arrival rate λ and service rate $\mu(1-p)$ (provided that we start both processes from the same initial distribution). A straightforward argument here is as follows. We may think that the customer returning to the queue goes straight back to service, simply continuing his previous service time. Then the total time $\tilde{S}$ in service for a single customer will be the sum of a random number N of IID exponential variables, where $(N-1) \sim \text{Geom}(1-p)$. This has the moment-generating function

$$\mathbb{E}\exp\left(\theta\widetilde{S}\right) \;=\; \mathbb{E}\left(\mathbb{E}\left[\exp\left(\theta\widetilde{S}\right)\big|N\right]\right)$$

$$=\; \sum_{n\geq 1}\mathbb{P}(N=n)\mathbb{E}\left[\exp\left(\theta S_{\text{tot}}\right)\big|N=n\right]$$

$$=\; \sum_{n\geq 1}(1-p)p^{n-1}\left(\frac{\mu}{\mu-\theta}\right)^n \;=\; \frac{1-p}{p}\sum_{n\geq 1}\left(\frac{p\mu}{\mu-\theta}\right)^n$$

$$=\; \frac{1-p}{p}\frac{p\mu}{\mu-\theta}\Big/\left(1-\frac{p\mu}{\mu-\theta}\right) \;=\; \frac{(1-p)\mu}{\mu-\theta}\Big/\frac{(1-p)\mu-\theta}{\mu-\theta},$$

$$=\; \frac{(1-p)\mu}{(1-p)\mu-\theta}.$$

which corresponds to the distribution $\text{Exp}\left[(1-p)\mu\right]$. Hence, positive recurrence holds when $\lambda < (1-p)\mu$, and in equilibrium, the probability $\mathbb{P}(Q_t=0)=1-\lambda/[(1-p)\mu]$. $\qquad\square$

> *Probable impossibilities are to be preferred*
> *to improbable possibilities.*
> Aristotle (384–322BC), Greek philosopher

In Volume 1 we touched on the theory of discrete-time *branching processes*. The basic model was where particles or living organisms divide and produce a number of descendents, or offspring. There exists a continuous-time counterpart of the theory; its basics are encapsulated below.

Suppose that a cell has been placed in a biological solution at time $t=0$. After an exponential time of rate μ it is divided producing k cells with probability p_k, $k=0,1,\dots$, with the mean value $\rho=\sum_{k\geq 1}kp_k$ ($k=0$ means that the cell dies). The same mechanism is applied to each of the living cells, independently.

Let M_t be the number of living cells in the solution by time $t>0$. We will prove that $\mathbb{E}M_t=\exp\left[t\mu(\rho-1)\right]$. To this end, set $g(t)=\mathbb{E}M_t$ and apply conditioning:

$$g(t+u) \;=\; \mathbb{E}\left[\mathbb{E}(M_{t+u}|M_u)\right]$$

$$=\; \sum_{k\geq 1}(k\,\mathbb{E}M_t)\mathbb{P}(M_u=k)$$

$$=\; (\mathbb{E}\,M_t)\sum_{k\geq 1}k\mathbb{P}(M_u=k)$$

$$=\; \mathbb{E}\left[M_t\right]\mathbb{E}\left[M_u\right] \;=\; g(t)g(u).$$

Next, for t near 0, again by conditioning:

$$g(t)=1-\mu t+\mu t\rho+o(t)=1+\mu t(\rho-1)+o(t).$$

We see that the function $g(t)$ is differentiable at $t = 0$, positive for $t > 0$ (since $g(t) > \mathbb{P}(\text{no division in } (0,t)) = e^{-\mu t}$), satisfies the multiplicativity equation

$$g(t+u) = g(t)g(u), \quad t, u \geq 0,$$

and is differentiable at zero. Then $g(t) = e^{\alpha t}, t \geq 0$, for some $\alpha \in \mathbb{R}$. Finally, $\alpha = \mu(\rho - 1)$.

In a similar fashion, one can derive the differential equation for $\phi_t(s) = \mathbb{E}s^{M_t}$, the probability generating function of M_t. In fact, $\phi_t(s)$ satisfies the following differential equation

$$\frac{d}{dt}\phi_t(s) = \mu\left(-\phi_t(s) + \sum_{k \geq 0} p_k[\phi_t(s)]^k\right), \quad \text{with} \quad \phi_0(s) = s. \tag{2.177}$$

Indeed,

$$\phi_{t+h}(s) = \phi_h[\phi_t(s)], \quad t, h \geq 0.$$

For h near 0:

$$\phi_h(s) = (1 - \mu h)s + \mu h \sum_{k \geq 0} p_k s^k + o(h),$$

and

$$\phi_{t+h}(s) = (1 - \mu h)\phi_t(s) + \mu h \sum_{k \geq 0} p_k[\phi_t(s)]^k + o(h),$$

i.e.

$$\phi_{t+h}(s) - \phi_t(s) = h\mu\left(-\phi_t(s) + \sum_{k \geq 0} p_k[\phi_t(s)]^k\right) + o(h).$$

Dividing by h and letting $h \to 0$ yields (2.177), the initial condition $\phi_0(s) = s$ being straightforward.

Branching processes produce examples of birth-and-death processes that are not Poisson. For instance, assume that each cell divides into two cells ($p_2 = 1$). Let $N_t = M_t - 1$ be the number of cells produced in the solution by time t. It turns out that N_t has a geometric distribution. In fact,

$$\begin{aligned}
\mathbb{P}(N_t = 0) &= \mathbb{P}(\text{no division in } (0,t)) = e^{-\mu t}, \\
\mathbb{P}(N_t = 1) &= \mathbb{P}(\text{a single division in } (0,t)) \\
&= \int_0^t \mu e^{-\mu s} e^{-2\mu(t-s)} ds = e^{-2\mu t}(e^{\mu t} - 1),
\end{aligned}$$

and in general

$$\mathbb{P}(N_t = n) = \mathbb{P}(n \text{ divisions in } (0, t))$$

$$= \int_0^t \int_{u_1}^t \cdots \int_{u_1}^t \mu e^{-\mu s_1}(2\mu)e^{-2\mu(s_2-s_1)}$$

$$\times \cdots$$

$$\times (n\mu)e^{-n\mu(s_n-s_{n-1})}\, e^{-(n+1)\mu(t-s_n)}\, ds_n \cdots ds_1$$

$$= n!e^{-(n+1)\mu t}\int_0^t \cdots \int_0^t \mu^n e^{\mu(s_1+\cdots+s_n)}$$

$$\times \mathbf{1}(s_1 < \cdots < s_n)\, ds_n \cdots ds_1$$

$$= n!e^{-(n+1)\mu t}\left(\int_0^t \mu e^{\mu s}\, ds\right)^n \Big/ n! = e^{-(n+1)\mu t}\left(e^{\mu t}-1\right)^n.$$

This indicates that, indeed, (N_t) does not constitute an inhomogeneous Poisson process. Alternatively, the infinitesimal probability of a jump

$$\mathbb{P}(N_{t+h}-N_t = 1 | N_t = k) = \mu k h + o(h)$$

depends on k, the value of N_t, whereas in an inhomogeneous Poisson process it should be of the form $\lambda(t)h + o(h)$ regardless of k.

Example 2.9.8 We continue the previous theme and find an explicit representation of $\phi_t(s)$ in the case of the quadratic function in (2.177). For $p_2 = 1$, integrating the equation

$$\frac{d\phi_t(s)}{dt} = -\mu\phi_t(s) + \mu\phi_t(s)^2, \quad \phi_0(s) = s,$$

we get

$$\phi_t(s) = \frac{s}{e^{\mu t} - (e^{\mu t}-1)s}, \quad -1 \le s \le 1.$$

For $p_0 = 1/3$, $p_2 = 2/3$, integrating the equation

$$\frac{d\phi_t(s)}{dt} = \frac{\mu}{3} - \mu\phi_t(s) + \frac{2\mu}{3}\phi_t(s)^2, \quad \phi_0(s) = s,$$

we get for $s \in (-1, 1)$

$$\phi_t(s) = \frac{(s-1)e^{\mu t/3} - (2s-1)}{2(s-1)e^{\mu t/3} - (2s-1)}, \quad \text{with } \phi_t(s) \to \frac{1}{2} \text{ as } t \to \infty.$$

In the case of a general quadratic polynomial, we write (2.177) in the form

$$\frac{d}{dt}\phi_t(s) = \mu p_2(\phi_t(s) - \lambda_1)(\phi_t(s) - \lambda_2), \quad \phi_0(s) = s, \quad -1 < s < 1,$$

λ_1 and λ_2 being the roots of

$$x = p_0 + p_1 x + p_2 x^2;$$

one root always equals 1. Without loss of generality we assume that $\lambda_1 \leq \lambda_2$.

Case (a): super-critical. Here $\lambda_2 = 1$ and $0 < \lambda_1 < 1$. We see that in this case $\lambda_1 = \pi_{\text{ext}}$, the probability of eventual extinction.

Case (b): critical. Here $\lambda_1 = \lambda_2 = 1$.

Case (c): sub-critical. Here $\lambda_1 = 1$, $\lambda_2 > 1$.

In cases (a) and (c)

$$\phi_t(s) = \frac{\lambda_1(s-\lambda_2) - \lambda_2(s-\lambda_1)e^{-\mu p_2(\lambda_2-\lambda_1)t}}{(s-\lambda_2) - (s-\lambda_1)e^{-\mu p_2(\lambda_2-\lambda_1)t}}, \quad t \geq 0, \ -1 < s < 1.$$

We see that

$$\lim_{t\to\infty} \phi_t(s) = \lambda_1 = \begin{cases} \pi_{\text{ext}}, & \text{in case (a),} \\ 1, & \text{in case (c),} \end{cases}$$

and convergence happens exponentially fast.

In case (b):

$$\phi_t(s) = 1 - \frac{1-s}{1 + (1-s)\mu t(1-p_1)/2}, \quad t \geq 0, \ -1 < s < 1.$$

Here,

$$\lim_{t\to\infty} \phi_t(s) = 1,$$

and convergence is inverse power-like: $(\phi_t(s) - 1) \approx O(1/t)$.

2.10 Examination questions on continuous-time Markov chains

> *The v-dity of the Bad, the π-ty of the Good*
> *(Or the Other way Around)*
> *(From the series 'Movies that never made it to the Big Screen'.)*

Question 2.10.1 (Markov chains, Part IB, 1994, 504D)
(i) Consider the four-state continuous-time Markov chain with states 1, 2, 3, 4 and the generator matrix

$$Q = \begin{pmatrix} -2 & 1 & 0 & 1 \\ 1 & -2 & 1 & 0 \\ 0 & 1 & -2 & 1 \\ 1 & 0 & 1 & -2 \end{pmatrix}.$$

Draw the figure associated with this chain. The chain begins at state 1. Find the probability that it is in state 1 at time t.

(ii) Consider now the five-state chain with states 1, 2, 3, 4 and 5, and the generator matrix

$$Q = \begin{pmatrix} -3 & 1 & 0 & 1 & 1 \\ 1 & -3 & 1 & 0 & 1 \\ 0 & 1 & -3 & 1 & 1 \\ 1 & 0 & 1 & -3 & 1 \\ 0 & 0 & 0 & 0 & 0 \end{pmatrix}.$$

The chain starts at state 1. By relating this matrix to the one previously considered, find the probability that it is in state 1 at time t.

Solution (i) The chain on four states behaves as a pair of independent Markov chains (X_t) and (Y_t), each with two states, say 0 and 1, and the generator matrix

$$Q = \begin{pmatrix} -1 & 1 \\ 1 & -1 \end{pmatrix}.$$

One can think about a pair (X_t, Y_t) moving around the corners of a square $[-1, 1]^2$ with equal chances of visiting each of the two neighbouring corners. Hence, the probability in question is written

$$(p_{11}(t))^2 = (A + B^{-2t})^2.$$

From the conditions at $t = 0$ and $t \to \infty$, $A = B = 1/2$. This yields the answer $(1 + e^{-2t})^2/4$.

(ii) The new chain adds the absorbing state 5, with absorption rate 1. Hence,

$$\mathbb{P}(\text{not in state 5 at time } t) = e^{-t},$$

regardless of the initial distribution λ. Then, by independence,

$$\mathbb{P}(\text{in the same state at time } t \text{ as at time } 0) = \left(\frac{e^{-t}}{4}\right)\left(1 + e^{-2t}\right)^2.$$

$\square$

Question 2.10.2 (Markov chains, Part IIA, 1994, A201K)
(i) Consider a random walk $(X_n)_{n \geq 0}$ on the graph shown in Figure 2.62.

At each step this process moves to a neighbouring vertex, with equal probability for each, and independently of past moves: thus from C it moves to A, B, D or E, each with probability $1/4$. Stating clearly any general theorems to which you appeal, show that $\mathbb{P}(X_n = A)$ converges as $n \to \infty$, and determine its limit.

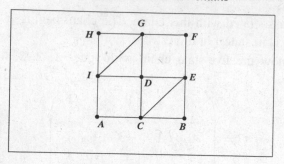

Fig. 2.62

(ii) Find for the random walk $(X_n)_{n\geq 0}$ defined in (i) the expected number of visits to C, starting from A, before the first return to A.

Let now $(Z_t)_{t\geq 0}$ be a continuous-time Markov chain on the graph of (i) with Q-matrix given by

$$q_{ij} = \begin{cases} 1 & \text{if } (i,j) \text{ is an edge,} \\ 0 & \text{otherwise.} \end{cases}$$

Let S denote the total time spent by $(Z_t)_{t\geq 0}$ in $\{C,E\}$, starting from B, before it first returns to B. Show that S follows an exponential distribution and determine its parameter.

Solution (i) For an irreducible aperiodic Markov chain with an invariant distribution π,

$$\mathbb{P}_\lambda(X_n = j) \to \pi_j, \quad n \to \infty,$$

for all states j, regardless of the initial distribution λ. In the example, the Markov chain represents the random walk on a graph. It is irreducible and aperiodic (all cycles being of co-prime lengths), and with equilibrium distribution π such that $\pi_i = v_i / \sum_k v_k$ where v_j is the valency of the vertex j. As the total valency $\sum_k v_k = 28$, $\lim_{n\to\infty} \mathbb{P}(X_n = A) = 2/28 = 1/14$.

(ii) For an irreducible aperiodic Markov chain with an ED π,

$$\mathbb{E}_i(\text{number of visits to } j \text{ before returning to } i) = \frac{\pi_j}{\pi_i},$$

which yields the answer $v_C/v_A = 2$.

Using symmetry, it helps to consider an aggregated chain with the states $B, (C,E), (A,F), D, (I,G), H$. Then all rates $(C,E) \to B$, $(C,E) \to (A,F)$ and $(C,E) \to D$ equal 1. Hence, for $(Z_t)_{t\geq 0}$, the time spent on each visit to (C,E) becomes exponential, of rate $\rho = 3$. On each visit, there arises a probability $1/3$ of returning to B. This means the number of visits to (C,E) before returning to B is

geometric, with parameter $q = 1/3$. The sum of a geometric number of independent exponential random variables takes an exponential distribution of rate qp, as,
e.g., the moment-generating function

$$\mathbb{E}_{\text{GM}} \left(\theta \sum_{i=1}^{N} X_i \right) = \sum_{n \geq 1} (1-q)^{n-1} q \left(\frac{\rho}{\mu - \theta} \right)^n = \frac{q\rho}{q\mu - \theta}.$$

Hence, S follows the exponential distribution, of rate 1.   □

Question 2.10.3 (Markov chains, Part IIA, 1994, A301K)
(i) A continuous-time Markov chain $(X_t)_{t \geq 0}$ with state space $\{1,2,3,4,5\}$ is
governed by the following generator matrix

$$Q = \begin{pmatrix} -3 & 1 & 0 & 1 & 1 \\ 1 & -3 & 1 & 0 & 1 \\ 0 & 1 & -3 & 1 & 1 \\ 1 & 0 & 1 & -3 & 1 \\ 0 & 0 & 0 & 0 & 0 \end{pmatrix}.$$

In the case $X_0 = 1$, find the probability that $X_t = 2$ for some $t \geq 0$.

(ii) For the chain described in (i), assuming that $X_0 \neq 5$, find the probability that
$(X_t)_{t \geq 0}$ eventually visits every state.

Solution   (i) From the description of the chain, the equations for $h_i = \mathbb{P}_i(\text{hit } 2)$ are

$$h_2 = 1, \quad h_1 = h_3 = \frac{1}{3} + \frac{1}{3}h_4, \quad h_4 = \frac{2}{3}h_1,$$

resulting in $h_1 = 3/7$ and $h_4 = 2/7$. By symmetry, $\mathbb{P}_i(\text{hit } j) = 3/7$ for any pair of
adjacent and $2/7$ for any pair of opposite corners i, j.

(ii) The condition on the initial distribution λ means $\lambda_5 = 0$. By symmetry, for
any such λ,

$$\mathbb{P}(\text{visit every state}) = \mathbb{P}_1(\text{visit every state}),$$

which in turn equals

$$\mathbb{P}_1(\text{hit } 2,3 \text{ and } 4) = 1 - \mathbb{P}_1(\{\text{avoid } 2\} \cup \{\text{avoid } 3\} \cup \{\text{avoid } 4\}).$$

By the inclusion–exclusion formula, the last term may be written

$$
\begin{aligned}
\mathbb{P}_1 \left(\bigcup_{i=2}^{4} \{\text{avoid } i\} \right) =\ & \mathbb{P}_1(\{\text{avoid } 2\}) + \mathbb{P}_1(\{\text{avoid } 3\}) + \mathbb{P}_1(\{\text{avoid } 4\}) \\
& - \mathbb{P}_1(\{\text{avoid } 2 \text{ and } 3\}) - \mathbb{P}_1(\{\text{avoid } 2 \text{ and } 4\}) \\
& - \mathbb{P}_1(\{\text{avoid } 3 \text{ and } 4\}) + \mathbb{P}_1(\{\text{avoid } 2, 3 \text{ and } 4\}).
\end{aligned}
$$

By (i), $\mathbb{P}_1(\{\text{avoid } j\}) = 4/7$ for $j = 2, 4$ and $5/7$ for $j = 3$.

Next,

$$\mathbb{P}_1(\text{hit 2 or 4}) = 2/3 \quad \text{and} \quad \mathbb{P}_1(\text{avoid 2 and 4}) = \frac{1}{3}.$$

Then for the hitting probability $h_i^{\{3,4\}} := \mathbb{P}_i(\text{hit 3 or 4})$ we obtain

$$h_1^{\{3,4\}} = \frac{1}{3} + \frac{1}{3}h_2^{3,4} \quad \text{and} \quad h_1^{\{3,4\}} = h_2^{\{3,4\}},$$

again, by symmetry. Hence $h_1^{\{3,4\}} = 1/2$ and

$$\mathbb{P}_1(\{\text{avoid 2 and 3}\}) = \frac{1}{2} = \mathbb{P}_1(\{\text{avoid 3 and 4}\}).$$

Finally,

$$\mathbb{P}_1(\{\text{avoid 2, 3 and 4}\}) = \mathbb{P}_1(\{\text{go straight to 5}\}) = 1/3.$$

Collecting all terms,

$$\mathbb{P}_1\left(\bigcup_{i=2}^{4}\{\text{avoid i}\}\right) = \frac{4}{7} + \frac{5}{7} + \frac{4}{7} - \frac{1}{3} - \frac{1}{2} - \frac{1}{2} + \frac{1}{3} = \frac{6}{7},$$

and so,

$$\mathbb{P}_1(\text{visit every state}) = 1/7.$$

$\square$

Question 2.10.4 (Markov chains, Part IB, 1995, 503G)

A spider climbs a vertical spout of height a at unit speed. At the times of a Poisson process, with rate λ per unit time, down comes the rain and washes the spider out. When rain begins, the spider is washed instantaneously to the bottom and straight away, starts to climb the spout again. Suppose that the spider initially sits at the bottom of the spout. Let T represent the time when it reaches the top and let N be the number of times it is washed back down before it gets to the top. For $\theta \geq 0$ and $0 \leq z \leq 1$, show that

$$\mathbb{E}\left(e^{-\theta T}z^N\right) = \frac{(\lambda + \theta)e^{-(\lambda+\theta)a}}{\lambda + \theta - \lambda z\left(1 - e^{-(\lambda+\theta)a}\right)}.$$

By first calculating $\mathbb{E}\left(e^{-\theta T}|N = n\right)$, or otherwise, determine $\mathbb{E}\left(T|N = n\right)$ for each $n = 0, 1, 2, \ldots$.

Solution Let J_1 be the time of the first rain. Write

$$g = \mathbb{E}\left(e^{-\theta T} z^N\right) = \int_0^\infty \lambda e^{-\lambda s} \mathbb{E}\left(e^{-\theta T} z^N | J_1 = s\right) ds$$

$$= \int_a^\infty \lambda e^{-\lambda s} e^{-\theta a} ds + \left(\int_0^a \lambda e^{-\lambda s} e^{-\theta s} z ds\right) g$$

$$= e^{-(\lambda+\theta)a} + z g \frac{\lambda}{\lambda+\theta}\left(1 - e^{-(\lambda+\theta)a}\right).$$

Thus,

$$g = e^{-(\lambda+\theta)a}\left(1 - \frac{\lambda z}{\lambda+\theta}\left(1 - e^{-(\lambda+\theta)a}\right)\right)^{-1}.$$

Then $\mathbb{E}\left(e^{-\theta T}\mathbf{1}(N=n)\right)$ is the coefficient at z^n in the expansion of g

$$\mathbb{E}\left(e^{-\theta T}\mathbf{1}(N=n)\right) = e^{-(\lambda+\theta)a}\left[\frac{\lambda}{\lambda+\theta}\left(1 - e^{-(\lambda+\theta)a}\right)\right]^n,$$

and $\mathbb{P}(N=n)$ is obtained when we set $\theta = 0$

$$\mathbb{P}(N=n) = e^{-\lambda a}\left(1 - e^{-\lambda a}\right)^n.$$

Next, define

$$g_n = \mathbb{E}\left(e^{-\theta T} | N=n\right) = e^{-\theta a}\left[\frac{\lambda\left(1 - e^{-(\lambda+\theta)a}\right)}{(\lambda+\theta)\left(1 - e^{-\lambda a}\right)}\right]^n.$$

Then

$$\frac{d}{d\theta} g_n = -a g_n + n e^{-\theta a}\left[\frac{\lambda\left(1 - e^{-(\lambda+\theta)a}\right)}{(\lambda+\theta)\left(1 - e^{-\lambda a}\right)}\right]^{n-1}$$

$$\times \left[-\frac{\lambda}{(\lambda+\theta)^2}\left(\frac{1 - e^{(\lambda+\theta)a}}{1 - e^{-\lambda a}}\right) + \frac{\lambda a e^{-(\lambda+\theta)a}}{(\lambda+\theta)\left(1 - e^{-\lambda a}\right)}\right].$$

To find $\mathbb{E}(T|N=n)$ we again set $\theta = 0$

$$\mathbb{E}(T|N=n) = -\frac{d}{d\theta} g_n \Big|_{\theta=0} = a + n\left(\frac{1}{\lambda} - \frac{a}{e^{\lambda a} - 1}\right).$$

$\square$

Question 2.10.5 (Markov chains, Part IB, 1995, 504G)

Between each pair of the cities A, B and C there is telephone line which may be put out of action by snowstorms. Snowstorms happen according to a Poisson process, with rate 8 per unit time, and when one occurs, each telephone line is put out of action independently, with probability $1/2$. When a line is out of action, it takes a random length of time to be repaired; the duration of this repair time has exponential distribution with mean $1/14$ and all repairs are carried out independently. Let $\{X_t, t \geq 0\}$ be a continuous-time Markov chain, where X_t represents the number of lines out of action at time t. Determine the expected holding times in each state and hence, or otherwise, determine the Q-matrix of $\{X_t, t \geq 0\}$.

Determine the long-run proportion of time when all pairs of cities may communicate, assuming that messages may be passed through the third city, if necessary.

Solution The states of the Markov chain are 0, 1, 2, 3 (the number of lines not in action), and the mean holding times are $1/7, 1/20, 1/32$ and $1/42$. The generator matrix is given by

$$Q = \begin{pmatrix} -7 & 3 & 3 & 1 \\ 14 & -20 & 4 & 2 \\ 0 & 28 & -32 & 4 \\ 0 & 0 & 42 & -42 \end{pmatrix}.$$

The invariant distribution $\pi = (\pi_0, \pi_1, \pi_3, \pi_4)$ is unique and satisfies $\pi Q = 0$, i.e.

$$\begin{aligned} -7\pi_0 + 14\pi_1 &= 0, \\ 3\pi_0 - 20\pi_1 + 28\pi_2 &= 0, \\ 3\pi_0 + 4\pi_1 - 32\pi_2 + 42\pi_3 &= 0, \\ \pi_0 + 2\pi_1 + 4\pi_2 - 42\pi_3 &= 0. \end{aligned}$$

This gives

$$\pi_0 = \frac{28}{51}, \ \pi_1 = \frac{14}{51}, \ \pi_2 = \frac{7}{51}, \ \pi_0 = \frac{2}{51}.$$

Thus the long-time proportion in question equals $\pi_0 + \pi_1 = 14/17$. $\qquad\square$

Question 2.10.6 (Part II, 314B, 1992, Stochastic Processes)
Consider a queueing system with one server in which there is a room for at most one customer to wait in addition to the customer being served. Arriving customers who find the waiting room full do not enter the system. The times between arrivals of customers are independent exponential random variables of parameter λ and the service times are independent exponential random variables of parameter μ. Write down the Q-matrix of the continuous-time Markov chain $X(t)$ where $X(t)$ is the number of customers in the system at time t.

Evaluate $\lim_{t \to \infty} \mathbb{P}(X(t) = j)$ for $j = 0, 1, 2$.

In the case $\lambda = \mu$, calculate $\mathbb{P}(X(t) = 0 | X(0) = 0)$ for all $t \geq 0$.

Solution Following common practice, we model the queue as a Markov chain with the following generator matrix

$$Q = \begin{pmatrix} -\lambda & \lambda & 0 \\ \mu & -\lambda - \mu & \lambda \\ 0 & \mu & -\mu \end{pmatrix}.$$

To find the invariant distribution, solve for $\pi = (\pi_0, \pi_1, \pi_2)$ with $\pi_i \geq 0$

$$\pi Q = 0, \quad \pi_0 + \pi_1 + \pi_2 = 1.$$

Then

$$\lambda \pi_0 = \mu \pi_1, \quad \lambda \pi_1 = \mu \pi_2,$$

and $\pi \propto (1, \lambda/\mu, \lambda^2/\mu^2)$, i.e.

$$\pi = \left(1, \frac{\lambda}{\mu}, \frac{\lambda^2}{\mu^2}\right) \Big/ \left(1 + \frac{\lambda}{\mu} + \frac{\lambda^2}{\mu^2}\right).$$

Further, by standard results, for $j = 0, 1, 2$,

$$\lim_{t \to \infty} \mathbb{P}(X(t) = j) = \pi_j = \left(\frac{\lambda}{\mu}\right)^j \Big/ \left(1 + \frac{\lambda}{\mu} + \frac{\lambda^2}{\mu^2}\right).$$

Next, in the case $\lambda = \mu$ consider first the case $\lambda = 1$. Here

$$Q = \begin{pmatrix} -1 & 1 & 0 \\ 1 & -2 & 1 \\ 0 & 1 & -1 \end{pmatrix}.$$

To find the eigenvalues solve

$$\begin{aligned} 0 &= \det(\nu \mathbf{I} - Q) \\ &= (\nu + 1)^2(\nu + 2) - 2(\nu + 1) \\ &= \nu(\nu + 1)(\nu + 3). \end{aligned}$$

Hence, obtain the eigenvalues $\nu_0 = 0$, $\nu_1 = -1$ and $\nu_2 = -3$. By the general result

$$p(t) := \mathbb{P}(X(t) = 0 | X(0) = 0) = \pi_0 + Ae^{\nu_1 t} + Be^{\nu_2 t}, \quad t \geq 0,$$

we have

$$p(t) = \frac{1}{3} + Ae^{-t} + Be^{-3t}.$$

But $p(0) = 1$ and $\frac{d}{dt} p(0) = -1$. So

$$1 = \frac{1}{3} + A + B, \quad -1 = -A - 3B,$$

$$1 = 2A, \quad A = \frac{1}{2}, \quad B = \frac{1}{6}.$$

Thus

$$p(t) = \frac{1}{3} + \frac{1}{2} e^{-t} + \frac{1}{6} e^{-3t}.$$

Alternatively, one can solve the backward or forward equations. $\square$

> *Man's unhappiness, as I construe, comes of his greatness;*
> *it is because there is Infinite in him,*
> *which with all his*
> *cunning he cannot quite bury under the Finite.*
> T. Carlyle (1795–1881), English poet and writer

Question 2.10.7 (Markov chains, Part IIA, 1995, A301M)
(i) Suppose that buses arrive at a bus-stop as a Poisson process $\{X_t\}_{t \geq 0}$ of parameter λ per hour, and that after an hour exactly n buses have arrived. Calculate the conditional probabilities $\mathbb{P}(X_t = k | X_1 = n)$ that exactly k buses, $0 \leq k \leq n$, have arrived at the bus-stop by time t, $0 \leq t \leq 1$, given that n have arrived by time 1.

(ii) Consider a continuous time Markov chain $\{X_t\}_{t \geq 0}$ with two states, 0 and 1. For each $t \geq 0$ and $i, j \in \{0, 1\}$, let $p_{ij}(t) = \mathbb{P}(X_t = j | X_0 = i)$. Suppose that for some $T > 0$ the matrix $P(T)$ has the form

$$P(T) = \begin{pmatrix} \alpha & 1 - \alpha \\ 1 - \alpha & \alpha \end{pmatrix}.$$

Prove that $1/2 < \alpha \leq 1$, and compute $P(t)$ for all $t \geq 0$.

Solution (i) The conditional probability $\mathbb{P}(X_t = k | X_1 = n)$ equals

$$\frac{\mathbb{P}(X_t = k, X_1 = n)}{\mathbb{P}(X_1 = n)} = \frac{\mathbb{P}(X_t = k)\mathbb{P}(X_1 = n | X_t = k)}{\mathbb{P}(X_1 = n)}$$

$$= \frac{\mathbb{P}(X_t = k)\mathbb{P}(Y_{1-t} = n - k)}{\mathbb{P}(X_1 = n)}$$

$$\text{(by the Markov property)}$$

$$= \frac{\left[e^{-\lambda t}(\lambda t)^k/k!\right]\left[e^{-\lambda(1-t)}(\lambda(1-t))^{n-k}/(n-k)!\right]}{e^{-\lambda}\lambda^n/n!}$$

$$= \binom{n}{k} t^k(1-t)^{n-k}.$$

That is, the conditional probabilities are binomial.

(ii) Write $e^{TQ} = P(T)$ where $Q = \begin{pmatrix} -\lambda_0 & \lambda_0 \\ \lambda_1 & -\lambda_1 \end{pmatrix}$ is the Q-matrix, with $\lambda_0, \lambda_1 \geq 0$. If $\lambda_0 = \lambda_1 = 0$, then Q becomes the zero matrix, e^{TQ} the unit matrix, and $\alpha = 1 > 1/2$.

Thus, we can assume that $\lambda_0 + \lambda_1 > 0$. Then the eigenvalues of Q equal 0 and $-(\lambda_0 + \lambda_1)$, and the equations for the diagonal entries read

$$\alpha = p_{00}(T) = \frac{\lambda_1}{\lambda_0 + \lambda_1} + \frac{\lambda_0}{\lambda_0 + \lambda_1} e^{-(\lambda_0 + \lambda_1)T};$$

and

$$\alpha = p_{11}(T) = \frac{\lambda_0}{\lambda_0 + \lambda_1} + \frac{\lambda_1}{\lambda_0 + \lambda_1} e^{-(\lambda_0 + \lambda_1)T}.$$

We obtain that $\lambda_0 = \lambda_1 = \lambda$, i.e.

$$\alpha = \frac{1}{2} + \frac{1}{2} e^{-2\lambda T},$$

whence $\alpha \in (1/2, 1)$.

For general $t > 0$

$$p_{00}(t) = \frac{1}{2} + \frac{1}{2} e^{-2\lambda t} = \frac{1}{2} + \frac{1}{2}(2\alpha - 1)^{t/T};$$

other entries being calculated similarly. $\square$

Question 2.10.8 (Markov chains, Part IB, 1996, 502G)
Patients arrive at a hospital department according to a Poisson process of rate λ. There are plenty of junior doctors, and an arriving patient is immediately seen by one of them. The doctor takes a random time (with mean μ^{-1}) to deal with the patient, after which the patient leaves the hospital with probability $1 - \alpha$, but with

probability α is referred to the consultant in charge of the department. The patients queue for the attention of the consultant, who takes a random time (with mean v^{-1}) to deal with each one. After being seen by the consultant, the patient leaves the hospital.

Show how, under appropriate assumptions (which should be stated), the situation may be formulated as a continuous-time Markov chain with state (r, s), where r is the number of junior doctors busy and s the number of patients referred to the consultant and still in the hospital. Describe the Q-matrix of the Markov chain.

Let $\pi(r, s)$ be the limit, as $t \to \infty$, of the probability of the chain being in state (r, s) at time t. Write down the equations which the $\pi(r, s)$ must satisfy, and show that these are satisfied by

$$\pi(r, s) = \frac{e^{-\beta} \beta^r}{r!} (1 - \gamma) \gamma^s,$$

for suitable positive values of β and γ, if $\lambda \alpha < v$.

Solution Assume that the times taken by doctors and the consultant to see patients are exponential and independent of each other and of the Poisson arrival process. Then, by the memoryless property of the exponential distribution, the pair (r, s) forms the state of a Markov chain whose generator matrix $Q = (q_{(r,s)(r',s')})$ contains the following non-zero off-diagonal entries

$$\begin{aligned}
q_{(r,s)(r+1,s)} &= \lambda, & r, s &\geq 0, \\
q_{(r,s)(r-1,s)} &= r\mu(1 - \alpha), & r &\geq 1, \ s \geq 0, \\
q_{(r,s)(r-1,s+1)} &= r\mu\alpha, & r &\geq 1, \ s \geq 0, \\
q_{(r,s)(r,s-1)} &= v, & r &\geq 0, \ s \geq 1.
\end{aligned}$$

The invariance equations $\pi Q = 0$, for $\pi = (\pi_{(r,s)})$, are specified as

$$\begin{aligned}
\pi_{(r,s)} \left[\lambda + r\mu + v\mathbf{1}(s \geq 1) \right] &= \pi_{(r-1,s)} \lambda \mathbf{1}(r \geq 1) \\
&\quad + \pi_{(r+1,s-1)} (r+1)\mu\alpha\mathbf{1}(s \geq 1) \\
&\quad + \pi_{(r+1,s)} (r+1)\mu(1 - \alpha) \\
&\quad + \pi_{(r,s+1)} v.
\end{aligned}$$

Substituting the suggested form yields

$$e^{-\beta} \frac{\beta^r}{r!} (1 - \gamma) \gamma^s \left(\lambda + r\mu + v\mathbf{1}(s \geq 1) \right)$$

$$= e^{-\beta} \frac{\beta^r}{r!} (1 - \gamma) \gamma^s$$

$$\times \left(\frac{r}{\beta} \lambda + \frac{\beta}{r+1} \frac{1}{\gamma} (r+1)\mu\alpha\mathbf{1}(s \geq 1) + \frac{\beta}{r+1} (r+1)\mu(1 - \alpha) + \gamma v \right).$$

We see that equality holds with $\beta = \lambda/\mu$ and $\gamma = \lambda\alpha/\nu$. To obtain an equilibrium distribution, it must be assumed that $\gamma < 1$.   □

Question 2.10.9 (Markov chains, Part IB, 1996, 503G)
Suppose that $(X(t), t \geq 0)$ is a continuous time Markov chain taking values in $\{0, 1, 2, \dots\}$. Define the jump chain. Describe a method of using the jump chain to construct a sample path of $X(t)$.

A population of individuals is subject to immigration and threat of total extermination. The number of individuals in the population at time t is a continuous-time Markov chain. For n individuals in the population at time t then, in the short time interval $(t, t+h)$:

(a) the probability that they are joined by a new member is $h/(n+2) + o(h)$;
(b) the probability that they are exterminated is $h/(n+2)(n+1) + o(h)$;
(c) the probability that more than one incident of either kind happens is $o(h)$.

Is state 0 recurrent? Are the other states recurrent? Justify your answers.

Solution The jump chain is made by the discrete time Markov chain observed at the times of jumps of $(X(t))$. If $(X(t))$ operates under the Q-matrix (q_{ij}) then the jump chain exhibits transition probabilities $\widehat{p}_{ij} = -q_{ij}/q_{ii}$, $j \neq i$. A sample path of $(X(t))$ is then constructed by iterating the following rule: the chain spends a random time $L_i \sim \text{Exp}(-q_{ii})$ in state i, independently of the history, then jumps to state $j \neq i$ with probability $-q_{ij}/q_{ii}$.

In the example given, the jump chain transition probabilities are given by

$$\begin{cases} \widehat{p}_{n,n+1} = \dfrac{n+1}{n+2}, \\ \widehat{p}_{n,0} = \dfrac{1}{n+2}. \end{cases}$$

A state $i \geq 0$ achieves recurrence in the continuous-time chain if and only if it does in the jump chain. In the jump chain this means that the return probability to state i is 1, i.e. $\mathbb{P}_i(T_i < \infty) = 1$. For $i = 0$

$$\mathbb{P}_0(n \text{ subsequent steps up}) = \frac{2}{3}\frac{3}{4}\cdots\frac{n+1}{n+2} = \frac{2}{n+2} \to 0 \text{ as } k \to \infty.$$

Hence,

$$\mathbb{P}_0(T_i < n) = 1 - \frac{2}{n+2} \to 1, \text{ as } n \to \infty,$$

and the return probability to 0 equals 1. So, 0 gets to be a recurrent state in the jump chain and hence also in the continuous-time one. Recurrence being a class property, when the chain is irreducible, every state is recurrent.   □

> *This method is, to define as the number of a class*
> *the class of all classes similar to the given class.*
> B. Russell (1872–1970), English mathematician and philosopher

Question 2.10.10 (Markov chains, Part IIA, 1996, A101E)
(i) Let S and T be independent exponential random variables of parameters α and β respectively. Set $M = \min\{S, T\}$. Determine the distribution of M and show that M is independent of the event $\{S < T\}$.

(ii) Customers enter a supermarket as a Poisson process of rate 2. There are two salesmen near the door who offer passing customers samples of a new product. Each customer takes an exponential time of parameter 1 to think about the new product, and during this time occupies the full attention of one salesman. Having tried the product, customers proceed into the store and leave by another door. When both salesmen are occupied, customers walk straight in. Assuming that both salesmen are free at time 0, find the probability that both end up busy at a later time t.

Solution (i) We see that

$$\mathbb{P}(M > t) = \mathbb{P}(S > t)\mathbb{P}(T > t) = e^{-(\alpha+\beta)t}, \ t \geq 0.$$

Hence, $M \sim \text{Exp}(\alpha + \beta)$. Next,

$$
\begin{aligned}
\mathbb{P}(S < T, M > t) &= \mathbb{P}(t < S < T) \\
&= \int_t^\infty \alpha e^{-\alpha x} e^{-\beta x} dx = \frac{\alpha}{\alpha + \beta} e^{-(\alpha+\beta)t}.
\end{aligned}
$$

On the other hand, at $t = 0$,

$$\mathbb{P}(S < T) = \frac{\alpha}{\alpha + \beta},$$

and so

$$\mathbb{P}(S < T, M > t) = \mathbb{P}(S < T)\mathbb{P}(M > t),$$

which implies independence.

(ii) There are three states: 0 (both salesmen are free), 1 (one busy, one free) and 2 (both busy). The non-zero jump rates are given as

$$q_{01} = \lambda, \ q_{10} = \mu, \ q_{12} = \lambda, \ q_{21} = 2\mu,$$

with $\lambda = 2$, $\mu = 1$. This leads to the generator matrix

$$Q = \begin{pmatrix} -2 & 2 & 0 \\ 1 & -3 & 2 \\ 0 & 2 & -2 \end{pmatrix},$$

with the eigenvalues $0, -2, -5$. It is easy to compute that $q_{00}^{(2)} = 6$. Then

$$p_{00}(t) = A + Be^{-2t} + Ce^{-5t}, \quad t \geq 0,$$

where

$$\begin{aligned} A + B + C &= 1, \\ -2B - 5C &= -2, \\ 4B + 25C &= 6. \end{aligned}$$

Hence, $A = 1/5$, $B = 2/3$, $C = 2/15$, and

$$\mathbb{P}_0(\text{both free at time } t) = \frac{1}{5} + \frac{2}{3}e^{-2t} + \frac{2}{15}e^{-5t},$$

$$\mathbb{P}_0(\text{both busy at time } t) = \frac{2}{5} - \frac{2}{3}e^{-2t} + \frac{4}{15}e^{-5t}.$$

$\square$

Question 2.10.11 (Markov chains, Part IIA, 1996, A201E)
(i) Let $(X_t)_{t\geq 0}$ be an irreducible non-explosive continuous-time Markov chain with
Q-matrix $Q = (q_{ij} : i, j \in I)$ and suppose that $(X_t)_{t\geq 0}$ has an invariant measure
$v = (v_i : i \in I)$. Denote the associated jump chain by $(Y_n)_{n\geq 0}$. Fix $h > 0$ and set
$Z_n = X_{nh}$. Explain how the transition matrices for the discrete time Markov chains
$(Y_n)_{n\geq 0}$ and $(Z_n)_{n\geq 0}$ are related to Q, and how their invariant measures are related
to v.

(ii): (a) In the case where $(X_t)_{t\geq 0}$ is recurrent, show that $(Z_n)_{n\geq 0}$ is also recurrent.

(b) In the case where the state space is $\mathbb{Z}_+$, and where $q_{ii} = -\lambda$, $q_{ii+1} = \lambda$, find
the transition probabilities for $(Z_n)_{n\geq 0}$.

(c) In the case where the state space is $\mathbb{Z}$, and where

$$q_{ii-1} = i^2 + 1, \quad q_{ii} = -2(i^2 + 1), \quad q_{ii+1} = i^2 + 1,$$

determine whether $(X_t)_{t\geq 0}$ is positive recurrent.

Solution (i) If v is an invariant measure for a non-explosive CTMC (X_t), then
$vQ = 0$. (Compare with Theorem 2.6.11.) Next,

- the transition matrix for (Y_n) is expressed as $\widehat{P} = (\widehat{p}_{ij})$, with $\widehat{p}_{ij} = -q_{ij}/q_{ii}$ for $j \neq i$ and $\widehat{p}_{ii} = 0$, and
- the transition matrix for (Z_n) gets to be $P(h) = e^{hQ}$. Then the IM for (Y_n) becomes $\mu = (\mu_i)$, with

$$\mu_i = -v_i q_{ii},$$

and that for (Z_n) is simply v.

In fact,

- $-q_{ii}(\widehat{p}_{ij} - \delta_{ij}) = q_{ij}$ for all states i, j, and hence

$$
\begin{aligned}
(\mu(\widehat{P} - I))_j &= \sum_i \mu_i (\widehat{p}_{ij} - \delta_{ij}) \\
&= \sum_i v_i q_i (\widehat{p}_{ij} - \delta_{ij}) \\
&= \sum_i v_i q_{ij} = (vQ)_j = 0,
\end{aligned}
$$

-

$$(vP(h))_j = \sum_{k \geq 0} \frac{h^k}{k!} (vQ^k)_j = v_j.$$

(ii)(a) As i is recurrent for (X_t), $\int_0^\infty dt\, p_{ii}(t) = \infty$. Next, if $nh \leq t < (n+1)h$ then, by the Markov property,

$$p_{ii}((n+1)h) \geq e^{-q_ih} p_{ii}(t) \text{ and so } \sum_{n \geq 1} p_{ii}(nh) \geq h^{-1} e^{-q_ih} \int_0^\infty p_{ii}(t)\, dt.$$

Hence, $\sum_{n \geq 1} p_{ii}(nh) = \infty$, with i recurrent for (Z_n).

(b) Here, Q forms the generator matrix of a Poisson process (N_t) of rate λ. Hence, the transition matrix $P(h) = e^{hQ}$ for Z_n turns into the upper-triangular matrix for increments of (N_t) in time h

$$P(h) = \begin{pmatrix} e^{-h\lambda} & (h\lambda)e^{-h\lambda} & (h\lambda)^2 e^{-h\lambda}/2! & (h\lambda)^3 e^{-h\lambda}/3! & \cdots \\ 0 & e^{-h\lambda} & (h\lambda)e^{-h\lambda} & (h\lambda)^2 e^{-h\lambda}/2! & \cdots \\ 0 & 0 & e^{-h\lambda} & (h\lambda)e^{-h\lambda} & \cdots \\ \ddots & \ddots & \ddots & \ddots & \cdots \end{pmatrix}.$$

(c) Here, (Y_n) constitutes the symmetric nearest-neighbour random walk on $\mathbb{Z}$ that is null-recurrent. All invariant measures for (Y_n) are proportional to $\mu = (\mu_i)$ with $\mu_i \equiv 1$. Then any invariant measure for (X_t) will be proportional to $v = (v_i)$ where $v_i = -\mu_i/q_{ii} = 1/2(i^2 + 1)$. As $\sum_{i \in \mathbb{Z}} v_i$ remains finite, the chain (X_t) is positive recurrent. $\qquad \square$

Remark 2.10.12 It is known that

$$\sum_{i=0}^{\infty} \frac{1}{i^2 + 1} = \frac{1}{2} + \frac{\pi}{2} \coth(\pi).$$

Hence, $\sum_{i \in \mathbb{Z}} v_i = \pi \coth \pi.$

Question 2.10.13 (Markov chains, Part IB, 1997, 502G)
Customers join a queue according to a Poisson process with rate λ. The service time of each customer is an exponential random variable with mean μ^{-1}, where $\lambda < \mu$; the service times are mutually independent, and independent of the arrival times.

(i) Quoting carefully any general theorems to which you appeal, show that the probability of there being n or more customers in the queue at time t converges to $(\lambda/\mu)^n$, as $t \to \infty$.

(ii) If the initial distribution is invariant, calculate the expected time until the queue becomes first empty.

Solution The number of customers in the above queue represents an irreducible birth-and-death process (X_t) with $q_{i\,i+1} = \lambda$, $q_{i\,i-1} = \mu$. From the detailed balance equations, an equilibrium distribution is geometric $\pi_i = (1 - \lambda/\mu)(\lambda/\mu)^i$ (since $\lambda < \mu$). Thus the chain is positive recurrent, with π the only equilibrium distribution. Theorem 2.8.1 states:

For an irreducible positive recurrent continuous-time Markov chain, $p_{ij}(t) \to \pi_j$ as $t \to \infty$.

(i) Hence, regardless of the initial distribution, for all $n = 1, 2, \ldots$,

$$\mathbb{P}(X_t \leq n-1) \to \left(1 - \frac{\lambda}{\mu}\right) \sum_{0 \leq i \leq n-1} \left(\frac{\lambda}{\mu}\right)^i = 1 - \left(\frac{\lambda}{\mu}\right)^n,$$

and

$$\mathbb{P}(X_t \geq n) \to \left(\frac{\lambda}{\mu}\right)^n.$$

(ii) Setting $k_i = \mathbb{E}_i(\text{hit } 0)$ and conditioning on the first jump, we find the following equations:

$$k_0 = 0, \quad k_i = \frac{1}{\lambda + \mu} + \frac{\lambda}{\lambda + \mu} k_{i+1} + \frac{\mu}{\lambda + \mu} k_{i-1}, \quad i \geq 1,$$

and we are interested in the minimal non-negative solution. It is given by $k_i = i/(\mu - \lambda)$, and hence

$$\sum_{i \geq 1} \pi_i k_i = \left(1 - \frac{\lambda}{\mu}\right) \sum_{i \geq 1} \left(\frac{\lambda}{\mu}\right)^i \frac{i}{\mu - \lambda} = \frac{\lambda}{(\mu - \lambda)^2}.$$

$\square$

Question 2.10.14 (Markov chains, Part IIA, 1997, A101J)
(i) Consider a birth-and-death process $X(t)$ with birth rates (λ_k) and death rates (μ_k) where $\mu_0 = 0$. Assume that the process does not explode. Write down the forward equations for the probabilities $p_k(t) = \mathbb{P}(X(t) = k)$, and deduce that the probability-generating function $g(s,t) = \mathbb{E}s^{X(t)}$ satisfies

$$\frac{\partial g}{\partial t} = (s - 1)\left(\Lambda(s) - \frac{1}{s} M(s)\right), \quad -1 < s \leq 1,$$

where $\Lambda(s) = \sum_k \lambda_k p_k(t)s^k$ and $M(s) = \sum_k \mu_k p_k(t)s^k$.

(ii) A community holds insufficient food resources to support more than N individuals. For k members at time t, a new member joins with probability $\lambda(N - k)h + o(h)$ during the time interval $(t, t + h)$, independently of the past history. Each member departs during this interval with probability $\mu h + o(h)$, independently of all other members and of the past history.

Write down the generator of the associated Markov chain $X(t)$, and show that

$$\frac{\partial g}{\partial t} = (s - 1)\left(\lambda N g - (\lambda s + \mu)\frac{\partial g}{\partial s}\right), \quad -1 < s \leq 1.$$

Assume that $X(0) = 0$. Find a function $h(t)$ such that $g(s,t) = (1 - h(t) + sh(t))^N$ solves the above equation, and hence find the distribution of $X(t)$.

Solution (i) The forward equations for the transition probabilities $p_k(t)$ are

$$\frac{d}{dt} p_k = -(\lambda_k + \mu_k)p_k + \lambda_{k-1}p_{k-1} + \mu_{k+1}p_{k+1}, \quad k \geq 0,$$

with $\lambda_{-1} = 0$.
For $g(s,t) = \mathbb{E}s^{X(t)} = \sum_k s^k p_k(t)$:

$$\sum_k s^k \frac{d}{dt} p_k = \sum_k \left(-(\lambda_k + \mu_k)p_k + \lambda_{k-1}p_{k-1} + \mu_{k+1}p_{k+1}\right)s^k,$$

whence

$$\frac{\partial g}{\partial t} = -\Lambda - M + s\Lambda + \frac{1}{s} M = (s - 1)\left(\Lambda - \frac{1}{s} M\right), \quad -1 < s \leq 1,$$

as required. If the process does not explode, the equations hold for all $t \geq 0$.

(ii) Here, the rates are

$$\lambda_k = \lambda(N-k), \quad \mu_k = \mu k, \quad k = 0, 1, \ldots, N.$$

Consequently,

$$\Lambda(s) = \sum_{k=0}^{N} \lambda(N-k)p_k s^k = \lambda N g - \lambda s \frac{\partial g}{\partial s},$$

and

$$M(s) = \mu \sum_{k=0}^{N} k p_k s^k = \mu s \frac{\partial g}{\partial s}.$$

Then from part (i)

$$\frac{\partial g}{\partial t} = (s-1)\left(\lambda N g - (\mu + \lambda s)\frac{\partial g}{\partial s}\right).$$

Substituting $g(s,t) = (1 - h(t) + sh(t))^N$, we obtain

$$\dot{h} = \lambda(1 - h + sh) - h(\mu + \lambda s),$$

or, subsequently,

$$\dot{h} + (\lambda + \mu)h - \lambda = 0, \quad h = \frac{\lambda}{\lambda + \mu} + ce^{-(\lambda+\mu)t}.$$

With $h(0) = 0$, we have

$$h = \frac{\lambda}{\lambda + \mu}\left(1 - e^{-(\lambda+\mu)t}\right).$$

We see that $X(t) \sim \text{Bin}(N, h(t))$, as

$$\mathbb{E}s^{X(t)} = \sum_{0 \leq r \leq N} \binom{N}{r} s^r h^r (1-h)^{N-r} = ((s-1)h(t) + 1)^N,$$

as suggested. □

Question 2.10.15 (Markov chains, Part IIA, 1998, A301E)
(i) The Russian ice-hockey team plays the Canadian team in the Olympic finals. Suppose that Russian goals occur according to a Poisson process with rate $r > 0$ and Canadian goals to a Poisson process with rate $c > 0$, independently of Russian goals. Let X be the number of goals scored by the Russians before Canada scores. Suppose that play continues forever. Find $\mathbb{P}(X = k)$, $k = 0, 1, \ldots$.
[*Hint.* Use the following formula, which we give without proof: $\int_0^\infty s^k e^{-s} ds = k!$]

(ii) (a) In the model of (i), now suppose that play continues only until time $t = 1$. What is $\mathbb{P}(X = 1)$?

(b) Suppose that by time $t = 1$ Russia has scored once. What is the expected value of the time at which the first goal of the game occurred?

Solution (i) If T gives the time of the first Canadian goal then $T \sim \text{Exp}(c)$, independently of $\{R(t),\ t \geq 0\}$, the Poisson process $\text{PP}(r)$ of Russian goals. Then, for $k = 0, 1, \ldots,$

$$
\begin{aligned}
\mathbb{P}(X = k) &= \mathbb{P}(R(T) = k) \\
&= c \int_0^\infty e^{-ct} \mathbb{P}(R(T) = k \mid T = t)\, dt = c \int_0^\infty e^{-ct} \mathbb{P}(R(t) = k)\, dt \\
&= \frac{cr^k}{k!} \int_0^\infty e^{-ct} t^k e^{-rt}\, dt = \frac{cr^k}{k!(r+c)^{k+1}} \int_0^\infty e^{-t} t^k\, dt \\
&= \frac{c}{(r+c)} \frac{r^k}{(r+c)^k}.
\end{aligned}
$$

(ii)(a) Write $P^{(1)}$ for the probability in question. Then

$$
\begin{aligned}
P^{(1)} &= \mathbb{P}(R(1) = 1,\ T > 1) + \mathbb{P}(R(1) = 1,\ T \leq 1) \\
&= \mathbb{P}(R(1) = 1)\mathbb{P}(T > 1) + c \int_0^1 e^{-ct} \mathbb{P}(R(t) = 1 \mid T = t)\, dt \\
&= re^{-r} e^{-c} + rc \int_0^1 t e^{-(r+c)t}\, dt \\
&= re^{-(r+c)} + \frac{rc}{(r+c)^2} \int_0^{r+c} t_1 e^{-t_1}\, dt_1 \quad (\text{where } t_1 = (r+c)t) \\
&= re^{-(r+c)} + \frac{rc}{(r+c)^2} \left[1 - (r+c+1)e^{-(r+c)} \right].
\end{aligned}
$$

(b) Let S stand for the time of the first Russian goal and, as before, T for the time of the first Canadian goal. Then, conditional on $R(1) = 1$, $S \sim U(0, 1)$. This implies

$$
\mathbb{P}(\min(S, T) \geq t \mid R(1) = 1) = (1 - t)\, e^{-ct},\ 0 < t < 1.
$$

Hence, the conditional density $f_{\min(S,T)\mid R(1)=1}$ of $\min(S, T)$ equals

$$
e^{-ct} + c(1 - t)\, e^{-ct} = e^{-ct}(1 + c(1 - t)),\ 0 < t < 1,
$$

and the conditional mean

$$\mathbb{E}\left[\min(S,T)|R(1)=1\right] = \int_0^1 te^{-ct}(1+c(1-t))\,dt = \frac{1}{c^2}\left(e^{-c}-1+c\right).$$

$\square$

The KGB's π's and the FBI's κ's

(From the series *'Movies that never made it to the Big Screen'*.)

Question 2.10.16 (Markov chains, Part IIA, 1999, A301E and Part IIB, 1999, B301E)

(i) Consider a *Poisson process* N with rate λ. Conditional on the event $\{N(t)=1\}$, show that the unique arrival time T of the process in the interval $(0,t]$ is uniformly distributed on this interval.

(ii) The rubbish bins of a certain Canadian campsite are renowned amongst bears for their tasty morsels. Bears arrive in the campsite at the times of a Poisson process with rate λ. After arrival, the mth bear spends a time R_m roaming for a bin, followed by a time S_m raiding it. The vectors (R_m, S_m), $m \geq 1$, are independent random vectors with the same (joint) distribution. Let $U(t)$ and $V(t)$ be the numbers of roaming and raiding bears, respectively, at time t, and assume that $U(0) = V(0) = 0$.

Let α (respectively β) be the probability that a bear arriving at some time T, chosen uniformly at random from the interval $(0,t)$, is roaming (respectively, raiding) at time t. Compute $\mathbb{P}(U(t)=u, V(t)=v)$ in terms of α and β, and hence show that $U(t)$ and $V(t)$ form independent random variables, each with a Poisson distribution.

Show that $\mathbb{E}(U(t)) \to \lambda \mathbb{E} R_1$ as $t \to \infty$.

[*Hint.* Conditional on $\{N(t)=m\}$, the first m arrival times have the same joint distribution as that of the order statistics of m independent random variables which are uniformly distributed on the interval $(0,t)$.]

Solution (i) The condition $N(t)=1$ means there was a single arrival in the interval $(0,t]$. Let the arrival time be T. Then, for all $0 \leq a \leq t$:

$$\begin{aligned}
\mathbb{P}(T \leq a | N(t) = 1) &= \mathbb{P}(N(a) = 1 | N(t) = 1) \\
&= \frac{\mathbb{P}(N(a) = 1, N(t) - N(a) = 0)}{\mathbb{P}(N(t) = 1)} \\
&= \frac{\lambda a e^{-\lambda a} e^{-\lambda(t-a)}}{\lambda t e^{-\lambda t}} = \frac{a}{t}.
\end{aligned}$$

So, $T \sim U(0,t)$.

(ii) The arrival times of bears in $(0, t]$, conditional on the event $\{N(t) = n\}$, make up independent, uniformly distributed points in $(0, t]$, the corresponding vectors (R_m, S_m), $1 \leq m \leq n$, being independent and identically distributed. Hence, at time t each bear is roaming with probability α, raiding with probability β or has left with probability $1 - \alpha - \beta$, independently. By definition,

$$\begin{cases} \alpha & = \mathbb{P}(R_1 \geq t - T), \\ \beta & = \mathbb{P}(R_1 < t - T \leq R_1 + S_1), \\ 1 - \alpha - \beta & = \mathbb{P}(t - T > R_1 + S_1), \end{cases}$$

where $T \sim \mathrm{U}(0, t)$, independently of (R_1, S_1). We see that, for all non-negative integers u, v with $u + v \leq n$, the conditional probability

$$\mathbb{P}(U(t) = u, \ V(t) = v | N(t) = n) = \frac{n! \alpha^u \beta^v (1 - \alpha - \beta)^{n-u-v}}{u! v! (n - u - v)!}.$$

Then the unconditional probability $\mathbb{P}(U(t) = u, V(t) = v)$ equals

$$\sum_{n \geq u+v} \mathbb{P}(U(t) = u, \ V(t) = v | N(t) = n) \mathbb{P}(N(t) = n)$$

$$= \sum_{n \geq u+v} \left(\frac{n! \alpha^u \beta^v (1 - \alpha - \beta)^{n-u-v}}{u! v! (n - u - v)!} \right) \left(\frac{(\lambda t)^n e^{-\lambda t}}{n!} \right)$$

$$= \left(\frac{(\alpha \lambda t)^u e^{-\alpha \lambda t}}{u!} \right) \left(\frac{(\beta \lambda t)^v e^{-\beta \lambda t}}{v!} \right)$$

$$\times \sum_{n \geq u+v} \frac{\left((1 - \alpha - \beta) \lambda t \right)^{n-u-v}}{(n - u - v)!} \, e^{-(1-\alpha-\beta)\lambda t}$$

$$= \left(\frac{(\alpha \lambda t)^u e^{-\alpha \lambda t}}{u!} \right) \left(\frac{(\beta \lambda t)^v e^{-\beta \lambda t}}{v!} \right).$$

Thus, $U(t)$ and $V(t)$ represent independent Poisson random variables, with means $\alpha \lambda t$ and $\beta \lambda t$, respectively.

Now,

$$\alpha = \int_0^t \mathbb{P}(R_1 \geq t - r) \, \mathrm{d}r \, \frac{1}{t} = \frac{1}{t} \int_0^t \mathbb{P}(R_1 \geq r) \, \mathrm{d}r.$$

So,

$$\lim_{t \to \infty} \alpha t = \int_0^\infty \mathbb{P}(R_1 \geq r) \, \mathrm{d}r = \mathbb{E} R_1,$$

and

$$\lim_{t \to \infty} \mathbb{E}U(t) = \lambda \mathbb{E}R_1.$$

$\square$

Question 2.10.17 (Part II, 213B, 1991, Stochastic Processes, modified)
Suppose that λ_i $(i \geq 0)$ and μ_i $(i \geq 0)$ are positive constants. Let

$$Q = (q_{ij} : i, j \geq 0)$$

be the generator matrix

$$Q = \begin{pmatrix} -\lambda_0 & \lambda_0 & 0 & 0 & 0 & \cdot \\ \mu_0 & -(M_0 + \lambda_1) & \lambda_1 & 0 & 0 & \cdot \\ \mu_0 & \mu_1 & -(M_1 + \lambda_2) & \lambda_2 & 0 & \cdot \\ \mu_0 & \mu_1 & \mu_2 & -(M_2 + \lambda_3) & \lambda_3 & \cdot \\ \cdot & \cdot & \cdot & \cdot & & \cdot \\ \cdot & \cdot & \cdot & & & \cdot \end{pmatrix}$$

where, for $i \geq 0$, $M_i = \mu_0 + \mu_1 + \cdots + \mu_i$. Prove that the minimal Markov chain associated with Q is regular (non-explosive) if and only if

$$\sum_{j \geq 1} \frac{1}{\lambda_j} \prod_{k=0}^{j-1} \left(1 + \frac{M_k}{\lambda_k} \right) = \infty.$$

You may use, without proof, the following fact: a continuous-time Markov chain (X_t) (or its generator matrix Q) is non-explosive if and only if the system

$$Qz = \theta z, \quad z = (z_i), \quad z_i \geq 0, \quad i = 0, 1, \ldots,$$

has no bounded non-trivial solution for some $\theta > 0$ (and hence for all $\theta > 0$).

[*Hint.* For $x > 0$ and $y > 0$, $1 + x + y \leq (1+x)e^y$.]

Solution A standard result says that $X(t)$ is regular if and only if the system

$$Qz = \theta z$$

exhibits no bounded non-trivial solution. So, solving

$$i = 0: \quad -\lambda_0 z_0 + \lambda_0 z_1 = \theta z_0,$$
$$i \geq 1: \quad \mu_0 z_0 + \mu_1 z_1 + \cdots + \mu_{i-1} z_{i-1} - (M_{i-1} + \lambda_i) z_i + \lambda_i z_{i+1} = \theta z_i,$$

i.e.

$$\sum_{i=0}^{i-1} \mu_i z_i - (M_{i-1} + \theta + \lambda_i) z_i + \lambda_i z_{i+1} = 0. \tag{2.178}$$

Suppose $i \geq 1$. Write

$$\sum_{j=0}^{i-1} \mu_j z_j - (M_{i-1} + \theta + \lambda_i) z_i + \lambda_i z_{i+1} = 0,$$

and

$$\sum_{j=0}^{i-2} \mu_j z_j - (M_{i-2} + \theta + \lambda_{i-1}) z_{i-1} + \lambda_{i-1} z_i = 0.$$

Subtracting yields

$$(\mu_{i-1} + M_{i-2} + \theta + \lambda_{i-1}) z_{i-1} - (M_{i-1} + \theta + \lambda_i) z_i + \lambda_i z_{i+1} - \lambda_{i-1} z_i = 0,$$

and after re-arranging

$$(M_{i-1} + \theta + \lambda_{i-1})(z_{i-1} - z_i) + \lambda_i(z_{i+1} - z_i) = 0.$$

Then

$$
\begin{aligned}
z_{i+1} - z_i &= \frac{M_{i-1} + \theta + \lambda_{i-1}}{\lambda_i}(z_i - z_{i-1}) \\
&= \cdots \\
&= \prod_{k=1}^{i} \frac{M_{k-1} + \theta + \lambda_{k-1}}{\lambda_k}(z_1 - z_0).
\end{aligned}
$$

Using (2.178), $z_1 - z_0 = \dfrac{\theta z_0}{\lambda_0}$, and hence

$$z_{i+1} - z_i = \frac{\theta z_0}{\lambda_0} \prod_{k=1}^{i} \frac{M_{k-1} + \theta + \lambda_{k-1}}{\lambda_k}.$$

Then

$$z_{i+1} = z_0 \left(1 + \frac{\theta}{\lambda_0} \sum_{j=1}^{i} \prod_{k=1}^{j} \frac{M_{k-1} + \theta + \lambda_{k-1}}{\lambda_k}\right).$$

The solution is bounded and non-trivial if and only if

$$\frac{1}{\lambda_0} \sum_{j \geq 1} \prod_{k=1}^{j} \frac{M_{k-1} + \lambda_{k-1} + \theta}{\lambda_k} < \infty,$$

that is,

$$\sum_{j \geq 1} \frac{1}{\lambda_j} \prod_{k=0}^{j-1} \frac{M_k + \lambda_k + \theta}{\lambda_k} < \infty.$$

If this holds for some $\theta > 0$ then, clearly,

$$\sum_j \frac{1}{\lambda_j} \prod_{k=0}^{j-1} \left(1 + \frac{M_k}{\lambda_k}\right) < \infty.$$

Conversely, if the last inequality holds then $\sum_j \frac{1}{\lambda_j} < \infty$.

Finally, as $1 + x + y < (1+x)e^y$, this implies that

$$\sum_j \frac{1}{\lambda_j} \prod_{k=0}^{j-1} \left(1 + \frac{M_k + \theta}{\lambda_k}\right) \leq \sum_j \frac{1}{\lambda_j} \prod_{k=0}^{j-1} \left(1 + \frac{M_k}{\lambda_k}\right) e^{\theta \sum_i 1/\lambda_i} < \infty.$$

$\square$

The fact that was quoted in the above problem is as follows.

Theorem 2.10.18 *Let* (X_t) *be a CTMC with a generator matrix* Q *and write* $T = T_{\text{explo}}$ *for the explosion time of* (X_t). *Fix* $\theta > 0$ *and set* $z_i = \mathbb{E}_i \left(e^{-\theta T}\right)$. *Then the column vector* $\mathbf{z}$ *with entries* z_i, $i \in I$, *satisfies:*

  (i) $|z_i| < 1$ *for all* i,
  (ii) $Q\mathbf{z} = \theta \mathbf{z}$.

Moreover, $\mathbf{z}$ *gives a maximal solution; that is, if* $\widetilde{\mathbf{z}} = (\widetilde{z}_i, i \in I)$ *is any solution to* (i) *and* (ii), *then*

$$\widetilde{z}_i \leq z_i, \text{ for all } i \in I.$$

This theorem implies that for each $\theta > 0$ the following conditions are equivalent:

  (a) Q is non-explosive;
  (b) $Q\mathbf{z} = \theta \mathbf{z}$ and $|z_i| < 1$ for all i implies $z = 0$.

Proof By the Markov property of the jump chain at time $n = 1$, conditional on $X_{J_1} = k$

$$\mathbb{E}_i \left(e^{-\theta T_{\text{explo}}} \big| X_{J_1} = k\right) = \int_0^\infty e^{-\theta u} q_i e^{-q_i u} \, du \, \mathbb{E}_k \left(e^{-\theta T_{\text{explo}}}\right) = \frac{q_i z_k}{q_i + \theta},$$

and

$$z_i = \sum_{k \neq i} \frac{q_i \widehat{p}_{ik} z_k}{q_i + \theta}. \tag{2.179}$$

Recall that $q_i = -q_{ii}$ and $q_i \widehat{p}_{ik} = q_{ik}$. Then (2.179) is equivalent to

$$\theta z_i = \sum_i q_{ik} z_k.$$

Now suppose that $\widetilde{\mathbf{z}}$ also satisfies (i) and (ii). Then the induction argument implies

$$\widetilde{z}_i \leq \mathbb{E}_i\left(e^{-\theta J_n}\right). \tag{2.180}$$

Indeed, (i) implies (2.180) for $n = 0$ and using (2.180) for n one gets it for $n+1$:

$$\widetilde{z}_i = \sum_{k \neq i} \frac{q_i \widehat{p}_{ik} \widetilde{z}_k}{q_i + \theta} \leq \sum_{k \neq i} \frac{q_i \widehat{p}_{ik}}{q_i + \theta} \mathbb{E}_k\left(e^{-\theta J_n}\right) = \mathbb{E}_i\left(e^{-\theta J_{n+1}}\right).$$

By the monotone convergence theorem,

$$\lim_{n \to \infty} \mathbb{E}_i\left(e^{-\theta J_{n+1}}\right) = \mathbb{E}_i\left(e^{-\theta T_{\text{explo}}}\right).$$

So $\widetilde{z}_i \leq z_i$. $\qquad\qquad\qquad\qquad\qquad\qquad\qquad\qquad\qquad\qquad\qquad\square$

Question 2.10.19 (Markov chains, Part IIA, 2000, A201E, part (ii))
The office of the Chair of the Department contains computer-controlled equipment which behaves erratically. If the window blind is down at time n, the computer raises it at time $n+1$ with probability β_1; if the blind is up at time n, it is lowered with probability β_2. If the light is off at time n, the computer switches it on at time $n+1$ with probability λ_1; if it is on, it switches off with probability λ_2. Changes in the states of blind and light are independent of one another and of earlier states.

The Chairman enters the room at time 0 and finds the blind down and light off. What is the probability that the blind and light occupy the same states on his departure at time n?

Determine the long run average amount of time for which both the blind is down and the light is off.

Solution For the blind, the states are written as D (down) and U (up), and the transition matrix

$$\begin{pmatrix} 1 - \beta_1 & \beta_1 \\ \beta_2 & 1 - \beta_2 \end{pmatrix}.$$

Hence, the n-step transition probability

$$p_{\text{DD}}^{(n)}(n) = \frac{\beta_2}{\beta_1 + \beta_2} + \frac{\beta_1}{\beta_1 + \beta_2}(1 - \beta_1 - \beta_2)^n;$$

similar formulas hold for $p_{\text{DU}}^{(n)}$, $p_{\text{UD}}^{(n)}$ and $p_{\text{UU}}^{(n)}$. The equilibrium distribution is given by $\pi^{\text{bl}} = (\pi_{\text{D}}^{\text{bl}}, \pi_{\text{U}}^{\text{bl}})$ is given by

$$\pi_{\text{D}}^{\text{bl}} = \frac{\beta_2}{\beta_1 + \beta_2}, \quad \pi_{\text{U}}^{\text{bl}} = \frac{\beta_1}{\beta_1 + \beta_2}.$$

Similar formulas hold for the ED $\pi^{li} = (\pi^{li}_{on}, \pi^{li}_{off})$ for the light (with λ instead of β). Finally, because of independence, we find that the joint transition probabilities are the products, viz.:

$$r^{(n)}_{(D,Off)(D,Off)} - \left[\frac{\beta_2}{\beta_1 + \beta_2} + \frac{\beta_1}{\beta_1 + \beta_2} (1 - \beta_1 - \beta_2)^n \right]$$
$$\times \left[\frac{\lambda_2}{\lambda_1 + \lambda_2} + \frac{\lambda_1}{\lambda_1 + \lambda_2} (1 - \lambda_1 - \lambda_2)^n \right],$$

as well as joint equilibrium probabilities, viz.:

$$\pi^{bl,li}_{U,Off} = \frac{\beta_1}{\beta_1 + \beta_2} \frac{\lambda_1}{\lambda_1 + \lambda_2}.$$

Similarly, for the long run average amount, the answer becomes

$$\frac{\beta_2}{(\beta_1 + \beta_2)} \frac{\lambda_2}{(\lambda_1 + \lambda_2)}.$$

$\square$

Question 2.10.20 (Markov chains, Part IIA, 2000, A301E and Part IIB, 2000, B301E)

The Quality Assurance Agency for Higher Education has sent a team to investigate the teaching of mathematics at University of Camford. As the visit progresses, the team keeps a count of the number of complaints which it has received. We assume that, during the time interval $(t, t + h)$, a new complaint is made with probability $\lambda h + o(h)$, while any given existing complaint is found groundless and removed from the list with probability $\mu h + o(h)$. Under reasonable conditions to be stated, show that the number $C(t)$ of active complains at time t constitutes a birth-and-death process with birth rates $\lambda_n = \lambda$ and death rates $\mu_n = n\mu$.

Derive, but do not solve, the forward system of equations for the probabilities $p_n(t) = P(C(t) = n)$. Show that $m(t) = \mathbb{E}C(t)$ satisfies the differential equation $m'(t) = \lambda - \mu m(t)$, and find $m(t)$ subject to the initial condition $m(0) = 1$.

Find the invariant distribution of the process.

Solution Assume that individual departures are independent of each other and that arrivals and departures independent from each other. Then the conditional probability

$$\mathbb{P}\big(C(t + h) - C(t) = -1 \mid C(t) = n \big) = \mu n h + o(h),$$

as the first departure corresponds to the minimum of n independent exponential random variables, i.e. occurs at an exponential time with rate μn. Then the forward equations $\frac{d}{dt} P(t) = P(t)Q$ become

$$\frac{d}{dt} p_0 = -\lambda p_0 + \mu p_1,$$

$$\frac{d}{dt} p_n = \lambda p_{n-1} - (\lambda + \mu n) p_n + \mu(n+1) p_{n+1}, \quad n \geq 1,$$

with the generator matrix

$$Q = \begin{pmatrix} -\lambda & \lambda & 0 & 0 & \cdots \\ \mu & -(\lambda + \mu) & \lambda & 0 & \cdots \\ 0 & 2\mu & -(\lambda + 2\mu) & \lambda & \cdots \\ \cdots & \cdots & & \cdots & \cdots \end{pmatrix},$$

and the equation for the invariant distribution π is $\pi Q = 0$.

We can also try the detailed balance equations

$$\pi_{i-1} \lambda = \pi_i i \mu, \quad i \geq 1,$$

giving

$$\pi_i = \left(\frac{\lambda}{\mu i} \right) \pi_{i-1} = \cdots = \left(\frac{\lambda}{\mu} \right)^i \frac{\pi_0}{i!}.$$

That is, $\pi_0 = e^{-\lambda/\mu}$ and thus $\pi \sim \mathrm{Po}\,(\lambda/\mu)$. For $m(t)$, the equation is

$$\frac{d}{dt} m(t) = \lambda - \mu m(t), \quad m(0) = 1,$$

whence

$$m(t) = \frac{\lambda}{\mu} + \left(1 - \frac{\lambda}{\mu} \right) e^{-\mu t}.$$

$\square$

> *Markov processes specialists like to do it in a transient state.*
> (From the series *How they do it.*)

Question 2.10.21 (Markov chains, Part IIA, 2001, A301D and Part IIB, 2001, B301D)

Let X be a continuous-time Markov chain on the state space $I = \{1, 2\}$ with generator matrix

$$Q = \begin{pmatrix} -\beta & \beta \\ \gamma & -\gamma \end{pmatrix}, \quad \text{where } \beta, \gamma > 0.$$

Show that the transition semigroup $P(t) = \exp(tQ)$ is given by

$$P(t) = (\beta + \gamma)^{-1} \begin{pmatrix} \gamma + \beta h(t) & \beta(1 - h(t)) \\ \gamma(1 - h(t)) & \beta + \gamma h(t) \end{pmatrix},$$

where $h(t) = e^{-t(\beta + \gamma)}$.

Solution The shortest way is to check that the matrix $P(t)$ satisfies the equations $P'(t) = P(t)Q = QP(t)$ and invoke the theorem of uniqueness of the solution. $\square$

Question 2.10.22 (Stochastic Processes, Part II, 213G, 1993)
Jobs arrive according to a Poisson process of rate $\lambda > 0$. They get processed individually, by a single processor, the processing times being independent random variables, each with the exponential distribution of parameter $v > 0$. After processing, a job either leaves the system with probability p, $0 < p < 1$, or, with probability $1 - p$, it is split into two separate jobs which are both sent to join the queue for processing again. Let $X(t)$ denote the number of jobs in the system at time t.
  In the case $1 + \lambda/v < 2p$, evaluate $\lim_{t \to \infty} \mathbb{P}(X(t) = j)$, $j = 0, 1, \ldots$, and find the expected time that the processor remains busy between two successive idle periods. What happens if $1 + \lambda/v \geq 2p$?

Solution In this situation,

$$\lambda_0 = \lambda \text{ and } \lambda_i = \lambda + qv, \ \mu_i = pv, \ i \geq 1,$$

where $q = 1 - p$, so $\gamma_i = \dfrac{\lambda}{pv} \alpha^{i-1}$, $i \geq 1$, where $\alpha = \dfrac{\lambda + qv}{pv}$.
  If $1 + \lambda/v < 2p$ then $\alpha < 1$, so

$$m = 1 + \frac{\lambda}{pv(1 - \alpha)} < \infty \text{ and } \mathbb{P}(X(t) = j) \to \gamma_j/m, \text{ as } t \to \infty.$$

  The expected return time to 0 in this case is $1/(\lambda \pi_0) = m/\lambda$ so the mean length of the busy periods will be

$$\frac{(m-1)}{\lambda} = \frac{1}{pv(1 - \alpha)} = \frac{1}{(2p-1)v - \lambda}.$$

If $1 + \lambda/v \geq 2p$, the chain turns either null-recurrent or transient, so $\mathbb{P}(X(t) = i) \to 0$ and the mean length of the busy period becomes infinite. $\square$

Question 2.10.23 (Applied Probability, Part II, B208D, 1996)
(a) Let $W(t)$ be the number of wasps which have landed in a bowl of soup during the time interval $(0, t]$, and assume that the chance of an arrival during the interval $(u, u + h)$ is $\lambda(u)h + o(h)$ for some given function λ. Give a clear statement of

any extra assumptions needed in order to set up W as an inhomogeneous Poisson process with rate function λ. Show that $W(t)$ exhibits the Poisson distribution with mean $\int_0^t \lambda(u)\mathrm{d}u$.

(b) Offers X_1, X_2, ... are received in sequence for the purchase of a house. Assume that the X_i are independent random variables with common density function f and distribution function F. Declare X_n to be a *record value* if either $n = 1$ or $X_n > X_i$ for all $i < n$. Find the probability that X_n represents a record value. Find also an estimate for the probability of such a record value lying in $(u, u+h)$ where h is small. Neglecting all terms which are $o(h)$, for h small, find the probability that $(u, u+h)$ contains a record value. Deduce that the number $R(t)$ of record values in $(0, t]$ has mean $-\ln(1 - F(t))$, if $F(t) < 1$.

Solution (a) Suppose that λ is a 'nice' function (e.g. bounded and integrable on every interval $(0, t)$). The assumptions on the process $(W(t))$ which we will use are: $W(0) = 0$, and, for all $t > 0$:

(i) the probability $\mathbb{P}(W(u+h) - W(u) = 1)$ of a single arrival in $(u, u+h)$ is $\lambda(u)h + o(h)$, uniformly in $u \in (0, t)$;

(ii) the probability $\mathbb{P}(W(u+h) - W(u) \geq 2)$ of a multiple arrival in $(u, u+h)$ is $o(h)$ uniformly in $u \in (0, t)$;

(iii) the increments $W(t_1) - W(t_0)$, ..., $W(t_n) - W(t_{n-1})$ are independent, for all time points $0 = t_0 < t_1 < \cdots < t_n = t$; i.e., for all $k_1, \ldots, k_n = 0, 1, \ldots$,

$$\mathbb{P}\big(W(t_j) - W(t_{j-1}) = k_j, \, 1 \leq j \leq n\big) = \prod_{j=1}^n \mathbb{P}\big(W(t_j) - W(t_{j-1}) = k_j\big).$$

Under these assumptions, $W(t) \sim \mathrm{Po}(\Lambda(t))$ where $\Lambda(t) = \int_0^t \lambda(u)\,\mathrm{d}u$. In fact, the moment-generating function $M_t(\theta) = \mathbb{E}e^{\theta W(t)}$ is represented as

$$
\begin{aligned}
M_t(\theta) &= \mathbb{E}\exp\big(\theta[W(t_1) - W(t_0)] + \cdots + \theta[W(t_n) - W(t_{n-1})]\big) \\
&= \prod_{j=1}^n \mathbb{E}\exp\big(\theta[W(t_j) - W(t_{j-1})]\big),
\end{aligned}
$$

and as $n \to \infty$ and $\max[t_j - t_{j-1}] \to 0$, for any given θ,

$$\prod_{i=1}^{n} \mathbb{E} \exp \left(\theta \left[W(t_j) - W(t_{j-1}) \right] \right)$$

$$= \prod_{j=1}^{n} \left[1 - \lambda(t_{j-1})(t_j - t_{j-1}) + e^{\theta} \lambda(t_{j-1})(t_j - t_{j-1}) + o(t_j - t_{j-1}) \right]$$

$$= \prod_{j=1}^{n} \left[1 + (e^{\theta} - 1) \lambda(t_{j-1})(t_j - t_{j-1}) + o(t_j - t_{j-1}) \right]$$

$$= \prod_{j=1}^{n} \exp \left[(e^{\theta} - 1) \lambda(t_{j-1})(t_j - t_{j-1}) + o(t_j - t_{j-1}) \right]$$

$$\to \exp \left[(e^{\theta} - 1) \int_0^t \lambda(u) \, du \right].$$

Hence, $M_t(\theta) = \exp \left[(e^{\theta} - 1) \Lambda(t) \right]$, and $W(t)$ is a Poisson random variable $\text{Po}(\Lambda(t))$ for all $t > 0$. In a similar fashion, one can show that $W(t+s) - W(s) \sim \text{Po}(\Lambda(t+s) - \Lambda(s))$. Then the family $(W(t), t \geq 0)$ is an IPP $(\lambda(t))$.

(b) For $n = 1$, $\mathbb{P}(X_1 \text{ is a record value}) = 1$, by definition. For $n > 1$, use conditional expectation, given a value of X_n:

$$\begin{aligned}
\mathbb{P}(X_n \text{ is a record value}) &= \mathbb{P}(X_n > X_i \text{ for all } i = 1, \dots, n-1) \\
&= \mathbb{E} \mathbf{1}(X_n > X_i \text{ for all } i = 1, \dots, n-1) \\
&= \mathbb{E}(\mathbb{E}[\mathbf{1}(X_n > X_i \text{ for all } i = 1, \dots, n-1) | X_n]) \\
&= \int_0^{+\infty} f(x) F(x)^{n-1} dx.
\end{aligned}$$

Next,

$$\mathbb{P}(X_n \text{ is a record value and } X_n \in (u, u+h))$$

$$= \int_u^{u+h} f(x) F(x)^{n-1} dx = f(u) F(u)^{n-1} h + o(h),$$

and

$$h \min \left[f(x) F(x)^{n-1} : u \leq x \leq u+h \right]$$
$$\leq \mathbb{P}(X_n \text{ is a record value and } X_n \in (u, u+h))$$
$$\leq h \max \left[f(x) F(x)^{n-1} : u \leq x \leq u+h \right].$$

To find the main contributing term in the probability that $(u, u+h)$ contains a record value, write:

$$\mathbb{P}\big((u, u+h) \text{ contains a record value }\big) = \mathbb{P}(u < X_1 < u+h)$$
$$+ \sum_{n>1} \mathbb{P}(X_n \text{ is a record value and } u < X_n < u+h) + o(h)$$

$$= h\big[f(u) + \sum_{n>1} f(u)F(u)^{n-1}\big] + o(h) = \frac{hf(u)}{1 - F(u)} + o(h).$$

We conclude that the process of records $(R(t))$ is IPP$(\lambda(t))$ where rate $\lambda(t) = f(u)/(1 - F(u))$.

Then the expectation

$$\mathbb{E}R(t) \;=\; \text{the mean number of record values in } (0, t)$$
$$=\; \int_0^t \lambda(u)\, du = \int_0^t \frac{f(u)}{1 - F(u)}\, du = \int_0^t \frac{dF(u)}{1 - F(u)}$$
$$=\; \int_0^t d\ln[1 - F(u)] = -\ln[1 - F(t)]$$

provided that $F(t) < 1$. We use the fact that $F(0) = 0$, hence $\ln[1 - F(0)] = 0$, because F has a density f concentrated on $(0, +\infty)$. $\qquad\square$

Question 2.10.24 (Applied Probability, Part II, B309D, 1996)
Customers arrive in a cake shop in the manner of a Poisson process with rate λ. The single server is ambidextrous and can serve people in batches of two (one with each hand). That is, at the beginning of each service period, she attends the next two people waiting; if there is only one waiting, this customer is served alone. Service periods S are independent random variables having common moment generating function $M(\theta) = \mathbb{E}\exp(\theta S)$. Let Q_n be the number of people in shop immediately after the completion of the nth service period. Express Q_{n+1} in the form

$$Q_{n+1} = A_n + Q_n - h(Q_n)$$

for an appropriate random variable A_n to be defined, where $h(x) = \min\{2, x\}$. Show that $Q = (Q_n : n \geq 1)$ is a Markov chain, and find an expression for $\mathbb{E}\, s^{A_n}$, $|s| \leq 1$.

If Q has stationary distribution $\pi = (\pi_i : i \geq 0)$ with generating function $G(s) = \sum_i \pi_i s^i$, show that

$$s^2 G(s) = M(\lambda(s-1))\left\{ G(s) + \sum_{i=0}^{2} (s^2 - s^i)\pi_i \right\}, \qquad |s| \leq 1.$$

When service times are exponentially distributed with parameter 1, show that

$$G(s) = (1-\alpha)/(1-\alpha s),$$

where $\alpha = 2/(1+\sqrt{5})$.

Solution Let A_n be the number of customers arriving in the nth service period and S_n be the duration of the nth service period, with the cumulative distribution function $F_S(t) = \mathbb{P}(S_n < t)$ and moment-generating function $M(\theta) = \mathbb{E}e^{\theta S_n}$. Then, conditional on $S_n = t$, the variable A_n has the Poisson distribution $\text{Po}(\lambda t)$. Thus, the probability-generating function

$$
\begin{aligned}
\phi_{A_n}(s) &= \mathbb{E}\, s^{A_n} = \mathbb{E}\left[\mathbb{E}\left(s^{A_n}|S_n\right)\right]\\
&= \int_0^{+\infty} \sum_{m=0}^{+\infty} s^m \frac{(\lambda t)^m}{m!} e^{-\lambda t} dF_S(t)\\
&= \int_0^{+\infty} e^{\lambda t(s-1)} dF_S(t) = M(\lambda(s-1)),
\end{aligned}
$$

which determines the distribution of A_n uniquely. Further, the random variables $A_1, A_2, \ldots$ are IID. Next, from the description of the queue,

$$Q_{n+1} = A_n + Q_n - h(Q_n), \quad \text{with } h(Q_n) = \min[2, Q_n],$$

where A_n is independent of Q_n (in fact, of the whole sequence $Q_1, \ldots, Q_{n-1}$). Therefore, (Q_n) is a discrete-time Markov chain.

Moreover, if $\pi = (\pi_n)$ is a stationary distribution then, in equilibrium,

$$
\begin{aligned}
G(s) &= \mathbb{E}\, s^{Q_{n+1}} = \mathbb{E}\, s^{A_n} \mathbb{E}\, s^{Q_n - h(Q_n)}\\
&= M(\lambda(s-1))\left(\pi_0 + \pi_1 + \pi_2 + \sum_{i=3}^{+\infty} \pi_i s^{i-2}\right),
\end{aligned}
$$

and

$$
\begin{aligned}
s^2 G(s) &= M(\lambda(s-1))\left(\pi_0 s^2 + \pi_1 s^2 + \sum_{i=2}^{+\infty} \pi_i s^i\right)\\
&= M(\lambda(s-1))\left[(s^2-1)\pi_0 + (s^2-s)\pi_1 + (s^2-s^2)\pi_2 + G(s)\right]\\
&= M(\lambda(s-1))\left\{G(s) + \sum_{i=0}^{2}(s^2-s^i)\pi_i\right\},
\end{aligned}
$$

as required.

Now assume that $S_n \sim \text{Exp}(\mu)$, with $M(\theta) = \dfrac{\mu}{\mu-\theta}$. Then, with $\rho = \dfrac{\lambda}{\mu}$,

$$M(\lambda(s-1)) = \frac{\mu}{\mu+\lambda-\lambda s} = \frac{1}{1+\rho-\rho s}.$$

Therefore,

$$G(s) = \frac{1}{(1+\rho-\rho s)s^2}\left[G(s)+(s^2-1)\pi_0+(s^2-s)\pi_1\right].$$

Following the suggested form of $G(s)$, set $G(s) = \dfrac{\pi_0}{1-\alpha s}$ with $\pi_1 = \alpha\pi_0$. In fact, this corresponds to a geometric equilibrium distribution, with $\pi_i = \alpha^i\pi_0,\ i\geq 1$, and $\pi_0 = 1-\alpha$. Then we obtain

$$\frac{\pi_0}{1-\alpha s} = \frac{1}{(1+\rho-\rho s)s^2}\left[\frac{\pi_0}{1-\alpha s}+(s^2-1)\pi_0+(s^2-s)\alpha\pi_0\right],$$

or

$$\begin{aligned}
1 &= \frac{1}{(1+\rho-\rho s)s^2}\left\{1+(1-\alpha s)\left[s^2-1+\alpha(s^2-s)\right]\right\}\\
&= \frac{1}{1+\rho-\rho s}\left(1-\alpha s+\alpha-\alpha^2 s+\alpha^2\right).
\end{aligned}$$

Equating coefficients in the 0th and 1st order terms in s yields a pair of (identical) relations

$$1+\rho = 1+\alpha+\alpha^2, \quad\text{and}\quad \rho s = \alpha s+\alpha^2 s,$$

whence

$$\alpha = \frac{-1+\sqrt{1+4\rho}}{2}.$$

Clearly, we need $\alpha < 1$, that is, $\rho < 2$ (which is a necessary and sufficient condition for existence (and uniqueness) of an ED). For $\lambda = \mu$, we have $\rho = 1$, and

$$\alpha = \frac{\sqrt{5}-1}{2}, \quad\text{and}\quad \pi_0 = 1-\alpha.$$

$\square$

Question 2.10.25 (Applied Probability, Part II, B411D, 1996)
A single bar contains N stools. Individuals enter the bar in the manner of a Poisson process with rate λ. If a stool does remains free, then an arriving customer sits on it. If none do, then the individual goes elsewhere. Customers stay in the bar for independent times exponentially distributed with parameter μ. Calculate the probability that an arriving customer finds no available stool, when the system is in equilibrium.

When the system is in equilibrium, describe the departure process of customers (not counting those who failed to find a vacant stool).

Solution The system is described by a birth-and-death process on $\{0, 1, \ldots, N\}$, which is a reversible CTMC. The detailed balance equations are

$$\pi_0 \lambda = \pi_1 \mu, \ldots, \pi_{N-1} \lambda = \pi_N \mu,$$

and are solved recurrently:

$$\pi_{N-1} = \frac{\mu}{\lambda} \pi_N, \ldots, \pi_0 = \left(\frac{\mu}{\lambda}\right)^N \pi_N.$$

Then

$$\pi_N = \left[1 + \frac{\mu}{\lambda} + \cdots + \left(\frac{\mu}{\lambda}\right)^N\right]^{-1}.$$

The departure process from the system is Poisson, of rate λ. This is Burke's Theorem for the loss M/M/1/N system. $\qquad\square$

> *All argument is against it, but all belief is for it.*
> *(On the appearance of the spirit of a person after death)*
> S. Johnson (1709–1784), English lexicographer and playwright

Question 2.10.26 (Applied Probability, Part II, B412J, 1997)
Describe the application of both discrete-time and continuous-time Markov chains to the single-server queue. Include discussions of the existence of an equilibrium queue length, together with calculation of quantities pertaining to the queue in equilibrium.

Solution (1) The M/M/1 queue. This is the simplest example: the inter-arrival times are IID $(\text{Exp} \lambda)$ and the service times IID $(\text{Exp} \mu)$. Here X_t, the number of customers in the system at time t forms a continuous-time Markov chain on $\mathbb{Z}_+ = \{0, 1, \ldots\}$ (a birth-death process). The rates are $q_{ii+1} = \lambda$, $i \geq 0$, and $q_{ii-1} = \mu$, $i \geq 1$. If

$$\rho = \frac{\lambda}{\mu} \begin{cases} > \\ = \\ < \end{cases} 1, \quad \text{the chain is} \quad \begin{cases} \text{transient,} \\ \text{null-recurrent,} \\ \text{positive-recurrent.} \end{cases}$$

If $\rho < 1$, the equilibrium distribution is geometric:

$$\pi_i = \rho^i (1 - \rho), \quad i \geq 0.$$

(2) The M/G/1 queue. Here, the inter-arrival times A_n are IID $\mathrm{Exp}\,(\lambda)$ and service times S_n are IID with a given distribution. The number X_n of customers in the queue at the time just after the nth departure forms a discrete-time Markov chain:

$$X_{n+1} = X_n + Y_{n+1} - \mathbf{1}(X_n \geq 1).\qquad(2.181)$$

Here, Y_n is the number of arrivals during the nth service time; Y_n is independent of X_n and, conditional on $S_n = s$, $Y_n \sim \mathrm{Po}\,(\lambda s)$. The PGF

$$\phi_Y(z) = \mathbb{E}z^Y = \mathbb{E}\left[\mathbb{E}\left(z^Y|S\right)\right] = \mathbb{E}\left(e^{\lambda(z-1)S}\right) = M_S(\lambda(z-1)).$$

This yields that $\mathbb{E}Y = \lambda\mathbb{E}S$ which is again denoted by ρ.

Lemma 2.10.27 *If $\rho = \mathbb{E}Y < 1$, the chain (X_n) is positive recurrent.*

Proof Iterating (2.181), we have:

$$X_n = X_0 + Y_1 + Y_2 + \cdots + Y_n - n + Z_n$$

where Z_n is the number of visits to 0 by time n. Assuming $X_0 = 0$,

$$\begin{aligned}
\mathbb{E}\left(Z_n/n\right) &= 1 - \mathbb{E}\left(\sum_{i=1}^{n}Y_i\right)\Big/ n\\
&= 1 - \rho + \mathbb{E}\left(X_n/n\right) \geq 1 - \rho > 0.
\end{aligned}$$

We see that $\mathbb{E}Z_n/n \to 1/m_0$ where

$$m_0 = \mathbb{E}_0(\text{return time to } 0) < \infty.$$

Hence, state 0 is positive recurrent. As the chain (X_n) is irreducible, it is positive recurrent. Hence, (X_n) has a unique invariant distribution. $\square$

Lemma 2.10.28 *In equilibrium, $\pi_0 = 1 - \rho$ and*

$$G_X(z) = \frac{(1-\rho)(1-z)G_Y(z)}{G_Y(z) - z} = \frac{(1-\rho)(1-z)M_S(\lambda(z-1))}{M_S(\lambda(z-1)) - z}.$$

Proof In equilibrium, $X_{n+1} \sim X_n$. Using this fact and the above equation, write

$$zG_X(z) = z\mathbb{E}z^X = \mathbb{E}z^{X+1} = \mathbb{E}\left(z^Y z^{X+\mathbf{1}(X=0)}\right)$$

$$= G_Y(z)\mathbb{E}z^{X+\mathbf{1}(X=0)} = G_Y(z)(\pi_0 z + G_X(z) - \pi_0),$$

whence

$$G_X(z) = \frac{\pi_0(1-z)G_Y(z)}{G_Y(z) - z}.$$

As $z \to 1$,

$$1 = \pi_0 \lim_{z \to 0} \frac{1-z}{G_Y(z) - z} = \frac{\pi_0}{1 - \mathbb{E}Y}.$$

By L'Hopital's rule, this equals $\pi_0 / (1 - \pi_0)$. Therefore, $\pi_0 = 1 - \rho$. $\qquad\square$

(3) The G/M/1 queue. Now suppose inter-arrival times A_n are IID with a given distribution, and service times S_n are IID (Exp μ). Let X_n be the number of customers in the queue just before the nth arrival. Then (X_n) is a Markov chain:

$$X_{n+1} = \max[X_n - Y_n + 1, 0],$$

where Y_n stands for the number of customers served during the nth interarrival time. Again Y_n is independent of X_n. Next, conditional on $A_n = t$, $Y_n \sim \mathrm{Po}(\mu t)$.

Using the same argument as before, one has:

$$G_Y(z) = M_A(\mu(z-1)), \quad \text{and} \quad \mathbb{E}Y = \mu\mathbb{E}A = 1/\rho.$$

Theorem 2.10.29 If $\rho < 1$, X_n is positive recurrent with equilibrium distribution $\pi_i = (1 - \eta)\eta^i$ where η is the unique root in $(0,1)$ of $\eta = G_Y(\eta)$.

Proof We have $G_Y(0) = \mathbb{P}(Y = 0) > 0$ and $G_Y(1) = 1$. Next, $G_Y'(1) = \mathbb{E}Y = 1/\rho > 1$. Also, $G''(z) > 0$, i.e., G_Y is a convex function. Hence, equation $\eta = G_Y(\eta)$ has a unique solution in $(0,1)$. Further, $\mathbb{P}(X_{n+1} = k) = \mathbb{P}(X_n - Y_n = k - 1)$. If $\pi = (\pi_i)$ is an equilibrium distribution then

$$\pi_k = \sum_{i \geq 0} \pi_{k+i-1} p_i,$$

where

$$p_i = \mathbb{P}(i \text{ services during a typical inter-arrival time});$$

substituting $\pi_i = \eta^i(1 - \eta)$ gives

$$(1 - \eta)\eta^k = \sum_{i \geq 0} (1 - \eta)\eta^{k+i-1} p_i.$$

Equivalently,

$$\eta^k = \sum_{i \geq 0} \eta^{k+i-1} p_i, \quad \text{i.e.} \quad \eta = \sum_{i \geq 0} \eta^i p_i = G_Y(\eta).$$

$\qquad\square$

$\qquad\square$

Question 2.10.30 (Applied Probability, Part II, B209E, 1998) (modified)
The following assertion is known as the Key Renewal Theorem for a discrete time renewal process.

Let $S_1, S_2, \ldots$ be IID positive integer-valued random variables with $p_k = \mathbb{P}(S_1 = k)$ and set

$$J_0 = 0, \quad J_n = S_1 + \cdots + S_n, \quad n \geq 1,$$

and

$$X_n = k \text{ if } J_k \leq n < J_{k+1}, \quad A_n = \{n = J_k \text{ for some } k\}, \quad n = 0, 1, \ldots.$$

That is, $A_n = \{n$ represents a time of renewal$\}$. Assume that the greatest common divisor gcd $(k : p_k > 0) = 1$. Then

$$\lim_{n \to \infty} \mathbb{P}(A_n) = \frac{1}{\mathbb{E}[S_1]}.$$

(a) Prove the above assertion.

(b) How long on average must one wait to see the pattern 000100 in a random sequence of binary digits?

Solution (a) Consider the Markov chain (Y_n) on $\{0, 1, \ldots\}$

$$\begin{aligned} Y_n &= \inf\{m \geq 0 : m + n = J_k \text{ for some } k\}, \\ &= \text{time from } n \text{ until next renewal}, \quad n = 1, 2, \ldots. \end{aligned}$$

The transition probabilities are given by

$$\begin{aligned} \mathbb{P}(Y_{n+1} = i | Y_n = 0) &= p_{i+1}, \quad i \geq 0, \\ \mathbb{P}(Y_{n+1} = i - 1 | Y_n = i) &= 1, \quad i \geq 1. \end{aligned}$$

The chain becomes irreducible and aperiodic, owing to the condition that gcd $(k : p_k > 0) = 1$. It follows a unique equilibrium distribution

$$\pi_0 = \frac{1}{\mathbb{E}[S_1]}, \quad \pi_k = \frac{1}{\mathbb{E}[S_1]} \sum_{i > k} p_i, \, k \geq 1,$$

and hence is positive recurrent. Then, as $n \to \infty$,

$$\mathbb{P}(A_n) = \mathbb{P}(Y_n = 0) \to \pi_0 = \frac{1}{\mathbb{E}[S_1]}.$$

(b) Let renewals occur when non-overlapping strings 000100 are produced. The probability that such a string appears by time $n \geq 6$ equals $1/2^6$. On the other hand,

$$\frac{1}{2^6} = \mathbb{P}(A_n) + \mathbb{P}(A_{n-4}) \frac{1}{2^4} + \mathbb{P}(A_{n-5}) \frac{1}{2^5}.$$

According to the Key Renewal Theorem, as $n \to \infty$, the right-hand side tends to

$$\frac{1}{\mathbb{E}[S_1]} \left(1 + \frac{1}{2^4} + \frac{1}{2^5}\right),$$

whence $\mathbb{E}[S_1] = 70$. In a general case where 1 occurs with probability p and 0 with probability q,

$$\mathbb{E}[S_1] = \frac{1}{q^5 p}(1 + q^3 p + q^4 p).$$

For more results on similar problems, see: G. Blom, D. Thorburn. How many random digits are required until given sequences are obtained? *Journ. Appl. Probab.*, **19** (1982), 518–531.

□

Question 2.10.31 (Applied Probability, Part II, B412E, 1998)
Consider an epidemic model $(S_t, I_t, R_t)_{t \geq 0}$ in a large population of size $N = S_t + I_t + R_t$, where S_t denotes the number of susceptible individuals, I_t denotes the number infected, and R_t denotes the number which have recovered or died. Suppose that $(S_t, I_t)_{t \geq 0}$ evolves as a Markov chain for which the non-zero transition rates are given by

$$q_{(s,i)(s-1,i+1)} = \lambda_{(s,i)} > 0, \quad \text{for } s \geq 1, \, i \geq 1,$$
$$q_{(s,i)(s,i-1)} = \mu_i > 0, \quad \text{for } s \geq 1, \, i \geq 1,$$

and assume that $S_0 = N - 1$, $I_0 = 1$.

(i) Show that $(R_t)_{t \geq 0}$ eventually becomes constant and, for the final value R_∞, that $\mathbb{P}(R_\infty > r)$ increases as the infection rate $\lambda_{(s,i)}$ rises, for all $r \geq 0$.

(ii) In the standard epidemic model one takes

$$\lambda_{(s,i)} = i\lambda \frac{s}{N}, \quad \mu_i = i\mu \text{ for some constants } \lambda, \mu > 0.$$

Give a justification for this choice of rates.

(iii) Show, in precise terms which you should make explicit, that in the limit as $N \to \infty$, a standard epidemic beginning with one infected individual affects a positive proportion of the population if and only if $\lambda > \mu$.

Solution (i) The transition rates guarantee that the sum $S_t + I_t$ remains non-increasing in t. Hence, R_t will be non-decreasing. In addition, R_t is bounded from above: $R_t \leq N$. Hence, $R_t \nearrow R_\infty$, almost surely.

Lemma 2.10.32 *If* $\tilde{\lambda}_{(s,i)} \geq \lambda_{(s,i)}$ *for all* (s,i) *then*

$$\mathbb{P}\left(\tilde{R}_\infty \geq r\right) \geq \mathbb{P}(R_\infty \geq r) \text{ for all } r > 0.$$

Proof Consider the jump chain on $\mathbb{Z}_+ \times \mathbb{Z}_+$ (a lattice quarter-plane), with a sample trajectory like that in Figure 2.63.

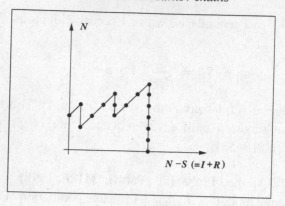

Fig. 2.63

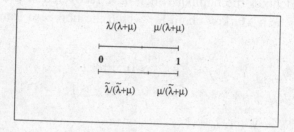

Fig. 2.64

Here, with probability $\lambda_{(s,i)}/(\lambda_{(s,i)} + \mu_i)$, the particle jumps one unit up and one to the right. With probability $\mu_i/(\lambda_{(s,i)} + \mu_i)$ it drops one unit down.

We can run the two chains corresponding to $\{\lambda_{(s,i)}\}$ and $\{\widetilde{\lambda}_{(s,i)}\}$ jointly, by using $U(0,1)$-IID random variables to determine transitions as shown in Figure 2.64. (This gives yet another example of coupling of random processes.)

Then the tilde-chain always jumps up and to the right whenever the original chain does. Hence, the trajectories of the tilde-chain lie above those of the original one. As $\widetilde{I}_\infty = I_\infty = 0$, the ultimate state $\widetilde{R}_\infty \geq R_\infty$. Hence, the assertion of the lemma. □

(ii) The choice

$$\lambda_{(s,i)} = i\lambda \frac{s}{N}, \ \mu_i = i\mu$$

means that each infected individual contacts every other at rate λ and recovers or dies at rate μ.

(iii) We begin with

Theorem 2.10.33 (a) *If $\lambda \leq \mu$ then for all $\varepsilon > 0$*

$$\mathbb{P}(R_\infty > \varepsilon N) \to 0 \text{ as } N \to \infty.$$

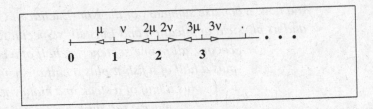

Fig. 2.65

Fig. 2.66

(b) If $\lambda > \mu$ and we choose $\varepsilon, \delta > 0$ so that $\lambda(1-\varepsilon)(1-\delta) = \mu$ then

$$\mathbb{P}(R_\infty > \varepsilon N) \geq \delta, \quad \text{for all } N.$$

Proof (a) Suppose $\lambda \leq \mu$. Take $\widetilde{\lambda}_{(s,i)} = \lambda i \geq \lambda si/N$ in the above lemma. The continuous-time chain $(\widetilde{I}_t)$ represents a birth-and-death process on $\{0,\ldots,N\}$ with the rates presented in Figure 2.64.

That is, the jump tilde-chain constitutes a random walk with a non-positive drift. Then, for all $\varepsilon > 0$, as $N \to \infty$,

$$\begin{aligned}
\mathbb{P}(R_\infty \geq \varepsilon N) &\leq \mathbb{P}\left(\widetilde{R}_\infty \geq \varepsilon N\right) \\
&\leq \mathbb{P}(\text{number of jumps to the right} \geq \varepsilon N) \to 0.
\end{aligned}$$

(b) Now suppose that $\lambda(1-\varepsilon)(1-\delta) = \mu$ for some $\varepsilon, \delta > 0$. Set

$$\widetilde{\lambda}_{(s,i)} = \min\left\{\frac{s}{N}, 1-\varepsilon\right\} \lambda i \leq \lambda \frac{s}{N} i.$$

Then, as long as $\widetilde{S}_t \geq N(1-\varepsilon)$, $(\widetilde{I}_t)$ forms a birth-and-death process with the rates presented in Figure 2.66 where $v = \lambda(1-\varepsilon)$. So, for all N,

$$\begin{aligned}
\mathbb{P}(R_\infty \leq \varepsilon N) &\leq \mathbb{P}\left(\widetilde{R}_\infty \leq \varepsilon N\right) \\
&\leq \mathbb{P}\left(\widetilde{S}_\infty \geq (1-\varepsilon)N, \text{ number of deaths} \leq \varepsilon N\right) \\
&\leq \mathbb{P}_1(\text{hit } 0) = \mu/v.
\end{aligned}$$

Thus,

$$\mathbb{P}\left(R_\infty > \varepsilon N\right) \geq 1 - \frac{\mu}{\nu} = \frac{\lambda(1-\varepsilon) - \lambda(1-\varepsilon)(1-\delta)}{\lambda(1-\varepsilon)} = \delta.$$

□

□

And now, a concise summary of the Non-Euclidean Geometry and its physical application and Einstein's Special Relativity Theory, all in one sentence. A half of a cottage π plus a half of a fish π plus a half of shepherd's π plus a half of a steak and kidney π will not make a full turn, but four pints might.
(From the series 'Thus spoke Superviser'.)

3

Statistics of discrete-time Markov chains

3.1 Introduction

> *Where are the weapons of math distraction?*
> (From the series *'When they go political'*.)

In this chapter we present some important facts about the statistics of discrete-time Markov chains (DTMC) with finitely many states. The basic question arises from observing a sample

$$\mathbf{x} = \mathbf{x}_n = \begin{pmatrix} x_0 \\ \vdots \\ x_n \end{pmatrix} \in I^{n+1}, \tag{3.1}$$

from a DTMC (X_m), with an unknown distribution, over $n+1$ subsequent time points $0, \ldots, n$: what we can say about this distribution? Typically, we are interested in *parameter estimation*, where the distribution $\mathbb{P}^\theta$ of the chain depends on a (scalar, or multi-dimensional) parameter θ, varying within a given (discrete or continuous) set Θ (in the continuous case, a subset on the line $\mathbb{R}$ or in a Euclidean space of a higher dimension). More precisely, in the DTMC setting, the transition probability matrix $\mathbb{P}^\theta$, and, in some cases, also the initial probability vector λ^θ, depend on θ, in the sense that the transition probabilities p_{ij}^θ, and initial probabilities λ_j^θ, are functions of $\theta \in \Theta$. For definiteness, we assume the (finite) state space I of the chain as fixed, and s will stand for the number of states $|I|$. Frequently, λ^θ coincides with an equilibrium distribution π^θ, so that $\mathbb{P}^\theta$ describes the chain in equilibrium. To further simplify the matter, we often take P^θ as irreducible and aperiodic for all $\theta \in \Theta$, so that the chain follows the unique equilibrium distribution π^θ, with $\pi^\theta = \pi^\theta P^\theta$, and exhibits a geometric convergence to π^θ, regardless of the initial probability vector (see Section 1.9, in particular, Theorems 1.9.2 and 1.9.3).

349

To this end, we introduce the special notation $\mathscr{P}_{IA}$ (see (3.10)). The assumption of irreducibility and aperiodicity will become particularly useful when we look at large samples (with $n \to \infty$).

As in the case of independent samples, we want to estimate θ from $\mathbf{x}$, i.e. to produce a function $\widehat{\theta}(\mathbf{x})$ (or a sequence of functions $\widehat{\theta}_n(\mathbf{x}_n)$), called an *estimator*, which provides a good approximation for θ. Here, we expect to be able to improve the quality of approximation when n grows to infinity. However, if we get restricted to a 'small', or 'moderate' sample size, asymptotical methods should be replaced by more appropriate ones.

For example, in the *hypothesis testing* approach, we wish to make a judgement about whether θ takes a particular value θ^0 (or is close to it), where θ^0 has been singled out from set Θ as a result of, say, some external information. This specifies a *simple null hypothesis*, $H_0 : \theta = \theta^0$. Again, in the simplest case, we compare it to a *simple alternative*, $H_1 : \theta = \theta^1$ where θ^1 represents another singled out value. Conveniently, the Neyman–Pearson Lemma is applicable in the case of a Markov chain, and we can work with the *likelihood ratio*

$$\frac{f_{\mathbf{X}}(\mathbf{x}, \theta^1)}{f_{\mathbf{X}}(\mathbf{x}, \theta^0)}.$$

Here $f_{\mathbf{X}}(\mathbf{x}, \theta)$ stands for the probability mass assigned to (or the *likelihood* of) the sample $\mathbf{x}$. We will consider two types of likelihood:

$$f_{\mathbf{X}}(\mathbf{x}, \theta) = L_{\mathbf{X}}(\mathbf{x}, \theta) \text{ (a full likelihood)},$$

and

$$f_{\mathbf{X}}(\mathbf{x}, \theta) = l_{\mathbf{X}}(\mathbf{x}, \theta) \text{ (a reduced likelihood)}.$$

More precisely,

$$L_{\mathbf{X}}(\mathbf{x}, \theta) = \mathbb{P}^\theta(X_0 = x_0, \ldots, X_n = x_n) = \lambda_{x_0}^\theta p_{x_0 x_1}^\theta \cdots p_{x_{n-1} x_n}^\theta, \tag{3.2}$$

which corresponds to a chain with an initial distribution λ^θ, and

$$l_{\mathbf{X}}(\mathbf{x}, \theta) = \mathbb{P}^\theta(X_1 = x_1, \ldots, X_n = x_n \mid X_0 = x_0) = p_{x_0 x_1}^\theta \cdots p_{x_{n-1} x_n}^\theta, \tag{3.3}$$

which corresponds to a chain starting from the state x_0. Here $\mathbf{X}\,(= \mathbf{X}^{(n)})$ stands for a random sample from the chain (X_m) observed between times 0 and n:

$$\mathbf{X} = \begin{pmatrix} X_0 \\ \vdots \\ X_n \end{pmatrix}. \tag{3.4}$$

So, the likelihood function (3.2) corresponds to the probability distribution $\mathbb{P}^\theta$ of the DTMC with an initial vector λ^θ, whereas (3.3) gives the conditional probability

$\mathbb{P}^{\theta}(X_1 = x_1, \ldots, X_n = x_n \mid X_0 = x_0)$. As explained, we will often assume that in the case of likelihood (3.2), the chain sets in equilibrium, i.e. λ^{θ} coincides with an equilibrium distribution π^{θ}, for which $\pi^{\theta} = \pi^{\theta} P^{\theta}$.

Then the likelihood ratio takes the form

$$\frac{l_{\mathbf{X}}(\mathbf{x}, \theta^1)}{L_{\mathbf{X}}(\mathbf{x}, \theta^0)} \quad \text{or} \quad \frac{l_{\mathbf{X}}(\mathbf{x}, \theta^1)}{l_{\mathbf{X}}(\mathbf{x}, \theta^0)}.$$

The Neyman–Pearson Lemma says that, for all $k > 0$, the test with the critical region

$$\mathscr{C}_k = \{\mathbf{x}: f_{\mathbf{X}}(\mathbf{x}, \theta^1) > k f_{\mathbf{X}}(\mathbf{x}, \theta^0)\}$$

forms the most powerful among all tests of the null hypothesis $H_0: \theta = \theta^0$ against the alternative $H_1: \theta = \theta^1$, of size

$$\alpha_k = \sum_{\mathbf{x} \in \mathscr{C}_k} \mathbb{P}^{\theta^0}(\mathbf{x}).$$

That is, for any test $\mathscr{C}^*$ with

$$\alpha^* = \sum_{\mathbf{x} \in \mathscr{C}^*} \mathbb{P}^{\theta^0}(\mathbf{x}) \leq \alpha_k,$$

the power $\beta^* \leq \beta_k$, where

$$\beta^* = \sum_{\mathbf{x} \in \mathscr{C}^*} \mathbb{P}^{\theta^1}(\mathbf{x}), \quad \beta_k = \sum_{\mathbf{x} \in \mathscr{C}_k} \mathbb{P}^{\theta^1}(\mathbf{x}).$$

Observe the insensitivity of the Neyman–Pearson Lemma to the nature of the parameter(s) θ. For instance, θ may be identified with the matrix of transition probabilities $P = (p_{ij})$; see (3.7). In this situation we would test the null hypothesis H_0 that the chain possesses a given transition probability matrix P^0 against the alternative H_1 that another transition matrix P^1 applies.

In a more general situation, with a simple null hypothesis: $\theta = \theta^0$, but a *composite* alternative hypothesis (viz., $\theta \in \Theta_0 \subset \Theta$ where $\theta^0 \in \Theta_0$), we may hope to benefit from a *generalised likelihood ratio* test, which belongs to the category of *goodness-of-fit* tests; see Volume 1, page 256. Here, one considers the ratio

$$\frac{\max\left[f_{\mathbf{X}}(\mathbf{x}, \vartheta): \vartheta \in \Theta_0\right]}{f_{\mathbf{X}}(\mathbf{x}, \theta^0)}$$

and rejects the null hypothesis when it gets large. Equivalently, passing to logarithms, we are interested in the difference

$$\max \ln\left[f_{\mathbf{X}}(\mathbf{x}, \vartheta): \vartheta \in \Theta_0\right] - \ln f_{\mathbf{X}}(\mathbf{x}, \theta^0).$$

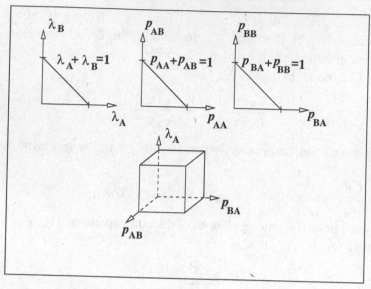

Fig. 3.1

To reach a correct conclusion, we would like to know the distribution of this *statistic*; in Volume 1 it was stated that in the case of IID samples this distribution is asymptotically χ^2 (Wilks' Theorem). But with Markov chains, an additional investigation into this matter is needed.

An important case arises where θ constitutes the whole pair (λ, P) (an initial vector and a transition matrix). In a sense, this case falls into the category of *non-parametric estimation*. For example, if the chain can take two states, say A and B, then $\lambda = (\lambda_A, \lambda_B)$ and $P = \begin{pmatrix} p_{AA} & p_{AB} \\ p_{BA} & p_{BB} \end{pmatrix}$, where λ_A and $\lambda_B = 1 - \lambda_A$ lie in $[0, 1]$, as well as $p_{AA}, p_{AB} = 1 - p_{AA}$ and $p_{BA}, p_{BB} = 1 - p_{BA}$. We may think that (i) the vector $\lambda = (\lambda_A, \lambda_B)$ runs over a segment Σ of a straight line in a non-negative quadrant of the plane $\mathbb{R}^2$, (ii) the transition matrix P expands over a Cartesian product of two segments, Σ_A and Σ_B in a non-negative orthant in $\mathbb{R}^4$. On the whole, the locus of the parameter (λ, P) traces a three-dimensional cube in $\mathbb{R}^6$, and (say) $\lambda_A, p_{AA}, p_{BA} \in [0, 1]$ make up independent co-ordinates on Θ. See Figure 3.1.

In general, $\theta = (\lambda, P)$ will be a multi-dimensional parameter. Assume, for definiteness, that the state space I is the set $\{1, \ldots, s\}$; then pairs (λ, P) lie in a subset $\mathscr{R}$ of the non-negative orthant in the Euclidean space $\mathbb{R}^M$ where the dimension $M = s + s^2$ (s for the number of entries λ_j and s^2 for the number of entries p_{ij}). More precisely, $\mathscr{R}(=\mathscr{R}_s)$ forms a (closed) set, of dimension $s-1+s(s-1)=s^2-1$, as we must take into account the linear constraints $\sum_{k=1}^{s}\lambda_k = \sum_{k=1}^{s}p_{ik} = 1$:

$$\mathscr{R} = \mathscr{R}_s = \left\{ (\lambda, P) : \lambda = (\lambda_j), P = (p_{ij}), \right.$$
$$\left. \sum_{k=1}^{s}\lambda_k = 1, \lambda_j \geq 0, \sum_{k=1}^{s}p_{ik} = 1, p_{ij} \geq 0, \text{ for all } i, j = 1, \ldots, s \right\}. \tag{3.5}$$

Geometrically, $\mathscr{R}$ is produced by the Cartesian product of $s+1$ simplexes, each simplex being of dimension $(s-1)$. A simplex of dimension s is a set of points $\mathbf{y} \in \mathbb{R}^s$ forming the locus for the inequalities $y_1 \geq 0, \ldots, y_s \geq 0$, and $(y_1 + \cdots + y_s) \leq a$, for some $a > 0$.

The interior of the set $\mathscr{R}$ is denoted by $\mathscr{R}^{\text{int}}$ ($= \mathscr{R}_s^{\text{int}}$) and is obtained when all inequalities are strict:

$$\mathscr{R}^{\text{int}} \ (= \mathscr{R}_s^{\text{int}}) = \left\{ (\lambda, P) : \lambda = (\lambda_j), P = (p_{ij}), \right.$$
$$\left. \sum_{k=1}^{s}\lambda_k = 1, \lambda_j > 0, \sum_{k=1}^{s}p_{ik} = 1, p_{ij} > 0, \text{ for all } i, j = 1, \ldots, s \right\}. \tag{3.6}$$

If we discard the initial distribution λ, the dimension of the set is reduced to $s^2 - s$; in this case we will use the notation $\mathscr{P} = \mathscr{P}_s$ and $\mathscr{P}^{\text{int}} = \mathscr{P}_s^{\text{int}}$:

$$\mathscr{P}(= \mathscr{P}_s) = \left\{ P = (p_{ij}) : \sum_{l=1}^{s}p_{il} = 1, p_{ij} \geq 0, \text{ for all } i, j = 1, \ldots, s \right\} \tag{3.7}$$

and

$$\mathscr{P}^{\text{int}}(= \mathscr{P}_s^{\text{int}}) = \left\{ P = (p_{ij}) : \sum_{l=1}^{s}p_{il} = 1, p_{ij} > 0, \text{ for all } i, j = 1, \ldots, s \right\}. \tag{3.8}$$

So, $\mathscr{P}$ is given by the Cartesian product of s simplexes each of dimension $s-1$. The structure of the above set $\mathscr{R}$ can be illustrated by means of the two-fold Cartesian product representation

$$\mathscr{R} = \mathbb{S} \times \mathscr{P}$$

where $\mathbb{S}$ ($= \mathbb{S}_s$) represents the simplex of stochastic vectors in $\mathbb{R}^s$

$$\mathbb{S}_s = \left\{ \lambda = (\lambda_j) : \lambda_j \geq 0, j = 1, \ldots, s, \sum_{l=1}^{s}\lambda_l = 1 \right\}.$$

Both sets $\mathscr{R}$ and $\mathscr{P}$ can be endowed with a distance generated by the Euclidean metrics in $\mathbb{R}^{s^2-1}$ and $\mathbb{R}^{s^2-s}$, correspondingly:

$$\text{dist}\left((\lambda,P),(\lambda',P')\right) = \left(\left[\text{dist}\left(\lambda,\lambda'\right)\right]^2 + \left[\text{dist}\left(P,P'\right)\right]^2\right)^{1/2},$$

where

$$\text{dist}\left(\lambda,\lambda'\right) = \left[\sum_{j=1}^{s}(\lambda_j - \lambda'_j)^2\right]^{1/2}, \quad \text{dist}\left(P,P'\right) = \left[\sum_{i,j=1}^{s}(p_{ij} - p'_{ij})^2\right]^{1/2},$$

and $\lambda = (\lambda_j)$, $\lambda' = (\lambda'_j)$, and $P = (p_{ij})$, $P' = (p'_{ij})$.

Remark 3.1.1 Observe that if $(\lambda,P) \in \mathscr{R}^{\text{int}}$ (or $P \in \mathscr{P}^{\text{int}}$) then $P = (p_{ij})$ becomes irreducible and aperiodic and hence features a unique equilibrium distribution $\pi = (\pi_j)$, with $\pi = \pi P$. Furthermore, in this case the matrix $P^n = \left(p_{ij}^{(n)}\right)$ converges to

$$\begin{pmatrix} \pi_1 & \cdots & \pi_s \\ \vdots & \cdots & \vdots \\ \pi_1 & \cdots & \pi_s \end{pmatrix}, \text{ as } n \to \infty. \text{ Moreover,}$$

$$\left|p_{ij}^{(n)} - \pi_j\right| \le (1-\rho)^{n-1}, \tag{3.9}$$

where $\rho = \min\left[p_{ij} : i,j = 1,\ldots,s\right]$, with $0 < \rho < 1$. See (1.84). We can express this fact as

$$\mathscr{P}^{\text{int}} \subset \mathscr{P}_{\text{IA}}, \text{ where } \mathscr{P}_{\text{IA}} = \left\{P \in \mathscr{P} : P \text{ irreducible and aperiodic}\right\}.$$

The set $\mathscr{P}_{\text{IA}}$ will remain of considerable interest in this chapter. It can be specified as follows:

$$\mathscr{P}_{\text{IA}} = \left\{P \in \mathscr{P} : \min\left[p_{ij}^{(m)} : i,j = 1,\ldots,s\right] > 0 \text{ for some } m \ge 1\right\}. \tag{3.10}$$

Remark 3.1.2 Depending on the particular problem, we may need to deal with other subsets of sets $\mathscr{R}$ and $\mathscr{P}$. This will be particularly relevant in Sections 3.7 and 3.8. For example, we may know *a priori* that our Markov chain cannot preserve its current state, i.e. that it always jumps away from it. In other words, the transition matrix P contains zeros on its main diagonal. In this case we can restrict ourselves to matrices $P \in \mathscr{P}_{\text{off-diag}}$ where $\mathscr{P}_{\text{off-diag}} \subset \mathscr{P}$ makes up a closed set of dimension $s^2 - 2s$:

$$\mathscr{P}_{\text{off-diag}} = \left\{P \in \mathscr{P} : p_{ii} = 0, \text{ for all } i = 1,\ldots,s\right\}. \tag{3.11}$$

In the case $s = 2$ (a two-state DTMC), we see that

$$\mathscr{P}^{\text{int}} = \mathscr{P}_{\text{IA}}.$$

On the other hand, the set $\mathscr{P}_{\text{off-diag}}$ from (3.11) is reduced to a single point. In fact, in this case the whole set $\mathscr{P}$ becomes a square, of relatively simple structure, as Worked Example 3.1.3 shows.

Another type of stochastic matrix we will refer to later in this chapter are the Hermitian, or symmetric, transition matrices $P = (p_{ij})$, with $p_{ij} = p_{ji}$. Here we use the notation $\mathscr{P}_{\text{symm}} \subset \mathscr{P}$:

$$\mathscr{P}_{\text{symm}} = \{P = (p_{ij}) \in \mathscr{P} : p_{ij} = p_{ji}, \ i, j = 1, \ldots, s\}. \qquad (3.12)$$

The matrix $P \in \mathscr{P}_{\text{symm}}$ presents an obvious equilibrium distribution π with entries $\pi_i = 1/s, \ i = 1, \ldots, s$.

Worked Example 3.1.3 Let I be $\{1, 2\}$ and P be a 2×2 transition matrix:

$$P = \begin{pmatrix} 1-p & p \\ q & 1-q \end{pmatrix},$$

where $0 \leq p, q \leq 1$. Prove that P is irreducible and aperiodic if and only if $0 < pq < 1$. In other words, the set of irreducible and aperiodic matrices is described by the condition $0 < p, q < 1$. Give a description of the communicating classes of the matrix P when the product $pq = 0$ or $pq = 1$. When is P periodic?

Solution By definition, P will be irreducible for $0 < p, q < 1$, as both states communicate and hence form a single (closed) communicating class. Aperiodicity arises from all transition probabilities being strictly positive.

On the other hand, suppose that the product $pq = 0$, so either p vanishes, or q, or both. If $p = q = 0$ then $P = \mathbf{I}$, and each state forms a closed communicating class. If $p = 0$ but $q > 0$ then state 1 becomes a closed and state 2 an open communicating class. If $q = 0$ but $p > 0$ then the converse applies. If $p = q = 1$, the chain is irreducible but periodic, with period 2. Thus, P proves to be irreducible and aperiodic only if $0 < pq < 1$. $\qquad \square$

It is convenient to fix a *Lebesgue measure* on the sets $\mathscr{R}$ and $\mathscr{P}$ from (3.5), (3.7), which will allow us to integrate and consider probability density functions on $\mathscr{R}$ and $\mathscr{P}$. By the nature of the Lebesgue measure, it will be supported by the interiors $\mathscr{R}^{\text{int}}$ and $\mathscr{P}^{\text{int}}$ specified in (3.6), (3.8). We denote an integral with respect to this measure by $\int d\lambda \times dP$ (or $\int dP$ when λ is discarded); it will be given by the natural volume (or area) on $\mathscr{R}$ and $\mathscr{P}$.

To be more specific, we should fix independent coordinates on $\mathcal{R}$ and $\mathcal{P}$. For example, we can write

$$d\lambda \times dP = \prod_{1 \leq j \leq s-1} d\lambda_j \times \prod_{1 \leq i \leq s} \prod_{1 \leq j \leq s-1} dp_{ij} \text{ or } dP = \prod_{1 \leq i \leq s} \prod_{1 \leq j \leq s-1} dp_{ij}, \quad (3.13)$$

which corresponds to omitting the 'last' entries λ_s and p_{is} as they linearly depend on the rest. Of course, here a purely subjective choice is made. In a similar way, the Lebesgue measure can be defined on the set $\mathcal{P}_{\text{off-diag}}$ (see (3.11)) and other subsets of $\mathcal{R}$ and $\mathcal{P}$.

Before we proceed further, it would be appropriate to comment on the contents of the current chapter. Its character differs from Chapters 1 and 2. Our original idea was to focus mainly on statistical methods specific to Markov chains. In our view, the statistics of Markov chains have been shaped during recent years by the theory of so-called hidden Markov models. They proved extremely successful in a variety of applications such as genome analysis. However, as some of the topics and methods discussed in Chapter 3 extend far beyond Markov chain theory, their role in this book extends to providing a 'training' ground for further chapters.

Also, formally speaking, the material of this chapter has not so far been taught in Cambridge. We have taught various parts of this material elsewhere or followed lectures and example classes given by other colleagues. Consequently, most problems in this chapter are not from the Cambridge Mathematical Tripos. However, we wished to follow the same plan as in other chapters, by setting problems and examples at the level which, in our opinion, would correspond to Cambridge standards.

Sometimes, in the process of studying the statistical properties of Markov chains we can directly rely on facts and/or methods from the basic Statistics course which focused on IID samples; see Chapters 3 and 4 of Volume 1. But more often than not we will face the task of developing more general methods. This forms the principal line of presentation in this chapter. Moreover, most of the methods emerging for Markov chains will be applicable in much wider contexts in *Statistics of Random Processes and Fields*. This discipline finds a broad range of applications, including economics, finance, telecommunication and image processing, to name a few. Learning these methods in the relatively simple case of Markov chains should prepare the interested reader to move further, by working on specialised books and articles. On the other hand, at our present level we will be able to prove formally some important facts which were used without proof in Volume 1: the amount of work for writing these proofs in the IID case comes to almost the same as in the Markov (or even a more general) case. We hope this will bring additional benefits to the interested reader.

3.2 Likelihood functions, 1. Maximum likelihood estimators

The Age of Inference

(From the series 'Movies that never made it to the Big Screen'.)

The structure of the likelihood functions (3.2) and (3.3) is clarified when we introduce the statistic $n_{ij}(\mathbf{x})$, for all $i, j \in I = \{1, \ldots, s\}$, the occurrence number of transition $i \to j$ in a sample $\mathbf{x}$:

$$n_{ij}(\mathbf{x}) = \sum_{m=1}^{n} \mathbf{1}(x_{m-1} = i, x_m = j), \tag{3.14}$$

the number of pairs (x_{m-1}, x_m) with $x_{m-1} = i$, $x_m = j$.

Then the likelihood functions $L^\theta(\mathbf{x})$ and $l^\theta(\mathbf{x})$ in (3.2) and (3.3) are written as

$$L(\mathbf{x}, \theta) = \lambda_{x_0}^\theta \prod_{1 \leq i, j \leq s} \left(p_{ij}^\theta\right)^{n_{ij}} \text{ (full), and } l(\mathbf{x}, \theta) = \prod_{1 \leq i, j \leq s} \left(p_{ij}^\theta\right)^{n_{ij}} \text{ (reduced). (3.15)}$$

From now on, the argument $\mathbf{x}$ in $n_{ij}(\mathbf{x})$ and subscript $\mathbf{X}$ in $L_\mathbf{X}(\mathbf{x}, \theta)$ and $l_\mathbf{X}(\mathbf{x}, \theta)$ will be omitted. Observe that $\sum_{1 \leq i, j \leq s} n_{ij} = n$.

In this regard, the following definition seems natural. We call a function $T(\mathbf{x})$ (in general, with vector values) a *sufficient statistic* (for θ) if, for all samples $\mathbf{x}$, the conditional probability $\mathbb{P}^\theta(\mathbf{X} = \mathbf{x} | T(\mathbf{X}) = t)$ does not depend on $\theta \in \Theta$, where t stands for the value of the statistic T at $\mathbf{x}$: $t = T(\mathbf{x})$. Formally,

$$\mathbb{P}^{\theta_1}(\mathbf{X} = \mathbf{x} \mid T(\mathbf{X}) = t) = \mathbb{P}^{\theta_2}(\mathbf{X} = \mathbf{x} \mid T(\mathbf{X}) = t), \text{ for all } \theta_1, \theta_2 \in \Theta. \tag{3.16}$$

Then the factorisation criterion holds: *T is sufficient for θ if and only if the likelihood $f_\mathbf{X}(\mathbf{x}, \theta)$ can be written as a product $g(T(\mathbf{x}), \theta)h(\mathbf{x})$ for some functions g and h where $h(\mathbf{x})$ does not depend on θ.* The proof repeats that given for (discrete) IID samples (see Volume 1, page 211).

Example 3.2.1 A sufficient statistic (actually a vector), in the case of likelihoods (3.2) is given by $(x_0, \{n_{ij}\})$ (or $\{n_{ij}\}$ and in the case of (3.3)) where $\{n_{ij}\}$ is the collection of occurrence numbers in $\mathbf{x}$.

Another property worth mentioning is unbiasedness. An estimator $\widehat{\theta}(\mathbf{x})$ of the parameter θ is called *unbiased*, if, for all $\theta \in \Theta$, the expected value $\mathbb{E}^\theta \widehat{\theta}(\mathbf{X})$ under the probability distribution $\mathbb{P}^\theta$ coincides with θ. This definition applies equally when θ is a scalar or a multi-dimensional parameter.

Example 3.2.2 Consider the likelihood (3.15), where $\lambda^\theta = \pi^\theta$, the equilibrium distribution for the transition matrix P^θ. We will assume that the whole matrix P forms the parameter $\theta = P$, and omit the superscript θ from the notation $\mathbb{P}^\theta$ and

$\mathbb{E}^\theta$. Then n_{ij}/n provides an unbiased estimator for the probability $\mathbb{P}(X_{k-1} = i, X_k = j) = \pi_i p_{ij}$. In fact,

$$
\begin{aligned}
\mathbb{E}\left[\frac{n_{ij}(\mathbf{X})}{n}\right] &= \left(\frac{1}{n}\right) \mathbb{E}\left[\sum_{k=1}^{n} \mathbf{1}(X_{k-1} = i, X_k = j)\right] \\
&= \frac{1}{n}\sum_{k=1}^{n} \mathbb{E}\left[\mathbf{1}(X_{k-1} = i, X_k = j)\right] \\
&= \frac{1}{n}\sum_{k=1}^{n} \mathbb{P}(X_{k-1} = i, X_k = j) \\
&= \pi_i p_{ij}.
\end{aligned}
\tag{3.17}
$$

We know, from the statistics of IID samples, that a powerful technique lies in considering *maximum likelihood estimators* (MLEs). This also works for Markov chain samples. However, one should be careful: even the definition of an MLE may require a subtle analysis.

The definition of a MLE for Markov samples proves similar to that for IID samples: a MLE $\theta^*(\mathbf{x})$ for a parameter θ satisfies

$$
\theta^*(\mathbf{x}) = \begin{cases} \underset{\theta}{\operatorname{argmax}}\, L(\mathbf{x}, \theta) & \text{(full likelihood)}, \\ \underset{\theta}{\operatorname{argmax}}\, l(\mathbf{x}, \theta) & \text{(reduced likelihood)}. \end{cases}
\tag{3.18}
$$

As in the case of IID samples, it often turns out more convenient to maximise the log-likelihoods $\mathscr{L}(\mathbf{x}, \theta) = \ln L(\mathbf{x}, \theta)$ and $\ell(\mathbf{x}, \theta) = \ln l(\mathbf{x}, \theta)$. In this section, the parameter θ is taken to be the whole transition matrix P, and the initial vector λ coincides with π, the equilibrium distribution. Thus, the full likelihood is defined by

$$
L(\mathbf{x}; P) = \pi_{x_0} \prod_{k=1}^{n} p_{x_{k-1}x_k},
\tag{3.19}
$$

and the corresponding log-likelihood by

$$
\mathscr{L}(\mathbf{x}, P) = \ln \pi_{x_0} + \sum_{k=1}^{n} \ln p_{x_{k-1}x_k},
\tag{3.20}
$$

where $\pi = (\pi_i, i \in I)$ constitutes the equilibrium distribution for the matrix P. The reduced likelihood is given by

$$
l(\mathbf{x}, P) = \prod_{k=1}^{n} p_{x_{k-1}x_k},
\tag{3.21}
$$

and its log-likelihood as

$$\ell(\mathbf{x}, P) = \sum_{k=1}^{n} \ln p_{x_{k-1} x_k}. \tag{3.22}$$

For instance, in the content of Worked Example 3.1.3, the transition matrix is

$$P = \begin{pmatrix} 1-p & p \\ q & 1-q \end{pmatrix}, \text{ so } p_{ii} = \begin{cases} 1-p, & i=0, \\ 1-q, & i=1, \end{cases} \text{ where } 0 \le p, q \le 1, \tag{3.23}$$

and the equilibrium distribution $\pi = (\pi_0, \pi_1)$ is

$$\pi_0 = \frac{q}{p+q}, \quad \pi_1 = \frac{p}{p+q}. \tag{3.24}$$

Sample vectors $\mathbf{x}$ show all entries $x_j = 0$ or $1, 0 \le j \le n$. Then (3.19) becomes

$$L = \left(\frac{1}{p+q}\right) p^{x_0+n_{01}} (1-p)^{n_{00}} q^{1-x_0+n_{10}} (1-q)^{n_{11}} \tag{3.25}$$

while (3.21) reads

$$l = p^{n_{01}} (1-p)^{n_{00}} q^{n_{10}} (1-q)^{n_{11}}. \tag{3.26}$$

For brevity, the reference to argument $\mathbf{x}$ and/or parameter P in $L(\mathbf{x}, P)$, and $l(\mathbf{x}, P)$ is often omitted.

Worked Example 3.2.3 (i) Let (X_m) be a two-state DTMC in equilibrium, with stochastic matrix P of transition probabilities (see (3.23)), where $0 < q, p < 1$. Calculate the covariance $\text{Cov}(X_j, X_{j+1}) = \mathbb{E}(X_j X_{j+1}) - \mathbb{E}(X_j)\mathbb{E}(X_{j+1})$ and the correlation coefficient

$$\rho = \text{Corr}(X_j, X_{j+1}) = \frac{\text{Cov}(X_j, X_{j+1})}{\sqrt{\text{Var} X_j} \sqrt{\text{Var} X_{j+1}}}, \tag{3.27}$$

as functions of q and p.

In some applications, parameters of interest could be $p+q$ and $\sigma = |p-q|$; re-write matrix P in terms of parameters $p+q$ and σ.

(ii) Prove that $\text{Cov}(X_j, X_{j+k}) = \rho^k$, for all $k \ge 1$.

Solution (i) The equilibrium distribution π takes the form

$$\pi_0 = \frac{q}{p+q}, \quad \pi_1 = \frac{p}{p+q},$$

which can be readily verified. Then

$$
\begin{aligned}
\mathrm{Cov}\,(X_j, X_{j+1}) &= \mathbb{E}(X_j X_{j+1}) - \mathbb{E}(X_j)\mathbb{E}(X_{j+1}) \\
&= \mathbb{P}(X_j = X_{j+1} = 1) - \mathbb{P}(X_j = 1)\,\mathbb{P}(X_{j+1} = 1) \\
&= \pi_1 p_{11} - \pi_1^2 \\
&= \frac{p}{p+q}\,(1-q) - \frac{p^2}{(p+q)^2} = \frac{pq}{(p+q)^2}\,(1-p-q).
\end{aligned}
$$

Next,

$$
\begin{aligned}
\mathrm{Var}\,X_j &= \mathbb{E}(X_j^2) - \big(\mathbb{E}(X_j)\big)^2 = \mathbb{E}(X_j) - \big(\mathbb{E}(X_j)\big)^2 = \pi_1 - \pi_1^2 \\
&= \frac{p}{p+q} - \frac{p^2}{(p+q)^2} = \frac{pq}{(p+q)^2},
\end{aligned}
$$

which is also equal to $\mathrm{Var}\,X_{j+1}$ as the chain sits in equilibrium. So,

$$
\sqrt{\mathrm{Var}\,X_j}\,\sqrt{\mathrm{Var}\,X_{j+1}} = pq/(p+q)^2 = \pi_1 - \pi_1^2,
$$

and

$$
\rho = \mathrm{Corr}\,(X_j, X_{j+1}) = 1 - p - q.
$$

Then: (a) if $p > q$ then $\sigma = p - q$, and

$$
P = \begin{pmatrix} \dfrac{1+\rho-\sigma}{2} & \dfrac{1-\rho+\sigma}{2} \\[2mm] \dfrac{1-\rho-\sigma}{2} & \dfrac{1+\rho+\sigma}{2} \end{pmatrix} \quad \text{and} \quad \pi = \left(\frac{1-\rho-\sigma}{2(1-\rho)}, \frac{1-\rho+\sigma}{2(1-\rho)} \right);
$$

(b) if $p < q$ then $\sigma = q - p$, and the matrix P is obtained from the former by swapping the rows and π by exchanging the entries.

(ii) Such direct calculations happen to be cumbersome and most often lead to hard-to-detect errors. To circumvent this obstacle, observe that the characteristic polynomial $\det\,(P - \kappa \mathbf{I})$ of matrix P from (3.23) reads

$$
(1-p-\kappa)(1-q-\kappa) - pq
$$

with roots $\kappa = 1$ and $\kappa = 1 - p - q = \rho$, being the eigenvalues of P. In addition, we know that $\pi = (\pi_0, \pi_1)$ forms the row eigenvector of P corresponding to the eigenvalue 1. In other words, the expression

$$
\mathrm{Corr}\,(X_j, X_{j+1}) = \frac{\pi_1 p_{11} - \pi_1^2}{\pi_1 - \pi_1^2} = 1 - p - q
$$

involving (a) the bottom right entry p_{11} of matrix P and (b) the right entry π_1 of the row eigenvector π with eigenvalue 1, yields the other eigenvalue of P. This holds for any 2×2 stochastic matrix P.

Now, obviously,

$$\text{Corr}(X_j, X_{j+k}) = \frac{\text{Cov}(X_j, X_{j+k})}{\text{Var}\, X_j} = \frac{\pi_1 p_{11}^{(k)} - \pi_1^2}{\pi_1 - \pi_1^2},$$

where $p_{11}^{(k)}$ makes up the bottom right entry of the stochastic matrix P^k, the vector π again being the eigenvector of P^k with the eigenvalue 1. Thus, $\text{Corr}(X_j, X_{j+k})$ must be equal to the other eigenvalue of P^k, the latter being nothing but ρ^k. □

We proceed with a discussion of the correctness of the definition of the MLE as a solution to the *maximum likelihood equation* (3.18). For definiteness, assume that the parameter θ is P, the transition matrix. It is represented by a point in the set $\mathscr{P}$ determined in (3.7). Recall the assumption that the matrix P is irreducible and aperiodic: $P \in \mathscr{P}_{\text{IA}}$.

We have seen earlier that the MLE p_{ij}^* of the transition probability p_{ij} in reduced likelihood $l(\mathbf{x}, P)$ is easily derived:

$$p_{00}^* = \frac{n_{00}}{n_{00} + n_{01}}, \; p_{01}^* = \frac{n_{01}}{n_{00} + n_{01}}, \; p_{10}^* = \frac{n_{10}}{n_{10} + n_{11}}, \; p_{11}^* = \frac{n_{11}}{n_{10} + n_{11}}.$$

However, the analysis of the MLE in full likelihood $L(\mathbf{x}, P)$ turns out to be far more involved, and at present has been formally completed only for a DTMC with two states. We present the related calculations in Example 3.2.4.

Example 3.2.4 Consider a two-state DTMC in equilibrium, with transition matrix P as in (3.23), where $q + p > 0$. This leads to the unique equilibrium distribution specified in (3.24). We assume that $n > 1$: that is, we consider at least three observations. In terms of the occurrence numbers, $n_{00} + n_{01} + n_{10} + n_{11} > 1$.

To find a MLE of $\theta = P$ for the likelihood L in (3.25), we need to find a maximum point P of the RHS in (3.25) in p and q, $0 \leq p, q \leq 1$, $p + q > 0$. We begin with a degenerate case where $n_{00}(x_0 + n_{01})(1 - x_0 + n_{10})n_{11} = 0$: that is, at least one transition did not occur in $\mathbf{x}$.

First, assume that $x_0 + n_{01} = 0$. Then, clearly, $x_j = 0$ for all $j = 0, \ldots, n$, hence $n_{00} = n$, $1 - x_0 + n_{10} = 1$, $n_{11} = 0$, and

$$L^\theta = \frac{q}{p+q}(1-p)^n.$$

Obviously, $p = 0$ yields a maximum value $L^\theta = 1$, for all $0 < q < 1$. In other words, the MLE turns out non-unique with the form

$$P = \begin{pmatrix} 1 & 0 \\ q & 1-q \end{pmatrix}, \quad 0 < q < 1, \tag{3.28}$$

which specifies 0 as an absorbing state. The symmetric case $1 - x_0 + n_{10} = 0$ is treated in the same fashion.

If $(x_0 + n_{01})(1 - x_0 + n_{10}) > 0$, but $n_{00} = n_{11} = 0$, then the (unique) MLE is found at $p = q = 1$

$$P = \begin{pmatrix} 0 & 1 \\ 1 & 0 \end{pmatrix} \tag{3.29}$$

(a deterministic jump to the other state).

Next, assume that $(x_0 + n_{01})(1 - x_0 + n_{10}) > 0$, but $n_{00} = 0$ and $n_{11} \geq 1$. Then

$$L^\theta = \frac{1}{p+q} \, p^{x_0 + n_{01}} \, q^{1 - x_0 + n_{10}} (1-q)^{n_{11}}.$$

We easily see that to get a maximum of L^θ, we should take the greatest possible value of p, i.e. $p = 1$. Next, the maximum in q does not occur at the boundary value $q = 0$ or 1, where L^θ vanishes, but at an interior point $q \in (0,1)$. To find this point, take the logarithm $\ln L^\theta$, substitute $p = 1$ and differentiate with respect to q:

$$\frac{\partial}{\partial q} \mathscr{L}^\theta \Big|_{p=1} = \frac{\partial}{\partial q} \ln L^\theta \Big|_{p=1} = -\frac{1}{1+q} + \frac{1 - x_0 + n_{10}}{q} - \frac{n_{11}}{1-q}. \tag{3.30}$$

The equation $\partial \ln L^\theta / \partial q \big|_{p=1} = 0$ yields a quadratic equation for q:

$$-(n_{10} + n_{11} - x_0)q^2 - (1 + n_{11})q + 1 - x_0 + n_{10} = 0,$$

with the solutions

$$q^\pm = \frac{-(n_{11} + 1) \pm \sqrt{(n_{11} + 1)^2 + 4(n_{10} + n_{11} - x_0)(1 - x_0 + n_{10})}}{2(n_{10} + n_{11} - x_0)}, \tag{3.31}$$

of which we take q^+, with the $+$ sign in front of the square root. So, in the case under consideration, the MLE becomes unique:

$$P = \begin{pmatrix} 0 & 1 \\ q^+ & 1 - q^+ \end{pmatrix}. \tag{3.32}$$

The symmetric case where $n_{11} = 0$ and $n_{00} \geq 1$ is analysed similarly.

Now consider the general case where $n_{00}(x_0 + n_{01})(1 - x_0 + n_{10})n_{11} > 0$. The full likelihood function

$$(p, q) \mapsto \text{the RHS of (3.25)}$$

is continuous on $[0,1] \times [0,1] \setminus \{(0;0)\}$, the closed unit square less the origin, and is extended by continuity to the value 0 at the origin. On the boundary $\partial([0,1] \times [0,1]) = \{0 \leq p, q \leq 1 : p(1-p)q(1-q) = 0\}$, this function drops to 0, hence its maximum lies in the open square $(0,1) \times (0,1)$.

Furthermore, if we show that the stationarity equations

$$\frac{\partial}{\partial p} \mathscr{L}^\theta = \frac{\partial}{\partial q} \mathscr{L}^\theta = 0$$

exhibit a unique solution (in $(0,1) \times (0,1)$) then they will yield a (unique) global maximum, i.e. the (uniquely determined) MLE.

So, take the logarithm and differentiate. The stationarity equations read

$$\frac{x_0 + n_{01}}{p} - \frac{n_{00}}{1 \quad p} = \frac{1}{p \mid q} = \frac{1 - x_0 + n_{10}}{q} - \frac{n_{11}}{1 \quad q}, \tag{3.33}$$

or, equivalently,

$$(x_0 + n_{01} - 1 + n_{00})p^2 - \left[x_0 + n_{01} - 1 \right.$$
$$\left. - (x_0 + n_{01} + n_{00})q \right]p - (x_0 + n_{10})q = 0, \tag{3.34}$$

$$(-x_0 + n_{10} + n_{11})q^2 - \left[-x_0 + n_{10} \right.$$
$$\left. - (1 - x_0 + n_{10} + n_{11})p \right]q - (1 - x_0 + n_{10})p = 0. \tag{3.35}$$

Writing these equations in the standard notation

$$Ap^2 + Bpq + Cq^2 + Dp + Eq + F = 0,$$

one observes that $B^2 > 4AC$ (which equals 0 in both equations); this fact indicates that each of the equations defines a hyperbola in the (p, q)-plane. It suffices to establish that these hyperbolas cannot have more than one common point in the open square $(0,1) \times (0,1)$ (they exhibit at least one point of intersection as the maximum of $\mathscr{L}$ is achieved in $(0,1) \times (0,1)$).

As a first useful remark notice that only one branch of each of the hyperbolas crosses $(0,1) \times (0,1)$. In fact, assuming that $x_0 + n_{01} + n_{00} > 1$, the equation (3.34) is solved, for p, by

$$p^{\pm} = \frac{1}{2(x_0 + n_{01} - 1 + n_{00})} \left[x_0 + n_{01} - 1 - (x_0 + n_{01} + n_{00})q \right.$$
$$\pm \left(\left(x_0 + n_{01} - 1 - (x_0 + n_{01} + n_{00})q \right)^2 \right.$$
$$\left. + 4(x_0 + n_{01} - 1 + n_{00})(x_0 + n_{10})q \right)^{1/2} \right]$$

$$= \frac{-b \pm \sqrt{b^2 - 4ac}}{2a}, \tag{3.36}$$

where

$$\left. \begin{aligned} a &= x_0 + n_{01} - 1 + n_{00}, \\ b &= -\left(x_0 + n_{01} - 1 - (x_0 + n_{01} + n_{00})q \right), \\ c &= -(x_0 + n_{10})q. \end{aligned} \right\} \tag{3.37}$$

To check this remark, suppose that $0 < q < 1$. Then the coefficients a, b, c in (3.36), (3.37) satisfy the following inequalities.

(a) The discriminant $b^2 - 4ac$ is non-negative. This holds because $n_{10} + n_{01} \geq 1$.

(b) The plus-solution $p^+ = (-b + \sqrt{b^2 - 4ac})/(2a)$ from (3.36) lies in $(0,1)$, or, equivalently, $0 < -b + \sqrt{b^2 - 4ac} < 2a$. The left bound here is true as $4ac < 0$, while the right bound holds because q is chosen to be > 0.

(c) The minus-solution $p^- = (-b - \sqrt{b^2 - 4ac})/(2a)$ from (3.36) is ≤ 0, which is straightforward to show.

This yields a unique solution in p for any $q \in (0,1)$ for equation (3.34) when $x_0 + n_{01} + n_{00} > 1$, meaning that the first hyperbola features a single branch crossing the open square $(0,1) \times (0,1)$. A similar argument works for the second hyperbola, under the assumption that $-x_0 + n_{10} + n_{11} > 0$.

Thus, we set

$$g(q) = \frac{-b + \sqrt{b^2 - 4ac}}{2a}, \quad \text{with } g(q) \in (0,1) \text{ for } 0 < q < 1,$$

and define $h(p)$ in a symmetric fashion, via the equation (3.35). We are interested in the solution (p^*, q^*), $0 < p^*, q^* < 1$, to

$$p^* = g(q^*), \quad q^* = h(p^*), \quad \text{or } p^* = g \circ h(p^*), \quad q^* = h \circ g(q^*). \tag{3.38}$$

It is possible to specify two open sub-intervals $J, K \subset (0,1)$, where the values of p^* and q^* can be confined in terms of the coefficients in (3.34)–(3.37):

$$J = \left(\frac{x_0 + n_{01} - 1}{x_0 + n_{01} - 1 + n_{00}}, \frac{x_0 + n_{01}}{x_0 + n_{01} + n_{00}} \right)$$

and

$$K = \left(\frac{-x_0 + n_{10}}{-x_0 + n_{10} + n_{11}}, \frac{1 - x_0 + n_{10}}{1 - x_0 + n_{10} + n_{11}} \right).$$

So, we claim that

(i) if $p \in (0,1)$, then $h(p) \in K$,

(ii) if $q \in (0,1)$, then $g(q) \in J$, which implies that solutions to (3.38) must obey

$$(p^*, q^*) \in J \times K. \tag{3.39}$$

To verify (i), pick $\widetilde{p} \in (0,1)$ and set $\widetilde{q} = h(\widetilde{p})$ and $r = \widetilde{p}/(\widetilde{p} + \widetilde{q})$, with $0 < \widetilde{q} < 1$ and $0 < r < 1$. Observe that $\widetilde{p}, \widetilde{q}$ and r obey the RHS of (3.33), viz.

$$\frac{1 - r}{\widetilde{q}} = \frac{1 - x_0 + n_{10}}{\widetilde{q}} - \frac{n_{11}}{1 - \widetilde{q}},$$

whence

$$\widetilde{q} = \frac{-x_0 + n_{10} + r}{-x_0 + n_{10} + n_{11} + r} \in K.$$

A similar argument yields (ii).

Next, we should check that, unless $n_{01} = 1 - x_0$ or $n_{10} = x_0$, the following two assertions hold true:

(iii) if $p \in (0,1)$, then $h'(p) \in (0,1)$,

(iv) if $q \in (0,1)$, then $g'(q) \in (0,1)$.

To check (iv), we differentiate, in q, (3.34):

$$g'(q) = \frac{x_0 + n_{01} - (x_0 + n_{01} + n_{00})g(q)}{2g(q)(x_0 + n_{01} + n_{00} - 1) + (x_0 + n_{01} + n_{00})q - (x_0 + n_{01} - 1)}.$$

Then pick $\widetilde{q} \in (0,1)$ and assume that $g'(\widetilde{q}) \notin (0,1)$; that is, $1/g'(\widetilde{q}) \leq 1$, or, equivalently,

$$2g(\widetilde{q})(x_0 + n_{01} + n_{00} - 1) + (x_0 + n_{01} + n_{00})\widetilde{q} - (x_0 + n_{01} - 1)$$
$$\leq x_0 + n_{01} - (x_0 + n_{01} + n_{00})g(\widetilde{q}).$$

Owing to the fact that $g(\widetilde{q}) \in J$, we see that

$$\frac{2(x_0 + n_{01} - 1)(x_0 + n_{01} + n_{00} - 1)}{x_0 + n_{01} + n_{00} - 1} + (x_0 + n_{01} + n_{00})\widetilde{q} - (x_0 + n_{01} - 1)$$
$$\leq x_0 + n_{01} - \frac{(x_0 + n_{01} + n_{00})(x_0 + n_{01} - 1)}{x_0 + n_{01} + n_{00} - 1},$$

or equivalently,

$$(x_0 + n_{01} + n_{00})\widetilde{q} + (x_0 + n_{01} - 1) \leq \frac{n_{00}}{x_0 + n_{01} + n_{00} - 1}.$$

So, unless $x_0 + n_{01} = 1$, the LHS remains > 1, which leads to a contradiction. Thus, assertion (iv) holds whenever $n_{01} > 1 - x_0$. In a similar way, assertion (iii) holds whenever $1 - x_0 + n_{10} > 1$, i.e. $n_{10} > x_0$.

One can now finish the analysis of (3.33), (3.34) and (3.35). To start with, let $x_0 + n_{01}$ and $1 - x_0 + n_{10}$ be > 1. Assume two distinct solutions of (3.38) are found, (p_1^*, q_1^*) and (p_2^*, q_2^*), both from $(0,1) \times (0,1)$: in fact, from $J \times K$ (cf. (3.39)). Suppose, for example, that $p_1^* < p_2^*$. Then applying Rolle's Theorem yields that there exists $\widetilde{p} \in J$ with

$$g'(h(\widetilde{p}))h'(\widetilde{p}) = 1.$$

But this contradicts the fact that both $h'(\widetilde{p})$ and $g'(h(\widetilde{p}))$ are < 1. The case where $q_1^* < q_2^*$ is treated in a similar fashion. Thus, in the case where $x_0 + n_{01} > 1$ and $1 - x_0 + n_{10} > 1$, (3.33), (3.34) and (3.35) have a unique solution in $(0,1) \times (0,1)$.

It remains to consider border cases where $x_0 + n_{01}$ or $1 - x_0 + n_{10}$ equal 1. For $x_0 + n_{01} = 1$, the value $1 - x_0 + n_{10}$ equals 1 or 2 (the possibility $1 - x_0 + n_{10} = 0$ was considered earlier). The above argument then should be modified and becomes somewhat longer and technically more involved, although follows the same idea. We omit this case from the discussion.

Unfortunately, there exists no convenient formula for the MLE

$$P^\theta = \begin{pmatrix} 1 - p^* & p^* \\ q^* & 1 - q^* \end{pmatrix}$$

in a general situation. A number of authors have produced results of numerical calculations, for different examples of sample vector **x**. For instance, with $n = 15$ and $\mathbf{x}^T = (0111001010000101)$, the MLE in the full likelihood is $p^* = 0.5329$, $q^* = 0.6893$, whereas the MLE in the reduced likelihood $p^* = 5/9 = 0.5556$, $q^* = 2/3 = 0.6667$.

We want to stress that, although the example of a two-state DTMC remains very basic, the above analysis of the MLEs appeared in the literature only recently; see S. Bisgaard and L.E. Travis. "Existence and uniqueness of the solution of the likelihood equations for binary Markov chains", *Statistics and Probability Letters*, **12** (1991), 29–35. Furthermore, prior to this publication, several conflicting statements had been made in the literature about the nature of the MLEs for a DTMC with two states. A quotation from the above paper reads: "It is ... perhaps disheartening, that this simplest example ... requires a very careful enumeration of special cases to establish the existence and uniqueness of [a solution to] the likelihood equations. More complicated Markov chains may be even more mischievous."

3.3 Consistency of estimators. Various forms of convergence

Consistency is too weak a property
to be of much interest in itself.
E.H. Lehmann (1917–), American statistician

This section of the book deals with an issue that has a flavour that is definitely more probabilistic than statistical, and that will reappear in a number of chapters in a later volume. Nevertheless, we believe we should introduce the relevant concepts here. One nice property of MLEs which we mentioned before is consistency. An estimator $\widehat{\theta}_n(\mathbf{x}_n)$ of parameter θ is called *consistent* if it converges to the 'true' value of the parameter as the size of the sample goes to infinity:

$$\lim_{n \to \infty} \widehat{\theta}_n(\mathbf{X}_n) = \theta. \tag{3.40}$$

There lies a subtle point about the limit here. The sample $\mathbf{X}_n = \begin{pmatrix} X_0 \\ \vdots \\ X_n \end{pmatrix}$ is random,

and its distribution $\mathbb{P}^\theta$ depends on $\theta \in \Theta$ (and is defined by the transition matrix P^θ or by a pair $(\lambda^\theta, \Gamma^\theta)$). So, we need to specify the form of convergence. Two specific forms of convergence will be discussed in this section: convergence in probability and convergence almost surely, or with probability 1. They prove popular in several areas of probability and measure theory and constantly appear in various chapters of the dynamical systems, random processes and fields, theoretical statistics and functional analysis literature.

Correspondingly, one calls an estimator $\widehat{\theta}_n(\mathbf{x}_n)$ of the parameter θ (i) consistent in probability, for a set $\Theta_0 \subseteq \Theta$, if, for all $\theta \in \Theta_0$, the limit in (3.40) holds in probability, and (ii) consistent almost surely, or with probability 1, if the limit holds almost surely, or with probability 1. As a set Θ_0 we will take all values of θ for which the transition matrix $\mathbb{P}^\theta$ is irreducible and aperiodic.

The basic definitions here are as follows:

Definition 3.3.1 A sequence of random variables U_n converges in probability, as $n \to \infty$, to a constant v, if for all $\varepsilon > 0$,

$$\lim_{n \to \infty} \mathbb{P}\left(|U_n - v| \geq \varepsilon\right) = 0, \ \text{i.e.} \ \lim_{n \to \infty} \mathbb{P}\left(|U_n - v| < \varepsilon\right) = 1. \tag{3.41}$$

More generally, U_n converge in probability to a random variable V if, for all $\varepsilon > 0$,

$$\lim_{n \to \infty} \mathbb{P}\left(|U_n - V| \geq \varepsilon\right) = 0, \ \text{i.e.} \ \lim_{n \to \infty} \mathbb{P}\left(|U_n - V| < \varepsilon\right) = 1. \tag{3.42}$$

Convergence in probability is denoted by $U_n \xrightarrow{P} v$ and $U_n \xrightarrow{P} V$.

In Volume 1, Section 1.6, we saw that the weak Law of Large Numbers corresponds to convergence in probability

$$\frac{1}{n} \sum_{k=1}^{n} X_k \xrightarrow{P} \mu$$

for a sum of IID RVs $X_1, X_2, \ldots$ with finite mean $\mu = \mathbb{E}X_k$ and finite variance $\operatorname{Var} X_k = \sigma^2$. This follows from the Chebyshev inequality:

$$
\mathbb{P}\left(\left|\frac{1}{n}\sum_{1\leq k\leq n}X_k-\mu\right|\geq\varepsilon\right) = \mathbb{P}\left(\frac{1}{n}\left|\sum_{1\leq k\leq n}X_k-n\mu\right|\geq\varepsilon\right)
$$

$$
\leq \left(\frac{1}{n^2\varepsilon^2}\right)\mathbb{E}\left(\sum_{1\leq k\leq n}X_k-n\mu\right)^2
$$

$$
= \left(\frac{1}{n^2\varepsilon^2}\right)\mathrm{Var}\left(\sum_{1\leq k\leq n}X_k\right)
$$

$$
= \left(\frac{1}{n^2\varepsilon^2}\right)\sum_{1\leq k\leq n}\mathrm{Var}\,X_k = \frac{\sigma^2}{n\varepsilon^2}. \qquad (3.43)
$$

Worked Example 3.3.2 Let (X_m) be a Markov chain. Subject to suitable assumptions, state and prove the weak Law of Large Numbers for $\sum_{1\leq k\leq n}X_k$. Your assumptions should include, but not be reduced to, the case where (X_n) are independent random variables.

Solution It is tempting to build an argument similar to the above when (X_m) is a Markov chain. Assume that the chain runs over a finite state space I with the total number of states $|I|=s$, transition matrix $P=(p_{ij})$ and initial distribution $\lambda=(\lambda_i)$. What shall we take as the value of μ? As a constant independent of m, it should be the mean value $\mathbb{E}(X_k)$ in equilibrium:

$$
\mu = \mu_{\mathrm{eq}} = \sum_{j\in I}j\pi_j, \qquad (3.44)
$$

where $\pi=(\pi_i)$ represents an equilibrium distribution for the chain. It helps to assume that the chain is irreducible and aperiodic. Then the equilibrium distribution becomes unique and the chain approaches it with time: $\mathbb{P}(X_n=j\mid X_0=i)=p_{ij}^{(n)}\to\pi_j$ and $\mathbb{P}(X_n=j)=(\lambda P^n)_j\to\pi_j$ as $n\to\infty$, for all $i,j\in I$ and initial probability vector λ. Moreover, the convergence happens geometrically (exponentially) fast:

$$
p_{ij}^{(n)} = \pi_j + \psi_{i,j}(n), \quad\text{where } |\psi_{i,j}(n)| \leq (1-\rho)^{n-1}; \qquad (3.45)
$$

cf. (1.84) in Theorem 1.9.3.

This of course covers the case of independent random variables (where the matrix P simply consists of repeated rows coinciding with π), but is not reduced to this case alone.

As before, by Chebyshev,

$$
\mathbb{P}\left(\left|\frac{1}{n}\sum_{1\leq k\leq n}X_k-\mu\right|\geq\varepsilon\right) \leq \left(\frac{1}{n^2\varepsilon^2}\right)\mathbb{E}\left[\sum_{1\leq k\leq n}X_k-n\mu\right]^2, \qquad (3.46)
$$

and it suffices to check that the expectation of the square of the sum in the RHS is $\leq \alpha^2 n + \beta$, where the constants α and β do not depend on n (in the case of IID RVs we find equality, with $\alpha = \sigma$).

Write

$$\mathbb{E}\left[\sum_{1 \leq k \leq n} X_k - n\mu\right]^2 = \mathbb{E}\left[\sum_{1 \leq k \leq n} (X_k - \mu)\right]^2$$

and expand:

$$\mathbb{E}\left[\sum_{1 \leq k \leq n} (X_k - \mu)\right]^2 = \mathbb{E}\left[\sum_{1 \leq k \leq n} (X_k - \mu)^2\right]$$

$$+ \mathbb{E}\left[\sum_{1 \leq k_1, k_2 \leq n} \mathbf{1}(k_1 \neq k_2)(X_{k_1} - \mu)(X_{k_2} - \mu)\right]. \quad (3.47)$$

The first sum is represented as

$$\Sigma_1 = \sum_{1 \leq k \leq n} \mathbb{E}(X_k - \mu)^2.$$

When the chain sits in equilibrium, $\lambda = \pi$ and $\mathbb{E}(X_k - \mu)^2$ does not depend on k and gives the variance of X_k. Then

$$\Sigma_1 = n\sigma_{\text{eq}}^2, \quad \text{where } \sigma_{\text{eq}}^2 = \operatorname{Var} X_k = \sum_{j \in I} (j - \mu)^2 \pi_j.$$

In the general case, it will still be $O(n)$. Indeed,

$$\mathbb{E}(X_k - \mu)^2 = \sum_{j \in I} (j - \mu)^2 \mathbb{P}(X_k = j) = \sum_{j \in I} (j - \mu)^2 (\lambda P^k)_j$$

$$= \sum_{j \in I} (j - \mu)^2 \sum_{i \in I} \lambda_i p_{ij}^{(k)} = \sum_{j \in I} (j - \mu)^2 \sum_{i \in I} \lambda_i [\pi_j + \psi_{i,j}(k)]$$

$$= \sum_{j \in I} (j - \mu)^2 \pi_j \sum_{i \in I} \lambda_i + \sum_{j \in I} (j - \mu)^2 \sum_{i \in I} \lambda_i \psi_{i,j}(k)$$

$$\leq \sigma_{\text{eq}}^2 + s(1 - \rho)^{k-1} A_1^2,$$

with

$$A_1 = \max[|j - \mu| : j \in I].$$

Denote $\alpha_1^2 = \sigma_{\text{eq}}^2$, and

$$\beta = A_1^2 s \sum_{k \geq 1} (1 - \rho)^{k-1} = \frac{A_1^2 s}{\rho}.$$

Then

$$\Sigma_1 \leq \alpha_1^2 n + \beta. \quad (3.48)$$

Next, the second sum

$$\Sigma_2 = \sum_{1 \le k_1, k_2 \le n} \mathbf{1}(k_1 \ne k_2) \mathbb{E}\left[(X_{k_1} - \mu)(X_{k_2} - \mu)\right]$$

$$= 2 \sum_{1 \le k \le n} \sum_{l \ge 1} \mathbf{1}(1 \le k + l \le n) \mathbb{E}\left[(X_k - \mu)(X_{k+l} - \mu)\right].$$

The general summand here is expressed as

$$\mathbb{E}\left[(X_k - \mu)(X_{k+l} - \mu)\right] = \sum_{i,j \in I} (i - \mu)(j - \mu) \mathbb{P}(X_k = i, X_{k+l} = j)$$

$$= \sum_{i,j \in I} (i - \mu)(j - \mu)(\lambda P^k)_i P_{ij}^{(l)}$$

$$= \sum_{i,j \in I} (i - \mu)(j - \mu)(\lambda P^k)_i [\pi_j + \psi_{ij}(l)]$$

$$= \sum_{i \in I} (i - \mu)(\lambda P^k)_i \sum_{j \in I} (j - \mu)\pi_j$$

$$+ \sum_{i,j \in I} (i - \mu)(j - \mu)(\lambda P^k)_i \psi_{ij}(l).$$

Observe that $\sum_{j \in I}(j - \mu)\pi_j = \sum_{j \in I} j\pi_j - \mu = 0$, by the choice of μ. Thus,

$$\left| \mathbb{E}\left[(X_k - \mu)(X_{k+l} - \mu)\right] \right| \le A_1^2 (1 - \rho)^{l-1},$$

and

$$|\Sigma_2| \le 2n \sum_{l > 1} \left| \mathbb{E}\left[(X_k - \mu)(X_{k+l} - \mu)\right] \right| \le n\alpha_2^2, \tag{3.49}$$

where $\alpha_2^2 = 2A_1^2 s / \rho$. Hence,

$$\mathbb{E}\left(\sum_{1 \le k \le n} X_k - n\mu \right)^2 \le (\alpha_1^2 + \alpha_2^2)n + \beta,$$

as required, where α_1, α_2 and β have been specified above. This establishes the weak Law of Large Numbers $\sum_{1 \le k \le n} X_k / n \xrightarrow{\mathbb{P}} \mu$. $\qquad \square$

Definition 3.3.3 Random variables U_n converge almost surely, or with probability 1, as $n \to \infty$, to a constant v, if

$$\mathbb{P}\left(\lim_{n \to \infty} U_n = v \right) = \mathbb{P}\left(\lim_{n \to \infty} |U_n - v| = 0 \right) = 1. \tag{3.50}$$

In other words, the set of outcomes where convergence $U_n \to v$ fails has probability zero.

As before, this definition is immediately extended to a more general situation of convergence to a random variable. Namely, U_n converge almost surely (AS) to a random variable V, if

$$\mathbb{P}\left(\lim U_n = V\right) = \mathbb{P}\left(\lim |U_n - V| = 0\right) = 1. \qquad (3.51)$$

Recall that AS convergence is denoted by $U_n \xrightarrow{\text{a.s.}} v$ and $U_n \xrightarrow{\text{a.s.}} V$.

A straightforward example where convergence holds not everywhere but only almost surely is provided by the sequence of functions $U_n(x) = (-x)^n$, where x forms a point of the unit segment $[0, 1]$, i.e. $0 \le x \le 1$. As $n \to \infty$, $U_n(x) \to 0$ for $0 \le x < 1$, but not for $x = 1$. So, if we consider a uniform distribution on the unit segment, then convergence $U_n \to 0$ happens with probability 1, but not everywhere. Clearly, the uniform distribution can be replaced here by any probability distribution on $[0, 1]$ provided that it is determined by a PDF $f(x), 0 \le x \le 1$.

In fact, if the probability distribution $\mathbb{P}$ is concentrated on a finite or a countable set of outcomes, the concept of AS convergence is not needed. In this case, from the very beginning we can assume that all outcomes under consideration arise with strictly positive probabilities, and AS convergence reduces to convergence everywhere. The concept of AS convergence becomes important when the set of outcomes constitutes a continuum, although the variables U_n under consideration may take finitely many values (for instance, only 0 and 1). From this point of view, the above example of a unit segment with a uniform distribution turns out particularly convenient. This distribution corresponds to the famous *Lebesgue* measure on $[0, 1]$, an object attentively studied in analysis.

This book does not assume any knowledge of measure theory, although clearly, such knowledge, however partial, would definitely help the reader. In other words, we avoid explicit references to such concepts as measurability and integrability. Instead, we declare that all 'abstract' events and random variables we work with happen to be (Lebesgue) measurable, and ditto their complements and their (countable) unions and intersections. Moreover, we will not mention this in future (as in fact we have not in the past). A helpful result, given without proof, is the so-called Luzin theorem: *A measurable function (on $[0, 1]$) is one which can be converted into a continuous function by alteration on a set of points carrying an arbitrarily small probability. In particular, a measurable subset (of $[0, 1]$) is one whose indicator function can be converted into a continuous function by altering it on a set of points carrying an arbitrarily small probability.* (Nikolai Luzin was a leader of the Moscow school of real analysis in the 1910–1930s, where a number of prominent mathematicians started their careers, including Kolmogorov.)

Physically, the concept of AS convergence can be difficult to grasp, especially at the first attempt. This results from the concept involving two evasive objects: a sequence of RVs (U_n) which is usually not given explicitly (because it may be cumbersome) and an event of probability zero where convergence $U_n \to v$ or $U_n \to V$ fails. Here, a helpful theorem is named after Yegorov (another patriarch of the Moscow mathematics school of the 1900–1920s): *AS convergence* $U_n \to V$ *holds if and only if for all* $\delta > 0$, *there exists an event* A_δ, *of probability at least* $1 - \delta$, *such that* $U_n - V$ *converges to 0 uniformly on* A_δ: $\sup \left[|U_n(x) - V(x) : x \in A_\delta| \right] \to 0$, *as* $n \to \infty$.

Yegorov's life ended in a tragedy. In 1930 he was dismissed from his position of the Head of the Institute of Mathematics of the Moscow State University. Shortly after Yegorov was arrested by the Soviet authorities and spent several months in harsh conditions in jail. (He often showed his opposition to the official ideology and was an active member of a spiritual minority movement in Russian Orthodoxy, which did not go well with the authorities at a time of fierce anti-religious campains.) Yegorov was lucky to receive a relatively mild sentence: he was exiled to Kazan, a town on the Volga river some 250 miles from Moscow. Kazan was then (as now) the capital of the Tatar Republic, a part of the Russian Federation, and got its fame from its University where, in the 19th century, Lobachevsky taught geometry and Lenin studied law (and was expelled from, after a student revolt). However, he continued to remonstrate with the authorities and died in 1931 as a result of a hunger strike which aggravated already serious health problems. Yet local mathematics enthusiasts managed to have him buried in a central cemetery, next to Lobachevsky.

We provide a short proof of Yegorov's theorem, starting with the if part: assume that the above property is satisfied for all $\delta > 0$ and take a sequence $\delta_n = 1/n^2$. Then the complement A^c_{1/n^2} of the event A_{1/n^2} happens with probability $\mathbb{P}\left(A^c_{1/n^2} \right) \leq 1/n^2$. Hence, the union $B_m = \bigcup_{n \geq m} A^c_{1/n^2}$ occurs with probability $\mathbb{P}(B_m) \leq \sum_{n \geq m} 1/n^2 \to 0$, as $m \to \infty$. In addition, events B_m decrease in probability as n increases: $B_{m+1} \subseteq B_m$. So, the intersection $B = \bigcap_m B_m$ must have $\mathbb{P}(B) = 0$, since $\mathbb{P}(B) \leq \mathbb{P}(B_m)$, for all m. This intersection consists of points belonging to infinitely many events A^c_{1/n^2}: therefore the complement B^c is formed by points belonging to only finitely many events A^c_{1/n^2}. In other words, B^c consists of points belonging to A_{1/n^2} for all n large enough. Thus, for all point x from B^c,

$$|U_n(x) - V(x)| \leq \frac{1}{n^2} \text{ for all } n, \text{ beginning with some } n_0(x).$$

Therefore, on B^c,

$$\lim_{n \to \infty} U_n = V.$$

But B^c arises with probability $1 - \mathbb{P}(B) = 1$, so the convergence $U_n \to V$ happens almost surely.

The proof of the only if part is more subtle; it is an epitome of AS convergence. Assume that $U_n \to V$ almost surely. Let A_n^m be the event where $|U_i - V| \leq 1/m$ for all $i > n$. By definition, events A_n^m increase in probability with n (for m fixed): $A_n^m \subseteq A_{n+1}^m$. Hence, $\mathbb{P}(A_n^m) \leq \mathbb{P}(A_{n+1}^m)$. The union $A^m = \bigcup_{n \geq 1} A_n^m$ represents the event where $|U_n - V| \leq 1/m$ for all n large enough, and probability $\mathbb{P}(A^m) = \lim_{n \to \infty} \mathbb{P}(A_n^m)$. Therefore, for all $m \geq 1$ and $\delta > 0$, there exists $n_0(m, \delta)$ such that $\mathbb{P}\left(A^m \setminus A_{n_0(m,\delta)}^m\right) < \delta/2^m$. Then we set

$$A_\delta = \bigcap_{m \geq 1} A_{n_0(m,\delta)}^m.$$

If $x \in A_\delta$ then, by definition, for all $m \geq 1$, $|U_i(x) - V(x)| \leq 1/m$ when $i \geq n_0(m, \delta)$. That is, $|U_n - V|$ converges to 0 uniformly on A_δ. Next, $\mathbb{P}(A^m) = 1$ as $U_n \to V$ almost surely (for the first time using this condition in the current proof). Hence, for the complement $B_{n_0(m,\delta)}^m = \left(A_{n_0(m,\delta)}^m\right)^c$ we can write

$$\mathbb{P}\left(B_{n_0(m,\delta)}^m\right) = \mathbb{P}\left(A^m \setminus A_{n_0(m,\delta)}^m\right) < \frac{\delta}{2^m}.$$

But we want to assess $\mathbb{P}(A_\delta)$, so, for the complement $B_\delta = A_\delta^c$,

$$\mathbb{P}(B_\delta) = \mathbb{P}\left(\bigcup_{m \geq 1} B_{n_0(m,\delta)}^m\right) \leq \sum_{m \geq 1} \mathbb{P}\left(A^m \setminus A_{n_0(m,\delta)}^m\right) < \sum_{m \geq 1} \frac{\delta}{2^m} = \delta.$$

This completes the proof.

The spirit of Yegorov's theorem clarifies the relation between convergence in probability and AS convergence. Namely, the theorem implies that *if $\{U_n\}$ converges to V almost surely then it does so in probability*. For the proof, let B be the set of probability 0 where convergence fails. Given $\delta > 0$, set:

$$B_k(\delta) = \text{the event where } |U_k - V| > \delta,$$

$$C_n(\delta) = \bigcup_{k \geq n} B_k(\delta) = \text{the event where } |U_k - V| > \delta, \text{ for some } k \geq n,$$

$$R(\delta) = \bigcap_{n \geq 1} C_n(\delta) = \text{the event where } |U_k - V| > \delta, \text{ for infinitely many } k.$$

Then we see:

(i) $R(\delta) \subseteq B$, and so $\mathbb{P}(R(\delta)) = 0$;
(ii) $\mathbb{P}(R(\delta)) = \lim_{n \to \infty} \mathbb{P}(C_n(\delta))$, as $C_{n+1}(\delta) \subseteq C_{n+1}(\delta)$. Hence, $\mathbb{P}(C_n(\delta)) \to 0$;

(iii) $B_n(\delta) \subseteq C_n(\delta)$, so $\mathbb{P}(B_n(\delta)) \to 0$, as $n \to \infty$, which establishes convergence in probability.

However, the inverse implication does not hold; see Examples 3.3.4 and 3.3.5 below.

Example 3.3.4 The following popular class of examples shows that a sequence of random variables converging in probability does not necessarily converge almost surely. Consider the unit segment $[0, 1)$ with a uniform distribution, and let

$$U_{k,i} = \begin{cases} 1, & \text{if } \left(\dfrac{i-1}{k}\right) \leq x < \dfrac{i}{k}, \\ 0, & \text{otherwise}, \end{cases} \quad \text{where } i = 1, \ldots, k, \, k = 1, 2, \ldots.$$

Next, consider the sequence

$$U_{1,1}, U_{2,1}, U_{2,2}, U_{3,1}, U_{3,2}, U_{3,3}, U_{4,1}, \ldots.$$

This sequence converges to 0 in probability, since for all $0 < \varepsilon < 1$,

$$\mathbb{P}(|U_{k,i}| > \varepsilon) = \mathbb{P}(U_{k,i} = 1) = \frac{1}{k} \to 0 \text{ as } k \to \infty.$$

However, the sequence does not converge to 0 almost surely. In fact, it does not converge to 0 at any given point x, and it does not converge almost surely at all. Indeed, for all $x \in [0, 1)$ and $k \geq 1$, there exists i such that $U_{k,i}(x) = 1$.

This example admits an interesting and far-reaching interpretation. Consider a fair coin-flipping: after an initial toss we observe two outcomes: 1 (heads) and 0 (tails), after two tosses four outcomes: 11, 10, 01 and 00, after three - eight, and so on. Then set

$$Y_1 \equiv 1,$$
$$Y_2 = \mathbf{1}(\text{the first flip is 1}), \, Y_3 = \mathbf{1}(\text{the first flip is 0}),$$
$$Y_4 = \mathbf{1}(\text{the first 2 flips are 11}), \, Y_5 = \mathbf{1}(\text{the first 2 flips are 10}),$$
$$Y_6 = \mathbf{1}(\text{the first 2 flips are 01}), \, Y_7 = \mathbf{1}(\text{the first 2 flips are 00}),$$
$$Y_8 = \mathbf{1}(\text{the first 3 flips are 111}), \, Y_9 = \mathbf{1}(\text{the first 3 flips are 110}),$$

and so on. In other words, we first list all possible outcomes of arbitrarily (but finitely) many flips in a certain order and then set Y_n to be the indicator of the outcome with the order number n. We order as follows. (i) We place the outcomes of n flips (that is, strings of 1s and 0s of length n) before the outcomes of $n+1$ tosses. (ii) For a fixed length n, we order the outcomes of n flips lexicographically, beginning with $11 \ldots 11$ followed by $11 \ldots 10$ then by $11 \ldots 01$, and so on.

Thus, a general member of the sequence, Y_n, gives the indicator of the $(n - 2^{m(n)} + 1)$st outcome of $\lceil \log_2 n \rceil$ flips, where $m(n) = \lceil \log_2 n \rceil$ stands for the integer part of $\log_2 n$. In other words, we partition the natural numbers $n = 1, 2, \ldots$ into disjoint 'blocks', the mth block beginning with 2^m and ending with $2^{m+1} - 1$ (both included), $m = 1, 2, \ldots$. Next, we assign to number the n from the mth block the $(n - 2^m + 1)$st binary string of length m and set Y_n to be the indicator of exactly the outcome assigned. Then the sequence (Y_n) converges to 0 in probability. Indeed, for all $\varepsilon > 0$,

$$\mathbb{P}(|Y_n| \geq \varepsilon) = \mathbb{P}(Y_n = 1) = \frac{1}{2^{\lceil \log_2 n \rceil}} \to 0, \text{ as } n \to \infty.$$

However, the random variables Y_n do not converge to 0 almost surely. In fact, Y_n should be formally considered as a function of an outcome of infinitely many flips, which depends only on the first $m(n) = \lceil \log_2 n \rceil$ tosses. An outcome of infinitely many flips would be an infinite string of 1s and 0s, and these strings (there are a continuum of them) fill a unit segment $[0, 1]$, endowed with a uniform probability distribution (which is the Lebesgue measure on $[0, 1]$). The correspondence $x \leftrightarrow (\alpha_1 \alpha_2 \ldots)$ between a point $x \in [0, 1]$ and a binary sequence $(\alpha_1 \alpha_2 \ldots)$, with $\alpha_j = 0$ or 1, is established via the binary representation (or binary decomposition) of x: $x = \sum_{j \geq 1} \alpha_j / 2^j$. Then $Y_n(x) = 1$ for x lying in a segment of length $2^{-m(n)}$ between the points $(n - m(n)) 2^{-m(n)}$ and $(n - m(n) + 1) 2^{-m(n)}$, and $Y_n(x) = 0$ for $x \in [0, 1]$ lying outside this segment. See Figure 3.2.

More precisely and to be formally correct, we have to add to the unit segment countably many points, as we face a problem with *diadic* numbers $x \in (0, 1)$, such as $x = 1/2$ or $x = 3/4$, or in general $x = k/2^m$, where $1 \leq k \leq 2^m - 1$, $m = 1, 2, \ldots$ (precisely the points where vertical bits appear on the graphs in the diagram.) Thus, if x is diadic then there exists j_0 such that coefficient $\alpha_{j_0} = 1$, but $\alpha_j \equiv 0$

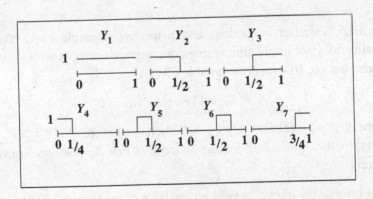

Fig. 3.2

for all $j > j_0$. However, such an x can also be represented as a sum $= \sum_{j \geq 1} \alpha'_j / 2^j$ where $\alpha'_{j_0} = 0$ and $\alpha'_j \equiv 1$ for all $j > j_0$. But this ambiguity does not affect the construction, as every single point carries Lebesgue measure 0.

Pictorially, we take a step-function over an interval $\left(0, \dfrac{1}{2^m}\right)$ of length $1/2^m$, then move the interval to the right, with step $1/2^m$, until it occupies the rightmost position (it takes exactly 2^m moves). Then we divide the length by two and repeat the motion.

Anyway, in this picture convergence almost surely means that, apart from a set of Lebesgue measure 0, $Y_n(x) = 0$ for all n large enough. But actually quite the opposite comes to be true: if $x \in [0,1]$ is not a diadic number then $Y_n(x) = 1$ for (some) indefinitely large n. This happens because, for all m, the point x will sooner or later be covered by an open interval $\left(\dfrac{k}{2^m}, \dfrac{k+1}{2^m}\right)$ of length 2^{-m}.

Coin-tossing is an example of a DTMC, with two states and all transition probabilities $1/2$; hence we can identify the Lebesgue measure on $[0,1]$ with $\mathbb{P}_{1/2,1/2}$. But a similar construction works if we consider a general two-state DTMC (X_n) in equilibrium, with transition matrix (3.9), where $0 < p, q < 1$. The difference lies in that, if $p \neq 1/2 \neq q$, then, instead of a 'nice' uniform distribution on the unit segment, we find a 'singular' measure $\mathbb{P}_{p,q}$ generated by the Markov chain. The 'singularity' means that there exists a partition of the unit segment into disjoint sets A and B (i.e. a representation $[0,1] = A \cup B$, with $A \cap B = \emptyset$) such that both probabilities $\mathbb{P}_{1/2,1/2}(B)$ and $\mathbb{P}_{p,q}(A)$ vanish. This picture is not made easier by assuming that $p + q = 1$ (in which case (X_n) forms an IID sequence). In fact, the measures $\mathbb{P}_{p,q}$ and $\mathbb{P}_{p',q'}$ become mutually singular in the above sense whenever $(p,q) \neq (p',q')$.

However, measures $P_{p,q}$ on $[0,1]$ prove no less 'physical' than the Lebesgue measure $\mathbb{P}_{1/2,1/2}$: they arise in a growing variety of situations in theory and applications.

Example 3.3.5 A shorter, but perhaps less instructive, example where convergence in probability does not imply almost sure convergence emerges with a sequence of independent, but not IID, random variables, $X_1, X_2, \ldots$. Set

$$p_n = \mathbb{P}(X_n = 1) = 1 - \mathbb{P}(X_n = 0)$$

and assume that $p_n \to 0$ but $\sum_n p_n = \infty$. Then, as before, $\mathbb{P}(|X_n| \geq \varepsilon) = p_n \to 0$. However, the Borel–Cantelli lemma guarantees that X_n does not converge to 0 almost surely.

We will involve the above facts in an analysis of maximum-likelihood estimators. Recall, an MLE $\theta^* = \theta_n^*(\mathbf{x})$ is defined as the value maximising $L(\mathbf{x}, \theta)$ or

$l(\mathbf{x}, \theta)$ in $\theta \in \Theta$ (see (3.18)). Equivalently, we will maximise the corresponding log-likelihoods $\mathscr{L}(\mathbf{x}, \theta) = \ln L(\mathbf{x}, \theta)$ or $\ell(\mathbf{x}, \theta) = \ln l(\mathbf{x}, \theta)$:

$$\widehat{\theta}^* = \mathrm{argmax}\left[\mathscr{L}(\mathbf{x}, \theta), \, \theta \in \Theta\right], \quad \text{or} \quad \widehat{\theta}^* = \mathrm{argmax}\left[\ell(\mathbf{x}, \theta), \, \theta \in \Theta\right]. \qquad (3.52)$$

Till the end of this section, Θ will stand for an arbitrary domain in $\mathbb{R}^M$, where the dimension $M \leq s^2 - 1$ (cf. (3.5)). Particular cases to note will be where $\Theta = \mathscr{R}$, a set of dimension $s^2 - 1$, as in (3.5), or $\Theta = \mathscr{P}$, a set of dimension $s^2 - s$, as in (3.7). Assume that probabilities p_{ij}^θ depend upon $\theta \in \Theta$ smoothly, and consider the equations

$$\frac{\partial}{\partial \theta} \mathscr{L}(\mathbf{x}, \theta) = \left(\frac{1}{\pi_{x_0}^\theta}\right) \frac{\partial}{\partial \theta} \pi_{x_0}^\theta + \sum_{k=1}^n \left(\frac{1}{p_{x_{k-1}x_k}^\theta}\right) \frac{\partial}{\partial \theta} p_{x_{k-1}x_k}^\theta$$

$$= \left(\frac{1}{\pi_{x_0}^\theta}\right) \frac{\partial}{\partial \theta} \pi_{x_0}^\theta + \sum_{i,j} n_{ij}(\mathbf{x}) \frac{1}{p_{ij}^\theta} \frac{\partial}{\partial \theta} p_{ij}^\theta = 0, \qquad (3.53)$$

and

$$\frac{\partial}{\partial \theta} \ell(\mathbf{x}, \theta) = \sum_{k=1}^n \left(\frac{1}{p_{x_{k-1}x_k}^\theta}\right) \frac{\partial}{\partial \theta} p_{x_{k-1}x_k}^\theta = \sum_{i,j} n_{ij}(\mathbf{x}) \left(\frac{1}{p_{ij}^\theta}\right) \frac{\partial}{\partial \theta} p_{ij}^\theta = 0. \qquad (3.54)$$

Equation (3.53) and (3.54) are often called, somewhat misleadingly, the maximum likelihood equations; in reality their solutions may provide a point of minimum not maximum of the likelihood functions. The latter cases should be excluded (see below), though in practice they seldom occur. Here $\partial/\partial\theta$ means the partial derivative in θ (the gradient vector for a multi-dimensional parameter).

Suppose that (3.53) or (3.54) gives a unique solution $\widehat{\theta} = \widehat{\theta}(\mathbf{x}) \in \Theta$ and $\widehat{\theta}$ determines a local maximum of $\mathscr{L}(\mathbf{x}, \theta)$ or $\ell(\mathbf{x}, \theta)$, i.e. the second derivative

$$\frac{\partial^2}{\partial \theta^2} \mathscr{L}(\mathbf{x}, \theta)\bigg|_{\theta=\widehat{\theta}} \leq 0 \quad \text{or} \quad \frac{\partial^2}{\partial \theta^2} \ell(\mathbf{x}, \theta)\bigg|_{\theta=\widehat{\theta}} \leq 0.$$

If θ is multi-dimensional, the operation $\partial^2/\partial\theta^2$ produces the matrix of second derivatives. In this case, the above matrices will be non-positive. Under these conditions, either $\widehat{\theta}^* = \widehat{\theta}$ or $\widehat{\theta}^*$ is attained at the boundary of the set Θ.

The analysis of the MLE for the log-likelihood ℓ^θ in (3.22) is straightforward.

Worked Example 3.3.6 (i) Let (X_m) be a DTMC, with state space $I = \{1, \ldots, s\}$ and unknown transition matrix $P = (p_{ij})$. Show that the MLE of $\theta = P$ for the likelihood l^θ is given by the normalised occurrence number

$$p_{ij}^* = \frac{n_{ij}}{\sum\limits_{1 \leq j \leq s} n_{ij}}, \tag{3.55}$$

where n_{ij} is defined as in (3.16).

(ii) Making the necessary assumptions on the matrix P and its equilibrium distribution π, state the *consistency property* of an estimator as in (3.55), in the sense of convergence in probability. Relating this convergence with a weak Law of Large Numbers, prove the above consistency property with the help of the Chebyshev or Markov inequality.

Solution (i) In this example, $\Theta = \mathscr{P}$ where set $\mathscr{P}$ is defined as in (3.7). We will work with (3.54), towards a solution of the maximisation problem

$$\max_{P=(p_{ij})} \ell^\theta(\mathbf{x}, P) = \sum_{1 \leq i,j \leq s} n_{ij} \ln(p_{ij})$$

subject to

$$p_{ij} \geq 0, \quad \sum_{1 \leq k \leq s} p_{ik} = 1, \quad \text{for all } 1 \leq i, j \leq s.$$

The Lagrangian reads

$$\sum_{i,j} n_{ij} \ln(p_{ij}) + \sum_{i} \lambda_i \left(\sum_j p_{ij} - 1 \right),$$

to be maximised in p_{ij} for $p_{ij} \geq 0$ (here $\lambda_1, \ldots, \lambda_s$ are Lagrange multipliers to be adjusted to satisfy the constraints $\sum_{1 \leq k \leq s} p_{ik} = 1, 1 \leq i \leq s$.) To find the maximisers of the Lagrangian, differentiate in p_{ij} and equate the derivative with zero:

$$\frac{n_{ij}}{p_{ij}} + \lambda_i = 0, \quad \text{whence } p_{ij} = -\frac{n_{ij}}{\lambda_i}, \quad 1 \leq i, j \leq s.$$

Adjusting the constraints yields

$$\lambda_i = - \sum_{1 \leq k \leq s} n_{ik}, \quad \text{i.e. } p_{ij}^* = \frac{n_{ij}}{\sum\limits_{1 \leq k \leq s} n_{ik}},$$

as required.

(ii) The consistency property in question is written as

$$n_{ij} \Big/ \sum\nolimits_{1 \leq k \leq s} n_{ik} \xrightarrow{\mathbb{P}} p_{ij}$$

and means that, for all $\varepsilon > 0$,

$$\lim_{n \to \infty} \mathbb{P} \left(\left| n_{ij} \Big/ \sum\nolimits_{1 \leq k \leq s} n_{ik} - p_{ij} \right| \geq \varepsilon \right) \to 0.$$

This will follow if we can prove that, for all $1 \leq i, j \leq s$, $n_{ij}/n \xrightarrow{\text{P}} \pi_i p_{ij}$. More precisely, for all $\varepsilon > 0$

$$\lim_{n \to \infty} \mathbb{P}\left(\left|\frac{n_{ij}}{n} - \pi_i p_{ij}\right| \geq \varepsilon\right) \to 0. \tag{3.56}$$

So, we need a unique equilibrium distribution $\pi = (\pi_1, \ldots, \pi_s)$, with all entries $\pi_i > 0$. We may naturally assume that P is irreducible and aperiodic, in which case the above properties hold, and also

$$\lim_{n \to \infty} P^n \to \Pi, \quad \text{where} \quad \Pi = \begin{pmatrix} \pi_1 & \cdots & \pi_s \\ \vdots & \cdots & \vdots \\ \pi_1 & \cdots & \pi_s \end{pmatrix}.$$

Moreover, taking, for simplicity, that $\rho := \min[p_{ij}] > 0$, the convergence happens geometrically (exponentially) fast: see (3.45).

Write

$$n_{ij}(\mathbf{X}) = \sum_{k=1}^{n} \mathbf{1}(X_{k-1} = i, X_k = j), \tag{3.57}$$

where $\mathbf{X} = \begin{pmatrix} X_0 \\ X_1 \\ \vdots \\ X_n \end{pmatrix}$ constitutes a random sample of the chain. Then convergence

in probability $n_{ij}(\mathbf{X})/n \xrightarrow{\text{P}} \pi_i p_{ij}$ becomes a weak Law of Large Numbers. First, write it in the equivalent form $\frac{1}{n}[n_{ij}(\mathbf{X}) - n\pi_i p_{ij}] \xrightarrow{\text{P}} 0$, or

$$\mathbb{P}\left(\left|\frac{1}{n}\sum_{k=1}^{n} I_k\right| \geq \varepsilon\right) \to 0, \quad \text{as } n \to \infty, \text{ for all } \varepsilon > 0. \tag{3.58}$$

Here, for short,

$$I_k = \mathbf{1}(X_{k-1} = i, X_k = j) - \pi_i p_{ij}, \tag{3.59}$$

with

$$\mathbb{E}[I_k] = \sum_{x_{k-1}, x_k = 1}^{s} [\delta_{x_{k-1}, i} \delta_{x_k, j} - \pi_i p_{ij}] \pi_{x_{k-1}} p_{x_{k-1} x_k} = 0, \tag{3.60}$$

where $\delta_{\cdot,\cdot}$ is the Kronecker delta. In fact, due to the stationarity of the chain (X_m), the random variables $I_1, \ldots, I_n$ are identically distributed, although not independent. We re-write (3.60) as

$$\sum_{x_{k-1}, x_k = 1}^{s} J_{x_{k-1}, x_k} \pi_{x_{k-1}} p_{x_{k-1} x_k} = 0, \quad \text{where} \quad J_{uv} = \delta_{u,i} \delta_{v,j} - \pi_i p_{ij}. \tag{3.61}$$

Then we make use of the Markov inequality: for all random variable $Y \geq 0$ and for all $\varepsilon > 0$, the probability $\mathbb{P}(Y \geq \varepsilon) \leq \mathbb{E}Y^2/\varepsilon^2$. Substituting

$$Y = \left| \sum_{1 \leq k \leq m} (I_k - \pi_i p_{ij}) \right|$$

yields

$$\mathbb{P}\left(\left| \left(\frac{1}{n}\right) n_{ij}(\mathbf{X}) - \pi_i p_{ij} \right| \geq \varepsilon \right) = \mathbb{P}\left(\frac{1}{n} \left| \sum_{1 \leq k \leq n} I_k \right| \geq \varepsilon \right)$$

$$\leq \left(\frac{1}{n^2 \varepsilon^2} \right) \mathbb{E}\left[\sum_{1 \leq k \leq n} I_k \right]^2. \qquad (3.62)$$

Suppose we manage to establish an inequality

$$\mathbb{E}\left[\sum_{1 \leq k \leq n} I_k \right]^2 \leq Cn, \qquad (3.63)$$

with C a constant independent of n. This would yield

$$\text{RHS of (3.62)} \leq \frac{C}{n\varepsilon^2} \to 0, \text{ as } n \to \infty, \text{ for all } \varepsilon > 0,$$

i.e. (3.58).

So, we wish to prove (3.63). Expand the square of the sum in the parentheses in the RHS of (3.63), by grouping together the terms I_k^2 and cross-products $I_{k_1} I_{k_2}$. Using additivity of expectation, we find

$$\mathbb{E}\left[\sum_{1 \leq k \leq n} I_k \right]^2 = \sum_{1 \leq k \leq n} \mathbb{E}[I_k^2] + \sum_{1 \leq k_1, k_2 \leq n} \mathbf{1}(k_1 \neq k_2) \mathbb{E}\left(I_{k_1} I_{k_2} \right). \qquad (3.64)$$

The first sum in the RHS of (3.64) equals $n\mathbb{E}[I_1^2]$, because the random variables I_k are identically distributed. In the second sum, the term $\mathbb{E}[I_{k_1} I_{k_2}]$ depends on the difference $k_1 - k_2$ only. This suggests using a summation over $k = k_1$ and $l = |k_1 - k_2|$. The second sum is re-written as

$$2 \sum_{1 \leq k \leq n} \sum_{0 < l \leq n-k} \mathbb{E}[I_k I_{k+l}].$$

and, in the absolute value, turns out less than or equal to

$$2n \sum_{1 < l < \infty} \left| \mathbb{E}[I_1 I_l] \right|. \qquad (3.65)$$

Thus, we aim to verify that the series in (3.65) converges. The bound in (3.45) plays the instrumental rôle here.

The idea is to check that, for l large, the expected value of the product gets close to the product of the expected values

$$\mathbb{E}[I_1 I_l] \approx \mathbb{E}[I_1]\mathbb{E}[I_l] = \left(\mathbb{E}[I_1]\right)^2 = 0, \tag{3.66}$$

as the factor $\mathbb{E}[I_1]$ vanishes; see (3.60).

To make this approximate equality precise, write down the general term $\mathbb{E}[I_1 I_l]$ as

$$\Sigma_l = \sum_{x_0, x_1, x_{l-1}, x_l} \pi_{x_0} \, p_{x_0 x_1} \, p_{x_1 x_{l-1}}^{(l-1)} \, p_{x_{l-1} x_l} \, J_{x_0, x_1} \, J_{x_{l-1}, x_l},$$

where J_{uv} was determined in (3.61). By (3.57)

$$\begin{aligned}
\Sigma_l &= \sum_{x_0, x_1, x_{l-1}, x_l} \pi_{x_0} \, p_{x_0 x_1} \left[\pi_{x_{l-1}} + \psi_{x_1, x_{l-1}}(l-1)\right] p_{x_{l-1} x_l} \, J_{x_0, x_1} \, J_{x_{l-1}, x_l} \\
&= \sum_{x_0, x_1} \pi_{x_0} \, p_{x_0 x_1} \, J_{x_0, x_1} \sum_{x_{l-1}, x_l} \pi_{x_{l-1}} \, p_{x_{l-1} x_l} \, J_{x_{l-1}, x_l} \\
&\quad + \sum_{x_0, x_1, x_{l-1}, x_l} \pi_{x_0} \, p_{x_0 x_1} \, \psi_{x_1, x_{l-1}}(l-1) \, p_{x_{l-1} x_l} \, J_{x_0, x_1} \, J_{x_{l-1}, x_l}. \tag{3.67}
\end{aligned}$$

According to (3.60), the sum

$$\sum_{x_0, x_1} \pi_{x_0} \, p_{x_0 x_1} \, J_{x_0, x_1} = \sum_{x_0, x_1} \pi_{x_0} \, p_{x_0 x_1} \, \delta_{x_0, i} \delta_{x_1, j} - \pi_i p_{ij} = 0.$$

Hence, in the RHS of (3.67) only the second term survives, with, according to (3.45), an absolute value less than or equal to

$$(1-\rho)^{l-1} \sum_{x_0, x_1, x_{l-1}, x_l} \pi_{x_0} \, p_{x_0 x_1} \, J_{x_0 x_1} \, p_{x_{l-1} x_l} \, J_{x_{l-1} x_l} \le A^2 s (1-\rho)^{l-1}. \tag{3.68}$$

Here, and below,

$$A = \left[(1 - \pi_i p_{ij}) \vee (\pi_i p_{ij})\right]. \tag{3.69}$$

Altogether, this implies that

$$\left|\mathbb{E}[I_1 I_l]\right| \le \text{the RHS of (3.68)},$$

which justifies (3.66). Then for the series in (3.65) we obtain

$$\sum_{1 < l < \infty} \left| \mathbb{E}[I_1 I_l] \right| \leq \frac{A^2 s}{\rho}.$$

Hence, we can bound the LHS in (3.64):

$$\mathbb{E}\left[\sum_{1 \leq k \leq n} I_k \right]^2 \leq n \frac{2A^2 s}{\rho}.$$

This proves (3.63) and completes the derivation of (3.56). Hence, the MLE (3.55) achieves consistency in the sense of convergence in probability. $\qquad \square$

Worked Example 3.3.7 Making the necessary assumptions on a DTMC (X_m) on $I = \{1, \ldots, s\}$, state and prove that the MLE (3.55) is consistent in the sense of AS convergence.

Solution The consistency property means that, as $n \to \infty$,

$$\frac{n_{ij}}{\displaystyle\sum_{1 \leq j \leq s} n_{ij}} \xrightarrow{\text{a.s.}} p_{ij},$$

the 'true' value of the parameter. This will follow if we prove that

$$\frac{1}{n} n_{ij} \to \pi_i p_{ij}, \quad \frac{1}{n} \sum_{1 \leq j \leq s} n_{ij} \to \sum_{1 \leq j \leq s} \pi_i p_{ij} = \pi_i, \qquad (3.70)$$

where π_i represents the stationary (invariant) probability. Various forms of convergence exist; we choose almost sure convergence, or convergence with probability 1, with respect to π, the equilibrium probability distribution of the Markov chain (X_m) with the (unknown) transition matrix P.

So, we seek a unique equilibrium distribution $\pi = (\pi_1, \ldots, \pi_s)$, with all entries $\pi_i > 0$. Naturally, we assume that P is irreducible and aperiodic, in which case the above properties hold, and also

$$\lim_{n \to \infty} P^n \to \Pi, \quad \text{where } \Pi = \begin{pmatrix} \pi_1 & \cdots & \pi_s \\ \vdots & \cdots & \vdots \\ \pi_1 & \cdots & \pi_s \end{pmatrix}.$$

Moreover, assuming, for simplicity, that $\rho := \min[p_{ij}] > 0$, the convergence will be geometrically (exponentially) fast: the entries $p_{ij}^{(n)}$ of P^n satisfy

$$p_{ij}^{(n)} = \pi_j + \psi_{i,j}(n), \quad \text{where } |\psi_{i,j}(n)| \leq (1 - \rho)^{n-1}. \qquad (3.71)$$

Write

$$n_{ij}(\mathbf{X}) = \sum_{m=1}^{n} \mathbf{1}(X_{m-1} = i, X_m = j),$$

where $X_0, \ldots, X_n$ form the entries of a random sample vector $\mathbf{X}$. Then almost sure convergence $n_{ij}(\mathbf{X})/n \to \pi_i p_{ij}$ becomes a strong Law of Large Numbers, and the first step is to write it in an equivalent form $\dfrac{1}{n}\left[n_{ij}(\mathbf{X}) - n\pi_i p_{ij}\right] \to 0$, or

$$\frac{1}{n} \sum_{m=1}^{n} I_m \to 0, \text{ almost surely.}$$

Here, for short,

$$I_m = \mathbf{1}(X_{m-1} = i, X_m = j) - \pi_i p_{ij}, \tag{3.72}$$

with

$$\mathbb{E}[I_m] = \sum_{x_{m-1},x_m=1}^{s} \left[\delta_{x_{m-1},i}\,\delta_{x_m,j} - \pi_i p_{ij}\right]\pi_{x_{m-1}}\, p_{x_{m-1}x_m} = 0, \tag{3.73}$$

where $\delta_{\cdot,\cdot}$ is the Kronecker delta. In fact, due to the stationarity of the chain (X_m), the RVs $I_1, \ldots, I_n$ are identically distributed, although not independent.

Then we make use of the Markov inequality: for all random variables $Y \geq 0$ and for all $\varepsilon > 0$, the probability $\mathbb{P}(Y \geq \varepsilon) \leq \mathbb{E}Y^4/\varepsilon^4$ (note the power 4, instead of the 'traditional' 2, with $\mathbb{P}(Y \geq \varepsilon) \leq \mathbb{E}Y^2/\varepsilon^2$). Substituting

$$Y = \left| \sum_{1 \leq m \leq n} (I_m - \pi_i p_{ij}) \right|$$

yields

$$\mathbb{P}\left(\left|\left(\frac{1}{n}\right) n_{ij}(\mathbf{X}) - \pi_i p_{ij}\right| \geq \varepsilon\right) = \mathbb{P}\left(\frac{1}{n}\left|\sum_{1 \leq k \leq n} I_k\right| \geq \varepsilon\right)$$

$$\leq \frac{1}{n^4 \varepsilon^4}\mathbb{E}\left[\sum_{1 \leq k \leq n} I_k\right]^4. \tag{3.74}$$

Suppose we manage to establish an inequality

$$\text{the RHS of } (3.74) \leq Cn^2, \tag{3.75}$$

with C a constant independent of n. Then, with $\varepsilon = 1/n^{1/8}$, we will find that

$$\mathbb{P}\left(\left|\left(\frac{1}{n}\right) n_{ij}(\mathbf{X}) - \pi_i p_{ij}\right| \geq \frac{1}{n^{1/8}}\right) \leq \frac{Cn^2}{n^4\, n^{-1/2}} = C\left(\frac{1}{n^{3/2}}\right).$$

The series $\sum_n n^{-3/2}$ converges. Then, by the Borel–Cantelli Lemma (see Volume 1, page 132), with probability 1, the event

$$\left\{ \left| \left(\frac{1}{n} \right) n_{ij}(\mathbf{X}) - \pi_i p_{ij} \right| \geq \frac{1}{n^{1/8}} \right\}$$

occurs for only finitely many n. In other words, with probability 1, the inequality

$$\left| \left(\frac{1}{n} \right) n_{ij}(\mathbf{X}) - \pi_i p_{ij} \right| < \frac{1}{n^{1/8}}$$

holds for all n large enough. The latter fact yields the almost sure convergence in (3.70).

To prove (3.75), expand the fourth power of the sum in the square brackets in the RHS of (3.74), by grouping together the terms I_k^4 and cross-products $I_{k_1} I_{k_2}^3$ and $I_{k_1}^2 I_{k_2}^2$. Using additivity of expectation, we may write

$$
\mathbb{E}\left[\sum_{1 \leq k \leq n} I_k \right]^4
$$

$$
= \sum_{1 \leq k \leq n} \mathbb{E}[I_k^4] + \sum_{1 \leq k_1, k_2 \leq n} \mathbf{1}(k_1 \neq k_2) \mathbb{E}\left[I_{k_1} I_{k_2}^3 \right]
$$

$$
+ \sum_{1 \leq k_1, k_2 \leq n} \mathbf{1}(k_1 \neq k_2) \mathbb{E}\left[I_{k_1}^2 I_{k_2}^2 \right]
$$

$$
+ \sum_{1 \leq k_1, k_2, k_3 \leq n} \mathbf{1}(k_1 \neq k_2 \neq k_3 \neq k_1) \mathbb{E}\left[I_{k_1} I_{k_2} I_{k_3}^2 \right]
$$

$$
+ \sum_{1 \leq k_1, k_2, k_3, k_4 \leq n} \mathbf{1}(k_\alpha \neq k_\beta, \text{ for all } \alpha \neq \beta) \mathbb{E}\left[I_{k_1} I_{k_2} I_{k_3} I_{k_4} \right] \qquad (3.76)
$$

for all $\alpha, \beta \in \{1, 2, 3, 4\}$. Anticipating the result of the forthcoming argument, the first and second sums in the RHS are $O(n)$, but the third, fourth and fifth amount to $O(n^2)$. The bound (3.71) will be instrumental in assessing the sums in the second, fourth and fifth lines in (3.76). Indeed, the first sum in the RHS of (3.76) equals $n\mathbb{E}[I_1^4]$, because the random variables I_k are identically distributed. In the next two sums, the terms $\mathbb{E}\left[I_{k_1} I_{k_2}^3 \right]$ and $\mathbb{E}\left[I_{k_1}^2 I_{k_2}^2 \right]$ depend on the difference $k_1 - k_2$ only. This suggests using summations over $k = k_1$ and $l = |k_1 - k_2|$. The second sum is rewritten as

$$
\sum_{1 \leq k \leq n} \sum_{0 < l \leq n-k} \left[\mathbb{E}\left(I_k I_{k+l}^3 \right) + \mathbb{E}\left(I_k^3 I_{k+l} \right) \right]
$$

and, in the absolute value, comes out less than or equal to

$$
n \sum_{1 < l < \infty} \left| \mathbb{E}\left(I_1 I_l^3 + I_1^3 I_l \right) \right|. \qquad (3.77)
$$

Thus, our aim with second sum in the RHS of (3.76) will be to check that the series in (3.77) converges.

The third sum in the RHS of (3.76) is non-negative, being formed by non-negative summands. It will at most be

$$n(n-1) \max\left[\mathbb{E}\left(I_1^2 I_l^2\right); l > 1\right] \leq n^2 \left[(1 - \pi_i p_{ij})^2 \vee (\pi_i p_{ij})^2\right], \qquad (3.78)$$

which gets as good as can be assessed in terms of a power of n. Here, $a \vee b = \max(a,b)$, and the latter bound in (3.78) follows from the formula

$$\mathbb{E}\left(I_1^2 I_l^2\right) = \sum_{x_0,x_1,x_{l-1},x_l} \pi_{x_0} \, p_{x_0 x_1} \, p_{x_1 x_{l-1}}^{(l-1)} \, p_{x_{l-1} x_l} \, J_{x_0,x_1}^2 \, J_{x_{l-1},x_l}^2, \qquad (3.79)$$

where $J_{u,v} = \delta_{ui}\delta_{vj} - \pi_i p_{ij}$.

The argument for the fourth and the fifth lines in the RHS of (3.76) elaborates that used to bound the series in (3.77). So, we deal with (3.77) first. The idea is to check that, for l large, the expected value of the product approximates the product of the expected values

$$\mathbb{E}[I_1 I_l^3] \approx \mathbb{E}[I_1]\mathbb{E}[I_l^3] \quad \text{and} \quad \mathbb{E}[I_1^3 I_l] \approx \mathbb{E}[I_1^3]\mathbb{E}[I_1]. \qquad (3.80)$$

Of course, both products of the expected values coincide:

$$\mathbb{E}[I_1]\mathbb{E}[I_l^3] = \mathbb{E}[I_1^3]\mathbb{E}[I_l] = \mathbb{E}[I_1]\mathbb{E}[I_1^3],$$

and becomes 0 as the factor $\mathbb{E}[I_1]$ vanishes; see (3.73).

To make the first relation in (3.80) precise, write down the general term $\mathbb{E}\left[I_1 I_l^3\right]$ in the series (3.77) as

$$\sum_{x_0,x_1,x_{l-1},x_l} \pi_{x_0} \, p_{x_0 x_1} \, p_{x_1 x_{l-1}}^{(l-1)} \, p_{x_{l-1} x_l} \, J_{x_0,x_1} \, J_{x_{l-1},x_l}^3$$

and use (3.71) to specify it in the form

$$\sum_{x_0,x_1,x_{l-1},x_l} \pi_{x_0} \, p_{x_0 x_1} \left[\pi_{x_{l-1}} + \psi_{x_1,x_{l-1}}(l-1)\right] p_{x_{l-1} x_l} \, J_{x_0,x_1} \, J_{x_{l-1},x_l}^3$$

$$= \sum_{x_0,x_1} \pi_{x_0} \, p_{x_0 x_1} \, J_{x_0,x_1} \sum_{x_{l-1},x_l} \pi_{x_{l-1}} \, p_{x_{l-1} x_l} \, J_{x_{l-1},x_l}^3$$

$$+ \sum_{x_0,x_1,x_{l-1},x_l} \pi_{x_0} \, p_{x_0 x_1} \, \psi_{x_1,x_{l-1}}(l-1) \, p_{x_{l-1} x_l} \, J_{x_0,x_1} \, J_{x_{l-1},x_l}^3. \qquad (3.81)$$

According to (3.73), the sum

$$\sum_{x_0,x_1} \pi_{x_0} \, p_{x_0 x_1} \, J_{x_0,x_1} = \sum_{x_0,x_1} \pi_{x_0} \, p_{x_0 x_1} \, \delta_{x_0,i} \, \delta_{x_1,j} - \pi_i p_{ij} = 0.$$

Hence, in the RHS of (3.81) only the second term survives, and, according to (3.71), its absolute value is bounded by

$$(1-\rho)^{l-1} \sum_{x_0,x_1,x_{l-1},x_l} \pi_{x_0} p_{x_0x_1} J_{x_0x_1} p_{x_{l-1}x_l} J_{x_{l-1},x_l} \leq A^2 s(1-\rho)^{l-1}. \qquad (3.82)$$

Here, and below,

$$A = \left[(1 - \pi_i p_{ij}) \vee (\pi_i p_{ij})\right]. \qquad (3.83)$$

Altogether, this implies that

$$\left| \mathbb{E}\left(I_1 I_l^3\right) \right| \leq \text{the RHS of (3.68)},$$

which justifies the first relation in (3.80). Similarly,

$$\left| \mathbb{E}\left(I_1^3 I_l\right) \right| \leq \text{the RHS of (3.82)},$$

which explains the second relation in (3.80). For future use, an analogous bound holds, of course, for $\left| \mathbb{E}\left(I_1^3 I_l\right) \right|$:

$$\left| \mathbb{E}\left(I_1^3 I_l\right) \right| \leq A^2 s(1-\rho)^{l-1}. \qquad (3.84)$$

For the expression (3.77) we obtain an upper bound as follows:

$$(3.77) \leq \left(\frac{A^2 s}{\rho} n\right), \qquad$$

confirming that the second sum in the RHS of (3.76) is $O(n)$.

As was said, the two last lines, the fourth and fifth, in the RHS of (3.76) are assessed by a similar argument. We discuss in detail the bound for the latter sum, which is technically more involved. The term $\mathbb{E}\left[I_{k_1} I_{k_2} I_{k_3} I_{k_4}\right]$ is not changed under permutation of the indexes k_α. Furthermore, it depends only on the pair-wise differences $k_\alpha - k_\beta$, and so we set $k_1 = k$, $k_2 = k_1 + l_1$, $k_3 = k_2 + l_2$, $k_4 = k_3 + l_3$ and evaluate the fifth line as

$$4! \left(\sum_{1 \leq k_1 < k_2 < k_3 < k_4 \leq n} \mathbb{E}\left[I_{k_1} I_{k_2} I_{k_3} I_{k_4}\right] \right)$$

$$= 4! \sum_{1 \leq k < n} \sum_{l_1,l_2,l_3 \geq 1} \mathbf{1}(k + l_1 + l_2 + l_3 \leq n) \mathbb{E}\left(I_k I_{k+l_1} I_{k+l_1+l_2} I_{k+l_1+l_2+l_3}\right). \qquad (3.85)$$

The idea being to write a representation similar to (3.80): if l_α equates to the maximum distance among l_1, l_2 and l_3, and l_α large, then

$$\mathbb{E}\left(I_{k_1} I_{k_2} I_{k_3} I_{k_4}\right) \approx \mathbb{E}\left[\prod_{\beta \leq \alpha} I_{k_\beta}\right] \mathbb{E}\left[\prod_{\beta > \alpha} I_{k_\beta}\right]. \qquad (3.86)$$

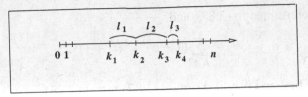

Fig. 3.3

We will use the fact that if the maximum distance l_α coincides with l_1 or l_3 then one of the factors $\mathbb{E}\left[\prod_{\beta \le \alpha} I_{k_\beta}\right]$ or $\mathbb{E}\left[\prod_{\beta > \alpha} I_{k_\beta}\right]$, reduces to a 'single' expectation $\mathbb{E}I_{k_1}$, and hence vanishes. Then, of course, the product in the RHS of (3.86) becomes zero, the 'worst' case being where $l_\alpha = l_2$ (see below).

More precisely, assuming $\max[l_1, l_2, l_3]$ is large, then

$$\mathbb{E}\left[I_{k_1} I_{k_2} I_{k_3} I_{k_4}\right] \approx \begin{cases} \mathbb{E}[I_{k_1}]\,\mathbb{E}[I_{k_2} I_{k_3} I_{k_4}] = 0, & \text{if } l_1 > l_2, l_3, \\ \mathbb{E}[I_{k_1} I_{k_2}]\,\mathbb{E}[I_{k_3} I_{k_4}], & \text{if } l_2 > l_1, l_3, \\ \mathbb{E}[I_{k_1} I_{k_2} I_{k_3}]\,\mathbb{E}[I_{k_4}] = 0, & \text{if } l_3 > l_1, l_2. \end{cases}$$

Correspondingly, the sum in the RHS of (3.85) splits into seven sums

$$\sum_{1 \le k < n} \left(\sum_{l_1 > l_2, l_3 \ge 1} + \sum_{l_2 > l_1, l_3 \ge 1} + \sum_{l_3 > l_1, l_2 \ge 1} \right)$$
$$+ \sum_{1 \le k < n} \left(\sum_{l_1 = l_2 > l_3 \ge 1} + \sum_{l_1 = l_3 > l_2 \ge 1} + \sum_{l_2 = l_3 > l_3 \ge 1} + \sum_{l_1 = l_2 = l_3 \ge 1} \right). \tag{3.87}$$

The sums in the second line turn out of a lesser order, and we mainly focus on the sums from the first line. The general term in every sum from (3.87) is written as

$$\sum \pi_{x_{k_1}-1}\, p_{x_{k_1}-1 x_{k_1}}\, J_{x_{k-1}, x_{k_1}}\, p_{x_{k_1} x_{k_2}-1}^{l_1-1}\, p_{x_{k_2}-1 x_{k_2}}\, J_{x_{k_2}-1, x_{k_2}}$$
$$p_{x_{k_2} x_{k_3}-1}^{l_2-1}\, p_{x_{k_3}-1 x_{k_3}}\, J_{x_{k_3}-1, x_{k_3}}\, p_{x_{k_3} x_{k_4}-1}^{l_3-1}\, p_{x_{k_4}-1 x_{k_4}}\, J_{x_{k_4}-1, x_{k_4}}. \tag{3.88}$$

Here the summation is accumulated over the states $x_{k_1}-1, x_{k_1}, x_{k_2}-1, x_{k_2}, x_{k_3}-1,$ $x_{k_3}, x_{k_4}-1, x_{k_4}$ running between 1 and s. When the maximum distance l_α is achieved for $\alpha = 1$ or $\alpha = 3$, we write $p_{x_{k_\alpha}-1 x_{k_\alpha}-1}^{(l_\alpha-1)} = \pi_{x_{k_\alpha}} + \psi_{x_{k_\alpha}-1, x_{k_\alpha}-1}(l_\alpha - 1)$, and split the sum (3.88) as was done in (3.81).

Then for the the first and the third sum in (3.87) we obtain the bound

$$\left| \sum_{1 \le k < n}\, \sum_{l_1 > l_2, l_3 \ge 1} \right|, \left| \sum_{1 \le k < n}\, \sum_{l_3 > l_1, l_2 \ge 1} \right| \le n A^4 s B_2$$

where A was determined in (3.83), and

$$B_2 = \sum_{l_1 \geq 2} (1-\rho)^{l_1-1}(l_1-1)^2.$$

We see that the first and the third sum in (3.87) remain $O(n)$.

In the second sum, writing $p_{x_{k_1} x_{k_2-1}}^{(l_1-1)} = \pi_{x_{k_1}} + \psi_{x_{k_1},x_{k_2-1}}(l_1-1)$ and $p_{x_{k_3} x_{k_4-1}}^{(l_3-1)} = \pi_{x_{k_4-1}} + \psi_{x_{k_3},x_{k_4-1}}(l_3-1)$ gives only that

$$\left| \sum_{1 \leq k < n} \sum_{l_1 > l_2, l_3 \geq 1} \mathbf{1}(k+l_1+l_2+l_3 \leq n) \mathbb{E}\left(I_k I_{k+l_1} I_{k+l_1+l_2} I_{k+l_1+l_2+l_3} \right) \right|$$

$$\leq A s^2 \sum_{1 \leq k < n} \sum_{l_1 > l_2, l_3 \geq 1} \mathbf{1}(k+l_1+l_2+l_3 \leq n)(1-\rho)^{l_1-1}(1-\rho)^{l_3-1}$$

$$\leq A \left(\frac{s^2}{\rho^2} \right) \sum_{1 \leq k < n} (n-k) = A \left(\frac{s^2}{\rho^2} \right) \frac{n(n-1)}{2}.$$

This confirms the claim that the last line in the RHS of (3.76) comes to $O(n^2)$. The fourth line in the RHS of (3.76) is estimated in a similar fashion. This yields (3.75) and completes the proof of (3.70). □

We conclude that the setting for Markov chains turns out in many aspects similar to that for IID observations; the big difference being of course that the likelihood contains a product of factors connecting pairs of subsequent states:

$$\begin{aligned} l^\theta(\mathbf{x}, \theta) &= \mathbb{P}^\theta(X_1 = x_1, \dots, X_n = x_n | X_0 = x_0) \\ &= p_{x_0 x_1}^\theta \cdots p_{x_{n-1} x_n}^\theta, \qquad \mathbf{x} = \begin{pmatrix} x_0 \\ \vdots \\ x_n \end{pmatrix}. \end{aligned} \qquad (3.89)$$

For the remainder of this section we assume the following condition.

Condition 3.3.8 If $p_{ij}^\theta = 0$ for some states $i, j \in I$ then this equality holds for all $\theta \in \Theta$. Therefore, if, for a given $\mathbf{x}$, the (reduced) likelihood $l(\mathbf{x}, \theta)$ happens to be strictly positive for some value $\theta \in \Theta$, it will remain so for every $\theta \in \Theta$. Then the set of samples $\mathbf{x}$ with $l(\mathbf{x}, \theta) = 0$ can be discarded once and for all. It obviously occurs with probability 0 under the Markov chain distribution $\mathbb{P}^\theta$, for all $\theta \in \Theta$. The remaining set of pairs (i, j) for which $p_{ij}^\theta > 0$, $\theta \in \Theta$, is denoted by D; we assume that for all $(i, j) \in D$, the transition probabilities p_{ij}^θ will be continuously differentiable in $\theta \in \Theta$.

We finish this section with a result showing that the log-likelihood ratios themselves satisfy the strong Law of Large Numbers.

Theorem 3.3.9 *Assume that the transition matrix* $P^\theta = (p_{ij}^\theta, i, j \in I)$, $\theta \in \Theta$, *satisfies Condition 3.3.8. Then, for all $\theta \in \Theta$ and for all initial distributions λ, as $n \to \infty$, the following convergence takes place with $\mathbb{P}^{\lambda,\theta}$-probability 1:*

$$\left(\frac{1}{n}\right) l^\theta(\mathbf{X}) \xrightarrow{\text{a.s.}} \mathbb{E}_{eq}^\theta \left[\ln\left(p_{X_0 X_1}^\theta \right) \right] = \sum_{(i,j) \in D} \pi_i^\theta \, p_{ij}^\theta \ln(p_{ij}^\theta). \tag{3.90}$$

Here and below, $\mathbb{P}^{\lambda,\theta}$ stands for the probability distribution of the DTMC with initial vector λ and transition matrix $\mathbb{P}^\theta$, and $\mathbb{E}_{eq}^\theta$ for the expectation with respect to the equilibrium measure π^θ.

Moreover, a similar fact holds for a general function $g(i,j)$, $(i,j) \in D$. Define the random variables $G_k = g(X_{k-1}, X_k)$. Then, as $n \to \infty$, for all $\theta \in \Theta$ and for all initial distributions λ, the sum $\sum_{k=1}^n G_k / n$ converges with $\mathbb{P}^{\lambda,\theta}$-probability 1:

$$\frac{1}{n} \sum_{k=1}^n G_k \xrightarrow{\text{a.s.}} \mathbb{E}_{eq}^\theta[G_1] = \sum_{i,j \in D} \pi_i^\theta \, p_{ij}^\theta \, g(i,j). \tag{3.91}$$

The sum in the RHS of (3.91) can be extended to all $i, j \in I$, with the standard agreement that $p_{ij}^\theta \ln(p_{ij}^\theta) = 0$ when $p_{ij}^\theta = 0$. The quantity

$$- \sum_{i,j \in I} \pi_i^\theta \, p_{ij}^\theta \ln(p_{ij}^\theta)$$

is called the *entropy* (or *entropy rate*) of the DTMC (X_m) and plays an important rôle in many applications. Under our assumption that (X_m) is irreducible and aperiodic, it remains strictly positive, which yields that the expectation in the RHS of (3.91) is strictly negative for the function $g(i,j) = \ln p_{ij}^\theta$.

Another example of the sum $\sum_{k=1}^n g(X_{k-1}, X_k)$ would be the sum of indicators $\mathbf{1}(X_{k-1} = i, X_k = j)$ in (3.59). But the function g may depend on a single variable: e.g. for $I \subset \mathbb{R}$ and $g(i,j) = j$ we observe that $G_k = X_k$, the value of the chain at time k. Equations (3.90) and (3.91) in this case become a *strong Law of Large Numbers* for the DTMC $(X_m, 0 \le m \le n)$: as $n \to \infty$,

$$\frac{1}{n} \sum_{k=1}^n X_k \xrightarrow{\text{a.s.}} \mathbb{E}_{eq}^\theta[X_1] = \sum_{x \in I} x \, \pi_x^\theta. \tag{3.92}$$

It is worth noting that in general, the sequence of random variables $G_k = g(X_{k-1}, X_k)$ does not form a Markov chain or a higher-order Markov chain. However, it represents a function of a DTMC (X_m) which exhibits an exponential decay of correlations: this property plays a key part in the proof. We will not give here the proof of Theorem 3.3.9, as it follows the same argument(s) as earlier in this section.

3.4 Likelihood functions, 2. Whittle's formula

A general approach to maximum-likelihood estimation of the transition probabilities of a DTMC is based on the so-called *Whittle formula*, which forms the subject of this section. To avoid confusion, we change some parts of the notation used so far. We will also follow terminology introduced in earlier works on this topics.

Let $\mathbf{x} = \begin{pmatrix} x_0 \\ \vdots \\ x_n \end{pmatrix} \in I^{n+1}$ be a sample from a finite-state DTMC (X_m), with transition matrix $P = (p_{ij})$ and initial distribution $\lambda = (\lambda_i)$.

Let $f_{ij}(\mathbf{x})$ be the number of transitions $i \to j$ in $\mathbf{x}$, i.e. the the number of times m, $0 \leq m < n$, for which $x_m = i$ and $x_{m+1} = j$. (In preceding sections, this quantity was denoted by n_{ij}.) Set $f_{i+} = \sum_j f_{ij}$ and $f_{+i}(\mathbf{x}) = \sum_j f_{ji}(\mathbf{x})$; the values f_{i+} and f_{+i} give the number of entrances to, and exits from, state i, in $\mathbf{x}$. We therefore get a matrix-valued function $\mathbf{x} \mapsto F(\mathbf{x}) = (f_{ij}(\mathbf{x}))$. The matrix $F(\mathbf{x})$ is called the *transition count* (in a given sample $\mathbf{x}$). For a random sample $\mathbf{X}$, we obtain a random matrix, $F(\mathbf{X}) = (f_{ij}(\mathbf{X}))$. It is also useful to adopt a reversed view:

Definition 3.4.1 Let $x, y \in I$ be a pair of distinct states $(x \neq y)$. Let $F = (f_{ij}, i, j \in I)$ be a matrix, with non-negative integer entries. We say that F gives an n-step *transition count*, for an initial state x and terminal state y, if (i) the sum $\sum_{i,j=1}^{s} f_{ij} = n$, and, (ii) with the previous notation $f_{i+} = \sum_j f_{ij}$ and $f_{+i} = \sum_j f_{ji}$,

$$f_{i+} - f_{+i} = 0, \quad \text{for all } i \in I, \text{ except for } i = x \text{ and } i = y,$$
$$\text{where } f_{x+} - f_{+x} = 1 \text{ and } f_{y+} - f_{+y} = -1. \tag{3.93}$$

For $x = y$, this definition is slightly modified: $f_{i+} - f_{+i} = 0$ for all $i \in I$, and in addition, $f_{x+} \geq 1$ and $f_{+x} \geq 1$. See Figure 3.4. So, excluding the initial and final states, transitions from any given state match the transitions to that state. However, for distinct initial and final states, moves from the initial state exceed those to it by 1 and moves from the final state amount to one fewer than moves to it. When the chain finishes back at its original state, transitions from any state equate those to it.

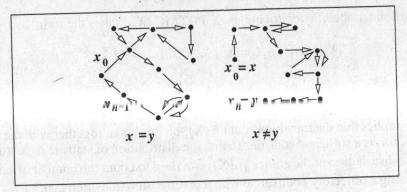

Fig. 3.4

Worked Example 3.4.2 Let $x \in I$ be fixed. Suppose that the non-negative integer matrix $F = (f_{ij}, \ i,j \in I)$ satisfies the property

$$f_{i+} - f_{+i} = \delta_{ix} - \delta_{iy}, \ i = 1, \ldots, s, \tag{3.94}$$

for some given $y = 1, \ldots, s$. Prove that y is the only point from I for which (3.94) holds.

Solution We provide a proof by contradiction. Assume that y and w both satisfy (3.94). If $x \neq y$, then

$$f_{y+} - f_{+y} = \delta_{yx} - \delta_{yy} = -1, \ \text{and} \ f_{y+} - f_{+y} = \delta_{yx} - \delta_{yw} = -\delta_{yw},$$

which implies $y = w$. If $x = y$, then

$$f_{i+} - f_{+i} = \delta_{ix} - \delta_{ix} = 0, \ \text{and} \ f_{i+} - f_{+i} = \delta_{ix} - \delta_{iw}, \ i = 1, \ldots, s.$$

Thus,

$$\delta_{ix} = \delta_{iw},$$

$\square$

and $w = x = y$.

We are interested in determining exactly how many trajectories prove compatible with a given (consistent) transition count.

Example 3.4.3 Let $I = \{1, 2, 3\}$ and consider the matrix

$$F = \begin{pmatrix} 0 & 1 & 1 \\ 0 & 0 & 1 \\ 1 & 0 & 0 \end{pmatrix}.$$

This gives a 4-step transition count, for an initial state 1 and a terminal state 3.

Recall the probability distributions of a DTMC (X_m) takes the form

$$\mathbb{P}(X_1 = x_1, \ldots, X_n = x_n \mid X_0 = x_0) = \prod_{i,j \in I}^{s} p_{ij}^{f_{ij}},$$

$$\mathbb{P}(X_0 = x_0, \ldots, X_n = x_n) = \lambda_{x_0} \prod_{i,j \in I} p_{ij}^{f_{ij}},$$

which implies that the transition count $F(\mathbf{X})$, on its own or together with the initial state x_0, forms a sufficient statistic. To find the distribution of statistic $F(\mathbf{X})$ (that is, the joint distribution of the entries $f_{ij}(\mathbf{X})$), we need to count the number of samples $\mathbf{x}$ following a trajectory compatible with a given n-step transition count F.

To this end, given $x, y \in I$, let $F = (f_{ij})$ be an n-step transition count with $x_0 = x$ and $x_n = y$, and denote by $N_{xy}^{(n)}(F)$ the number of samples $\mathbf{x} \in I^{n+1}$ such that $F = F(\mathbf{x})$. Some straightforward (although tedious) combinatorics performed below results in the formula

$$N_{xy}^{(n)}(F) = \left(\frac{\prod_i f_{i+}!}{\prod_{ij} f_{ij}!} \right) \varphi_{(x,y)}^-. \tag{3.95}$$

Here $\varphi_{(x,y)}^-$ stands for the (x,y)th cofactor in the matrix $F^- = (f_{ij}^-)$ which we determine below:

$$\varphi_{(x,y)}^- = (-1)^{x+y} \left(\det(f_{ij}^-)_{i \neq x, \, j \neq y} \right). \tag{3.96}$$

Entries f_{ij}^- of F^- are given by

$$f_{ij}^- = \begin{cases} \delta_{ij} - f_{ij}/f_{i+}, & \text{if } f_{i+} > 0, \\ \delta_{ij}, & \text{if } f_{i+} = 0, \end{cases} \quad i, j = 1, \ldots s. \tag{3.97}$$

Equation (3.95) means that, given an n-step transition count $F = (f_{ij})$ with $x_0 = x$, $x_n = y$, the conditional probability that $F(\mathbf{X}) = F$, given that $X_0 = x$ and $X_n = y$, becomes

$$\mathbb{P}\Big(F(\mathbf{X}) = F \mid X_0 = x, X_n = y \Big) = \prod_i f_{i+}! \prod_{ij} \left(\frac{p_{ij}^{f_{ij}}}{f_{ij}!} \right) \varphi_{(x,y)}^-; \tag{3.98}$$

likewise for the conditional probability

$$\mathbb{P}\Big(X_n = y, F(\mathbf{X}) = F \mid X_0 = x \Big) = p_{xy}^{(n)} \prod_i f_{i+}! \prod_{ij} \left(\frac{p_{ij}^{f_{ij}}}{f_{ij}!} \right) \varphi_{(x,y)}^-, \tag{3.99}$$

and for the unconditional probability

$$\mathbb{P}\Big(X_0 = x, X_n = y, F(\mathbf{X}) = F \Big) = \lambda_x p_{xy}^{(n)} \prod_i f_{i+}! \prod_{ij} \left(\frac{p_{ij}^{f_{ij}}}{f_{ij}!} \right) \varphi_{(x,y)}^-. \tag{3.100}$$

Here $p_{xy}^{(n)}$ represents the n-step transition probability, from x to y. Equations (3.98)–(3.100) are called *Whittle's formulas*.

Example 3.4.4 For the matrix F from Example 3.4.3, Whittle's formula gives

$$N_{13}^{(4)} = 2! \det \begin{pmatrix} 1/2 & 1/2 \\ 1 & -1 \end{pmatrix} = 2.$$

On the other hand, $N_{23}^{(4)} = N_{33}^{(4)} = 0$ because the transition count F is not compatible with any of the pairs $(2,3)$ and $(3,3)$. Hence, Whittle's formula cannot apply for these pairs.

This brings us to the following

Definition 3.4.5 Let $x, y \in I$ be given states (not necessarily distinct), and n an integer ≥ 1. Denote by $\mathscr{B}(n, x, y)$ the set of matrices $F = (f_{ij})$ with integer elements f_{ij}, $i, j \in I$, such that

(i) the sum $\sum_{i,j \in I} f_{ij} = n$,

(ii) $f_{i+} - f_{+i} = \delta_{ix} - \delta_{iy}$, $i \in I$, and, if $x = y$, $f_{x+} \geq 1$ and $f_{+x} \geq 1$.

In other words, the matrices $F \in \mathscr{B}(n, x, y)$ give n-step transition counts with $x_0 = x$ and $x_n = y$. Next, set $\mathscr{B}(n, x) = \bigcup_{y \in I} \mathscr{B}(n, x, y)$, which gives all n-step transition counts with $x_0 = x$. (Observe that, given x, the sets $\mathscr{B}(n, x, y)$ must be disjoint for different $y \in I$.)

Now let $P = (p_{ij}, i, j \in I)$ be a transition matrix. A *Whittle distribution* with parameters (P, n, x, y) is a probability distribution on $\mathscr{B}(n, x, y)$ which assigns to $F \in \mathscr{B}(n, x, y)$ the probability

$$\upsilon(F) = \prod_i f_{i+}! \prod_{ij} \left(\frac{p_{ij}^{f_{ij}}}{f_{ij}!} \right) \varphi_{(x,y)}^-. \tag{3.101}$$

Next, a Whittle distribution with parameters (P, n, x) constitutes a probability distribution on $\mathscr{B}(n, x)$ assigning to $F \in \mathscr{B}(n, x, y)$ the probability

$$\Upsilon(F) = p_{xy}^{(n)} \prod_i f_{i+}! \prod_{ij} \left(\frac{p_{ij}^{f_{ij}}}{f_{ij}!} \right) \varphi_{(x,y)}^-. \tag{3.102}$$

That is, the conditional probability $\Upsilon(F \mid \mathscr{B}(n, x, y))$ equals $\upsilon(F)$, for $F \in \mathscr{B}(n, x, y)$.

Proof of formula (3.95)   The proof goes by induction in n. Equation (3.95) holds for $n = 1$ (in this case both sides of (3.95) give 1). Assume it holds for $n - 1$.

Denote by $G(k,m)$ the matrix obtained from $G = (G_{ij})$ when its (k,m)th entry is diminished by 1: $G(k,m) = G - E(k,m)$ where $(E(k,m))_{ij} = \delta_{k,i}\delta_{j,m}$. Then clearly, for $F = (f_{ij})$,

$$N_{kl}^{(n)}(F) = \sum_{m=1}^{s} \mathbf{1}(f_{km} > 0)N_{ml}^{(n-1)}(F(k,m)). \tag{3.103}$$

Hence it suffices to show that the expressions $\varphi_{(k,l)}^{-}$ in the right-hand side of (3.95) satisfy a similar relation, viz.,

$$\varphi_{(k,l)}^{-} = \sum_{m=1}^{s} \mathbf{1}(f_{km} > 0)\left(\frac{f_{km}}{f_{k+}}\right)\left(\varphi^{-}(k,m)\right)_{(m,l)}. \tag{3.104}$$

Here and below, $\left(\varphi^{-}(k,m)\right)_{(m,l)}$ stands for the (m,l)th cofactor in matrix $F^{-}(k,m)$. Since $F^{-}(k,m)$ and F^{-} agree outside the mth column, the cofactors coincide: $\left(\varphi^{-}(k,m)\right)_{(m,l)} = \varphi_{(m,l)}^{-}$. This fact, together with definition (3.96), implies that (3.104) is equivalent to

$$\sum_{m=1}^{s}\left[\delta_{k,m} - \left(\frac{f_{k,m}}{f_{k+}}\right)\mathbf{1}(f_{km} > 0)\right]\varphi_{(m,l)}^{-} = 0. \tag{3.105}$$

Since $\sum_{m=1}^{s} f_{km}^{-}\varphi_{(m,l)}^{-} = \delta_{k,l}\det F^{-}$, equation (3.105) follows immediately for the case where $k \neq l$. Thus, we need only show that $\det F^{-} = 0$ if $k = l$. Suppose for notational convenience that $f_{i+} = f_{+i}$, is positive for $i \leq r$ and zero for $i > r$. Then F presents the form

$$F = \begin{pmatrix} A & 0 \\ 0 & 0 \end{pmatrix}, \tag{3.106}$$

where A forms an $r \times r$ matrix. By Definition (3.97),

$$F^{-} = \begin{pmatrix} A^{-} & 0 \\ 0 & I \end{pmatrix}, \tag{3.107}$$

where the rows of A^{-} sum to 0. Thus $\det F^{-} = \det A^{-} = 0$. (If $k \neq l$, F^{-} may become non-singular.) □

An interesting question arises in calculating the moments of the Whittle distribution. Take $I = \{1, \ldots, s\}$. Let $B = (b_{ij})$ be an $s \times s$ matrix with eigenvalues $\mu_1, \ldots, \mu_s$, which we assume at the moment to be distinct. Let $g(x)$ be an arbitrary polynomial of degree n and let $g(B)$ be the corresponding matrix polynomial. The well-known Sylvester theorem states that

$$g(B) = \sum_{k \in I} g(\mu_k) B(k), \tag{3.108}$$

where matrices $B(k)$ are defined by the expression

$$B(k) = \frac{\prod\limits_{i \in I:\, i \neq k} \left(\mu_i \mathbf{I} - B\right)}{\prod\limits_{i \in I:\, i \neq k} (\mu_i - \mu_k)}, \quad k \in I. \tag{3.109}$$

The matrix $B(k)$ forms the rank 1 (non-orthogonal) projection onto the 1-dimensional subspace generated by the eigenvector of B corresponding to the kth eigenvalue μ_k; the projection is performed along the hyperplane spanned by the remaining eigenvectors. The matrices $B(k)$ are *idempotent*, with

$$\left(B(k)\right)^2 = B(k), \text{ and } B(k)B(k') = 0, \ k \neq k', \ k, k' \in I.$$

Observe that the matrix $B(1)$ is an $s \times s$ matrix each row of which is the invariant vector $\underline{\pi}$ defined by the relation $\underline{\pi} = \underline{\pi} P$. An analogue of (3.108) also holds in the case of multiple eigenvalues, although (3.109) will then include the derivatives of polynomial g. We omit the details.

Theorem 3.4.6 *If an $s \times s$ random matrix $\mathbf{F}$ follows the Whittle distribution with parameters (P, n, x), then the expected value of its (α, β)th entry is given by*

$$E_{\alpha\beta}(n, x) = \sum_{m=0}^{n-1} p_{x\alpha}^{(m)} P_{\alpha\beta}, \quad \alpha, \beta, x \in I, \ n = 1, 2, 3, \ldots \tag{3.110}$$

where $p_{x\alpha}^{(m)}$ appears as the (x, α)th entry of the m-step transition matrix P^m. Furthermore, suppose the matrix P is irreducible and aperiodic, with distinct eigenvalues, say $1 = \mu_1 > |\mu_2|, \ldots, |\mu_s|$. Then the expected value $E_{\alpha\beta}(n, x)$ admits the representation

$$E_{\alpha\beta}(n, x) = p_{\alpha\beta} \left(n\pi_\alpha + \sum_{2 \leq k \leq s} \left(\frac{1 - \mu_k^n}{1 - \mu_k}\right) (p(k))_{x\alpha} \right). \tag{3.111}$$

Here, $(p(k))_{x\alpha}$ makes up the (x, α)th entry of the matrix $P(k)$ defined by (3.109), for $B = P$:

$$P(k) = \frac{\prod_{i \in I:\, i \neq k} \left(\mu_i \mathbf{I} - P\right)}{\prod_{i \neq k} (\mu_i - \mu_k)}, \quad k = 1, \ldots, s.$$

Proof Let $N_{\alpha,\beta}(n, x)$ be the random number of transitions $\alpha \to \beta$ in the random matrix $\mathbf{F}$, or, equivalently, in a sample $\mathbf{X}$ with distribution $\mathbb{P}(\,\cdot\, | X_0 = x)$.

Let the first transition in $\mathbf{X}$ be $x \to k$, $k \in I$. Then $N_{\alpha\beta}(n,x)$ satisfies the equations

$$N_{\alpha\beta}(n,x) = \delta_{x,\alpha}\,\delta_{k,\beta} + N_{\alpha,\beta}(n-1,k), \;\; n \geq 2, \tag{3.112}$$

and

$$N_{\alpha\beta}(1,x) = \delta_{x,\alpha}\,\delta_{k,\beta}, \;\; k \in I. \tag{3.113}$$

Thus, $E_{\alpha\beta}(n,x)$ satisfies the equations

$$E_{\alpha\beta}(n,x) = p_{x\beta}\,\delta_{x,\alpha} + \sum_{k=1}^{s} p_{xk}\,E_{\alpha\beta}(n-1,k), \;\; n \geq 2,$$

and

$$E_{\alpha\beta}(1,x) = p_{x\beta}\,\delta_{x,\alpha}. \tag{3.114}$$

We shall prove inductively that

$$E_{\alpha\beta}(n,x) = \sum_{m=0}^{n-1} p_{x\alpha}^{(m)}\,p_{\alpha\beta}. \tag{3.115}$$

Since $\delta_{x,\alpha}\,p_{\alpha\beta} = \delta_{x,\alpha}\,p_{x\beta}$, equation (3.115) is satisfied by (3.114) for $n = 1$. Then, using (3.115)

$$
\begin{aligned}
E_{\alpha\beta}(n+1,x) &= p_{x\beta}\,\delta_{x,\alpha} + \sum_{k=1}^{s} p_{xk} \sum_{m=0}^{n-1} p_{k\alpha}^{(m)}\,p_{\alpha\beta} \\
&= p_{x\beta}\,\delta_{x,\alpha} + \sum_{m=0}^{n-1} p_{x\alpha}^{(m+1)}\,p_{\alpha\beta} \\
&= \sum_{m=0}^{n} p_{x\alpha}^{(m)}\,p_{\alpha\beta}, \tag{3.116}
\end{aligned}
$$

establishing the induction step.

If all the eigenvalues μ_k of P are distinct, Sylvester's theorem yields

$$p_{x\alpha}^{(m)} = \sum_{k=1}^{s} \mu_k^m\,(p(k))_{x\alpha}, \;\; m = 0,1,\dots. \tag{3.117}$$

Arranging the eigenvalues so that $1 = \mu_1 > |\mu_2|, \dots, |\mu_s|$ allows to write (3.116) as

$$
\begin{aligned}
E_{\alpha\beta}(n,x) &= \sum_{m=0}^{n-1} \sum_{k=1}^{s} \mu_k^m (p(k))_{x\alpha}\,p_{\alpha\beta} \\
&= p_{\alpha\beta}\left(n\pi_\alpha + \sum_{k=2}^{s}\left(\frac{1-\mu_k^n}{1-\mu_k}\right)(p(k))_{x\alpha}\right).
\end{aligned}
$$

$\square$

Theorem 3.4.7 *If a random matrix* **F** *adheres to the Whittle distribution with parameter* (P, n, x), *then, for all* $\alpha, \beta, \gamma, \delta \in I$, *the covariance between the* (α, β)th *and the* (γ, δ)th *entries of* **F** *is given by*

$$C_{\alpha,\beta;\gamma,\delta}(n,x) = F_{\alpha\beta}(1,x)\left[\delta_{\alpha,\gamma}\,\delta_{\beta,\delta} - F_{\gamma\delta}(1,x)\right], \quad n=1.$$

$$C_{\alpha,\beta;\gamma,\delta}(n,x) = E_{\alpha\beta}(n,x)\left[\delta_{\alpha,\gamma}\,\delta_{\beta,\delta} - E_{\gamma\delta}(n,x)\right]$$

$$+ \sum_{m=1}^{n-1}\left[p_{x\alpha}^{(n-1-m)}\, p_{\alpha\beta}\, E_{\gamma\delta}(m,\beta)\right.$$

$$\left. + p_{x\gamma}^{(n-m-1)}\, p_{\gamma\delta}\, E_{\alpha\beta}(m,\delta)\right], \quad n \geq 2. \quad (3.118)$$

If P *happens to be irreducible and aperiodic and with distinct eigenvalues* $\mu_1, \ldots, \mu_s$, *then the covariance between the* (α, β)th *and the* (γ, δ)th *entries of* **F** *from the second equation in (3.118) above admits the representation*

$$C_{\alpha,\beta;\gamma,\delta}(n,x)$$

$$= p_{\alpha\beta}\,\delta_{\alpha,\gamma}\,\delta_{\beta,\delta}\left(n\pi_\alpha + \sum_{k=2}^{s}\left(\frac{1-\mu_k^n}{1-\mu_k}\right)(p(k))_{x\alpha}\right)$$

$$- p_{\alpha\beta}\,p_{\gamma\delta}\left\{n\left(\sum_{k=2}^{s}\left(\frac{1-\mu_k^n}{1-\mu_k}\right)\left[\pi_\alpha\,(p(k))_{x\gamma} + \pi_\gamma\,(p(k))_{x\alpha}\right]\right)\right.$$

$$+ \sum_{k=2}^{s}\sum_{k'=2}^{s}\left(\frac{1-\mu_k^n}{1-\mu_k}\right)\left(\frac{1-\mu_{k'}^n}{1-\mu_{k'}}\right)(p(k))_{x\alpha}\,(p(k'))_{x\gamma}$$

$$- \left(-n\pi_\alpha\,\pi_\gamma + \sum_{k=2}^{s}\left(\frac{n-1-n\mu_k+\mu_k^2}{(1-\mu_k)^2}\right)\right.$$

$$\left. \times\left[\pi_\alpha\,((p(k))_{x\gamma} + (p(k))_{\beta\gamma}) + \pi_\gamma\,((p(k))_{x\alpha} + (p(k))_{\delta\alpha})\right]\right)$$

$$+ \sum_{k=2}^{s}\left(\frac{1-n\mu_k^{n-1}+(n-1)\mu_k^2}{(1-\mu_k^2)}\right)$$

$$\left[(p(k))_{x\alpha}\,p(k))_{\beta\gamma} + (p(k))_{x\gamma}(p(k))_{\delta\alpha}\right]$$

$$+ \sum_{k=2}^{s}\sum_{k'=2,k'\neq k}^{s}\left(\frac{(p(k))_{x\alpha}\,(p(k'))_{\beta\gamma} + (p(k))_{x\gamma}\,(p(k'))_{\delta\alpha}}{1-\mu_{k'}}\right)$$

$$\left. \times\left[\left(\frac{1-\mu_k^{n-1}}{1-\mu_k}\right) - \left(\frac{\mu_{k'}(\mu_k^{n-1}-\mu_{k'}^{n-1})}{\mu_k-\mu_{k'}}\right)\right]\right\}. \quad (3.119)$$

Proof Step 1. Let, as before, $N_{\alpha,\beta}(n,x)$ be the number of transitions from state $\alpha \to \beta$ in $\mathbf{X}$, and assume that the first transition is $x \to k$. Then

$$N_{\alpha,\beta}(n,x)N_{\gamma,\delta}(n,x) = \delta_{\alpha,\gamma}\,\delta_{\beta,\delta}\,\delta_{\alpha,x}\,\delta_{\beta,k}, \text{ for } n = 1.$$

$$N_{\alpha,\beta}(n,x)N_{\gamma,\delta}(n,x)$$
$$= \delta_{\alpha,\gamma}\,\delta_{\beta,\delta}\,\delta_{\alpha,x}\,\delta_{\beta,k} + \delta_{x,\gamma}\,\delta_{k,\delta}N_{\alpha,\beta}(n-1,k)$$
$$+\delta_{x,\alpha}\,\delta_{k,\beta}N_{\gamma,\delta}(n-1,k) + N_{\alpha,\beta}(n-1,k)N_{\gamma,\delta}(n-1,k), \text{ for } n \geq 2.$$

Furthermore, the 'mixed' second moment

$$\sigma_{\alpha,\beta;\gamma,\delta}(n,x) := \mathbb{E}\left[N_{\alpha,\beta}(n,x)N_{\gamma,\delta}(n,x)\right]$$

satisfies the relations

$$\sigma_{\alpha,\beta;\gamma,\delta}(n,x) = \delta_{\alpha,\gamma}\,\delta_{\beta,\delta}\,\delta_{\alpha,x}\,p_{x\beta} + \delta_{x,\gamma}\,p_{x,\delta}\,E_{\alpha,\beta}(n-1,\delta)$$
$$+\delta_{x,\alpha}\,p_{x\beta}\,E_{\gamma\delta}(n-1,\beta) + \sum_{k=1}^{s}p_{xk}\,\sigma_{\alpha,\beta;\gamma,\delta}(n-1,k), \text{ for } n \geq 2,$$
$$\sigma_{\alpha\beta\gamma\delta}(n,x) = \delta_{\alpha\gamma};\delta_{\beta\delta}\,\delta_{\alpha x}\,p_{x\beta}, \text{ for } n = 1.$$

Step 2. Next, we show that

$$\sigma_{\alpha,\beta;\gamma,\delta}(n,x) = \delta_{\alpha,\gamma}\,\delta_{\beta,\delta}\,E_{\alpha,\beta}(n,x)$$
$$+\sum_{k=1}^{n-1}\left[p_{x\alpha}^{(n-1-k)}\,p_{\alpha\beta}\,E_{\gamma,\delta}(k,\beta) + p_{x\gamma}^{(n-1-k)}\,p_{\gamma\delta}E_{\alpha,\beta}(k,\delta)\right], \, n \geq 2, \left.\begin{array}{c} \\ \\ \\ \end{array}\right\} \quad (3.120)$$
$$\sigma_{\alpha,\beta;\gamma,\delta}(n,x) = \delta_{\alpha\gamma}\,\delta_{\beta,\delta}\,E_{\alpha,\beta}(1,s), \, n = 1,$$

after which (3.118) follows.

The case $n \geq 2$ will again be proved by induction. For $n = 2$, equation (3.120) is easily verified. To make the inductive step from n to $n+1$, write

$$\sigma_{\alpha,\beta;\gamma,\delta}(n+1,x)$$
$$= \delta_{\alpha,\gamma}\,\delta_{\beta,\delta}\,\delta_{\alpha,x}\,p_{x\beta} + \delta_{x,\gamma}\,p_{x\delta}\,E_{\alpha,\beta}(n,\delta) + \delta_{x,\alpha}\,p_{x\beta}\,E_{\gamma,\delta}(n,\beta)$$
$$+\sum_{k=1}^{s}p_{xk}\left(\delta_{\alpha,\gamma}\,\delta_{\beta,\delta}\,\sum_{m=1}^{n-1}p_{k\alpha}^{(m)}\,p_{\alpha\beta}\right.$$

$$+ \sum_{m=1}^{n-1} \left[p_{k\alpha}^{(n-1-m)} \, p_{\alpha\beta} E_{\gamma,\delta}(m,\beta) + p_{k\gamma}^{(n-1-m)} \, p_{\gamma\delta} \, E_{\alpha,\beta}(m,\delta) \right] \Bigg)$$

$$= \delta_{\alpha,\gamma}\delta_{\beta,\delta} \, \delta_{\alpha,x} \, p_{x\beta} + \delta_{x,\gamma} \, p_{x\delta} \, E_{\alpha,\beta}(n,\delta) + \delta_{x,\alpha} \, p_{x\beta} \, E_{\gamma,\delta}(n,\beta)$$

$$+ \, \delta_{\alpha,\gamma} \, \delta_{\beta,\delta} \sum_{m=0}^{n-1} p_{x\mu}^{(m+1)} \, r_{\mu\rho}$$

$$+ \sum_{m=0}^{n-1} \left[p_{x\alpha}^{(n-m)} \, p_{\alpha\beta} \, E_{\gamma,\delta}(m,\beta) + p_{x\gamma}^{(n-m)} \, p_{\gamma\delta} \, E_{\alpha,\beta}(m,\delta) \right]$$

$$= \delta_{\alpha,\gamma} \, \delta_{\beta,\delta} E_{\alpha,\beta}(n+1,x)$$

$$+ \sum_{m=1}^{n} \left[p_{x\alpha}^{(n-m)} \, p_{\alpha\beta} \, E_{\gamma\delta}(m,\beta) + p_{x\gamma}^{(n-m)} \, p_{\gamma\delta} \, E_{\alpha,\beta}(m,\delta) \right].$$

Step 3. Finally, suppose matrix P is irreducible and aperiodic and has distinct eigenvalues. Then we have for the covariance $C_{\alpha,\beta,\gamma,\delta}(n,x)$:

$$C_{\alpha,\beta;\gamma,\delta}(n,x) = p_{\alpha\beta} \left[n\pi_\alpha + \sum_{k=2}^{s} \left(\frac{1-\mu_k^n}{1-\mu_k} \right) (p(k))_{x\alpha} \right]$$

$$\times \left(\delta_{\alpha,\gamma}\delta_{\beta,\delta} - p_{\gamma\delta} \left[n\pi_\gamma + \sum_{k=2}^{s} \left(\frac{1-\mu_k^n}{1-\mu_k} \right) (p(k))_{x\gamma} \right] \right)$$

$$+ p_{\alpha\beta} \, p_{\gamma\delta} \sum_{m=1}^{n-1} \sum_{k=2}^{s} \mu_k^{(n-1-m)} \left[(p(k))_{x\alpha} \left(m\pi_\gamma + \sum_{k'=2}^{s} \left(\frac{1-\mu_{k'}^m}{1-\mu_{k'}} \right) p_{\beta\gamma}(k') \right) \right.$$

$$\left. + (p(k))_{x\gamma} \left(m\pi_\alpha + \sum_{k'=2}^{s} \left(\frac{1-\mu_{k'}^m}{1-\mu_{k'}} \right) p_{\delta\alpha}(k') \right) \right].$$

At the end, we invoke four relations:

$$\sum_{m=1}^{n-1} m\mu_k^{(n-1-m)} = \frac{n-1-n\mu_k+\mu_k^n}{(1-\mu_k)^2}, \ k=2,\ldots,s,$$

$$\sum_{m=1}^{n-1} (1-\mu_k^m) = \frac{n-1-n\mu_k+\mu_k^n}{1-\mu_k}, \ k=2,\ldots,s,$$

$$\sum_{m=1}^{n-1} \mu_k^{(n-1-m)}(1-\mu_{k'}^m) = \frac{1-\mu_k^{n-1}}{1-\mu_k} - \frac{\mu_{k'}(\mu_k^{n-1}-\mu_{k'}^{n-1})}{\mu_k-\mu_{k'}}, \tag{3.121}$$

$$k,k'=2,\ldots,s, \ k'\neq k,$$

$$\sum_{m=1}^{n-1} \mu_k^{(n-1-m)}(1-\mu_k^m) = \frac{1-n\mu_k^{n-1}+(n-1)\mu_k^n}{1-\mu_k}, \ k=2,\ldots,s.$$

and deduce (3.118). $\qquad\qquad\qquad\qquad\qquad\qquad\qquad\qquad\qquad\qquad\quad\square$

We note that (3.111) implies that as $n \to \infty$,

$$E_{\alpha,\beta}(n,x) \approx p_{\alpha\beta} \left[n\pi_\alpha + \sum_{k=2}^{s} \frac{(p(k))_{x\alpha}}{1-\mu_k} \right]. \tag{3.122}$$

Similarly, from (3.118) one deduces that as $n \to \infty$,

$$C_{\alpha,\beta,\gamma,\delta}(n,x)$$

$$\approx n \left[\pi_\alpha \, p_{\alpha\beta} \, \delta_{\alpha,\gamma} \, \delta_{\beta,\delta} - \pi_\alpha \, p_{\alpha\beta} \, \pi_\gamma \, p_{\gamma\delta} \right.$$

$$\left. + p_{\alpha\beta} \, p_{\gamma\delta} \sum_{k=2}^{s} \frac{\pi_\alpha (p(k))_{\beta\gamma} + \pi_\gamma (p(k))_{\delta\alpha}}{1-\mu_k} \right]$$

$$+ p_{\alpha\beta} \, \delta_{\alpha\gamma} \, \delta_{\beta\delta} \sum_{k=2}^{s} \frac{(p(k))_{x\alpha}}{1-\mu_k}$$

$$- p_{\alpha\beta} p_{\gamma\delta} \left[\sum_{k=2}^{s} \frac{\pi_\alpha ((p(k))_{x\gamma} + (p(k))_{\beta\gamma}) + \pi_\gamma ((p(k))_{x\alpha} + (p(k))_{\delta\alpha})}{(1-\mu_k)^2} \right.$$

$$\left. - \sum_{k,k'=2}^{s} \frac{(p(k))_{x\alpha}(p(k))_{\beta\gamma} + (p(k))_{x\gamma}(p(k))_{\delta\alpha} - (p(k))_{x\alpha}(p(k))_{x\gamma}}{(1-\mu_k)(1-\mu_{k'})} \right]. \tag{3.123}$$

Worked Example 3.4.8 Check that for the 2×2 transition matrix

$$P = \begin{pmatrix} 1-p & p \\ q & 1-q \end{pmatrix}, \quad 0 \leq p, q \leq 1,$$

Sylvester's theorem leads to the spectral decomposition

$$P = \begin{pmatrix} q/(p+q) & p/(p+q) \\ q/(p+q) & p/(p+q) \end{pmatrix} + (1-p-q) \begin{pmatrix} p/(p+q) & -p/(p+q) \\ -q/(p+q) & q/(p+q) \end{pmatrix}.$$

Derive expressions for entries of the m-step transition matrix P^m:

$$p_{12}^{(m)} = p \sum_{k=1}^{m-1} (1-p-q)^k, \quad p_{21}^{(m)} = q \sum_{k=1}^{m-1} (1-p-q)^k, \tag{3.124}$$

and similar formulas for $p_{11}^{(m)}$ and $p_{22}^{(m)}$. Here $m = 1, 2, \ldots$.

Solution We immediately get

$$\pi = \left(\frac{q}{p+q}, \frac{p}{p+q} \right).$$

Then for the matrix P^m:

$$P^m = \begin{pmatrix} q/(p+q) & p/(p+q) \\ q/(p+q) & p/(p+q) \end{pmatrix} + (1-p-q)^m \begin{pmatrix} p/(p+q) & -p/(p+q) \\ -p/(p+q) & q/(p+q) \end{pmatrix}.$$

This yields, in addition to (3.124),

$$p_{12}^{(m)} = p\left(\frac{1-(1-p-q)^m}{1-(1-p-q)}\right), \quad \text{and} \quad p_{21}^{(m)} = q\left(\frac{1-(1-p-q)^m}{1-(1-p-q)}\right).$$

$\square$

A detailed account of properties of Whittle's distribution can be found in P. Billingsley, "Statistical methods in Markov chains", *Annals Math. Statist.*, **32** (1961), 12–40.

3.5 Bayesian analysis of Markov chains: prior and posterior distributions

> *Bayesian Instincts, 2*
> *The Last of the Bayesians*
> *The Statsman Who Started as a Frequentist*
> *But Came Back as a Bayesian*
>
> (From the series '*Movies that never made it to the Big Screen*'.)

In the case of a Bayesian set-up, an unknown parameter θ is considered as *random*, with a given *prior* distribution Π^{prior}. Again, we will focus in this section on the extreme case where θ is a pair (λ, P) varying within the set $\mathscr{R}$ defined in (3.5) or θ is reduced to matrix P, varying within the set $\mathscr{P}$ defined in (3.7). The question is what we should take as a 'natural' probability distribution, Π^{prior}, of θ.

In many applications, one assumes that Π^{prior} is a product of *Dirichlet* distributions (or even more generally, a *Liouville* distributions). Formally, in the case $\theta = (\lambda, P)$, Π^{prior} is determined by the probability density function $\pi^{\text{prior}}(\lambda, P)$, relative to the Lebesgue measure $d\lambda \times dP$, on the set $\mathscr{R}$ from (3.5). See the first equation in (3.13). The PDF in question has the form of the product: $\pi^{\text{prior}}(\lambda, P) = \pi_{\text{in}}^{\text{prior}}(\lambda)\pi_{\text{tr}}^{\text{prior}}(P)$ where $\pi_{\text{in}}^{\text{prior}}(\lambda)$ is the joint PDF for entries λ_j of the initial vector λ, and $\pi_{\text{tr}}^{\text{prior}}(P)$ is the joint PDF for entries p_{ij} of the transition matrix P. Further,

$$\pi_{\text{in}}^{\text{prior}}(\lambda) = \Gamma\left(\sum_{k\in I} b_k\right)\prod_{j\in I}\frac{\lambda_j^{b_j-1}}{\Gamma(b_j)}\lambda_j^{b_j-1}, \quad \lambda = (\lambda_j), \tag{3.125}$$

$$\pi_{\text{tr}}^{\text{prior}}(P) = \prod_{i\in I}\Gamma\left(\sum_{k\in I} a_{ik}\right)\prod_{j\in I}\frac{p_{ij}^{a_{ij}-1}}{\Gamma(a_{ij})}, \quad P = (p_{ij}). \tag{3.126}$$

Here, parameters b_j and a_{ij} are positive numbers, $i, j \in I$. Formulas (3.125) and (3.126) should be treated with a reservation, as entries λ_j and p_{ij} satisfy the relations $\sum \lambda_j = 1$ and $\sum_j p_{ij} = 1$, for all $i \in I$, and hence are not linearly independent. This means that PDFs $\pi_{\text{in}}^{\text{prior}}(\lambda)$ and $\pi_{\text{tr}}^{\text{prior}}(P)$ should be considered in linearly independent variables (when one excludes one entry in the vector λ and one entry in each row of the matrix P). Recall that the same comment was made when we defined Lebesgue measures in (3.13).

As is plain to see, the PDF $\pi_{\text{in}}^{\text{prior}}(\lambda)$ in (3.125) is of the same type as the joint PDF of entries of a single row of the transition matrix P. Thus, we will focus on studying the PDF $\pi^{\text{prior}}(P) = \pi_{\text{tr}}^{\text{prior}}(P)$ from (3.126), omitting the subscript tr. In other words, we consider the case where $\theta = P$ varies within the set $\mathscr{P}$ defined in (3.7), or even within its interior $\mathscr{P}^{\text{int}}$ from (3.8). As was noted in Remark 3.1.1, if $P \in \mathscr{P}^{\text{int}}$ then it is irreducible and aperiodic and hence has a unique equilibrium distribution π.

For a detailed review of Liouville distributions, see R.D. Gupta and D.S.P. Richards, "Multivariate Liouville distributions", *Journ. Multivariate Anal.*, **23** (1987), 233–256.

Example 3.5.1 Recall (cf. Question 1.16.1), for a chain with two states $\{1, 2\}$, the matrix $P = \begin{pmatrix} 1-p & p \\ q & 1-q \end{pmatrix}$ is identified with the pair (p, q), and the set $\mathscr{P}$ can be thought of as a closed unit square $[0, 1] \times [0, 1]$. For the PDF (3.126) we then obtain

$$\pi^{\text{prior}}(p, q)$$
$$= \frac{\Gamma(a_{11} + a_{12})}{\Gamma(a_{11})\Gamma(a_{12})} \frac{\Gamma(a_{21} + a_{22})}{\Gamma(a_{21})\Gamma(a_{22})} (1-p)^{a_{11}-1} p^{a_{12}-1} q^{a_{21}-1} (1-q)^{a_{22}-1}. \quad (3.127)$$

Remembering that $1/B(\alpha, \beta) = \Gamma(\alpha + \beta)/\Gamma(\alpha)\Gamma(\beta)$, (3.127) is a product of two Beta-PDFs,

$$\frac{1}{B(a_{11}, a_{12})} (1-p)^{a_{11}-1} p^{a_{12}-1}, \quad 0 < p < 1,$$

and

$$\frac{1}{B(a_{21}, a_{22})} q^{a_{21}-1} (1-q)^{a_{22}-1}, \quad 0 < q < 1.$$

It is easy to compute all moments of matrix elements. Say,

$$\mathbb{E}[p_{11}] = \frac{\Gamma(a_{11} + a_{12})\Gamma(a_{21} + a_{22})}{\Gamma(a_{11})\Gamma(a_{12})\Gamma(a_{21})\Gamma(a_{22})} B(a_{11} + 1, a_{12}) B(a_{21}, a_{22})$$
$$= \frac{a_{11}}{a_{11} + a_{12}},$$

etc. See Worked Example 3.5.4 for details.

Reflecting the above fact, the distributions Π^{prior} with PDF $\pi^{\text{prior}}(P)$ as in (3.126) are sometimes called products of *multivariate* Beta-distributions.

An important role is played by the so-called *Dirichlet integral formula*. It states the following well-known fact from analysis:

$$\int \cdots \int_{A_n} x_1^{a_1-1} \cdots x_n^{a_n-1} \left(1 - \sum_{i=1}^{n} x_i\right)^{a_{n+1}-1} \, dx_1 \cdots dx_n$$

$$= \frac{\Gamma(a_1) \cdots \Gamma(a_{n+1})}{\Gamma(a_1 + \cdots + a_{n+1})}. \tag{3.128}$$

Here the domain of integration is

$$A_n = \left\{ (x_1, \ldots, x_n) : x_1, \ldots, x_n \geq 0, \ \sum_{i=1}^{n} x_i \leq 1 \right\} \subset \mathbb{R}^n,$$

and $a_1, \ldots, a_{n+1}$ are positive numbers. The analytic proof of (3.128) is rather tedious. A more transparent proof is provided by probabilistic methods.

Conside therefore IID RVs $Y_k \sim \Gamma(a_k, 1)$. The joint PDF $f_{Y_1, \ldots, Y_{n+1}}$ of $Y_1, \ldots, Y_{n+1}$ is a product:

$$f_{Y_1, \ldots, Y_{n+1}}(y_1, \ldots, y_{n+1}) = \frac{e^{-(y_1 + \cdots + y_{n+1})}}{\Gamma(a_1) \ldots \Gamma(a_{n+1})} y_1^{a_1-1} \cdots y_{n+1}^{a_{n+1}-1},$$

$$y_1, \ldots, y_{n+1} \geq 0. \tag{3.129}$$

It is convenient to use the following change of variables:

$$V_1 = Y_1, \ V_2 = Y_2, \ \ldots, \ V_n = Y_n, \ V_{n+1} = Y_1 + \cdots + Y_{n+1}$$

and

$$X_1 = \frac{V_1}{V_{n+1}}, \ \ldots, \ X_n = \frac{V_n}{V_{n+1}}, \ X_{n+1} = V_{n+1}.$$

Then the joint PDF $f_{V_1, \ldots, V_{n+1}}$ of $V_1, \ldots, V_{n+1}$ is calculated as

$$f_{V_1, \ldots, V_{n+1}}(v_1, \ldots, v_{n+1})$$

$$= \frac{e^{-v_{n+1}} v_1^{a_1-1} \cdots v_n^{a_n-1}}{\Gamma(a_1) \cdots \Gamma(a_{n+1})} \left(v_{n+1} - \sum_{i=1}^{n} v_i \right)^{a_{n+1}-1}$$

$$\times \mathbf{1}\left(v_{n+1} \geq \sum_{i=1}^{n} v_i \right), \ v_1, \ldots, v_n > 0. \tag{3.130}$$

The Jacobian

$$\frac{\partial(v_1,\ldots,v_{n+1})}{\partial(x_1,\ldots,x_{n+1})} = \det \begin{pmatrix} x_{n+1} & 0 & 0 & 0 & \cdots & 0 & x_1 \\ 0 & x_{n+1} & 0 & 0 & \cdots & 0 & x_2 \\ 0 & 0 & x_{n+1} & 0 & \cdots & 0 & x_3 \\ \vdots & \vdots & \vdots & \vdots & \cdots & \vdots & \vdots \\ 0 & 0 & 0 & 0 & \cdots & 0 & 1 \end{pmatrix}$$

equals x_{n+1}^n. We then obtain the following formula for the joint PDF $f_{X_1,\ldots,X_{n+1}}$ of variables $X_1, \ldots, X_{n+1}$:

$$f_{X_1,\ldots,X_{n+1}}(x_1,\ldots,x_{n+1})$$
$$= \frac{e^{-x_{n+1}} x_{n+1}^{a_1+\cdots+a_{n+1}-1} x_1^{a_1-1} \ldots x_n^{a_n-1}}{\Gamma(a_1)\ldots\Gamma(a_{n+1})} \left(1 - \sum_{i=1}^n x_i\right)^{a_{n+1}-1}. \tag{3.131}$$

Now, integrating in the variable x_{n+1} yields the joint PDF $f_{X_1,\ldots,X_n}(x_1,\ldots,x_n)$ of RVs $X_1, \ldots, X_n$. Direct calculation gives for $f_{X_1,\ldots,X_n}(x_1,\ldots,x_n)$ the expression

$$\frac{\Gamma(a_1+\cdots+a_{n+1})}{\Gamma(a_1)\cdots\Gamma(a_{n+1})} x_1^{a_1-1} \cdots x_n^{a_n-1} \left(1 - \sum_{i=1}^n x_i\right)^{a_{n+1}-1}. \tag{3.132}$$

The proof of (3.128) is concluded by observing that (3.132) defines a PDF (that is, a non-negative function, with integral 1).

Definition 3.5.2 Given $a_1, \ldots, a_{n+1} > 0$, the PDF

$$f(x_1,\ldots,x_n)$$
$$= \frac{\Gamma(a_1+\cdots+a_{n+1})}{\Gamma(a_1)\cdots\Gamma(a_{n+1})} x_1^{a_1-1} \cdots x_n^{a_n-1}$$
$$\times \left(1 - \sum_{i=1}^n x_i\right)^{a_{n+1}-1} \mathbf{1}\left(\sum_{i=1}^n x_i < 1\right), \quad x_1,\ldots,x_n > 0, \tag{3.133}$$

is called a *Dirichlet* PDF: we denote it by $f^{\mathrm{Dir}}(x_1,\ldots,x_n)$. A vector $\mathbf{X}$ of RVs $X_1, \ldots, X_n$ with joint PDF $f^{\mathrm{Dir}}(x_1,\ldots,x_n)$ is said to have a *Dirichlet distribution* with (vector) parameter $\mathbf{a}$, or briefly, $\mathrm{Dir}(\mathbf{a})$, and we write:

$$\mathbf{X} = \begin{pmatrix} X_1 \\ \vdots \\ X_n \end{pmatrix} \sim \mathrm{Dir}(\mathbf{a}), \quad \text{where } \mathbf{a} = \begin{pmatrix} a_1 \\ \vdots \\ a_n \\ a_{n+1} \end{pmatrix}.$$

Revisiting (3.126), we see that the joint PDF $\pi^{\mathrm{prior}}(P)$, with parameters $a_{ij}, i, j \in I$, is the product, over $i \in I$, of Dirichlet PDFs $\mathrm{Dir}(\mathbf{a}_i)$, with vectors $\mathbf{a}_i = (a_{ij}, j \in I)$. Furthermore, the factor

$$\Gamma\left(\sum_{k \in I} a_{ik}\right) \prod_{j \in I} \frac{p_{ij}^{a_{ij}-1}}{\Gamma(a_{ij})}$$

in this product describes the joint PDF of the entries $p_{ij}, j \in I$, of row i of the transition matrix $P = (p_{ij})$.

Dirichlet's formula implies a more general *Liouville* formula:

$$\int \cdots \int_{\{x_i \geq 0,\, x_1 + \cdots + x_n < h\}} g(x_1 + \cdots + x_n) x_1^{a_1-1} \cdots x_n^{a_n-1} dx_1 \cdots dx_n$$

$$= \frac{\Gamma(a_1) \cdots \Gamma(a_{n+1})}{\Gamma(a_1 + \cdots + a_{n+1})} \int_0^h g(t) t^{a_1 + \cdots + a_n - 1} dt \qquad (3.134)$$

valid for any function g for which the integral in the RHS is correctly defined.

Worked Example 3.5.3 (a) Consider the Liouville distribution, $\mathrm{Liouv}(g,h)$, with joint PDF

$$f^{\mathrm{Liouv}}(x_1, \ldots, x_n) = Cg(x_1 + \cdots + x_n) x_1^{a_1-1} \cdots x_n^{a_n-1}$$
$$\times \mathbf{1}(x_1 + \cdots + x_n \leq h), \quad x_1, \ldots, x_n \geq 0, \quad a_1, \ldots, a_n > 0. \qquad (3.135)$$

Here $g(s), s > 0$, is a given function, $h > 0$ is a parameter, and $C > 0$ is the normalizing constant, chosen so that $\int_{\mathbb{R}^n} f^{\mathrm{Liouv}}(x_1, \ldots, x_n) dx_1 \cdots dx_n = 1$. Check that PDF (3.135) coincides with Dirichlet's distribution, $\mathrm{Dir}(\mathbf{a})$, with PDF

$$f^{\mathrm{Dir}}(x_1, \ldots, x_n) = \frac{\Gamma\left(\sum_{j=1}^{n+1} a_j\right)}{\prod_{j=1}^{n+1} \Gamma(a_j)} x_1^{a_1-1} \cdots x_n^{a_n-1}$$

$$\times \left(1 - \sum_{i=1}^n x_i\right)^{a_{n+1}-1} \mathbf{1}\left(\sum_{j=1}^n x_j \leq 1\right), \quad x_1, \ldots, x_n \geq 0, \qquad (3.136)$$

if we set $h = 1$ and $g(s) = (1-s)^{a_{n+1}-1}$. Here, $\mathbf{a} = (a_1, \ldots, a_n, a_{n+1})$.

(b) Deduce Liouville's formula (3.134) from Dirichlet's formula (3.128).

Solution   (a) Equation (3.132) follows from (3.131), with h and g as indicated, by a direct substitution. The value of the corresponding constant C equals $\dfrac{\Gamma(a_1 + \cdots + a_{n+1})}{\Gamma(a_1) \cdots \Gamma(a_{n+1})}$ by the earlier calculation.

(b) The integral (3.134) equals

$$\int_0^h g(t) \int_{\{x_1 + \cdots + x_n = t\}} x_1^{a_1 - 1} \cdots x_n^{a_n - 1} dx_1 \cdots dx_{n-1} dt$$

$$= \int_0^h g(t) \int_{\left\{ \sum_{j=1}^{n-1} x_j \leq t \right\}} x_1^{a_1 - 1} \cdots x_{n-1}^{a_{n-1} - 1} \left(t - \sum_{j=1}^{n-1} x_j \right)^{a_n - 1} dx_1 \cdots dx_{n-1} dt.$$

In the variables

$$y_1 = \frac{x_1}{t}, \ldots, y_{n-1} = \frac{x_{n-1}}{t},$$

this integral takes the form

$$\int_0^h g(t) t^{a_1 + \cdots + a_n - 1}$$

$$\times \left[\int_{\left\{ \sum_{j=1}^{n-1} y_j \leq 1 \right\}} y_1^{a_1 - 1} \cdots y_{n-1}^{a_{n-1} - 1} \left(1 - \sum_{j=1}^{n-1} y_j \right)^{a_n - 1} dy_1 \cdots dy_{n-1} \right] dt.$$

By (3.128), the internal integral in the square brackets equals

$$\frac{\Gamma(a_1) \cdots \Gamma(a_n)}{\Gamma(a_1 + \cdots + a_n)},$$

which completes the solution. $\square$

Worked Example 3.5.4 The moments of the Dirichlet distribution are defined by

$$\mathbb{E}\left(X_1^{\alpha_1} \cdots X_n^{\alpha_n} \right) = \int_{A^n} x_1^{\alpha_1} \cdots x_n^{\alpha_n} f^{\mathrm{Dir}}(x_1, \ldots, x_n) \, dx_1 \cdots dx_n.$$

Prove that for all $\alpha_1 > -a_1, \ldots, \alpha_n > -a_n$

$$\mathbb{E}\left(X_1^{\alpha_1} \cdots X_n^{\alpha_n} \right) = \frac{\Gamma\left(\sum_{j=1}^{n+1} a_j \right)}{\Gamma\left(\sum_{j=1}^{n+1} a_j + \sum_{k=1}^n \alpha_k \right)} \prod_{i=1}^n \frac{\Gamma(a_i + \alpha_i)}{\Gamma(a_i)}. \tag{3.137}$$

In particular,

$$E(X_k) = \frac{a_k}{a}, \quad EX_k^2 = \frac{a_k(a_k+1)}{a(a+1)}, \quad \text{Var} \, X_k = \frac{a_k(a-a_k)}{a^2(a+1)} \tag{3.138}$$

$$E(X_1 \cdots X_n) = \prod_{i=1}^{n} \frac{a_i}{a_i+i-1}, \quad E(X_k X_l) = \frac{a_k a_l}{a(a+1)}. \tag{3.139}$$

where $a = a_1 + \cdots + a_{n+1}$.

Solution Write

$$E\left(X_1^{\alpha_1} \cdots X_n^{\alpha_n}\right) = \frac{\Gamma(a_1 + \cdots + a_{n+1})}{\Gamma(a_1) \cdots \Gamma(a_{n+1})}$$

$$\times \int \cdots \int_{A_n} x_1^{a_1+\alpha_1-1} \cdots x_n^{a_n+\alpha_n-1} \left(1 - \sum_{i=1}^{n} x_i\right)^{a_{n+1}-1} dx_1 \cdots dx_n.$$

with $A_n = \{x_i \geq 0, \sum_{i=1}^{n} x_i \leq 1\}$ and apply Dirichlet's integral formula. This yields

$$E\left(X_1^{\alpha_1} \cdots X_n^{\alpha_n}\right) = \frac{\Gamma\left(\sum_{j=1}^{n+1} a_j\right)}{\Gamma(a_1) \cdots \Gamma(a_{n+1})} \frac{\Gamma(a_1+\alpha_1) \cdots \Gamma(a_1+\alpha_n)\Gamma(a_{n+1})}{\Gamma\left(\sum_{j=1}^{n+1} a_j + \sum_{j=1}^{n} \alpha_j\right)}$$

$$= \frac{\Gamma\left(\sum_{j=1}^{n+1} a_j\right)}{\Gamma\left(\sum_{j=1}^{n+1} a_j + \sum_{j=1}^{n} \alpha_j\right)} \prod_{i=1}^{n} \frac{\Gamma(a_i+\alpha_i)}{\Gamma(a_i)}.$$

$\square$

Worked Example 3.5.5 Verify that the mean value $E_{i,j}$ and the variance $V_{i,j}$ of the (i,j) entry of the transition matrix, under the distribution with joint PDF (3.126) are given by

$$E_{i,j} = \frac{a_{ij}}{\sum_k a_{ik}}, \quad V_{i,j} = \frac{a_{ij}(\sum_k a_{ik} - a_{ij})}{(\sum_k a_{ik})^2 (\sum_k a_{ik} + 1)}. \tag{3.140}$$

Verify that the covariance of the (i,j) and (i,j') entries is

$$C_{i,j;i,j'} = -\frac{a_{ij} a_{ij'}}{(\sum_k a_{ik})^2 (\sum_k a_{ik} + 1)}. \tag{3.141}$$

Solution The result follows immediately from (3.138)–(3.139). For additional details see Martin, 1967.

$\square$

Worked Example 3.5.6 Let $S_{n+1} = Y_1 + \cdots + Y_{n+1}$, where $Y_k \sim \text{Gam}(a_k, 1)$, independently, and $a = a_1 + \cdots + a_{n+1}$.

(a) Prove that

$$X_k = \frac{Y_k}{S_{n+1}} \sim \text{Bet}(a_k, a - a_k).$$ (3.142)

(b) For the joint distribution prove that

$$\begin{pmatrix} X_k \\ X_l \end{pmatrix} \sim \text{Dir} \begin{pmatrix} a_k \\ a_l \\ a - a_k - a_l \end{pmatrix}.$$

(c) For a symmetric Dirichlet distribution $\text{Dir}(a, \ldots, a)$ prove that

$$X_i \sim \text{Bet}(a, na), \quad i = 1, \ldots, n.$$ (3.143)

Here $\text{Bet}(\alpha, \beta)$ stands for the Beta-distribution.

Solution (Sketch) For parts (a) and (b), make the following change of variables

$$X_l = \frac{Y_l}{S_{n+1}}, \quad l = 1, \ldots, n, \quad X_{n+1} = S_{n+1},$$

and integrate with respect to redundant variables. In this way the joint PDF of x_1 and x_2 can be written as

$$f_{X_1, X_2}(x_1, x_2) = C x_1^{a_1 - 1} x_2^{a_2 - 1} (1 - x_1 - x_2)^{a_{n+1} - 1}$$

$$\times \int_{A_{n-2}} x_3^{a_3 - 1} \cdots x_n^{a_n - 1} \left(1 - \frac{\sum_{i=3}^{n} x_i}{1 - x_1 - x_2} \right)^{a_{n+1} - 1} dx_3 \cdots dx_n$$

where

$$A_{n-2} = \left\{ x_3, \ldots, x_n \geq 0, \sum_{i=3}^{n} x_i \leq 1 - x_1 - x_2 \right\}$$

and C is a normalizing constant needed to make the integral of f_{X_1, X_2} in $dx_1 dx_2$ equal to 1. Introducing new variables $v_i = x_i / (1 - x_1 - x_2), i = 3, \ldots, n$ and computing the integral by Dirichlet formula (3.128), one obtains assertion (b).

(c) Similarly, the marginal PDF $f_{X_1}(x_1)$ of X_1 is

$$C x_1^{a-1} (1 - x_1)^{a-1} \int_{A_n} x_2^{a-1} \cdots x_n^{a-1} \left(1 - \frac{\sum_{i=2}^{n} x_i}{1 - x_1} \right)^{a-1} dx_2 \cdots dx_n$$

$$= \frac{1}{B(a, na)} x_1^{a-1} (1 - x_1)^{na-1}.$$

Here, as usual, $B(a, na) = \Gamma(a) \Gamma(na) / \Gamma((n+1)a)$ stands for the Beta-function.

$\square$

Worked Example 3.5.7 (a) Let $\begin{pmatrix} X_1 \\ \vdots \\ X_n \end{pmatrix} \sim \text{Dir} \begin{pmatrix} a_1 \\ \vdots \\ a_n \\ a_{n+1} \end{pmatrix}$. Prove that for the

sum $Y = X_1 + \cdots + X_n$,

$$Y \sim \text{Bet}(a_1 + \cdots + a_n, a_{n+1}). \tag{3.144}$$

(b) Prove that for the vector $\mathbf{X}_k = \begin{pmatrix} X_1 \\ \vdots \\ X_k \end{pmatrix}$ with $k < n$,

$$\mathbf{X}_k \sim \text{Dir} \begin{pmatrix} a_1 \\ \vdots \\ a_k \\ a_{k+1} + \cdots + a_{n+1} \end{pmatrix}. \tag{3.145}$$

(c) Set $Y_1 = X_1 + \cdots + X_{n_1}$, $Y_2 = X_{n_1+1} + \cdots + X_{n_1+n_2}$, ..., $Y_k = X_{n_1+\cdots+n_{k-1}+1} + \cdots + X_{n_1+\cdots+n_k}$. Then show

$$\mathbf{Y}_k = \begin{pmatrix} Y_1 \\ \vdots \\ Y_k \end{pmatrix} \sim \text{Dir} \begin{pmatrix} a(1) \\ \vdots \\ a(k) \\ a(k+1) \end{pmatrix}, \tag{3.146}$$

where

$$a(1) = a_1 + \cdots + a_{n_1},$$
$$\cdots \tag{3.147}$$
$$a(k) = a_{n_1+n_2+\cdots+n_{k-1}+1} + \cdots + a_{n_1+\cdots+n_k},$$
$$a(k+1) = a_{n_1+n_2+\cdots+n_k+1} + \cdots + a_{n+1}.$$

Solution (Sketch) For part (a), apply Dirichlet's formula (3.134) with $g(t) = (1 - t)^{a_{n+1}-1}$. For part (b), use the same calculation as in Worked Example 3.5.6. In part (c), the joint PDF $f_{Y_1,\ldots,Y_k}(t_1,\ldots,t_k)$ is proportional to

$$t_1^{a(1)-1} \cdots t_{k+1}^{a(k+1)-1}$$

where $t_1 + \cdots + t_{k+1} = 1$. That is, $\mathbf{Y}_k$ has the Dirichlet distribution with parameters $a(1),\ldots,a(k+1)$. $\qquad\square$

In Volume 1, we discussed the issue of conjugacy of a given family (or class) of distributions. The meaning of this concept is that if the prior distribution Π^{prior} is from a given class (described by one or several parameters), then the *posterior* distribution, given sample vector $\mathbf{x}$, is from the same family (class). In this case we only have to indicate how the parameters of the posterior distribution are calculated as functions of the sample vector and the parameters of the prior distribution. Recall that, if the prior distribution has PDF $\pi^{\text{prior}}(\theta)$, $\theta \in \Theta$, and the likelihood of a sample vector $\mathbf{x}$ is $L(\mathbf{x}; \theta)$ or $l(\mathbf{x}; \theta)$, then the posterior PDF is determined by

$$\pi^{\text{post}}(\theta|\mathbf{x}) \propto \pi^{\text{prior}}(\theta)L(\mathbf{x}; \theta), \quad \text{or} \quad \pi^{\text{post}}(\theta|\mathbf{x}) \propto \pi^{\text{prior}}(\theta)\, l(\mathbf{x}; \theta).$$

Here the proportionality coefficient is fixed by the condition that the integral of $\pi^{\text{post}}(\theta|\mathbf{x})$ equals 1. The parameter θ may be a scalar or a vector; the case of maximum uncertainty which we analysed at some length in previous sections is where $\theta = (\lambda, P) \in \mathscr{R}$ or $\theta = P \in \mathscr{P}$.

Worked Example 3.5.8 Let $X_1, \ldots, X_n$ be IID RVs with values $k \in \{1, \ldots, \kappa\}$ and common marginal probabilities

$$\theta_k = \mathbb{P}(X = k).$$

Suppose that the vector $\theta = \begin{pmatrix} \theta_1 \\ \vdots \\ \theta_\kappa \end{pmatrix}$ is random, with Dirichlet distribution

$\text{Dir}\begin{pmatrix} a_1 \\ \vdots \\ a_\kappa \end{pmatrix}$. Then, given a sample vector $\mathbf{x} = \begin{pmatrix} x_1 \\ \vdots \\ x_n \end{pmatrix}$, the posterior distribu-

tion of θ is $\text{Dir}\begin{pmatrix} n_1 + a_1 \\ \vdots \\ n_\kappa + a_\kappa \end{pmatrix}$, where n_k stands for the cardinality of $\{i : i = 1, \ldots, n, \ x_i = k\}$.

In particular, the posterior mean value of the RV θ_k equals the ratio $(n_k + a_k)/(n + a)$ where $a = \sum_{k=1}^{\kappa} a_k$.

Solution (Sketch) This follows immediately from (3.138). $\qquad\square$

Worked Example 3.5.9 Consider a DTMC on a given finite state space $I = \{1, \ldots, s\}$, where the transition matrix P is chosen randomly, with PDF $\pi^{\text{prior}}(P)$ for $P \in \mathscr{P}^{\text{int}}$, and $\mathscr{P}^{\text{int}}$ is the interior of the set of dimension $s(s-1)$, determined in (3.8). Verify that the family of Dirichlet PDFs given by (3.126) is *conjugate* relative to the reduced likelihood $l(\mathbf{x}; P) = \prod_{i,j=1}^{s} p_{ij}^{n_{ij}(\mathbf{x})}$. That is, check that if $\pi^{\text{prior}}(P)$

is of the form (3.126) with a given collection of values $a_{ij} > 0$, then the posterior density $\pi^{\text{post}}(P|\mathbf{x})$ defined by

$$\pi^{\text{post}}(P|\mathbf{x}) \propto l(\mathbf{x}, P)\pi^{\text{prior}}(P)$$

again has the form

$$\pi^{\text{post}}(P|\mathbf{x}) = \prod_{i \in I} \Gamma\left(\sum_{k \in I} a'_{ik}\right) \prod_{j \in I} \frac{p_{ij}^{d'_{ij}-1}}{\Gamma(a'_{ij})}. \tag{3.148}$$

Determine the value a'_{ij} as a function of a_{ij} and $\mathbf{x}$.

Solution (Sketch) Use the fact that $a'_{ik} = a_{ik} + n_{ik}$ where n_{ik} is the transition count defined in (3.14). □

Worked Example 3.5.10 Assume that a distribution of the 2×2 transition matrix $P = \begin{pmatrix} 1-p & p \\ q & 1-q \end{pmatrix}$ from Example 3.5.1 is the product of two Beta distributions, with the PDF

$$f(p,q) = \frac{p^{\alpha-1}(1-p)^{\beta-1}q^{\gamma-1}(1-q)^{\delta-1}}{B(\alpha,\beta)B(\gamma,\delta)}, \quad 0 < p,q < 1, \tag{3.149}$$

where $\alpha, \beta, \gamma, \delta > 0$. Write, alternatively, $P = \begin{pmatrix} 1-p_{12} & p_{12} \\ p_{21} & 1-p_{21} \end{pmatrix}$.

(a) Check that the mean $\mathbb{E}\left[p_{12}^{(m)}\right]$ of the matrix element $p_{12}^{(m)} = (P^m)_{12}$ equals

$$\frac{\alpha}{\beta+\alpha} \sum_{k=0}^{m-1} \sum_{l=0}^{k} \binom{k}{l} (-1)^l \frac{(\beta)_{k-l}(\gamma)_l}{(\beta+\alpha+1)_{k-l}(\gamma+\delta)_l}, \quad m = 1,2,\ldots, \tag{3.150}$$

where $(x)_k = \Gamma(x+k)/\Gamma(x) = x(x+1)\cdots(x+k-1)$ is the *Pochhammer symbol*. Next, verify that the mean value $\mathbb{E}\left[p_{21}^{(m)}\right]$ of the entry $p_{21}^{(m)} = (P^m)_{21}$ is given by

$$\mathbb{E}\left[p_{21}^{(m)}\right] = \sum_{k=0}^{m-1} \sum_{l=0}^{k} \binom{k}{l} (-1)^l \frac{(\beta)_{k-l}(\gamma)_{l+1}}{(\alpha+\beta)_{k-l}(\gamma+\delta)_{l+1}}, \quad m = 1,2,\ldots, \tag{3.151}$$

and that the mean value $\mathbb{E}\left[p_{12}^{(m)} p_{21}^{(m)}\right]$ equals

$$\frac{\alpha}{\beta+\alpha} \sum_{j,k=0}^{m-1} \sum_{l=0}^{j+k} \binom{j+k}{l} (-1)^l \frac{(\beta)_{j+k-l}(\gamma)_{l+1}}{(\alpha+\beta+1)_{j+k-l}(\gamma+\delta)_{l+1}}. \tag{3.152}$$

(b) The entries π_1 and π_2 of the equilibrium distribution π of the matrix P become random variables. Check that the mean values $\mathbb{E}[\pi_1]$ and $\mathbb{E}[\pi_2]$ are given by

$$
\begin{aligned}
\mathbb{E}[\pi_1] &= \sum_{k=0}^{\infty} \sum_{l=0}^{k} \binom{k}{l} (-1)^l \frac{(\beta)_{k-l}(\gamma)_{l+1}}{(\alpha+\beta)_{k-l}(\gamma+\delta)_{l+1}}, \\
\mathbb{E}[\pi_2] &= \frac{\alpha}{\beta+\alpha} \sum_{k=0}^{\infty} \sum_{l=0}^{k} \binom{k}{l} (-1)^l \frac{(\beta)_{k-l}(\gamma)_{l}}{(\alpha+\beta+1)_{k-l}(\gamma+\delta)_{l}}.
\end{aligned}
\tag{3.153}
$$

Solution (Sketch) (a) It has been proved in Worked Example 3.4.8 that

$$
p_{12}^{(m)} = p \frac{1-(1-p-q)^m}{1-(1-p-q)} = p \sum_{k=0}^{m-1} (1-p-q)^k.
$$

Expand the factor $(1-p-q)^k$ as $\sum_{l=0}^{m} \binom{k}{l} (-1)^l q^l (1-p)^{k-l}$, and use independence of p_{12} and p_{21} to obtain the representation

$$
\mathbb{E}\big[p_{12}^{(m)}\big] = \sum_{k=0}^{m-1} \sum_{l=0}^{k} \binom{k}{l} (-1)^l \mathbb{E}[q^l]\mathbb{E}[p(1-p)^{k-l}].
$$

Next, the product $\mathbb{E}[q^l]\mathbb{E}[p(1-p)^{k-l}]$ equals the ratio

$$
\frac{B(\alpha+1,\beta+k-l)B(\gamma+l,\delta)}{B(\alpha,\beta)B(\gamma,\delta)}.
$$

Substituting the corresponding Gamma-functions, we obtain (3.150). Similarly, we obtain the expression (3.151) for $\mathbb{E}[p_{21}^{(m)}]$. Next, expanding

$$
p^m q^m = pq \sum_{k,j=0}^{m-1} (1-p-q)^{j+k}
$$

we obtain the formula (3.152) for $\mathbb{E}\big[p_{12}^{(m)} p_{21}^{(m)}\big]$.

Equation (3.153) then emerges in the limit $m \to \infty$. Note an analytical remark: the series in (3.153) converge only conditionally, not absolutely. See again Martin, 1967.   □

Worked Example 3.5.11 Let (X_m) be a two-state DTMC, with two states. Suppose that the transition matrix $P = \begin{pmatrix} 1-p & p \\ q & 1-q \end{pmatrix}$ of the chain (X_m) is random and distributed with the product-PDF f as in (3.149), with positive parameters α, β, γ, δ. Next, assume that we are given a *reward matrix*

$$
R = (r_{ij}) = \begin{pmatrix} a & b \\ c & d \end{pmatrix}
$$

where the entries $r_{ij} = a, b, c, d \in \mathbb{R}$ indicate the reward earned when the DTMC (X_m) makes a transition from state i to state j.

Define the average *discounted reward vector* $\begin{pmatrix} V_1(P) \\ V_2(P) \end{pmatrix}$ with entries

$$V_i(P) = \sum_{n \geq 0} \rho^n \sum_{j,k=1}^{2} p_{ij}^{(n)} p_{jk} r_{jk}, \quad i = 1, 2,$$

where $\rho \in [0, 1/2)$ is a *discount factor*. As the transition matrix P is assumed to be random, the entries $V_1(P)$ and $V_2(P)$ are also random variables.

Prove the relations

$$
\begin{aligned}
\mathbb{E}[V_1] &= \frac{\beta a + \alpha b}{(1-\rho)(\alpha+\beta)} + \frac{\alpha \rho}{(1-\rho)(\alpha+\beta)} \\
&\quad \times \sum_{k \geq 0} \sum_{l=0}^{k} \binom{k}{l} \rho^k (-1)^l \frac{(\beta)_{k-l}(\gamma)_l}{(\alpha+\beta+1)_{k-l}(\gamma+\delta)_l} \\
&\quad \times \left[\frac{(\gamma+l)c + \delta d}{\gamma+\delta+l} - \frac{a(\beta+k-l) + b(\alpha+1)}{\alpha+\beta+1+k-l} \right], \quad (3.154)
\end{aligned}
$$

and

$$
\begin{aligned}
\mathbb{E}[V_2] &= \frac{\gamma c + \delta d}{(1-\rho)(\gamma+\delta)} + \frac{\rho}{1-\rho} \\
&\quad \times \sum_{k \geq 0} \sum_{l=0}^{k} \binom{k}{l} \rho^k (-1)^l \frac{(\beta)_{k-l}(\gamma)_{l+1}}{(\alpha+\beta)_{k-l}(\gamma+\delta)_{l+1}} \\
&\quad \times \left[\frac{a(\beta+k-l) + b\alpha}{\alpha+\beta+k-l} - \frac{c(\gamma+l+1) + d\delta}{\gamma+\delta+l+1} \right]. \quad (3.155)
\end{aligned}
$$

Solution Let M stand for the parameter matrix:

$$M = \begin{pmatrix} \alpha & \beta \\ \gamma & \delta \end{pmatrix}$$

and write $\mathbb{E}_M$ for the expectation with respect to the PDF $f(p, q)$ from (3.149), with parameters identified in matrix M. Next, set

$$S_{ij}(M) = \sum_{m \geq 1} \rho^m \mathbb{E}_M \left[p_{ij}^{(m)} \right], \quad i, j = 1, 2.$$

Then for $S_{ij}(M)$ one can write down series in terms of the entries of M, viz.,

$$S_{12}(M) = \frac{\alpha}{\alpha+\beta} \sum_{m \geq 1} \sum_{k=0}^{m} \rho^m \sum_{l=0}^{k} \binom{k}{l} (-1)^l \frac{(\beta)_{k-l}(\gamma)_l}{(\alpha+\beta+1)_{k-l}(\gamma+\delta)_l},$$

or, changing the order of the first two summations,

$$S_{12}(M) = \frac{\alpha\rho}{(1-\rho)(\alpha+\beta)} \sum_{k\geq 0}\sum_{l=0}^{k} \binom{k}{l} \rho^k (-1)^l \frac{(\beta)_{k-l}(\gamma)_l}{(\alpha+\beta+1)_{k-l}(\gamma+\delta)_l}. \tag{3.156}$$

Similarly,

$$S_{21}(M) = \frac{\rho}{1-\rho} \sum_{k\geq 0}\sum_{l=0}^{k} \binom{k}{l} \rho^k (-1)^l \frac{(\beta)_{k-l}(\gamma)_{l+1}}{(\alpha+\beta)_{k-l}(\gamma+\delta)_{l+1}}. \tag{3.157}$$

It can be shown that the series (3.156), (3.157) converge absolutely for $\rho < 1/2$. Moreover,

$$S_{11}(M) = \sum_{m\geq 1} \rho^m \left(1 - \mathbb{E}_M\left[p_{12}^m\right]\right) = \frac{\rho}{1-\rho} - S_{12}(M), \tag{3.158}$$

and similarly,

$$S_{22}(M) = \frac{\rho}{1-\rho} - S_{21}(M). \tag{3.159}$$

Further, denote by $T_{ij}(M)$, $i,j = 1,2$, the matrix obtained when the (i,j)th entry of M increases by 1:

$$T_{11}(M) = \begin{pmatrix} \alpha+1 & \beta \\ \gamma & \delta \end{pmatrix}, \quad T_{12}(M) = \begin{pmatrix} \alpha & \beta+1 \\ \gamma & \delta \end{pmatrix},$$

and so on, and write $\mathbb{E}_{T_{ij}(M)}$ for the expectation under the PDF as in (3.149), but with parameters identified in the matrix $T_{ij}(M)$. Then the following equality holds:

$$\mathbb{E}_M[V_i] = \sum_{k=1}^{2} \mathbb{E}_M[p_{ik}]r_{ik} + \sum_{j,k=1}^{2} S_{ij}(T_{jk}(M))\mathbb{E}_M[p_{jk}]r_{jk}, \quad i=1,2, \tag{3.160}$$

where $S_{ij}(T_{jk}(M))$ is determined by the same formulas (3.156)–(3.159), with M replaced by the matrix $T_{jk}(M)$.

This is the crucial equation. Substituting into (3.160) the expressions for $S_{ij}(T_{jk}(M))$ and re-arranging terms in a suitable manner eventually leads to (3.154) and (3.155). For instance,

$$\mathbb{E}_M[V_1] = \frac{\beta}{\alpha+\beta}a + \frac{\alpha}{\alpha+\beta}b + \left(\frac{\rho}{1-\rho} - S_{12}(T_{11}(M))\right)\frac{\beta}{\alpha+\beta}a$$

$$+ \left(\frac{\rho}{1-\rho} - S_{12}(T_{12}(M))\right)\frac{\alpha}{\alpha+\beta}b$$

$$+ S_{12}(T_{21}(M))\frac{\delta}{\gamma+\delta}c + S_{12}(T_{22}(M))\frac{\gamma}{\gamma+\delta}d$$

$$= \frac{\beta a + \alpha b}{(1-\rho)(\alpha+\beta)} + \frac{\alpha\rho}{(1-\rho)(\alpha+\beta)}$$

$$\times \sum_{k\geq 0}\sum_{l=0}^{k}\binom{k}{l}\rho^k(-1)^l\frac{(\beta)_{k-l}(\gamma)_l}{(\alpha+\beta+1)_{k-l}(\gamma+\delta)_l}$$

$$\times [A_l c + B_l d - C_l a - D_l b],$$

and a simple calculation shows that

$$A_l = \frac{\gamma+l}{\gamma+\delta+l}, \qquad B = \frac{\delta}{\gamma+\delta+l},$$

$$C = \frac{\beta+k-l}{\alpha+\beta+1+k-l}, \qquad D = \frac{\alpha+1}{\alpha+\beta+1+k-l},$$

which yields (3.154). □

3.6 Elements of control and information theory

We begin with two examples referring to the Secretary problem (see Section 1.11).

Worked Example 3.6.1 Let $X_1, \ldots, X_m$ be independent and identically distributed random variables, $X_j \sim U(0,1)$. (We may think that X_j represents a 'quality' of an object j drawn at random from an 'unlimited population', without replacement.) As in Worked Example 1.11.1, we consider a single-choice Secretary problem, aiming to select the object of highest quality, by comparing the currently emerging object with the preceding ones, with no possibility to return to previously rejected objects. Recall that in Section 1.11 the final (and relatively simple) answers for the probability emerged in the limit $m \to \infty$. For example, allowing two choices increases the probability of success from 0.3678 to 0.5910. Here we consider the single-choice case with fully known distribution and specify the optimal strategy. As we shall see in Worked Example 3.6.2, this information increases the probability of success to 0.5802, which is only marginally lower than 0.5910.

Solution It is not hard to convince oneself that for each $i = 1, \ldots, m$ there exists an optimal threshold value $b_i \in (0,1)$ such that at draw $m-i+1$ one should select the emerging object if $X_{m-i+1} = \max[X_l : 1 \leq l \leq m-i+1] > b_i$ and reject it if $X_{m-i+1} < b_i$ or $X_{m-i+1} < \max[X_l : 1 \le l \le m - i]$. (In the case where $X_{m-i+1} = \max[X_l : 1 \le l \le m - i + 1] = b_i$, any of the two decisions leads to the same probability of success). Indeed, $b_1 = 0$ (which means you take the last emerging object if it appears to be the global maximum and you haven't made the choice before), whereas $b_2 = 1/2$ (which is the median of the uniform distribution $U(0, 1)$). The remaining b_i will be $> 1/2$ (they will monotonically increase with i); to calculate them exactly we will use the above indifference condition.

Suppose that we have not made a selection by the $(m - i)$th draw and $X_{m-i+1} = \max[X_l : 1 \le l \le m - i + 1]$ (in which case we call object $m - i + 1$ a candidate). There are $i - 1$ draws left. Then $x = b_i$ is the solution to

$$x^{i-1} = \sum_{j=1}^{i-1} \binom{i-1}{j} \frac{1}{j} x^{i-1-j}(1-x)^j.$$

Indeed, if we stop (make a selection), the chance of success equals x^{i-1}. If we continue, then $j \le i - 1$ values bigger than x may appear. If we stop at the first appearance of such a value, the probability that it is the absolute maximum is $1/j$ due to symmetry. For $i = 2$ we get $x = 1 - x$, or $x = 1/2$. Thus $b_2 = 1/2$, as stated. For $i = 3$, after simplifying, we get $5x^2 - 2x - 1 = 0$, or

$$x = b_3 = (1 + \sqrt{6})/5 \approx 0.6899$$

For modest values of $i + 1$ one can find b_{i+1} numerically:

$i+1$	b_{i+1}	$i+1$	b_{i+1}
2	0.5000000	20	0.95891663
3	0.68989795	25	0.96727367
4	0.77584508	30	0.97280561
5	0.82458958	35	0.97672783
6	0.85594922	40	0.97967655
10	0.91604417	45	0.98195608
15	0.94482887	50	0.98377582

Note that the optimal threshold b_i does not depend on $m > i$. $\qquad\square$

The General Secretary Problem
(From the series '*Movies that never made it to the Big Screen*'.)

Worked Example 3.6.2 Continuing the previous example, let us fix a strategy (not necessary the optimal) with thresholds $d_1 \ge d_2 \ge d_3 \cdots \ge d_m$, where $0 \le d_i \le 1$.

That is, we select the first object whose quality X_j gives the maximum $\max[X_l : 1 \leq l \leq j]$ and exceeds d_j. Prove that the probability of success at the first draw equals

$$P(1) = \frac{1 - d_1^m}{m},$$

whereas the probability of success $P(r+1)$ at draw $r+1$ is given by

$$P(r+1) = \sum_{i=1}^{r} \frac{d_i^r}{r(m-r)} - \sum_{i=1}^{r} \frac{d_i^m}{m(m-r)} - \frac{d_{r+1}^m}{m}, \quad 1 \leq r \leq m-1.$$

Solution The expression for $P(1)$ is straightforward: $1 - d_1^m$ gives the probability that at least one of the objects has quality at least d_1 and $1/m$ the conditional probability that, given the above event, the best quality is X_1 (by symmetry). For a general r, we take i, $1 \leq i \leq r$, and consider the probability that the first r draws resulted in no selection, and that the (globally) best object is among the remaining $m - r$ draws. Equivalently, this the probability that, for all $i = 1, \ldots, r$ such that the quality X_i is the highest among $X_1, \ldots, X_r$, we have that $X_i < d_i$ and $X_i < \max[X_1, \ldots, X_m]$. As before, the probability of the event $X_i = \max[X_1, \ldots, X_r] \leq d_i$ equals d_i^r/r.

Within this event, it is possible that there will be no selection, when X_i is the global maximum $\max[X_1, \ldots, X_m]$. The probability that $X_i = \max[X_1, \ldots, X_m] < d_i$ equals d_i^m/m. Therefore, the difference $d_i^r/r - d_i^m/m$ gives the probability of the event that (i) $X_i < d_i$, (ii) $X_i = \max[X_1, \ldots, X_r]$ and (iii) $X_i < \max[X_1, \ldots, X_m]$. Since the thresholds $d_1, \ldots, d_m$ have been chosen monotone decreasing, within the last event, no X_l with $l < i$ could be larger than d_l. Summing the differences $d_i^r/r - d_i^m/m$ over $1 \leq i \leq r$ yields the probability that no selection has been made among the first r draws and that the best quality object is among the last $m - r$ draws.

Given this information, the probability that the $(r+1)$st quality X_{r+1} is the global maximum equals

$$\frac{1}{m-r} \sum_{i=1}^{r} (d_i^r/r - d_i^m/m).$$

This expression gives the probability that no selection occurred among the first r draws, and that $X_{r+1} = \max[X_1, \ldots, X_m]$; that is, X_{r+1} is the globally highest quality. If we choose the $(r+1)$st object when X_{r+1} yields the globally highest quality, we succeed. But there is a chance that we do not take the $(r+1)$st object, when it is globally the best, and this probability is equal to d_{r+1}^m/m. This must be subtracted, and we obtain the equation for $P(r+1)$. $\square$

Under the optimal strategy, where $d_i = b_{m-i+1}$, $i = 1, \ldots, m$, we obtain the optimal probability $P_{\text{opt}}(\text{success})$:

$$P_{\text{opt}}(\text{success}) = \frac{1 - b_m^m}{m}$$

$$+ \sum_{r=2}^{m} \left[\sum_{i=1}^{r-1} \frac{b_{m-i+1}^{r-1}}{(r-1)(m-r+1)} - \sum_{i=1}^{r-1} \frac{b_{m-i+1}^m}{m(m-r+1)} - \frac{b_{m-r+1}^m}{m} \right].$$

The table below shows these probabilities for a sample of values of m.

m	P_{opt}	m	P_{opt}
2	0.75000	20	0.594200
3	0.684293	30	0.589472
4	0.655396	40	0.587126
5	0.639194	50	0.585725
10	0.608699	∞	0.580164

Secretaries do it without any problem.
(From the series 'How they do it'.)

In the remaining part of this section we discuss connections between statistics and the information theory. Some of this material will be used in Section 3.8. We will abbreviate probability density function by PDF and probability mass function by PMF. The former refers to continuous RVs and the latter to discrete ones, but we will succeed in treating both simultaneously.

Definition 3.6.3 Let X be a RV with PMF/PDF $f(x; \theta)$, $x \in \mathbb{R}$, depending on a parameter $\theta \in \Theta$. Suppose that Θ is an interval of the real line $\mathbb{R}$ and all PMFs/PDFs $f(x; \theta)$ have the same support set $\mathbb{S} \subseteq \mathbb{R}$ which is a finite or countable discrete set in the case of a PMF or an interval in the case of a PDF. (That is, for all $\theta \in \Theta$, $f(x; \theta) > 0$ iff $x \in \mathbb{S}$.) Suppose that $f(x; \theta)$ depends smoothly on $\theta \in \Theta$. The *score* (of X) is the RV $V(X; \theta)$ dependent on the random argument X:

$$V(X; \theta) = \frac{\partial}{\partial \theta} \ln f(X; \theta). \tag{3.161}$$

Under mild assumptions, we have that

$$\mathbb{E}_\theta V(X; \theta) = \begin{cases} \displaystyle\sum_{x \in \mathbb{S}} f(x; \theta) \frac{\partial \ln f(x; \theta)}{\partial \theta} \\ \displaystyle\int_{\mathbb{S}} f(x; \theta) \frac{\partial \ln f(x; \theta)}{\partial \theta} \, dx \end{cases} = 0.$$

(It suffices to write $\partial[\ln f(x;\theta)]/\partial\theta = \left[\partial f(x;\theta)/\partial\theta\right]/f(x;\theta)$ and carry the derivative $\partial/\partial\theta$ outside the sum/integral.) The *Fisher information* (contained in X under the distribution with the PMF/PDF $f(x;\theta)$ is the value $J(\theta)$ defined by

$$J(\theta) = \mathbb{E}_\theta\left(V(X;\theta)\right)^2 = \begin{cases} \sum\limits_{x\in\mathbb{S}} f(x;0)\left(\dfrac{\partial}{\partial\theta}f(x;\theta)\right)^2, \\ \int\limits_{\mathbb{S}} f(x;\theta)\left(\dfrac{\partial}{\partial\theta}f(x;\theta)\right)^2 \, dx. \end{cases} \tag{3.162}$$

In other words, $J(\theta) = \mathrm{Var}_\theta V(X;\theta)$.

A similar definition can be introduced in a more general case where $\theta\in\Theta\subseteq\mathbb{R}^d$ and x is replaced by a vector $\mathbf{x}\in\mathbb{R}^n$. (For example, a multivariate normal density with unknown mean and covariance matrix corresponds with $\mathbb{S} = \mathbb{R}^n$ and $\theta = (\mu,\Sigma)\in\mathbb{R}^{n+n(n+1)/2}$.) Here, instead of a scalar quantity, we speak of a *Fisher information matrix* $J(\theta) = (J_{ij}(\theta))$, where

$$\begin{aligned} J_{ij}(\theta) &= \mathbb{E}\left[V_i(\mathbf{X};\theta)V_j(\mathbf{X};\theta)\right] \\ &= \begin{cases} \sum\limits_{x\in\mathbb{S}} f(\mathbf{x};\theta)\left(\dfrac{\partial}{\partial\theta_i}f(\mathbf{x};\theta)\dfrac{\partial}{\partial\theta_j}f(\mathbf{x};\theta)\right), \\ \int\limits_{\mathbb{S}} f(\mathbf{x};\theta)\left(\dfrac{\partial}{\partial\theta_i}f(\mathbf{x};\theta)\dfrac{\partial}{\partial\theta_j}f(\mathbf{x};\theta)\right) \, dx, \end{cases} \end{aligned} \tag{3.163}$$

for $i,j = 1,\ldots,d$. Here $\mathbf{V}(\mathbf{X};\theta) = \begin{pmatrix} V_1(\mathbf{X};\theta) \\ \vdots \\ V_d(\mathbf{X};\theta) \end{pmatrix}$ is a vector score:

$$V_i(\mathbf{X};\theta) = \frac{\partial}{\partial\theta_i}\ln f(\mathbf{X};\theta), \quad i = 1,\ldots,d. \tag{3.164}$$

As before, under mild assumptions, the mean values $\mathbb{E}V_i(\mathbf{X};\theta) = 0$, and the entry $J_{ij}(\theta)$ is identified as the covariance of $V_i(\mathbf{X};\theta)$ and $V_j(\mathbf{X};\theta)$: $J_{ij}(\theta) = \mathrm{Cov}\left[V_i(\mathbf{X};\theta), V_j(\mathbf{X};\theta)\right]$.

We will refer below to Definitions (3.161)–(3.162) as a 'scalar case' and to (3.163)–(3.164) as a 'vector case'.

Definition 3.6.4 Let f_0 and f_1 be two PMFs/PDFs, on $\mathbb{R}$ or $\mathbb{R}^n$. Set:

$$D(f_1\|f_0) = \begin{cases} \sum \mathbf{1}(f_1(x) > 0)f_1(x)\ln\dfrac{f_1(x)}{f_0(x)}, \\ \int \mathbf{1}(f_1(x) > 0)f_1(x)\ln\dfrac{f_1(x)}{f_0(x)} \, dx. \end{cases} \tag{3.165}$$

The quantity $D(f_1 \| f_0)$ is called variously the *Kullback* (or *Kullback–Leibler*) *distance* from f_1 to f_0 or the Kullback *divergence* (or *information divergence*) between f_1 and f_0. Yet another popular term is the *relative entropy* of f_1 relative to f_0. We will often call it briefly the divergence.

In this definition we set $f_1(x)\ln\left[f_1(x)/f_0(x)\right] = +\infty$ if $f_0(x) = 0$ and $f_1(x) > 0$, so $D(f_1 \| f_0)$ may take value $+\infty$. If f_0 and f_1 have the same support set $\mathbb{S} \subseteq \mathbb{R}$ or $\mathbb{R}^d$ (so that $f_0(x) > 0$ if and only if $x \in \mathbb{S}$ and $f_1(x) > 0$ if and only if $x \in \mathbb{S}$) then the summation/integration in the RHS of (3.163) is carried out precisely on $\mathbb{S}$. (The nature of the support set $\mathbb{S}$ is not important: the definition works when f_0 and f_1 are PMFs/PDFs on any given set.) The indicator function $\mathbf{1}(f_1(x) > 0)$ can be omitted if we adopt the standard agreement that $0\ln 0 = 0$ (extension by continuity).

The term 'distance' is rather misleading here: the quantity $D(f_1 \| f_0)$ does not satisfy the symmetry property or triangle inequality. That is, there are examples where $D(f_1 \| f_0) \neq D(f_0 \| f_1)$ and $D(f_2 \| f_0) > D(f_2 \| f_1) + D(f_1 \| f_0)$: see below. However, this concept has a profound geometric meaning, and the term 'distance' is widely used.

> *Divergence of Character and the Extinction of Less-improved Forms.*
> C.R. Darwin (1809–1892), English naturalist

The connection between the Fisher information and the Kullback–Leibler distance is established in the following

Lemma 3.6.5 *Assume Definition 3.6.3. Then, in the scalar case, the following property holds true: the divergence between PMFs/PDFs $f(\,\cdot\,;\widetilde{\theta})$ and $f(\,\cdot\,;\theta)$, $\theta, \widetilde{\theta} \in \Theta$, satisfies*

$$\lim_{\widetilde{\theta}\to\theta} \frac{D\big(f(\,\cdot\,;\widetilde{\theta})\,\|\,f(\,\cdot\,;\theta)\big)}{(\widetilde{\theta}-\theta)^2} = \frac{1}{2}J(\theta), \tag{3.166}$$

or, equivalently, with $\delta \to 0$, for all $\theta \in \Theta$,

$$D\big(f(\,\cdot\,;\theta+\delta)\,\|\,f(\,\cdot\,;\theta)\big) = \frac{1}{2}J(\theta)\delta^2 + o(\delta^2). \tag{3.167}$$

Similarly, in the vector case, where $\delta \in \mathbb{R}^d$ has $\|\delta\| \to 0$, for all $\theta \in \Theta$, the divergence obeys

$$D\big(f(\,\cdot\,;\theta+\delta)\,\|\,f(\,\cdot\,;\theta)\big) = \frac{1}{2}\langle \delta, J(\theta)\delta \rangle + o(\|\delta\|^2). \tag{3.168}$$

Proof (For the scalar PMF case with finitely many outcomes only.) Assume that set $\mathbb{S}$ is finite. Perform the standard Taylor expansion, using $\ln(1+\varepsilon) = \varepsilon + o(\varepsilon)$:

$$D(f(\cdot;\theta+\delta) \| f(\cdot;\theta)) = \sum_{x \in \mathbb{S}} f(x;\theta+\delta) \ln \frac{f(x;\theta+\delta)}{f(x;\theta)}$$

$$= \sum_{x \in \mathbb{S}} \left[f(x;\theta) + \delta \frac{\partial}{\partial \theta} f(x;\theta) + o(\delta) \right] \ln \frac{f(x;\theta) + \delta \frac{\partial}{\partial \theta} f(x;\theta) + o(\delta)}{f(x;\theta)}$$

$$= \sum_{x \in \mathbb{S}} \left[f(x;\theta) + \delta \frac{\partial}{\partial \theta} f(x;\theta) + o(\delta) \right]$$

$$\times \left[\delta \frac{\partial f(x;\theta)/\partial \theta}{f(x;\theta)} + \delta^2 \frac{\partial^2 f(x;\theta)/\partial \theta^2}{2 f(x;\theta)} - \delta^2 \frac{(\partial f(x;\theta)/\partial \theta)^2}{2 f(x;\theta)^2} + o(\delta^2) \right]$$

$$= \sum_{x \in \mathbb{S}} \left[\delta \frac{\partial f(x;\theta)}{\partial \theta} + \frac{\delta^2}{2} \frac{\partial^2 f(x;\theta)}{\partial \theta^2} + \frac{(\partial f(x;\theta)/\partial \theta)^2}{f(x;\theta)} \left(-\frac{\delta^2}{2} + \delta^2 \right) + o(\delta^2) \right].$$

The sums of $\partial f(x;\theta)/\partial \theta$ and $\partial^2 f(x;\theta)/\partial \theta^2$ vanish. (As before, the derivatives $\partial/\partial \theta$ and $\partial^2/\partial \theta^2$ can be taken out of the sum.) The term $\dfrac{(\partial f(x;\theta)/\partial \theta)^2}{f(x;\theta)}$ yields the result. $\qquad \square$

Lemma 3.6.6 (*Gibbs' inequality*) *The Kullback–Leibler distance $D(f_1 \| f_0)$ defined in (3.165) is non-negative:*

$$D(f_1 \| f_0) \geq 0. \tag{3.169}$$

The equality holds if and only if the two PMFs/PDFs coincide.

Proof We use an elementary inequality $\ln y \leq y - 1$, $y > 0$, with equality if and only if $y = 1$. Substituting $f_0(x)/f_1(x)$ for y yields

$$-D(f_1 \| f_0) \ \leq \ \begin{cases} \sum \mathbf{1}(f_1(x) > 0) f_1(x) \left(\dfrac{f_0(x)}{f_1(x)} - 1 \right) \\ \displaystyle\int \mathbf{1}(f_1(x) > 0) f_1(x) \left(\dfrac{f_0(x)}{f_1(x)} - 1 \right) \, \mathrm{d}x \end{cases}$$

$$= \ \begin{cases} \sum \mathbf{1}(f_1(x) > 0) \, (f_0(x) - f_1(x)) \\ \displaystyle\int \mathbf{1}(f_1(x) > 0) \, (f_0(x) - f_1(x)) \, \mathrm{d}x \end{cases}$$

$$\leq \ 1 - 1 = 0.$$

The equality occurs if and only if the term $\mathbf{1}(f_1(x) > 0) \dfrac{f_0(x)}{f_1(x)} \equiv \mathbf{1}(f_1(x) > 0)$ which means precisely that the two PMFs/PDFs coincide. $\qquad \square$

The Kullback–Leibler distance emerges naturally in the context of hypothesis testing. Let $\mathbf{X} = \begin{pmatrix} X_1 \\ \vdots \\ X_n \end{pmatrix}$ be a random vector with IID entries taking values from a finite set $\mathbb{S}$. Suppose that we test the null hypothesis that $X_m \sim f_0$ versus the alternative that $X_m \sim f_1$ where f_0 and f_1 are two given PMFs on $\mathbb{R}$. Given a sample vector $\mathbf{x} = \begin{pmatrix} x_1 \\ \vdots \\ x_n \end{pmatrix}$, we count the *empirical distribution* formed by frequencies

$$\widehat{p}_{\mathbf{x}}(b) = \frac{1}{n}\sum_{i=1}^{n} \mathbf{1}(x_i = b), \quad b \in \mathbb{S}. \tag{3.170}$$

Then the log-likelihood ratio can be written as

$$
\begin{aligned}
\ln \frac{f_1(x_1)\cdots f_1(x_n)}{f_0(x_1)\cdots f_0(x_n)} &= \sum_{i=1}^{n} \ln \frac{f_1(x_i)}{f_0(x_i)} \\
&= \sum_{b \in \mathbb{S}} n\widehat{p}_{\mathbf{x}}(b) \ln \frac{f_1(b)\widehat{p}_{\mathbf{x}}(b)}{f_0(b)\widehat{p}_{\mathbf{x}}(b)} \\
&= n\Big[D(\widehat{p}_{\mathbf{x}} \| f_0) - D(\widehat{p}_{\mathbf{x}} \| f_1) \Big]. \tag{3.171}
\end{aligned}
$$

Let us calculate some of the most frequent examples.

Example 3.6.7 (a) Let f_0 and f_1 be two Poisson PMFs, on $\mathbb{Z}_+ = \{0,1,\ldots\}$:

$$f_0(n) = \frac{\lambda_0^n e^{-\lambda_0}}{n!}, \quad f_1(n) = \frac{\lambda_1^n e^{-\lambda_1}}{n!}, \quad n \in \mathbb{Z}_+.$$

Then

$$
\begin{aligned}
D(f_1 \| f_0) &= \sum_{n \geq 0} \frac{\lambda_1^n e^{-\lambda_1}}{n!} \left(n\ln \lambda_1 - \lambda_1 - n\ln \lambda_0 + \lambda_0 \right) \\
&= \lambda_1 \ln \frac{\lambda_1}{\lambda_0} + (\lambda_0 - \lambda_1) \\
&= \lambda_0 (r\ln r + 1 - r), \quad r = \frac{\lambda_1}{\lambda_0}. \tag{3.172}
\end{aligned}
$$

(b) If f_0 and f_1 are geometric PMFs on $\mathbb{Z}_+$, with

$$f_0(n) = p_0(1-p_0)^n, \quad f_1(n) = p_1(1-p_1)^n, \quad n \in \mathbb{Z}_+,$$

then

$$D(f_1 \| f_0) = \sum_{n \geq 0} p_1(1-p_1)^n \left[n\ln\frac{1-p_1}{1-p_0} + \ln\frac{p_1}{p_0} \right]$$

$$= \frac{1-p_1}{p_1}\ln\frac{1-p_1}{1-r_0} + \ln\frac{p_1}{r_0}$$

$$= \frac{1}{p_1}D(p_1, 1-p_1 \| p_0, 1-p_0). \tag{3.173}$$

(c) Assume that f_0 and f_1 are binomial PMFs, on $\{0, 1, \ldots, n\}$:

$$f_0(k) = \binom{n}{k} p_0^k(1-p_0)^{n-k}, \quad f_1(k) = \binom{n}{k} p_1^k(1-p_1)^{n-k}, \quad k = 0, 1, \ldots, n.$$

Then

$$D(f_1 \| f_0) = \sum_{k=0}^{n} \binom{n}{k} p_1^k(1-p_1)^{n-k} \left[k\ln\frac{p_1}{p_0} + (n-k)\ln\frac{1-p_1}{1-p_0} \right]$$

$$= n \left[p_1\ln\frac{p_1}{p_0} + (1-p_1)\ln\frac{1-p_1}{1-p_0} \right]$$

$$= nD(p_1, 1-p_1 \| p_0, 1-p_0). \tag{3.174}$$

(d) Let f_0 and f_1 be two negative binomial PMFs on $\mathbb{Z}_+$: $f_0 \sim \text{NegBin}(p_0, k)$ and $f_1 \sim \text{NegBin}(p_1, k)$:

$$f_0(n) = \binom{n+k-1}{n} p_i^k(1-p_i)^n, \quad n = 0, 1, \ldots, \quad i = 0, 1.$$

Then

$$D(f_1 \| f_0) = \sum_{n \geq 0} \binom{n+k-1}{k-1} p_1^k(1-p_1)^n \left[k\ln\frac{p_1}{p_0} + n\ln\frac{1-p_1}{1-p_0} \right]$$

$$= k\ln\frac{p_1}{p_0} + \frac{k(1-p_1)}{p_1}\ln\frac{1-p_1}{1-p_0}$$

$$= \frac{k}{p_1}D(p_1, 1-p_1 \| p_0, 1-p_0). \tag{3.175}$$

(e) Now suppose that f_0 and f_1 are two (discrete) uniform PMFs: $f_0 \sim \text{U}[1, n_0]$ and $f_1 \sim \text{U}[1, n_1]$:

$$f_0(k) = \frac{1}{n_0}, \quad k = 1, \ldots, n_0, \quad f_0(k) = \frac{1}{n_1}, \quad k = 1, \ldots, n_1.$$

Then, according to the definition, $D(f_1 \| f_0) = +\infty$ if $n_1 > n_0$. For $n_1 \leq n_0$,

$$D(f_1 \| f_0) = \sum_{k=0}^{n_1} \frac{1}{n_1}\ln\frac{n_0}{n_1} = \ln\frac{n_0}{n_1}. \tag{3.176}$$

We proceed with continuous random variables:

Example 3.6.8 (a) Let f_0 and f_1 be two exponential PDFs, on $\mathbb{R}_+ = (0, +\infty)$:

$$f_0 = \lambda_0 e^{-\lambda_0 x} \mathbf{1}(x > 0), \quad f_1 = \lambda_1 e^{-\lambda_1 x} \mathbf{1}(x > 0).$$

Then

$$
\begin{aligned}
D(f_1 \| f_0) &= \int_0^\infty \lambda_1 e^{-\lambda_1 x} \left[(\lambda_0 - \lambda_1) x + \ln \frac{\lambda_1}{\lambda_0} \right] dx \\
&= \frac{\lambda_0 - \lambda_1}{\lambda_1} + \ln \frac{\lambda_1}{\lambda_0} \\
&= r - 1 - \ln r, \quad \text{where } r = \frac{\lambda_0}{\lambda_1}.
\end{aligned}
\tag{3.177}
$$

Extending this calculation to the case where $f_0 \sim \text{Gam}(\alpha, \lambda_0)$ and $f_1 \sim \text{Gam}(\alpha, \lambda_1)$ yields

$$D(f_1 \| f_0) = \alpha \left(\ln \frac{\lambda_1}{\lambda_0} + \frac{\lambda_0 - \lambda_1}{\lambda_1} \right). \tag{3.178}$$

(b) Assume that f_0 and f_1 are two normal PDFs. First, consider the simple case where $f_0 \sim N(\mu_0, \sigma^2)$ and $f_1 \sim N(\mu_1, \sigma^2)$ (different means but the same variance), $\mu_0, \mu_1 \in \mathbb{R}$, $\sigma^2 > 0$. Here,

$$
\begin{aligned}
D(f_1 \| f_0) &= \frac{1}{\sqrt{2\pi}\sigma} \int e^{-(x-\mu_1)^2/(2\sigma^2)} \frac{\left[(x - \mu_0)^2 - (x - \mu_1)^2 \right]}{2\sigma^2} dx \\
&= \frac{1}{\sqrt{2\pi}\sigma} \int e^{-(x-\mu_1)^2/(2\sigma^2)} \frac{\left[x - \mu_1 + (\mu_1 - \mu_0) \right]^2}{2\sigma^2} dx - \frac{1}{2\sigma^2} \\
&= \frac{1}{2\sigma^2} + \frac{(\mu_1 - \mu_0)^2}{2\sigma^2} - \frac{1}{2\sigma^2} \\
&= \frac{(\mu_1 - \mu_0)^2}{2\sigma^2}.
\end{aligned}
\tag{3.179}
$$

Note that in this case $D(f_1 \| f_0) = D(f_0 \| f_1)$.

Now suppose that f_0 and f_1 are general normal multivariate PDFs: $f_0 \sim N(\mu_0, \Sigma_0)$ and $f_1 \sim N(\mu_1, \Sigma_1)$ where $\mu_0, \mu_1 \in \mathbb{R}^n$, and Σ_0, Σ_1 are two $n \times n$ real positive-definite invertible matrices. Recall the form of the multivariate normal PDF:

$$f_i(\mathbf{x}) = \frac{\exp \left[-\frac{1}{2} \langle \mathbf{x} - \mu_i, \Sigma_i^{-1}(\mathbf{x} - \mu_i) \rangle \right]}{(2\pi)^{n/2} (\det \Sigma_i)^{1/2}}, \quad \mathbf{x} \in \mathbb{R}^n, \, i = 0, 1.$$

Then, following the same lines as before, after some calculations, one obtains

$$D(f_1 \| f_0) = \frac{1}{2} \left[\ln \frac{\det \Sigma_0}{\det \Sigma_1} + \operatorname{tr} \left(\Sigma_1 \Sigma_0^{-1} - \mathbf{I} \right) + \langle \mu_1 - \mu_0, \Sigma_0^{-1} (\mu_1 - \mu_0) \rangle \right], \quad (3.180)$$

where, as before, $\mathbf{I}$ is the $n \times n$ unit matrix. So, in the case where $\Sigma_0 = \Sigma_1 = \Sigma$, we have

$$D(f_1 \| f_0) = \frac{1}{2} \langle \mu_1 - \mu_0, \Sigma^{-1} (\mu_1 - \mu_0) \rangle, \qquad (3.181)$$

generalising (3.179). On the other hand, for $\mu_0 = \mu_1$ and general Σ_0, Σ_1,

$$D(f_1 \| f_0) = \frac{1}{2} \left[\operatorname{tr} \left(\Sigma_1 \Sigma_0^{-1} \right) - \ln \left(\det \left(\Sigma_1 \Sigma_0^{-1} \right) \right) - n \right]. \qquad (3.182)$$

(c) A more challenging example is of two Cauchy distributions: $f_0 \sim \mathrm{Ca}(\alpha_0, \tau)$ and $f_1 \sim \mathrm{Ca}(\alpha_1, \tau)$. Here

$$f_0(x) = \frac{\tau}{\pi \left[(x - \alpha_0)^2 + \tau^2 \right]}, \quad f_1(x) = \frac{\tau}{\pi \left[(x - \alpha_1)^2 + \tau^2 \right]}, \quad x \in \mathbb{R},$$

$$D(f_1 \| f_0) = \ln \left(1 + \frac{(\alpha_1 - \alpha_0)^2}{4 \tau^2} \right). \qquad (3.183)$$

In fact, the change of variables $x \mapsto x - \alpha_1$ leads to the representation

$$D(f_1 \| f_0) = \frac{\tau}{\pi} \int \frac{1}{x^2 + \tau^2} \ln \frac{x^2 + \tau^2}{(x - \alpha)^2 + \tau^2} \, dx := g(\alpha),$$

where $\alpha = \alpha_1 - \alpha_0$. Differentiating this integral in α yields

$$g'(\alpha) = -\frac{2\tau}{\pi} \int \frac{x - \alpha}{(x^2 + \tau^2) \left[(x - \alpha)^2 + \tau^2 \right]} \, dx.$$

The integrand in the RHS is a rational function, with two poles (zeroes of the denominator) in the upper complex half-plane, at $x = i\tau$ and $x = \alpha + i\tau$. A standard integration procedure then yields

$$g'(\alpha) = 4 i \tau \left[\frac{i\tau - \alpha}{2 i \tau (\alpha^2 - 2 i \alpha \tau)} + \frac{i\tau}{2 i \tau (\alpha^2 + 2 i \alpha \tau)} \right] = \frac{2\alpha}{\alpha^2 + 4 \tau^2}.$$

Integrating the last expression in α, with $g(0) = 0$, we obtain (3.183).

Worked Example 3.6.9 (The log-sum inequality) Let $a_1, a_2, \ldots$ and $b_1, b_2, \ldots$ be non-negative numbers, with $\sum_i b_i < \infty$. Prove that

$$\sum_i \mathbf{1}(a_i > 0) a_i \ln \frac{a_i}{b_i} \geq \left(\sum_i a_i \right) \ln \frac{\sum_i a_i}{\sum_i b_i}, \tag{3.184}$$

with equality if and only if $a_i \equiv b_i$.

Solution   Without loss of generality, we assume that all numbers are strictly positive. Use Jensen's inequality for a strictly convex function $\phi(t) = t \ln t$, $t > 0$:

$$\sum_i \lambda_i \phi(t_i) \geq \phi \left(\sum_i \lambda_i t_i \right)$$

with $\lambda_i = b_i / \left(\sum_j b_j \right)$ and $t_i = a_i / b_i$, to obtain

$$\sum_i \mathbf{1}(a_i > 0) \frac{a_i}{\sum_j b_j} \ln \frac{a_i}{b_i} \geq \sum_i \frac{a_i}{\sum_j b_j} \ln \sum_i \frac{a_i}{\sum_j b_j}.$$

Because of the strict convexity, equality holds iff $a_i \equiv b_i$.   □

The Gibbs inequality (Lemma 3.6.6) states that $D(f_1 \| f_0)$ is non-negative (see (3.169)). Lemma 3.6.10 gives a more precise bound. Define

$$\|f_1 - f_0\|_1 = \begin{cases} \sum |f_1(x) - f_0(x)|, \\ \int |f_1(x) - f_0(x)| \mathrm{d}x. \end{cases} \tag{3.185}$$

Lemma 3.6.10   *The Kullback–Leibler distance satisfies*

$$D(f_1 \| f_0) \geq \frac{1}{4} \|f_1 - f_0\|_1^2. \tag{3.186}$$

Proof   (For the discrete case only; the proof for the continuous case is a mere repetition.) The first step is to show that

$$D(f_1 \| f_0) \geq -2 \ln \left[\sum f_1(x)^{1/2} f_0(x)^{1/2} \right]. \tag{3.187}$$

(Here the summation is restricted to points $x = x_i$ where $f_1(x) > 0$, and the indicator $\mathbf{1}(f_1(x) > 0)$ is omitted. The same agreement is used in various summations below.) To this end, we write

$$D(f_1 \| f_0) = 2 \sum \left[f_1(x)^{1/2} f_1(x)^{1/2} \right] \ln \left[\frac{f_1(x)}{f_0(x)} \right]^{1/2}$$

$$= 2 \left[\sum f_1(x)^{1/2} \right] \sum f_1(x)^{1/2} f_1(x)^{1/2} \ln \left[(f_1(x)/f_0(x))^{1/2} \right] \frac{1}{\sum f_1(x)^{1/2}}.$$

Next, we use the log-sum inequality (3.184) with

$$a_i = \frac{f_1(x_i)^{1/2} f_1(x_i)^{1/2}}{\sum_j f_1(x_j)^{1/2}}$$

and

$$b_i = \frac{f_1(x_i)^{1/2} f_0(x_i)^{1/2}}{\sum_j f_1(x_j)^{1/2}}.$$

This yields (3.187).

Next, we use, as in the proof of Lemma 3.6.6, the inequality $\ln y \le y - 1$, $y > 0$, to show that

$$-2\ln\left[\sum f_1(x)^{1/2} f_0(x)^{1/2}\right] \ge \sum \left[f_1(x)^{1/2} - f_0(x)^{1/2}\right]^2.$$

Finally, we check that

$$\sum \left[f_1(x)^{1/2} - f_0(x)^{1/2}\right]^2 \ge \frac{1}{4}\left[\sum |f_1(x) - f_0(x)|\right]^2.$$

In fact, by the Cauchy–Schwarz inequality

$$\left[\sum |f_1(x) - f_0(x)|\right]^2$$
$$= \left(\sum \left|f_1(x)^{1/2} - f_0(x)^{1/2}\right| \left[f_1(x)^{1/2} + f_0(x)^{1/2}\right]\right)^2$$
$$\le \sum \left|f_1(x)^{1/2} - f_0(x)^{1/2}\right|^2 \sum \left[f_1(x)^{1/2} + f_0(x)^{1/2}\right]^2.$$

Then, by expanding the square, one checks that the second sum ≤ 4. This yields (3.186). $\qquad\square$

In fact, a more elaborate argument shows that the constant $1/4$ in (3.186) can be replaced by $1/(2\ln 2)$.

Lemma 3.6.11 (Additivity property of the Kullback–Leibler distance) (a) *Let*

$$\mathbf{X} = \begin{pmatrix} X_1 \\ \vdots \\ X_n \end{pmatrix} \text{ and } \mathbf{Y} = \begin{pmatrix} Y_1 \\ \vdots \\ Y_n \end{pmatrix} \text{ be two random vectors, each with independent}$$

entries, where $X_i \sim f_0^{(i)}$ *and* $Y_i \sim f_1^{(i)}$. *Then*

$$D\left(f_{\mathbf{Y}} \| f_{\mathbf{X}}\right) = \sum_{i=1}^{n} D\left(f_1^{(i)} \| f_0^{(i)}\right). \tag{3.188}$$

(b) *Let* (X_m) *and* (Y_m) *be two Markov chains on the same (finite) state space* I, *with transition matrices* $P^{(0)} = (p_{ij}^{(0)})$ *and* $P^{(1)} = (p_{ij}^{(1)})$, *respectively. Assume that* the chain (X_m) has initial probabilities $\lambda_i = \mathbb{P}(X_1 = i)$ whereas (Y_i) is in equilibrium, with $\mathbb{P}(Y_m = i) = \pi_i$ and $\pi_j = \sum_{i \in I} \pi_i p_{ij}^{(1)}$, $i, j \in I$. As above, let $f_{\mathbf{X}}$ and $f_{\mathbf{Y}}$ stand for the PMFs of the sample vectors $\mathbf{X}$ and $\mathbf{Y}$. Then

$$D\left(f_{\mathbf{Y}} \,\|\, f_{\mathbf{X}}\right) = D(\pi \,\|\, \lambda) + (n-1)E_\pi\left(P^{(1)} \,\|\, P^{(0)}\right), \qquad (3.189)$$

where $\pi = (\pi_i)$, $\lambda = (\lambda_i)$ and

$$E_\pi\left(P^{(1)} \,\|\, P^{(0)}\right) = \sum_{i,j \in I} \pi_i p_{ij}^{(1)} \ln \frac{p_{ij}^{(1)}}{p_{ij}^{(0)}}. \qquad (3.190)$$

Proof (a) Straightforward, by expanding the logarithm.

(b) Similarly, with $\mathbf{x} = \begin{pmatrix} x_1 \\ \vdots \\ x_n \end{pmatrix} \in I^n$:

$$
\begin{aligned}
D\left(f_{\mathbf{Y}} \,\|\, f_{\mathbf{X}}\right) &= \sum_{\mathbf{x}} \mathbb{P}(\mathbf{Y} = \mathbf{x}) \ln \frac{\mathbb{P}(\mathbf{Y} = \mathbf{x})}{\mathbb{P}(\mathbf{X} = \mathbf{x})} \\
&= \sum_{\mathbf{x}} \left(\pi_{x_1} \prod_{l=1}^{n-1} p_{x_l x_{l+1}}^{(1)} \right) \ln \frac{\pi_{x_1} \prod_{l=1}^{n-1} p_{x_l x_{l+1}}^{(1)}}{\lambda_{x_1} \prod_{l=1}^{n-1} p_{x_l x_{l+1}}^{(0)}} \\
&= \sum_{\mathbf{x}} \pi_{x_1} \prod_{l=1}^{n-1} p_{x_l x_{l+1}}^{(1)} \left[\ln \frac{\pi_{x_1}}{\lambda_{x_1}} + \sum_{l=1}^{n-1} \ln \frac{p_{x_l x_{l+1}}^{(1)}}{p_{x_l x_{l+1}}^{(0)}} \right] \\
&= \sum_{i \in I} \pi_i \ln \frac{\pi_i}{\lambda_i} + (n-1) \sum_{i,j \in I} \pi_i p_{ij}^{(1)} \ln \frac{p_{ij}^{(1)}}{p_{ij}^{(0)}},
\end{aligned}
$$

which yields (3.189). $\qquad \square$

Remark 3.6.12 The quantity $E_\pi\left(P^{(1)} \,\|\, P^{(0)}\right)$ in (3.190) can be written as the expected value:

$$
\begin{aligned}
E_\pi\left(P^{(1)} \,\|\, P^{(0)}\right) &= \sum_{i,j \in I} \mathbb{P}(Y_m = i, Y_{m+1} = j) \ln \frac{p_{ij}^{(1)}}{p_{ij}^{(0)}} \\
&= \mathbb{E}_{Y_m, Y_{m+1}} \ln \frac{p_{Y_m Y_{m+1}}^{(1)}}{p_{Y_m Y_{m+1}}^{(0)}}; \qquad (3.191)
\end{aligned}
$$

it does not depend on m as the chain (Y_m) is in equilibrium. Equivalently, let $\mathbf{p}_i^{(0)}$ and $\mathbf{p}_i^{(1)}$ denote the probability distributions on I represented by the ith row of matrices $P^{(0)}$ and $P^{(1)}$, respectively. Then the Kullback divergence $D(\mathbf{p}_i^{(1)}||\mathbf{p}_i^{(0)})$, considered as a function on I, is defined via

$$K : i \in I \mapsto D(\mathbf{p}_i^{(1)}||\mathbf{p}_i^{(0)}).$$

Then $E_\pi(P^{(1)}||P^{(0)})$ represents the expectation of K considered as a random variable with the probability distribution $\pi = (\pi_i)$:

$$E_\pi\left(P^{(1)}||P^{(0)}\right) = \sum_{i \in I} \pi_i D(\mathbf{p}_i^{(1)}||\mathbf{p}_i^{(0)}) = \mathbb{E}_\pi K, \tag{3.192}$$

which is simply another form of (3.191).

A useful fact is the chain rule: let p_{X_1,X_2} stand for the joint distribution of X_1 and X_2 and p_{Y_1,Y_2} for that of Y_1 and Y_2; in the notation of Lemma 3.6.11(b),

$$\begin{cases} p_{X_1,X_2}(i,j) &= \mathbb{P}(X_1 = i, X_2 = j) = \lambda_i p_{ij}^{(0)}, \\ p_{Y_1,Y_2}(i,j) &= \mathbb{P}(Y_1 = i, Y_2 = j) = \pi_i p_{ij}^{(1)}, \end{cases} \quad i, j \in I.$$

Then

$$\begin{aligned} D\left(p_{Y_1,Y_2}||p_{X_1,X_2}\right) &= \sum_{i,j \in I} p_{Y_1,Y_2}(i,j) \ln \frac{p_{Y_1,Y_2}(i,j)}{p_{X_1,X_2}(i,j)} \\ &= \sum_{i,j \in I} \pi_i p_{ij}^{(1)} \ln \frac{\pi_i p_{ij}^{(1)}}{\lambda_i p_{ij}^{(0)}} \\ &= \sum_{i,j \in I} \pi_i p_{ij}^{(1)} \left[\ln \frac{\pi_i}{\lambda_i} + \ln \frac{p_{ij}^{(1)}}{p_{ij}^{(0)}} \right] \\ &= D(\pi||\lambda) + E_\pi\left(P^{(1)}||P^{(0)}\right). \tag{3.193} \end{aligned}$$

We can write this in a general form:

Lemma 3.6.13 (The chain rule for the Kullback–Leibler distance) *Let X_1, X_2 and Y_1, Y_2 be two pairs of random variables, where X_1, Y_1 take values in a set $\mathbb{S}_1$ and X_2, Y_2 in a set $\mathbb{S}_2$. Let f_{X_1,X_2} and f_{Y_1,Y_2} stand for the joint PMFs/PDFs of X_1 and X_2 and of Y_1 and Y_2, and f_{X_1} and f_{Y_1} for the marginal PMFs/PDFs of X_1 and Y_1, respectively. Furthermore, let $f_{X_2|X_1}$ and $f_{Y_2|Y_1}$ denote the conditional PMFs/PDFs of X_2 given X_1 and of Y_2 given Y_1, respectively. Then*

$$D\left(f_{Y_1,Y_2}||f_{X_1,X_2}\right) = D\left(f_{X_1}||f_{Y_1}\right) + D_{f_{Y_1}}\left(f_{Y_2|Y_1}||f_{X_2|X_1}\right), \tag{3.194}$$

where

$$D_{f_{Y_1}}\left(f_{Y_2|Y_1}\|f_{X_2|X_1}\right)$$

$$=\begin{cases}\displaystyle\sum_{y_1\in\mathbb{S}_1}f_{Y_1}(y_1)\sum_{y_2\in\mathbb{S}_2}f_{Y_2|Y_1}(y_2|y_1)\ln\frac{f_{Y_2|Y_1}(y_2|y_1)}{f_{X_2|X_1}(y_2|y_1)}\\[2ex]\displaystyle\int_{\mathbb{S}_1}f_{Y_1}(y_1)\int_{\mathbb{S}_2}f_{Y_2|Y_1}(y_2|y_1)\ln\frac{f_{Y_2|Y_1}(y_2|y_1)}{f_{X_2|X_1}(y_2|y_1)}\mathrm{d}y_2\mathrm{d}y_1\end{cases}\geq 0,\qquad (3.195)$$

with equality if and only if $f_{Y_2|Y_1}=f_{X_2|X_1}$.

This brings us to a generalisation of Definition 3.6.4:

Definition 3.6.14 The quantity $D_{f_{Y_1}}\left(f_{Y_2|Y_1}\|f_{X_2|X_1}\right)$ in (3.195) is called the *conditional* Kullback divergence.

We can now extend (3.188) the case of general random vectors **X** and **Y**:

$$D\left(f_{\mathbf{Y}}\|f_{\mathbf{X}}\right)=D\left(f_{Y_1}\|f_{X_1}\right)+D_{f_{Y_1}}\left(f_{Y_2|Y_1}\|f_{X_2|X_1}\right)$$

$$+\cdots+D_{f_{Y_1,\dots,Y_{n-1}}}\left(f_{Y_n|Y_1,\dots,Y_{n-1}}\|f_{X_n|X_1,\dots,X_{n-1}}\right).\qquad (3.196)$$

Suppose that PMFs f_0 and f_1 are written as convex linear combinations:

$$f_0(x)=\lambda g_0(x)+(1-\lambda)h_0(x),\text{ and }f_1(x)=\lambda g_1(x)+(1-\lambda)h_1(x),\quad (3.197)$$

where $0<\lambda<1$ and g_i and h_i, $i=0,1$, are PMFs/PDFs, on the same set.

Lemma 3.6.15 (Joint convexity of the Kullback–Leibler distance) *The following inequality holds true:*

$$D\left(\lambda g_1+(1-\lambda)h_1\|\lambda g_0+(1-\lambda)h_0\right)\leq\lambda D(g_1\|g_0)+(1-\lambda)D(h_1\|h_0).$$

$$(3.198)$$

Proof Using the log-sum inequality we get

$$\left[\lambda g_1(x)+(1-\lambda)h_1(x)\right]\ln\frac{\lambda g_1(x)+(1-\lambda)h_1(x)}{\lambda g_0(x)+(1-\lambda)h_0(x)}d$$

$$\leq\lambda g_1(x)\ln\frac{g_1(x)}{g_0(x)}+(1-\lambda)h_1(x)\ln\frac{h_1(x)}{h_0(x)}.$$

Summing up/integrating yields (3.198). $\qquad\square$

Remark 3.6.16 The convex linear combinations

$$f_0(x)=\lambda g_0(x)+(1-\lambda)h_0(x)\text{ and }f_1(x)=\lambda g_1(x)+(1-\lambda)h_1(x)$$

have a transparent probabilistic meaning: consider a random variable U with two values, say 1 and 2, taken with probabilities λ and $1 - \lambda$. Then consider U jointly with a random variable X such that the PMF/PDF of X conditional on $U = 1$ is g_0 and conditional on $U = 2$ is h_0. The unconditional PMF/PDF of X will coincide with f_0. A similar coupling can be performed when we use g_1 instead of g_0 and h_1 instead of h_0; the emerging random variable Y will have the PMF/PDF f_1. Then (3.198) takes the form

$$D(f_Y \| f_X) \leq D_{f_U}(f_{Y|U} \| f_{X|U}) \tag{3.199}$$

and can be extended to the case of a general random variable U.

The next property of the Kullback–Leibler distance is called the data processing inequality. Suppose that random variables X and Y, with values in set $\mathbb{S}$ are transformed by a transition function with values $p(x, y)$; in the case of PMFs we talk about a transition matrix (p_{xy}). That is, we assume that

$$\sum_y p_{xy} = 1 \quad \text{and} \quad \int p(x, y) \, dy = 1$$

and pass from X and Y to random variables X' and Y', where the PMFs/PDFs $f_{X'}$ and $f_{Y'}$ are related to f_X and f_Y by

$$f_{X'}(y) = \begin{cases} \sum_{x \in \mathbb{S}} f_X(x) p_{xy}, \\ \int_{\mathbb{S}} f_X(x) p(x, y) \, dx, \end{cases} \quad f_{Y'}(y) = \begin{cases} \sum_{x \in \mathbb{S}} f_Y(x) p_{xy}, \\ \int_{\mathbb{S}} f_Y(x) p(x, y) \, dx. \end{cases} \tag{3.200}$$

This operation is termed 'processing' and includes 'merging' several values x_1, ..., x_l together (when, for a given y, $p_{xy} = 1$ for $x = x_1, \ldots, x_l$) and other types of 'massaging' data represented by X and Y. Lemma 3.6.17 below shows that any such operation cannot result in increasing the divergence.

Call Back and Libel'er

(From the series *'Movies that never made it to the Big Screen'*.)

Lemma 3.6.17 (Data processing inequality for the Kullback–Leibler distance) *Under the above transformation, the Kullback divergence decreases:*

$$D(f_{Y'} \| f_{X'}) \leq D(f_Y \| f_X) \tag{3.201}$$

Proof We use the chain rule (3.194):

$$\begin{aligned} D(f_{Y,Y'} \| f_{X,X'}) &= D(f_Y \| f_X) + D_{f_Y}(f_{Y'|Y} \| f_{X'|X}) \\ &= D(f_{Y'} \| f_{X'}) + D_{f_{Y'}}(f_{Y|Y'} \| f_{X|X'}) \end{aligned}$$

But $f_{X'|X}$ and $f_{Y'|Y}$ coincide, by construction:

$$f_{X'|X}(y|x) = f_{Y'|Y}(y|x) = \begin{cases} p_{xy}, \\ p(x,y). \end{cases}$$

So, the conditional divergence $D_{f_Y}(f_{Y'|Y} \| f_{X'|X})$ vanishes:

$$D_{f_Y}(f_{Y'|Y} \| f_{X'|X}) = 0.$$

At the same time, $D_{f_{Y'}}(f_{Y|Y'} \| f_{X|X'}) \geq 0$. This yields (3.201). $\qquad\square$

We also see when the equality in (3.201) is attained: this happens iff $D_{f_{Y'}}(f_{Y|Y'} \| f_{X|X'}) = 0$; that is,

$$f_{Y|Y'} = f_{X|X'}. \tag{3.202}$$

In words, data processing does not change the Kullback–Leibler distance if and only if the conditional PMF/PDF of Y, given that $Y' = y$, and that of X, given that $X' = y$, coincide (for almost all y under the PMF/PDF $f_{Y'}$). This can be expressed as a sufficiency property of the processing transformation, relative to the pair of variables X and Y, which is a generalisation of the concept of a sufficient statistic.

Worked Example 3.6.18 Let (X_m) be a DTMC with an initial distribution λ and a transition matrix P. Prove that $D(f_{X_m} \| \pi)$ decreases with m where π is an equilibrium distribution for P.

Solution More generally, let (X_m) and (Y_m) be a two DTMCs with the same transition matrix P. Then the distance between PMFs f_{Y_m} and f_{X_m} decreases with m:

$$D(f_{Y_{m+1}} \| f_{X_{m+1}}) \leq D(f_{Y_m} \| f_{X_m}).$$

This follows immediately from Lemma 3.6.17. $\qquad\square$

We conclude our discussion of properties of the Kullback–Leibler distance with monotonicity in the case of parametric families with a monotone likelihood ratio. A family of PMFs/PDFs $f(\,\cdot\,;\theta)$, $\theta \in \Theta$, is said to have a monotone likelihood ratio (MLR) if there exists an order $\prec$ on set Θ such that for $\theta_1 \prec \theta_2$ the ratio

$$\Lambda_{\theta_1,\theta_2} = \frac{f(x;\theta_1)}{f(x;\theta_2)} = g_{\theta_1,\theta_2}(T(x)). \tag{3.203}$$

where T is a real-valued statistic and $g_{\theta_1,\theta_2}(y)$ is a monotone non-decreasing function (of a real variable y). See Volume 1, page 249.

Lemma 3.6.19 *Suppose that PMFs/PDFs* $f(\,\cdot\,;\theta)$, $\theta \in \Theta$, *form a family with an MLR. Then, for all* $\theta_1, \theta_2, \theta_3 \in \Theta$ *with* $\theta_1 \prec \theta_2 \prec \theta_3$,

$$D(f(\,\cdot\,;\theta_3)\,\|\,f(\,\cdot\,;\theta_2)) \le D(f(\,\cdot\,\theta_3)\,\|\,f(\,\cdot\,;\theta_1)). \qquad (3.204)$$

The proof is based on the concept of *convex order* between random variables (or their distributions). This topic (important in a number of applications) will be discussed in a later volume.

Solomon Kullback (1903–1994) began his career as a high school teacher in his native New York but soon moved to the US Army's Signal Intelligence Service (SIS). He went on to a long and distinguished career at the SIS and its eventual successor, the National Security Agency (NSA). In the late 1950s Kullback became the Chief Scientist at the NSA until his retirement in 1962. After that he took a position at the Georgetown University. In 1942, Major Kullback was sent to Britain to learn how at Bletchley Park the British were producing intelligence by exploiting the German Enigma machine. He contributed to the Bletchley Park team efforts and after his return to the States was made the head of the Japanese section at the NSA. He was very much liked by colleagues both in academia and special services for being "totally guileless, you always knew where you stood with him."

Richard Leibler (1914–2003) was an American mathematician and cryptographer. He took part in World War II, in the Iwo Jima and Okinawa invasions. His most distinguished periods were with the Institute for Defense Analysis at Princeton and the NSA. He is credited with the programme that enabled the NSA team to solve previously undecipherable Soviet espionage messages in the project codenamed VENONA.

The paper, S. Kullback, R.A. Leibler, "On information and sufficiency", *Annals of Mathematical Statistics*, **22** (1951), 79–86, where the concept of information divergence was formulated, was perhaps the most famous of the authors' academic achievements. The paper appeared at the height of the Cold War and was immediately noticed by the Soviets who had their own powerful cryptography division associated with special services. It has to be said that the control of publications in the Soviet system was (apparently) much tighter, and a paper written by authors of status similar to that of Kullback and Leibler had a little chance of being published in an open academic source. However, the Soviets had a developed network of 'secret' journals and periodicals accessible only to members of certain departments (carefully vetted at, and through, the time of their employment). It was even possible to obtain PhD or DSci degrees or to get elected into the membership of the USSR Academy of Sciences with few or no publications accessible to the general public. (Such members were often called 'closed' or 'secret' Academicians; Sakharov was the most famous example of them.)

3.7 Hidden Markov models, 1. State estimation for Markov chains

We now begin discussing the topic of *hidden Markov models* (HMMs). Consider the following situation. There is a discrete-time Markov chain, (X_m), on a state space I, say $I = \{1, \ldots, s\}$, with a (fully or partially) unknown initial distribution $\lambda = (\lambda_i)$ and a (fully or partially) unknown transition matrix $P = (p_{ij})$, $i, j = 1, \ldots, s$. In addition, the chain is not fully observable. For example, one may observe values X_{n_k} only at some selected times $n_1, n_2, \ldots$, or one can only record values $b(X_1), b(X_2), \ldots$, where $b : I \to \mathcal{K}$ is an unknown function of a state, possibly random, with values in a new 'alphabet' $\mathcal{K} = \{1, \ldots, \kappa\}$ (an unknown we know we don't know). In applications, typically, $\kappa < s$, and the function b is a many-to-one. In an 'unrestricted' problem, the pair (λ, P) runs over the full set $\mathcal{R}$ (see (3.5)) or over its subset $\mathcal{R}^{int}$ (see (3.6)). If we discard the initial distribution λ (viz., consider a stationary DTMC, with an equilibrium disribution π) then it will be convenient to assume that $P \in \mathcal{P}_{IA}$. However, we may have a priori information about (λ, P), for instance, that the matrix P is off-diagonal (i.e. $P \in \mathcal{P}_{\text{off-diag}}$; cf. (3.11)) or P is Hermitian (i.e. $P \in \mathcal{P}_{\text{symm}}$; cf. (3.12)). The function b may also be specified to a certain degree, which extracts a (known) class of functions (e.g. for $s = \kappa$, b may be a permutation). In this case we will have a restricted problem.

Another example is where the chain is observed accurately, but not all the time: we only see its states at (integer) times $t_0, \ldots, t_m$ where $0 \le t_0 < \cdots < t_m$ and $t_m > m$. There may occur a situation where we have to combine the two problems, but for simplicity we will treat them separately.

The task is to estimate λ and P from a recorded string of observed values $\sigma_0 = b(X_0), \ldots, \sigma_n = b(X_n)$ or from a given sequence of states $x_{n_1}, x_{n_2}, \ldots, x_{n_m}$.

Example 3.7.1 You observe a string $\sigma = \begin{pmatrix} \sigma_0 \\ \vdots \\ \sigma_n \end{pmatrix}$ of 0s and 1s. You suspect that it gives a record of (a function of) a Markov chain (X_m) with three states, say A, B and C: $\sigma_m = b(x_m)$, with $\mathbf{X} = \begin{pmatrix} X_0 \\ \vdots \\ X_n \end{pmatrix}$, $\mathbf{x} = \begin{pmatrix} x_0 \\ \vdots \\ x_n \end{pmatrix}$. You think that the chain is symmetric, i.e. its transition matrix $P = (p_{ij})$, $i, j = A, B, C$, is

$$P = \begin{pmatrix} 1 - 2p & p & p \\ p & 1 - 2p & p \\ p & p & 1 - 2p \end{pmatrix}, \text{ with } 0 \le p \le \frac{1}{2}.$$

There are several possibilities you can think of as to what function b may be: (a) on a pair of states b equals 0, say, $b(A) = b(B) = 0$, and on the remaining state it equals 1: $b(C) = 1$, or vice versa;, (b) on a pair of states b is equal to 0 with probability q and 1 with probability $1 - q$, whereas on the remaining state b is 1 with probability 1 (or, alternatively, 0 with probability 1); (c) each of $b(A)$, $b(B)$ and $b(C)$ takes value 0 with probabilities q_A, q_B and q_C, or 1 with probability $1 - q_A$, $1 - q_B$ and $1 - q_C$, independently. Altogether you have 2 possibilities for a non-random model (option (a)), 4 possibilities for a semi-random model (option (b)), and (essentially) 1 possibility for a fully random model (option (c)). You also have a reason to believe that the chain is stationary, i.e. λ is the equilibrium distribution $\pi = (1/3, 1/3, 1/3)$.

In sum, the transition matrix P is specified by a parameter p running over the interval $[0, 1/2]$, and for q we have the above possibilities (a), (b) and (c). A triple (π, P, b) in the context of this example is considered as a 'model'; in the case of a random function b it may be convenient to speak of a triple $(\pi, P, \mathbf{Q})$ where $\mathbf{Q}$ represents the collection of probabilities for random variables $b(X_0), \ldots, b(X_n)$.

For example, the states A, B and C may correspond to some consonants in an (idealised) problem of automated speech recognition. Some of these consonants may be clearly recognised from their spectrograms while others are more difficult to separate from each other.

Suppose you want to compare two particular families of models:

(i) the family of models with a deterministic (i.e. non-random) function b, denoted by $Z_{\det} = (\pi, P, b_{\det})$, where

$$b_{\det}(A) = b_{\det}(B) = 1, \quad \text{and} \quad b_{\det}(C) = 0;$$

and

(ii) a fully random model, denoted by $Z_{\mathrm{ran}} = (\pi, P, \mathbf{Q})$, where

$$b_{\mathrm{ran}}(\star) = \begin{cases} 1, & \text{with probability } q_\star, \\ 0, & \text{with probability } 1 - q_\star, \end{cases}, \star = A, B, C, \text{ independently,}$$

and $q_A = q_B = q_1$, $q_C = q_2$.

So, we compute the *aggregated* likelihoods:

$$\mathbf{L}^{\det}(\sigma; Z_{\det})$$
$$= \sum_{\mathbf{x}} \prod_{i=1}^{n} p_{x_{i-1} x_i} \left[\mathbf{1}(\sigma_i = 1, x_i = A \text{ or } B) + \mathbf{1}(\sigma_i = 0, x_i = C) \right] \qquad (3.205)$$

and

$$\mathbf{L}^{\mathrm{ran}}(\sigma; Z_{\mathrm{ran}}) = \sum_{\mathbf{x}} \prod_{i=1}^{n} p_{x_{i-1} x_i} \Big(\mathbf{1}(\sigma_i = 0) \big[(1 - q_1) \mathbf{1}(x_i = A \text{ or } B)$$
$$+ (1 - q_2) \mathbf{1}(x_i = C) \big] + \mathbf{1}(\sigma_i = 1) \big[q_1 \mathbf{1}(x_i = A \text{ or } B) + q_2 \mathbf{1}(x_i = C) \big] \Big). \qquad (3.206)$$

(We dropped the factor π_{x_0} in the RHS of (3.205) and (3.206) as it equals $1/3$ and plays no role in the analysis.) For a given σ, the function $\mathbf{L}^{\text{det}}$ is a polynomial, of degree n, of the variable $p \in [0, 1/2]$, while $\mathbf{L}^{\text{ran}}$ is a polynomial in the variables $p \in [0, 1/2]$ and $q_1, q_2 \in [0, 1]$. Of course, if there is no additional restriction on q_1 and q_2, then the polynomial $\mathbf{L}^{\text{ran}}$ is a continuation of $\mathbf{L}^{\text{det}}$ (or, if you like, $\mathbf{L}^{\text{det}}$ is a restriction of $\mathbf{L}^{\text{ran}}$ at $q_1 = 1$, $q_2 = 0$). However, if there is an additional condition, for instance, $q_1, q_2 \in (q^-, q^+) \subset [0, 1]$, then comparing two polynomials becomes non-trivial.

So, we maximise both polynomials, obtaining optimal models

$$Z_{\text{det}}^*(\sigma) = \operatorname*{argmax}_{p} \mathbf{L}^{\text{det}}(\sigma; Z_{\text{det}}) \text{ and } Z_{\text{ran}}^*(\sigma) = \operatorname*{argmax}_{p, q_1, q_2} \mathbf{L}^{\text{ran}}(\sigma; Z_{\text{ran}}). \quad (3.207)$$

Then we compare optimal values

$$\mathbf{L}^{\text{det}}(\sigma; Z_{\text{det}}) \text{ and } \mathbf{L}^{\text{ran}}(\sigma; Z_{\text{ran}});$$

the maximum of the two specifies the better fit (for a given string σ). A similar procedure can be performed for any choice of the above types (a)–(c) of function b.

Remark 3.7.2 It is important to take into account that models with too many parameters (e.g. an arbitrary $s \times s$ transition matrix P and an arbitrary collection of probabilities $\mathbf{Q}$) may result in an 'overfitted' model $Z^*(\sigma)$; this may generate an unwanted instability, where $Z^*(\sigma)$ changes drastically with the string σ. This makes it desirable to use any 'side information' available on a possible model, to include it in the maximisation problems for the likelihoods.

Example 3.7.3 Consider a discrete-time Markov chain (X_m) with 3 states and 3×3 transition matrix $P = (p_{ij})$. It is known that diagonal transition probabilities vanish: $p_{ii} = 0$, $i = 1, 2, 3$. Further, suppose we know that at the initial time $X_0 = 1$, and at time 4, $X_4 = 3$, but do not know states at times 1, 2 and 3. Write down the aggregated likelihood as the sum of the likelihoods over sample vectors $\mathbf{x} = \begin{pmatrix} x_0 \\ \vdots \\ x_4 \end{pmatrix} \in \{1, 2, 3\}^5$ compatible with this restriction (i.e. with $x_0 = 1$, $x_4 = 3$):

$$\mathbf{L}(P | X_0 = 1, X_4 = 3) = p_{13}^{(4)}$$
$$= p_{12}^2 p_{21} p_{23} + p_{13} p_{31} p_{12} p_{23} + p_{12} p_{23} p_{31} p_{13} + p_{12} p_{23}^2 p_{32} + p_{13}^2 p_{32} p_{21}; \quad (3.208)$$

this is a polynomial function in the variables p_{ij}. Following the maximum likelihood philosophy, we would like to maximise $\mathbf{L}(P | X_0 = 1, X_4 = 3)$ in $P = (p_{ij})$ over the set $\mathscr{P}_{\text{off-diag}}$; see (3.11). The maximum likelihood estimator P_{ML}^* can be at an internal point or on the boundary. In general, the problem of finding the exact MLE becomes computationally difficult; a number of other factors also interfere which makes it desirable to develop 'reasonable' approximative methods.

As will be shown, the problem of constructing an approximation to estimators of transition probabilities p_{ij}, $1 \leq i, j \leq 3$, can be (reasonably) solved by iterations of a certain transformation. More precisely, set:

$$\widehat{p}_{ij} = \frac{p_{ij} \dfrac{\partial}{\partial p_{ij}} \mathbf{L}(P|X_0 = 1, X_4 = 3)}{\Xi(P|X_0 = 1, X_4 = 3),}, \quad 1 \leq i, j \leq 3. \tag{3.209}$$

where the denominator is given by

$$\Xi(P|X_0 = 1, X_4 = 3) = \sum_{k=1}^{3} p_{ik} \frac{\partial}{\partial p_{ik}} \mathbf{L}(P|X_0 = 1, X_4 = 3)$$

$$= 2p_{12}^2 p_{21} p_{23} + p_{12} p_{23}^2 p_{32} + 2p_{13} p_{31} p_{12} p_{23}$$

$$+ 2p_{12} p_{23} p_{31} p_{13} + 2p_{13}^2 p_{32} p_{21}. \tag{3.210}$$

In particular,

$$\widehat{p}_{12} = \frac{2p_{12}^2 p_{21} p_{23} + p_{12} p_{23}^2 p_{32} + p_{13} p_{31} p_{12} p_{23} + p_{12} p_{23} p_{31} p_{13}}{\Xi(P|X_0 = 1, X_4 = 3)},$$

$$\widehat{p}_{13} = \frac{p_{13} p_{31} p_{12} p_{23} + p_{12} p_{23} p_{31} p_{13} + 2p_{13}^2 p_{32} p_{21}}{\Xi(P|X_0 = 1, X_4 = 3)},$$

and so on. Iterations of this transformation provide a solution within a reasonable margin.

Examples 3.7.1 and 3.7.3 outline the main directions of our investigation. See also Koski, 2001. One direction is related to 'noisy' observations where we have records of all states subsequently taken by the chain, albeit subject to noise which results in an aggregation of states. We call this an HMM *filtration* problem. The second direction is when the chain is available for observation only at some selected times. We call it an HMM *interpolation* problem. The methods used in these cases bear some similarities, but also differ in some essential aspects.

Consider first a general set-up for the HMM filtration problem. We are given a vector of observed values $\sigma = \begin{pmatrix} \sigma_0 \\ \vdots \\ \sigma_n \end{pmatrix}$, called a *training sequence*, where $\sigma_0, \ldots, \sigma_n$ take values in $\{1, \ldots, \kappa\}$. This means we know that the event $\{\sigma_0 = b(X_0), \ldots, \sigma_n = b(X_n)\}$ occurred. However, b remains an unknown function $\{1, \ldots, s\} \to \{1, \ldots, \kappa\}$, possibly random. More precisely, we will assume that for all $\mathbf{x}$, the values $b(X_i)$ are conditionally independent, given that $\mathbf{X} = \mathbf{x}$, (3.211)

and set

$$\mathbb{P}(b(X_i) = k | X_i = j) = q_{jk}, \quad j = 1, \ldots, s, \quad k = 1, \ldots, \kappa, \quad (3.212)$$

where $q_{jk} \geq 0$, $\sum_{k=1}^{\kappa} q_{jk} = 1$ for all $j \in I$. The case of a non-random function b occurs when q_{jk} equals 0 or 1 (obviously, not more than one q_{jk} can equal 1 for a given j). In the case of a 'perfect' observation, we would have $s = \kappa$ and $q_{ij} = \delta_{ij}$.

In general, the collection of probabilities q_{jk} is denoted by $\mathbf{Q}$ (from the definition, it forms an $s \times \kappa$ stochastic matrix). We call them *noise probabilities* and refer to the triple $(\lambda, P, \mathbf{Q})$ as a (hidden Markov) model (with a memoryless noise) and denote it, as before, by Z.

We assume that there has been given a set $\mathscr{Z}$ of models (that is, triples $Z = (\lambda, P, \mathbf{Q})$), and all considerations are reduced to this set $\mathscr{Z}$. The 'largest' such set will correspond to an 'unrestricted' situation.

The HMM filtration problem (also known in the literature as a learning, or training, or estimation, problem for an HMM with noise) is to find a 'most likely' model, $Z^* = (\lambda^*, P^*, \mathbf{Q}^*)$, maximising the aggregated likelihood function $\mathbf{L}(\sigma; Z)$ in $Z \in \mathscr{Z}$ for given σ:

$$\mathbf{L}(\sigma; Z) := \mathbb{P}(\mathbf{b}(\mathbf{X}; Z) = \sigma) = \sum_{\mathbf{X}} \mathbb{P}(\mathbf{X} = \mathbf{x}; Z) \prod_{i=0}^{n} q_{x_i \sigma_i}$$

$$= \sum_{\mathbf{X}} \lambda_{x_0} q_{x_0 \sigma_0} \prod_{i=1}^{n} p_{x_{i-1} x_i} q_{x_i \sigma_i}. \quad (3.213)$$

Here and below, $\mathbb{P}(\cdot; Z)$ stands for the probability distribution generated by the model Z (that is, by a (λ, P) DTMC for $\mathbf{X}$, independent observations $b(X_j)$ with noise probabilities $\mathbf{Q} = (q_{jk})$). Sometimes we also use the alternative notation $\mathbb{P}_Z$.

Therefore, we are interested in a triple $Z_{\mathrm{ML}}^* = (\lambda_{\mathrm{ML}}^*, P_{\mathrm{ML}}^*, \mathbf{Q}_{\mathrm{ML}}^*)$ defined by

$$Z_{\mathrm{ML}}^* = \underset{Z \in \mathscr{Z}}{\mathrm{argmax}} \; \mathbb{P}(\mathbf{b}(\mathbf{X}) = \sigma; Z), \quad (3.214)$$

where $\mathbf{b}(\mathbf{X})$ stands for the (random) vector $\begin{pmatrix} b(X_0) \\ \vdots \\ b(X_n) \end{pmatrix}$. (The subscript ML refers to maximum likelihood.)

In practice, finding the minimiser Z^* in (3.214) is often difficult, particularly when the numbers s and κ are large and the set $\mathscr{Z}$ carries numerous constraints.

Unsuprisingly therefore, there is a substantial literature discussing various algorithmic methods giving approximations to the value Z^*. This is the subject of the current and next sections.

Example 3.7.4 An unrestricted HMM filtration problem arises when the pair (λ, P) runs over the set $\mathscr{R}$ defined in (3.5) and the matrix $\mathbf{Q}$ over the set $\mathscr{P}_{s,\kappa}$ of dimension $s(\kappa - 1)$

$$\mathscr{P}_{s,\kappa} = \left\{ \mathbf{Q} = (q_{jk}) : q_{jk} \geq 0, \; \sum_{k=1}^{\kappa} q_{jk} = 1 \right\}.$$

Correspondingly, we set

$$\mathscr{U} = \mathscr{R}_s \times \mathscr{P}_{s,\kappa} = \Lambda_s \times \mathscr{P}_s \times \mathscr{P}_{s,\kappa} \tag{3.215}$$

and bear in mind that the unrestricted problem corresponds to $Z \in \mathscr{U}$. The unrestricted stationary HMM filtration problem would correspond to the pair (π, P), with the matrix P running over set $\mathscr{P}^{\text{int}}$ from (3.8), and the matrix $\mathbf{Q}$ running over $\mathscr{P}_{s,\kappa}$.

In the unrestricted problem, the set $\mathscr{U}$ can be endowed with a distance, by setting

$$\text{dist}\,(Z, Z') = \left[\sum_j (\lambda_j - \lambda_j')^2 + \sum_{i,j} (p_{ij} - p_{ij}')^2 + \sum_{j,k} (q_{jk} - q_{jk}')^2 \right]^{1/2},$$

where $Z = (\lambda, P, \mathbf{Q})$, $Z' = (\lambda', P', \mathbf{Q}')$ and $\lambda = (\lambda_j)$, $P = (p_{ij})$, $\mathbf{Q} = (q_{jk})$, $\lambda' = (\lambda_j')$, $P' = (p_{ij}')$, $\mathbf{Q}' = (q_{jk}')$. In other words, this is the Euclidean distance in $\mathbb{R}^{s(s+1+\kappa)}$ restricted upon $\mathscr{U}$. We will use this distance in Section 3.9.

An example of a restricted HMM filtration problem arises when the matrix P runs over the set $\mathscr{P}_{\text{off-diag}}$ in (3.11) or the set $\mathscr{P}_{\text{symm}}$ in (3.12). In the first case, $\mathscr{Z} = \Lambda \times \mathscr{P}_{\text{off-diag}} \times \mathscr{P}_{s,\kappa}$, and in the second, $\mathscr{Z} = \Lambda \times \mathscr{P}_{\text{symm}} \times \mathscr{P}_{s,\kappa}$.

We will work with sample vectors $\mathbf{x} = \begin{pmatrix} x_0 \\ \vdots \\ x_n \end{pmatrix}$ which have positive probabilities

$\mathbb{P}(\mathbf{X} = \mathbf{x}; Z)$, under a given model Z; a natural assumption which we will follow throughout is that the set of these vectors, $\mathscr{X} \subseteq I^n$, is the same for all models $Z \in \mathscr{Z}$ under consideration. For instance, consider the above case of a restricted filtration problem, where $P \in \mathscr{P}_{\text{off-diag}}$, i.e. the transition matrix $P = (p_{ij})$ of the Markov chain does not permit a repetition of states (that is, features $p_{ii} = 0$ for all state $i = 1, \ldots, s$) but allows all other transitions for any model $Z = (\lambda, P, \mathbf{Q}) \in \mathscr{Z}$ (see 3.7.3). In this case, $\mathscr{X}$ consists of all vectors $\mathbf{x} \in I^n$ with $x_{i-1} \neq x_i$, $i = 1, \ldots, n$.

Moreover, we assume that for all $Z \in \mathscr{Z}$ and a training sequence σ occurring as a value $b(\mathbf{X})$ (i.e. with $\mathbb{P}\big(b(\mathbf{X}) = \sigma; Z\big) > 0$),

$$\mathbb{P}\big(\mathbf{X} = \mathbf{x} \big| b(\mathbf{X}) = \sigma; Z\big) > 0 \text{ if and only if } \mathbf{x} \in \mathscr{X}. \tag{3.216}$$

Next, let t be the cardinality of $\mathscr{X}$. It is convenient to enumerate strings $\mathbf{x} \in \mathbf{X}$ by $l = 1, \ldots, t$ (in any order) and write $\mathbf{x}(l) = \begin{pmatrix} x_0(l) \\ \vdots \\ x_n(l) \end{pmatrix}$ for the lth string. Then, given $Z = (\lambda, P, \mathbf{Q})$, set

$$\begin{aligned} u_l(\sigma; Z) &= \mathbb{P}\Big(b(\mathbf{X}) = \sigma, \mathbf{X} = \mathbf{x}(l); Z\Big) \\ &= \lambda_{x_0(l)} q_{x_0(l)\sigma_0} \prod_{j=1}^{n} P_{x_{j-1}(l)x_j(l)} q_{x_j(l)\sigma_j}. \end{aligned} \tag{3.217}$$

That is, $u_l(\sigma; Z)$ gives the probability of the intersection

$$\{\mathbf{X} = \mathbf{x}(l)\} \cap \{b(\mathbf{X}) = \sigma\},$$

in model Z.

Theorem 3.7.5 *Assume that we are given a model $Z \in \mathscr{U}$ and a training sequence σ such that $u_l(\sigma; Z) > 0$ for at least one l (that is, $\mathbb{P}\big(b(\mathbf{X}) = \sigma; Z\big) > 0$). Then, under the assumption (3.216), for all $\widehat{Z} \in \mathscr{U}$,*

$$\ln \frac{\mathbb{P}\big(b(\mathbf{X}) = \sigma; \widehat{Z}\big)}{\mathbb{P}\big(b(\mathbf{X}) = \sigma; Z\big)} \geq \frac{U(Z, Z; \sigma) - U(Z, \widehat{Z}; \sigma)}{\mathbb{P}\big(b(\mathbf{X}) = \sigma; Z\big)}, \tag{3.218}$$

where

$$U(Z, Z; \sigma) = \sum_{l=1}^{t} \big[-u_l(\sigma; Z) \ln u_l(\sigma; Z)\big], \tag{3.219}$$

and

$$U(Z, \widehat{Z}; \sigma) = \sum_{l=1}^{t} \big[-u_l(\sigma; Z) \ln u_l(\sigma; \widehat{Z})\big]. \tag{3.220}$$

Here we follow an agreement that

$$-u_l(\sigma; Z) \ln u_l(\sigma; Z) = 0 \text{ when } u_l(\sigma; Z) = 0,$$

and

$$-u_l(\sigma;Z)\ln u_l(\sigma;\widehat{Z}) = +\infty, \quad \text{when } u_l(\sigma;Z) > 0 \text{ but } u_l(\sigma;\widehat{Z}) = 0,$$

and so the sums in (3.219), (3.220) have all summands ≥ 0.

Proof of Theorem 3.7.5 The proof follows immediately from Example 3.6.7 with $n = t, a_l = u_l, b_l = \widehat{u}_l$. In fact, the assertion of the theorem is equivalent to

$$\ln \frac{\sum\limits_{l=1}^{t} \widehat{u}_l}{\sum\limits_{l=1}^{t} u_l} \geq \left[\sum_{l=1}^{t} (u_l \ln \widehat{u}_l - u_l \ln u_l) \right] \Bigg/ \sum_{l=1}^{t} u_l,$$

where $u_l = u_l(\sigma;Z)$ and $\widehat{u}_l = u_l(\sigma;\widehat{Z})$. $\qquad\square$

Thus if, for given σ and Z, we can find a model $\widehat{Z}$ for which the RHS of (3.218) is positive, then we obtain an 'improved' model, in a sense of a higher value for the likelihood.

Hence, we are interested in minimising the function $U(Z,\widehat{Z};\sigma)$ defined in (3.220), in the variable $\widehat{Z} \in \mathscr{Z}$, for a given model $Z \in \mathscr{Z}$ and a given training sequence σ. In general, the minimiser will of course depend on Z and σ (and upon the choice of the set $\mathscr{Z}$).

To this end, it will be convenient to use transition counts. As in Section 3.4, given $i, j = 1,\ldots,s$ and $l = 1,\ldots,t$, let $f_{ij}(l)$ be the number of transitions $i \to j$ in the string $\mathbf{x}(l)$:

$$f_{ij}(l) = \sum_{m=1}^{n} \mathbf{1}(x_{m-1}(l) = i, x_m(l) = j). \tag{3.221}$$

Next, set

$$r_j(l) = \mathbf{1}(x_0(l) = j). \tag{3.222}$$

Furthermore, given a training sequence σ and $k = 1,\ldots,\kappa$, denote by $n_{jk}(l)$ $(= n_{jk}(l,\sigma))$ the number of times the value k was recorded in state j in the string $\mathbf{x}(l)$:

$$n_{jk}(l) = \sum_{m=0}^{n} \mathbf{1}(x_m(l) = j, \sigma_m = k). \tag{3.223}$$

Finally, denote:

$$e_j = \sum_{l=1}^{t} u_l r_j(l), \quad c_{ij} = \sum_{l=1}^{t} u_l f_{ij}(l), \quad \text{and} \quad d_{jk} = \sum_{l=1}^{t} u_l n_{jk}(l), \tag{3.224}$$

where we again write $u_l = u_l(\sigma;Z)$ for model $Z = (\lambda, P, \mathbf{Q}) \in \mathscr{Z}$. Thus, e_j, c_{ij} and d_{jk} are all functions of Z and σ.

Going back to (3.220), we re-group summands in the expression for $U(Z, \widehat{Z}; \sigma)$, according to occurrences of initial states i, transitions $i \to j$ and recorded values k. As a result, we obtain that for $\widehat{Z} = (\widehat{\lambda}, \widehat{P}, \widehat{\mathbf{Q}})$

$$
\begin{aligned}
U(Z, \widehat{Z}; \sigma) \;=\; & \sum_{l=1}^{t} u_l \Bigg[-\ln \widehat{\lambda}_{x_0(l)} \\
& - \sum_{j=1}^{s} \sum_{k=1}^{\kappa} n_{jk}(l) \ln \widehat{q}_{jk} - \sum_{i=1}^{s} \sum_{j=1}^{s} f_{ij}(l) \ln \widehat{p}_{ij} \Bigg] \\
=\; & -\sum_{j=1}^{s} e_j \ln \widehat{\lambda}_j - \sum_{j=1}^{s} \sum_{k=1}^{\kappa} d_{jk} \ln \widehat{q}_{jk} - \sum_{i=1}^{s} \sum_{j=1}^{s} c_{ij} \ln \widehat{p}_{ij}.
\end{aligned}
\tag{3.225}
$$

The unique global minimum, in $\widehat{Z}$, of the expression in the RHS of (3.225) is attained at the point $\widehat{Z}^* = (\widehat{\lambda}^*, \widehat{P}^*, \widehat{\mathbf{Q}}^*)$ where $\widehat{\lambda} = (\widehat{\lambda}_i^*)$, $\widehat{P}^* = (\widehat{p}_{ij}^*)$ and $\widehat{\mathbf{Q}}^* = (\widehat{q}_{jk}^*)$ are given by

$$
\widehat{\lambda}_j^* = e_j \Bigg/ \sum_{i=1}^{s} e_i, \quad j = 1, \ldots, s,
\tag{3.226}
$$

$$
\widehat{p}_{ij}^* = c_{ij} \Bigg/ \sum_{m=1}^{s} c_{im}, \quad i, j = 1, \ldots, s,,
\tag{3.227}
$$

and

$$
\widehat{q}_{jk}^* = d_{jk} \Bigg/ \sum_{k=1}^{\kappa} d_{jk}, \quad j = 1, \ldots, s, \; k = 1, \ldots, \kappa.
\tag{3.228}
$$

The reader should bear in mind that the probabilities $\widehat{\lambda}_j^*$, $\widehat{p}_{ij}^*$ and $\widehat{q}_{jk}^*$ are functions of Z and σ.

Thus, if model $\widehat{Z}^* \in \mathscr{Z}$, it provides an 'improvement' upon Z. For instance, in Example 3.7.3, where the transition matrix P has all diagonal entries $p_{ii} = 0$, transition matrix $\widehat{P}^* = (\widehat{p}_{ij}^*)$ from (3.227) holds the same property.

The denominators in (3.226), (3.228) can be simplified. Indeed, observe that

$$
\sum_{j=1}^{s} e_j = \sum_{l=1}^{t} \sum_{j=1}^{s} r_j(l) = \sum_{l=1}^{t} u_l = \mathbb{P}(\mathbf{b}(\mathbf{X}) = \sigma; Z)
$$

and

$$
n_j := \sum_{k=1}^{\kappa} d_{jk} = \mathbb{E}_Z \big(\text{the number of visits to state } j \text{ prior to time } n \big),
$$

where $\mathbb{E}_Z$ stands for the expectation relative to $\mathbb{P}_Z$. Using these relations one can write (3.226)–(3.228) in the compact form

$$\widehat{\lambda}_j^* = e_j/\mathbb{P}_Z(\mathbf{b}(\mathbf{X}) = \sigma), \quad \widehat{p}_{ij}^* = c_{ij} \left/ \sum_{m-1}^s c_{im} \right., \quad \text{and } \widehat{q}_{jk}^* = d_{jk}/n_j. \qquad (3.229)$$

In general, we have to solve the constrained problem

$$\begin{aligned} \text{minimise} \quad & \text{the RHS of (3.225), in } \widehat{Z}, \text{ for a given } Z \\ \text{subject to} \quad & \widehat{Z} \in \mathscr{Z}. \end{aligned} \qquad (3.230)$$

This suggests the following 'learning' algorithm: given an initial model $Z^{(0)} \in \mathscr{Z}$ and a training sequence σ, solve problem (3.230) thereby obtaining an improved model, $Z^{(1)} = \widehat{Z}^* \in \mathscr{Z}$. Then repeat it for $Z^{(1)}$ and σ, and so on. Suppose that the minimiser $Z^{(N)}$ obtained in the course of N iterations converges to a limit:

$$\lim_{N \to \infty} Z^{(N)} = Z^{(\infty)} \in \mathscr{Z} \qquad (3.231)$$

then the model $Z^{(\infty)}$ can be considered as the 'best fit' for the algorithm.

The following questions then arise: (i) Does the limit $Z^{(\infty)}$ in (3.231) exist (for all or some initial models $Z^{(0)}$); (ii) if this limit does exist, does it coincide with a maximiser Z_{ML}^* in (3.214)? As was mentioned, these questions give rise to a substantial literature embracing a number of important applications. Some of the results in this direction are discussed in Section 3.9.

The above algorithm is called the *Baum–Welch learning algorithm*; its attraction lies in the simplicity (and therefore practicality) of the solution to problem (3.230) for various types of set $\mathscr{Z}$, as was demonstrated by formulas (3.226)–(3.229). It is important to associate with these relations a map $\Phi: \mathscr{U} \to \mathscr{U}$ of the set $\mathscr{U}$ (see (3.215)) to itself

$$\Phi: Z = (\lambda, P, \mathbf{Q}) \mapsto \widehat{Z}^* = (\widehat{\lambda}^*, \widehat{P}^*, \widehat{\mathbf{Q}}^*), \qquad (3.232)$$

which we call the *Baum–Welch transformation* for (unrestricted) filtration. (In the course of our presentation, we will re-write formulas (3.226)–(3.229) in a number of (equivalent) forms, clarifying various aspects of the map Φ.) The transformation Φ will be particularly useful for problem (3.230) when it takes the original set of models $\mathscr{Z}$ to itself:

$$\Phi(\mathscr{Z}) \subseteq \mathscr{Z}.$$

The above questions (i) and (ii) (in the unrestricted filtration problem) are about iterations Φ^N of the transformation Φ. A straightforward corollary of Theorem 3.7.5 is

Worked Example 3.7.6 Prove that any point Z_{ML}^* defined in (3.214) is a fixed point of the map Φ:

$$\Phi(Z_{\text{ML}}^*) = Z_{\text{ML}}^*. \tag{3.233}$$

Solution In fact, if $\Phi(Z_{\text{ML}}^*) \neq Z_{\text{ML}}^*$ then the probability $\mathbb{P}(b(X) = \sigma, \Phi(Z_{\text{ML}}^*))$ is $\geq \mathbb{P}(b(X) = \sigma, Z_{\text{ML}}^*)$.

$\square$

Chariots of Φ, Chariots of Π
(From the series 'Movies that never made it to the Big Screen'.)

We now pass to the HMM interpolation problems. Here, we work again with a DTMC (X_m) with state space $I = \{1, \dots, s\}$ and a transition matrix $P = (p_{ij})$. In our setting, the matrix P completely defines the model; we assume for simplicity that the initial distribution λ is known. Suppose the chain is observed at (integer) times $0 = t_0 < t_1 < \cdots < t_k \leq n$; denote $\mathbf{T} = \{t_1, \dots, t_k\}$. Correspondingly, denote:

$$\mathbf{X_T} = \begin{pmatrix} X_{t_0} \\ \vdots \\ X_{t_k} \end{pmatrix} \text{ and } \mathbf{x_T} = \begin{pmatrix} x_{t_0} \\ \vdots \\ x_{t_k} \end{pmatrix}. \text{ Next, given } \mathbf{y} = \begin{pmatrix} y_0 \\ \vdots \\ y_n \end{pmatrix} \in I^{n+1}, \text{ let } \mathbf{y}\big|_{\mathbf{T}} \text{ stand}$$

for the restriction $\begin{pmatrix} y_{t_0} \\ \vdots \\ y_{t_k} \end{pmatrix}$. Then define the aggregated likelihood:

$$\begin{aligned} \mathbf{L}(P|\mathbf{X_T} = \mathbf{x_T}) &= \sum_{\mathbf{y} \in I^{n+1}} \lambda_{y_0} \prod_{m=1}^{n} p_{y_{m-1}y_m} \mathbf{1}\left(\mathbf{y}\big|_{\mathbf{T}} = \mathbf{x_T}\right) \\ &= \sum_{\mathbf{y} \in I^{n+1}} \prod_{i,j=1}^{s} \lambda_i^{r_i} p_{ij}^{f_{ij}} \mathbf{1}\left(\mathbf{y}\big|_{\mathbf{T}} = \mathbf{x_T}\right), \end{aligned} \tag{3.234}$$

where

$$f_{ij} = f_{ij}(\mathbf{y}) = \sum_{m=1}^{n} \mathbf{1}(y_{m-1} = i, y_m = j), \ r_i = r_i(\mathbf{y}) = \mathbf{1}(y_0 = i);$$

cf. (3.221), (3.222). The HMM interpolation problem is to find the maximiser $P_{\text{ML}}^* (= P_{\text{ML}}^*(\mathbf{x_T}))$ over a given set $\mathscr{Y} \subseteq \mathscr{P}$ (see (3.7)):

$$P_{\text{ML}}^* = \underset{P \in \mathscr{Y}}{\text{argmax}} \ \mathbf{L}(P|\mathbf{X_T} = \mathbf{x_T}). \tag{3.235}$$

As before, when $\mathscr{Y} = \mathscr{P}$, we obtain an unrestricted problem, and if $\mathscr{Y}$ is a proper subset of $\mathscr{P}$, we speak of a restricted problem. A further generalisation (not considered here) would occur if we use a more general condition $X_0 \in A_0$, $X_{t_1} \in A_1$, ..., $X_{t_k} \in A_k$ where $A_1, \ldots, A_k$ are subsets of state space I.

We face here a kind of difficulty similar to that in the HMM filtration problem: the maximiser P_{ML}^* in (3.235) is hard to calculate, and it is sensitive to the choice of set $\mathscr{Y}$ specifying a priori information about the model. Therefore, an approximate solution is sought, based on a reasonably straightforward construction.

To this end, define the matrix $\widehat{P}$ with entries $\widehat{p}_{ij}$, $i, j = 1, \ldots$, given by

$$\widehat{p}_{ij} = \frac{p_{ij} \dfrac{\partial}{\partial p_{ij}} \mathbf{L}(P|\mathbf{X_T} = \mathbf{x_T})}{\displaystyle\sum_{k=1}^{s} p_{ik} \dfrac{\partial}{\partial p_{ik}} \mathbf{L}(P|\mathbf{X_T} = \mathbf{x_T})}, \tag{3.236}$$

where the aggregated likelihood $\mathbf{L}(P|\mathbf{X_T} = \mathbf{x_T})$ is given by (3.234). Clearly, $\widehat{P}$ depends on P and $\mathbf{x_T}$: $\widehat{P} = \widehat{P}(P, \mathbf{x_T})$. For a given sample $\mathbf{x_T}$, formula (3.236) defines a map $\Pi(= \Pi(\mathbf{x_T}))$ on the set $\mathscr{P}_s$:

$$\Pi: P = (p_{ij}) \mapsto \widehat{P} = (\widehat{p}_{ij}) \tag{3.237}$$

which we call the Baum–Welch transformation for the HMM interpolation problem.

Two observations are worthwhile here.

(I) Suppose that $t_0 = 0$, $t_1 = 1$, ..., $t_k = k$, i.e. we observe the chain at subsequent time points $0, \ldots, k$. Then $\mathbf{x_T}$ becomes the sample vector $\mathbf{x}_0^k = \begin{pmatrix} x_0 \\ \vdots \\ x_k \end{pmatrix} \in I^{k+1}$,

and the RHS in (3.236) yields a matrix which does not depend on P but on $\mathbf{x}_0^k$ only. More precisely, in this case formula (3.236) returns, as $\widehat{p}_{ij}$, the empirical (or relative) frequency $\widehat{f}_{ij}(= \widehat{f}_{ij}(\mathbf{x}_0^k))$ of the transition $i \to j$ in the sample $\mathbf{x}_0^k$:

$$\widehat{p}_{ij} = \widehat{f}_{ij} := \left(\sum_{l=1}^{s} f_{il}(\mathbf{x}_0^k) \right)^{-1} f_{ij}(\mathbf{x}_0^k), \quad i, j = 1, \ldots, s. \tag{3.238}$$

Geometrically, this means that the Baum–Welch transformation Π sends any $P \in \mathscr{P}$ to the matrix $\widehat{F} = (\widehat{f}_{ij})$ of empirical frequencies:

$$\Pi(P) = \widehat{F}, \quad P \in \mathscr{P}.$$

Hence, in this case the matrix $\widehat{F}$ forms a unique fixed point of the transformation (3.237), and if we repeat the procedure (3.236) (i.e. iterate the map (3.237)), then the result will be again the matrix $\widehat{F}$.

(II) If the DTMC (X_m) yields a sequence of IID RVs, then formula (3.236) returns, as $\widehat{p}_{ij}$, the empirical (relative) frequency $\widehat{g}_j = \widehat{g}_j(\mathbf{x_T})$ of visits to state j in sample $\mathbf{x_T}$. Formally:

$$\widehat{p}_{ij} = \widehat{g}_j := \left(\sum_{l=1}^{s} g_l \right)^{-1} g_j, \quad j = 1, \ldots, s, \tag{3.239}$$

where,

$$g_j = g_j(\mathbf{x_T}) = \sum_{l=0}^{k} \mathbf{1}(x_{t_l} = j).$$

In other words, we forget in this case about states visited during time intervals between points $t_0, \ldots, t_k$ and compute the frequencies of visits to each state $j = 1, \ldots, s$ based on the available data. In other words, every matrix $P = (p_{ij})$ whose rows are repetitions of a fixed stochastic vector (or, equivalently, whose entries $p_{ij} = p_j$ are constant along the columns) is taken by map Π to the matrix $\widehat{G}$ of the empirical frequencies $\widehat{g}_j$ (which, obviously, satisfies the same property). Geometrically, this means the matrices $\widehat{G} = (\widehat{g}_{ij})$ always form a family of fixed points for the Baum–Welch transformation Π.

Worked Example 3.7.7 Prove remarks (I) and (II).

Solution Both equalities (3.238) and (3.239) follow from (3.236) by differentiation.

$\square$

An remarkable fact is that iterating Π from (3.237) leads to an increase in the value of the aggregated likelihood $\mathbf{L}(P|\mathbf{X_T} = \mathbf{x_T})$ defined in (3.234).

Theorem 3.7.8 *For any transition matrix $P = (p_{ij})$, set of time points* $\mathbf{T} = \{t_0, t_1, \ldots, t_k\}$, *with* $0 = t_0 < t_1 < \cdots < t_k \leq n$, *and sample string* $\mathbf{x_T} = \begin{pmatrix} x_{t_0} \\ \vdots \\ x_{t_k} \end{pmatrix}$,

$$\mathbf{L}(\Pi(P)|\mathbf{X_T} = \mathbf{x_T}) \geq \mathbf{L}(P|\mathbf{X_T} = \mathbf{x_T}). \tag{3.240}$$

Moreover, equality in (3.240) is attained if and only if $\Pi(P) = P$.

Proof The main idea of the proof is algebraic. Given $\mathbf{x_T}$, the functions

$$P \mapsto \mathbf{L}(P|\mathbf{X_T} = \mathbf{x_T}) \text{ and } P \mapsto \mathbf{L}(\Pi(P)|\mathbf{X_T} = \mathbf{x_T})$$

are homogeneous polynomials in variables p_{ij}, in the sense that both $\mathbf{L}(P|\mathbf{X_T} = \mathbf{x_T})$ and $\mathbf{L}(\Pi(P)|\mathbf{X_T} = \mathbf{x_T})$ are sums of monomials of a fixed (total) degree equal to $t_k + 1$. Moreover, these monomials enter the sum with coefficients 0 or 1 (see (3.234)). Theorem 3.7.8 will follow from the more general Theorem 3.7.10 stated and proved below for such polynomials. □

Before we pass to Theorem 3.7.10, we would like to invoke the famous Euler theorem on homogeneous functions. A function of n real variables $f(x_1, x_2, \ldots, x_n)$ is said to be *homogeneous of degree d*, if for any real a

$$f(ax_1, ax_2, \ldots, ax_n) = a^d f(x_1, x_2, \ldots, x_n). \tag{3.241}$$

Euler's theorem says that for any differentiable homogeneous function,

$$\sum_{i=1}^{n} x_i \frac{\partial}{\partial x_i} f = df. \tag{3.242}$$

Worked Example 3.7.9 Assuming property (3.241), prove (3.242).

Solution (Sketch) Differentiate $f(ax_1, ax_2, \ldots, ax_n)$ with respect to a. Then (3.241) gives

$$\begin{aligned}
\frac{\mathrm{d}}{\mathrm{d}a} f(ax_1, ax_2, \ldots, ax_n) &= \sum_{i=1}^{n} x_i \frac{\partial}{\partial x_i} f(ax_1, \ldots, ax_n) \\
&= da^{d-1} f(x_1, x_2, \ldots, x_n).
\end{aligned}$$

Finally, set $a = 1$. □

We are now in a position to state and prove Theorem 3.7.10.

Theorem 3.7.10 *Suppose we are given positive integers q and q_i, where $i = 1, \ldots, q$. We will be working with arrays of (non-negative) variables p_{ij}, $i = 1, \ldots, q$, $j = 1, \ldots, q_i$, denoted by P. Consider a closed set $\mathscr{D}$ of dimension $\sum_{i=1}^{q}(q_i - 1)$ given by*

$$\mathscr{D} = \left\{ p_{ij} \geq 0, \sum_{l=1}^{q_i} p_{ij} = 1, i = 1, \ldots, q, j = 1, \ldots, q_i \right\}. \tag{3.243}$$

Next, let $P \mapsto Z(P)$, $P = (p_{ij})$, be a homogeneous polynomial of degree d in variables p_{ij}, $i = 1, \ldots, q$, $j = 1, \ldots, q_i$, with non-negative coefficients. Given $P = (p_{ij}) \in \mathscr{D}$, let $\Pi(P) = (\Pi(P)_{ij})$ denote the point in $\mathscr{D}$ with

$$\Pi(P)_{ij} = p_{ij}\frac{\partial Z}{\partial p_{ij}}\left(\sum_{j=1}^{q_i} p_{ij}\frac{\partial Z}{\partial p_{ij}}\right)^{-1}. \tag{3.244}$$

Then $Z(\Pi(P)) > Z(P)$ unless $\Pi(P) = P$.

Proof First, we establish some notation. Let $v = (v_{ij})$ be an array of non-negative integers v_{ij}, where $i = 1,\ldots,q$, $j = 1,\ldots,q_i$. Given an array $P = (p_{ij}) \in \mathscr{D}$, $[P]^v$ will stand for the product $\prod_{i=1}^{q}\prod_{j=1}^{q_i}p_{ij}^{v_{ij}}$. Next, $c_v \geq 0$ will denote the coefficient in $Z(P)$ in front of monomial $[P]^v$:

$$Z(P) = \sum_v c_v [P]^v.$$

Using this notation, we can write

$$\Pi(P)_{ij} = \frac{\sum\limits_v c_v v_{ij}[P]^v}{\sum\limits_{j=1}^{q_i}\sum\limits_v c_v v_{ij}[P]^v}. \tag{3.245}$$

We wish to prove that

$$\begin{aligned}
Z(P) &= \sum_v c_v [P]^v = \sum_v c_v \prod_{i=1}^{q}\prod_{j=1}^{q_i}p_{ij}^{v_{ij}}\\
&\leq \sum_v c_v \prod_{i=1}^{q}\prod_{j=1}^{q_i}[\Pi(P)_{ij}]^{v_{ij}} = \sum_v c_v \left[\Pi(P)\right]^v = Z(\Pi(P)), \tag{3.246}
\end{aligned}$$

and analyse the case of equality.

To this end, represent

$$\begin{aligned}
Z(P) &= \sum_v \left(c_v\prod_{i=1}^{q}\prod_{j=1}^{q_i}(\Pi(P)_{ij})^{v_{ij}}\right)^{1/(d+1)}\\
&\quad \times \left[c_v^{d/d+1}[P]^v\prod_{i=1}^{q}\prod_{j=1}^{q_i}\left(\frac{1}{\Pi(P)_{ij}}\right)^{v_{ij}/(d+1)}\right].
\end{aligned}$$

and apply the Hölder inequality

$$\left|\sum_v f_v g_v\right| \leq \left(\sum_v |f_v|^p\right)^{1/p}\left(\sum_v |g_v|^q\right)^{1/q},$$

with $p = d + 1$ and $q = (d+1)/d$, to obtain

$$Z(P)$$

$$\leq \left(\sum_{v} c_v \prod_{i=1}^{q} \prod_{j=1}^{q_i} (\Pi(P)_{ij})^{v_{ij}} \right)^{1/(d+1)}$$

$$\times \left(\sum_{v} c_v [P]^v \prod_{i=1}^{q} \prod_{j=1}^{q_i} \left(\frac{p_{ij}}{\Pi(P)_{ij}} \right)^{v_{ij}/d} \right)^{d/(d+1)}$$

$$= \left(Z(\Pi(P)) \right)^{1/(d+1)} \left(\sum_{v} c_v [P]^v \prod_{i=1}^{q} \prod_{j=1}^{q_i} \left(\frac{p_{ij}}{\Pi(P)_{ij}} \right)^{v_{ij}/d} \right)^{d/(d+1)} ; \qquad (3.247)$$

here, in the second factor we have used the fact that $([P]^v)^{d+1/d} = [P]^v \prod_{i=1}^{p} \prod_{j=1}^{q_i} p_{ij}^{v_{ij}/d}$.
Since the polynomial Z is homogeneous and

$$\sum_{i=1}^{p} \sum_{j=1}^{q_i} \frac{v_{ij}}{d} = 1,$$

we can use the inequality $\prod z_i^{\alpha_i} \leq \sum \alpha_i z_i$ between the geometric and arithmetic means with $\alpha_i \geq 0$, and $\sum_i \alpha_i = 1$ to conclude that

$$\sum_{v} c_v [P]^v \prod_{i=1}^{p} \prod_{j=1}^{q_i} \left(\frac{p_{ij}}{\Pi(P)_{ij}} \right)^{v_{ij}/d} \leq \sum_{v} c_v [P]^v \sum_{i=1}^{p} \sum_{j=1}^{q_i} \frac{v_{ij}}{d} \left(\frac{p_{ij}}{\Pi(P)_{ij}} \right).$$

We now substitute formula (3.245) to obtain

$$\sum_{v} c_v [P]^v \sum_{i=1}^{q} \sum_{j=1}^{q_i} \frac{v_{ij}}{d} \left(\frac{p_{ij}}{\Pi(P)_{ij}} \right)$$

$$= \frac{1}{d} \sum_{v} c_v \left([P]^v \sum_{i=1}^{q} \sum_{j=1}^{q_i} v_{ij} p_{ij} \right) \frac{\sum_{j'=1}^{q_i} \sum_{v'} c_{v'} v'_{ij'} [P]^{v'}}{\sum_{v'} c_{v'} v'_{ij} [P]^{v'}}$$

$$= \frac{1}{d} \sum_{i=1}^{q} \sum_{j=1}^{q_i} p_{ij} \left(\frac{\sum_{v} v_{ij} c_v [P]^v}{\sum_{v'} v'_{ij} c_{v'} [P]^{v'}} \right) \sum_{j'=1}^{q_i} \sum_{v'} c_{v'} v'_{ij'} [P]^{v'}. \qquad (3.248)$$

Here we have interchanged the order of finite summations. For every pair (i, j), the ratio within the brackets equals 1 and by (3.243) for each i, we have $\sum_{j=1}^{q_i} p_{ij} = 1$. Hence, the whole expression in the RHS of (3.248) reduces to

$$\frac{1}{d} \sum_{i=1}^{q} \sum_{j'=1}^{q_i} \sum_{v'} c_{v'} v'_{ij'} [P]^{v'} = \frac{1}{d} \sum_{ij'} p_{ij'} \frac{\partial P}{\partial p_{ij'}}. \qquad (3.249)$$

So, by Euler's theorem, expression (3.249) becomes $\sum_v c_v[P]^v = Z(P)$.

Thus, we obtain the following bound for the second factor in the RHS of (3.247):

$$\left(\sum_v c_v[P]^v \prod_{i=1}^{q}\prod_{j=1}^{q_i}\left(\frac{p_{ij}}{\Pi(P)_{ij}}\right)^{v_{ij}/d}\right) \leq Z(P).$$

Correspondingly, (3.247) becomes

$$Z(P) \leq \left(Z(\Pi(P))\right)^{1/(d+1)}(Z(P))^{d/(d+1)},$$

which is equivalent to (3.246).

Finally, $Z(\Pi(P)) > Z(P)$ if $\Pi(P) \neq P$, as follows from (3.247) and the fact that: (i) the inequality between geometric and arithmetic means becomes equality if and only if all numbers z_i are equal; (ii) the Hölder inequality becomes equality if and only if f_v and g_v are proportional. But equality of all z_i means that the ratio $\dfrac{p_{ij}}{\Pi(P)_{ij}}$ is a constant. But this constant must be 1 by (3.243). Then (ii) holds too. $\square$

We see that iterating the Baum–Welch transformation Π strictly increases the aggregated likelihood $\mathbf{L}(P|\mathbf{X_T} = \mathbf{x_T})$ unless we reach a fixed point. But the function $P \mapsto \mathbf{L}(P|\mathbf{X_T} = \mathbf{x_T})$ is uniformly bounded from above for $P \in \mathscr{P}$. So, suppose we start from a point $P^{(0)} \in \mathscr{P}$ and let $P^{(N)}$ be $\Pi^N(P^{(0)})$, the result of the N-fold application of transformation Π. Then the limit

$$\lim_{N\to\infty} \mathbf{L}(\Pi^N(P^{(0)}|\mathbf{X_T} = \mathbf{x_T}) \tag{3.250}$$

always exists. However, questions similar to (i) and (ii) for the transformation Φ (see above) remain open:

(i) Does the matrix $P^{(N)}$ itself converge to a limit $P^{(\infty)}$ as $N \to \infty$? If yes then $P^{(\infty)}$ must be a fixed point for Π, with value $\mathbf{L}(P^{(\infty)}|\mathbf{X_T} = \mathbf{x_T})$ coinciding with the limit in (3.250). In general, the sequence $\{P^{(N)}\}$ may have more than one limiting point in $\mathscr{P}$ (that is, the limits will exists along different subsequences $\{P^{(N_m)}\}$), but every limiting point will be a fixed point for Π.

(ii) Is the limit $P^{(\infty)}$ (or a limiting point) a point of maximum of $\mathbf{L}(P|\mathbf{X_T} = \mathbf{x_T})$ (local or global)?

(iii) For a given restricted problem, with $P \in \mathscr{Y}$, does the point $P^{(\infty)}$ lie in $\mathscr{Y}$? In general these questions do not have 'nice' (let alone straightforward) answers and require painstaking analysis.

Remark 3.7.11 Despite these nice properties, values $\hat{p}_{ij}^*$ and $\hat{q}_{jk}^*$ have a serious drawback: they are calculated for a given model Z, i.e. are not functions of training sequence σ only. Thus, we cannot call them unbiased and consistent estimators for λ_i, p_{ij} and q_{jk}.

We finish this section by noting that Theorem 3.7.10 enables us to establish that the transformation Φ (see (3.232)) also increases the aggregated likelihood:

Theorem 3.7.12 *For any initial distribution* $\lambda = (\lambda_j)$, *transition matrix* $P = (p_{ij})$, *and collection of noise probabilities* $\mathbf{Q} = (q_{jk})$ *determining a model* $Z = (\lambda, P, \mathbf{Q})$,

and any training sequence $\sigma = \begin{pmatrix} \sigma_0 \\ \vdots \\ \sigma_n \end{pmatrix}$,

$$\mathbf{L}(\sigma; \Phi(Z)) \geq \mathbf{L}(\sigma; Z). \tag{3.251}$$

Moreover, equality in (3.251) is attained if and only if $\Phi(Z) = Z$.

Worked Example 3.7.13 Prove Theorem 3.7.12.

Solution (Sketch) Two alternative proofs are possible: either by using Theorem 3.7.5 or via Theorem 3.7.10. $\square$

3.8 Hidden Markov models, 2. The Baum–Welch learning algorithm

> *Desperately Seeking Smoothness*
> (From the series '*Movies that never made it to the Big Screen*'.)

We start this section by discussing the *smoothing procedure*, in the HMM filtration problem. The philosophy behind this term is as follows. Prior to the procedure, we deal with the situation where there is an unknown model, represented by a point $Z = (\lambda, P, \mathbf{Q}) \in \mathscr{Z}$ or, figuratively speaking, by a function on $\mathscr{Z}$ vanishing 'outside Z' and having a peak at Z. Given a training sequence $\sigma = \begin{pmatrix} \sigma_0 \\ \vdots \\ \sigma_n \end{pmatrix}$, the procedure

enables us to consider a family of models $\widehat{Z}^* = (\widehat{\lambda}^*, \widehat{P}^*, \widehat{\mathbf{Q}}^*)$ compatible with σ, where $\widehat{\lambda}^* = \widehat{\lambda}^*(Z, \sigma)$, $\widehat{P}^* = \widehat{P}^*(Z, \sigma)$ and $\widehat{\mathbf{Q}}^* = \widehat{\mathbf{Q}}^*(Z, \sigma)$ vary with Z. In other words, we pass to 'distributed', or 'smoothed' objects represented by functions on the set $\mathscr{Z}$. Formally, we obtain the map $\Phi\colon Z \mapsto \widehat{Z}^*$; see (3.232). (Of course, a single application of this procedure does not yet solve the problem of assessing the unknown HMM, but it represents a step towards such an assessment. An important issue which we will discuss in the bulk of this section is the result of iterations of the transformation Φ.)

Thus, suppose that we have recorded a training sequence $\sigma = \begin{pmatrix} \sigma_0 \\ \vdots \\ \sigma_n \end{pmatrix}$, for a

random string $\mathbf{X} = \begin{pmatrix} X_0 \\ \vdots \\ X_n \end{pmatrix}$ generated by a DTMC (X_m). That is, we will work

with conditional probabilities, given that $\mathbf{b}(\mathbf{X}) = \sigma$, where $\mathbf{b}(\mathbf{X}) = \begin{pmatrix} b(X_0) \\ \vdots \\ b(X_n) \end{pmatrix}$.

Given $0 \leq m \leq n$, set

$$\widetilde{p}_{ij}(m,n) = \mathbb{P}_Z(X_m = i, X_{m+1} = j \mid \mathbf{b}(\mathbf{X}) = \sigma) \tag{3.252}$$

and

$$\widetilde{p}_i(m,n) = \mathbb{P}_Z(X_m = i \mid \mathbf{b}(\mathbf{X}) = \sigma) = \sum_{j=1}^{s} \widetilde{p}_{ij}(m,n), \quad \widetilde{\lambda}_i = \widetilde{p}_i(0,n). \tag{3.253}$$

With $u_l = \mathbb{P}_Z(\mathbf{b}(\mathbf{X}) = \sigma, \mathbf{X} = \mathbf{x}(l))$, $l = 1,\ldots,t$, denote the normalization constant for $u_1,\ldots,u_t$ by

$$C = \frac{1}{\mathbb{P}_Z(\mathbf{b}(\mathbf{X}) = \sigma)} = \frac{1}{\sum_{l=1}^{t} u_l}. \tag{3.254}$$

In Lemmas 3.8.1 and 3.8.2 we rewrite formulas (3.226)–(3.228) in an alternative, and more convenient, form.

Lemma 3.8.1 *Minimisers* $\widehat{q}_{jk}^* \left(= \widehat{q}_{jk}^*(\sigma) \right)$ *from (3.228) and (3.229) can be written in the form*

$$\widehat{q}_{jk}^* = \frac{\sum_{m=1}^{n} \mathbf{1}(\sigma_m = k)\, \widetilde{p}_j(m,n)}{\sum_{m=1}^{n} \widetilde{p}_j(m,n)}, \quad j = 1,\ldots,s, \ k = 1,\ldots,\kappa. \tag{3.255}$$

Proof We start with an obvious observation that by (3.253), for all $j = 1,\ldots,s$ and $k = 1,\ldots,\kappa$,

$$\sum_{m=1}^{n} \mathbf{1}(\sigma_m = k)\mathbb{P}_Z(X_m = j \mid \mathbf{b}(\mathbf{X}) = \sigma) = \sum_{m=1}^{n} \mathbf{1}(\sigma_m = k)\widetilde{p}_j(m,n). \tag{3.256}$$

Our next goal is to check that

$$d_{jk} = \mathbb{P}_Z(\mathbf{b}(\mathbf{X}) = \sigma) \sum_{m=1}^{n} \mathbf{1}(\sigma_m = k)\mathbb{P}_Z(X_m = j \mid \mathbf{b}(\mathbf{X}) = \sigma). \tag{3.257}$$

To prove (3.257), consider the function of the sample $\mathbf{x}(l)$, $l = 1, \ldots, t$:

$$l \mapsto z(l, j, k) := \sum_{m=0}^{n} \mathbf{1}(x_m(l) = j, \sigma_m = k),$$

giving the number of times one sees a match $i \to k$ in $\mathbf{x}(l)$, for given σ. Then

$$\sum_{m=1}^{n} \mathbb{P}_Z(X_m = j, b(X_m) = k \mid \mathbf{b}(\mathbf{X}) = \sigma) = C \sum_{l=1}^{t} z(l, j, k) u_l = C d_{jk}$$

with C defined in (3.254). To finish the proof, it is enough to note that the denominator

$$\sum_{m=1}^{n} \widetilde{p}_j(m, n) = C n_j,$$

which is obvious. Hence,

$$\widehat{q}_{jk}^* = \frac{C^{-1} \sum\limits_{m=1}^{n} \mathbf{1}(\sigma_m = k) \, \widetilde{p}_j(m, n)}{C^{-1} \sum\limits_{m=1}^{n} \widetilde{p}_j(m, n)}.$$

This implies (3.255). $\qquad\square$

In a similar way one establishes

Lemma 3.8.2 *The minimisers* $\widehat{p}_{ij}^* \left(= \widehat{p}_{ij}^*(\sigma) \right)$ *from (3.226) and (3.227) can be written in the form*

$$\widehat{p}_{ij}^* = \frac{\sum_{m=1}^{n-1} \widetilde{p}_{ij}(m, n)}{\sum_{m=1}^{n-1} \widetilde{p}_i(m, n)}, \quad i, j = 1, \ldots, s. \tag{3.258}$$

The minimisers $\widehat{\lambda}_j^* \left(= \widehat{\lambda}_j^*(\sigma) \right)$ *from (3.216) and (3.219) can be written in the form*

$$\widehat{\lambda}_j^* = \widetilde{p}_j(0, n), \quad j = 1, \ldots, s. \tag{3.259}$$

Unfortunately, formulas (3.255), (3.258), (3.259), (like (3.226)–(3.228)) are not very useful computationally. As was mentioned in Section 3.7, in practice the maximisation procedure is performed according to an algorithm, due to Baum and Welch (also called the Baum–Welch re-estimation), which performs subsequent 'improvements' of $\widetilde{p}_{ij}(m, n)$ and $\widetilde{p}_i(m, n)$ at each iteration.

Our immediate goal is to write the smoothed probabilities in terms of so-called forward and backward variables $\alpha_m(j)$ and $\beta_m(j)$; see (3.260). This gives a computationally effective form of the Baum–Welch re-estimation. Given $m = 0,\ldots,n$, define random strings

$$\mathbf{b}_{m\uparrow}(\mathbf{X}) = \begin{pmatrix} b(X_0) \\ \vdots \\ b(X_m) \end{pmatrix} \quad \text{and} \quad \mathbf{b}^{m\downarrow}(\mathbf{X}) = \begin{pmatrix} b(X_{m+1}) \\ \vdots \\ b(X_n) \end{pmatrix}.$$

Similarly,

$$\sigma_{m\uparrow} = \begin{pmatrix} \sigma_0 \\ \vdots \\ \sigma_m \end{pmatrix} \quad \text{and} \quad \sigma^{m\downarrow} = \begin{pmatrix} \sigma_{m+1} \\ \vdots \\ \sigma_n \end{pmatrix}.$$

Next, define

$$\alpha_m(j) = \mathbb{P}_Z(\mathbf{b}_{m\uparrow}(\mathbf{X}) = \sigma_{m\uparrow}, X_m = j),$$

$$\beta_m(j) = \mathbb{P}_Z(\mathbf{b}^{m\downarrow}(\mathbf{X}) = \sigma^{m\downarrow} | X_m = j). \tag{3.260}$$

Then we have

$$\alpha_m(j)\beta_m(j) = \mathbb{P}_Z(\mathbf{b}(\mathbf{X}) = \sigma, X_m = j). \tag{3.261}$$

By definition of conditional probability, the following recursion relations hold true:

$$\alpha_0(j) = \lambda_j q_{j\sigma_0}, \quad j = 1,\ldots,s, \tag{3.262}$$

$$\alpha_{m+1}(j) = \left[\sum_{i=1}^{s} \alpha_m(i) p_{ij} \right] q_{j\sigma_{m+1}}, \quad j = 1,\ldots,s, \ m = 1,\ldots,n-1, \tag{3.263}$$

$$\beta_n(j) = 1, \quad j = 1,\ldots,s, \tag{3.264}$$

and

$$\beta_m(j) = \sum_{i=1}^{s} \beta_{m+1}(i) q_{i\sigma_{m+1}} p_{ji}, \quad j = 1,\ldots,s, \ m = 0,\ldots,n-1. \tag{3.265}$$

Equations (3.262) and (3.263) yield the forward recursion for probabilities α, while (3.264), (3.265) the backward recursion for probabilities β.

Lemma 3.8.3 *The probability* $\widetilde{p}_i(m,n)$ *from (3.253) admits the following representation:*

$$\widetilde{p}_i(m,n) = \frac{\alpha_m(i)\beta_m(i)}{\mathbb{P}_Z(\mathbf{b}(\mathbf{X}) = \sigma)}. \tag{3.266}$$

Proof The result follows directly from (3.253). $\qquad\square$

Now, applying Bayes' formula, write

$$\sum_{m=1}^{n-1} \widetilde{p}_{ij}(m,n) = \sum_{m=1}^{n-1} \mathbb{P}_Z(X_m = i, X_{m+1} = j \mid \mathbf{b}(\mathbf{X}) = \sigma)$$

$$= C \sum_{m=1}^{n-1} \left[\mathbb{P}_Z(\mathbf{b}(\mathbf{X}) = \sigma \mid X_m = i, X_{m+1} = j) \mathbb{P}_\sigma(Y_m = i, Y_{m+1} = j) \right]$$

$$= C \sum_{m=1}^{n-1} \mathbb{P}_Z(\mathbf{b}(\mathbf{X}) = \sigma \mid X_m = i, X_{m+1} = j) \mathbb{P}_Z(X_m = i) p_{ij},$$

where C is the constant from (3.254). The next step is to check, by using the Markov property, that

$$\mathbb{P}_Z(\mathbf{b}(\mathbf{X}) = \sigma \mid X_m = i, X_{m+1} = j)$$
$$= \mathbb{P}_Z(\mathbf{b}_{m\uparrow}(\mathbf{X}) = \sigma_{m\uparrow} \mid X_m = i) q_{j\sigma_{m+1}} \mathbb{P}_Z(\mathbf{b}^{m\downarrow}(\mathbf{X}) = \sigma^{m\downarrow} \mid X_{m+1} = j).$$

Thus we obtain that

$$\mathbb{P}_Z(X_m = i) \mathbb{P}_Z(\mathbf{b}(\mathbf{X}) = \sigma \mid X_m = i, X_{m+1} = j) = \alpha_m(i) q_{j\sigma_{m+1}} \beta_{m+1}(j).$$

Summing up yields

$$\sum_{m=1}^{n-1} \widetilde{p}_{ij}(m,n) = \frac{\sum_{m=0}^{n-1} \alpha_m(i) p_{ij} q_{j\sigma_{m+1}} \beta_{m+1}(j)}{\mathbb{P}_Z(\mathbf{b}(\mathbf{X}) = \sigma)}. \tag{3.267}$$

Our next task is to provide a computationally efficient expression for p_{ij}^*. Combining the expression (3.266) with Lemma 3.8.2, one obtains the *smoothed transition probabilities*

$$\widehat{p}_{ij}^* = \frac{\sum_{m=1}^{n-1} \widetilde{p}_{i,j}(m,n)}{\sum_{m=1}^{n-1} \widetilde{p}_i(m,n)} = p_{ij} \frac{\sum_{l=0}^{n-1} \alpha_l(i) q_{j\sigma_{l+1}} \beta_{l+1}(i)}{\sum_{l=0}^{n-1} \alpha_l(i) \beta_l(i)}. \tag{3.268}$$

Next, the *smoothed initial probabilities* are given by

$$\widehat{\lambda}_j^* = \widetilde{p}_j(0,n) = \frac{\alpha_0(j) \beta_0(j)}{\mathbb{P}_Z(\mathbf{b}(\mathbf{X}) = \sigma)}. \tag{3.269}$$

Finally, for the *smoothed noise probabilities* we have the formula

$$\widehat{q}_{jk}^* = \frac{\sum_{m=1}^{n} \mathbf{1}(\sigma_m = k) \widetilde{p}_j(m,n)}{\sum_{m=1}^{n} \widetilde{p}_j(m,n)} = \frac{\sum_{m=1}^{n} \mathbf{1}(\sigma_m = k) \alpha_m(j) \beta_m(j)}{\sum_{m=1}^{n} \alpha_m(j) \beta_m(j)}. \tag{3.270}$$

Formulas (3.268)–(3.270) constitue the basis of the modern computational machinery extensively used in many applications; see, e.g. Rabiner & Juang, 1993. For an accessible introduction to HMMs in biology, see A. Krogh. "An introduction to hidden Markov models for biological sequences", in S.L, Salzberg, D.B. Searls and S. Kasif. *Computational Methods in Molecular Biology.* Amsterdam: Elsevier, 1999, pp. 45–63; see also Durbin *et al.*, 1998.

We want to repeat that (3.268)–(3.270) yield the point

$$\Phi(Z) = \widehat{Z}^* = (\widehat{\lambda}^*, \widehat{P}^*, \widehat{\mathbf{Q}}^*) \tag{3.271}$$

which depends on the variable Z (that is, on the initial choice of an attempted HMM). In this situation, the key step is to perform the N-fold iteration of the procedure (i.e. to consider the transformation Φ^N):

$$Z^{(N)} = \Phi(Z^{(N-1)}) = \Phi^N(Z^{(0)}), \quad N = 1, 2, \ldots, \tag{3.272}$$

and to identify the limit point(s) as $N \to \infty$. In a 'nice' situation one may hope that there exists a limit $Z^{(\infty)} = \lim_{N \to \infty} \Phi^N(Z^{(0)})$ which does not depend on the initial point $Z^{(0)}$ (or depends on $Z^{(0)}$ 'weakly', where $Z^{(\infty)}$ will vary only when we pass from one 'basin of attraction' to another). Assume in addition that $Z^{(\infty)}$ is a global maximiser for the likelihood $\mathbf{L}(\sigma; Z) = \mathbb{P}(\mathbf{b}(\mathbf{X}) = \sigma; Z)$; that is,

$$\mathbf{L}(\sigma; Z^{(\infty)}) = \max \left[\mathbf{L}(\sigma; Z) : Z = (\lambda, P, \mathbf{Q}) \in \mathscr{U} \right]. \tag{3.273}$$

Then we can treat the point $Z^{(\infty)}$ as an 'estimator' (in fact, as an MLE, Z^*_{ML}) yielding the 'best guess' of the HMM for the given training sequence σ.

Geometrically, the limit $Z^{(\infty)}$, if it exists, is represented by a fixed point of the map Φ. We see that the analysis of fixed points Z^* of Φ, and particularly the issue of convergence $Z^{(N)} = \Phi^N(Z) \to Z^*$, becomes a crucial question here. This issue is addressed in Section 3.9.

Before we proceed further, let us discuss the (straightforward) issue of maximising the expression (3.273) in the variables λ, P and $\mathbf{Q}$ constituting the model Z. The Lagrangian is

$$\mathbf{L}(\sigma; Z) + \gamma \left(\sum_{j=1}^s \lambda_j - 1 \right) + \sum_{i=1}^s v_i \left(\sum_{j=1}^s p_{ij} - 1 \right) + \sum_{j=1}^s l_j \left(\sum_{k=1}^\kappa q_{jk} - 1 \right),$$

where γ, v_i and l_j are Lagrange multipliers. The stationary point in the interior of the domain satisfies

$$\lambda_j > 0, \quad \sum_{j=1}^s \lambda_j = 1, \quad \frac{\partial}{\partial \lambda_j} \mathbf{L}(\sigma; Z) + \gamma = 0 \quad j = 1, \ldots, s,$$

$$p_{ij} > 0, \quad \sum_{j=1}^{s} p_{ij} = 1, \quad \frac{\partial}{\partial p_{ij}} \mathbf{L}(\sigma; Z) + v_i = 0, \quad i, j = 1, \dots, s$$

$$q_{ik} > 0, \quad \sum_{k=1}^{\kappa} q_{ik} - 1, \quad \frac{\partial}{\partial q_{jk}} \mathbf{L}(\upsilon, Z) + l_i = 0, \quad l = 1, \dots, s, \ k = 1, \dots, \kappa,$$

and is given by

$$\lambda_j = \left(\sum_{m=1}^{s} \lambda_m \frac{\partial}{\partial \lambda_m} \mathbf{L}(\sigma; Z) \right)^{-1} \lambda_j \frac{\partial}{\partial \lambda_j} \mathbf{L}(\sigma; Z), \qquad (3.274)$$

$$p_{ij} = \left(\sum_{m=1}^{s} p_{im} \frac{\partial}{\partial p_{im}} \mathbf{L}(\sigma; Z) \right)^{-1} p_{ij} \frac{\partial}{\partial p_{ij}} \mathbf{L}(\sigma; Z), \qquad (3.275)$$

and

$$q_{jk} = \left(\sum_{m=1}^{\kappa} q_{jm} \frac{\partial}{\partial q_{jm}} \mathbf{L}(\sigma; Z) \right)^{-1} q_{jk} \frac{\partial}{\partial q_{jk}} \mathbf{L}(\sigma; Z). \qquad (3.276)$$

These equations appeared before in (3.209), Example 3.7.3, and (3.226)–(3.229) in a general context (see also (3.236) and (3.244)). They form a coupled system of nonlinear equations, which in general cannot be solved analytically.

Since the maximisation can be done in each of these variables separately, we consider maximisation in variables p_{ij} only.

Lemma 3.8.4 *The following equation holds true:*

$$\frac{\partial}{\partial p_{ij}} \mathbf{L}(\sigma; Z) = \sum_{m=1}^{n-1} \alpha_m(i) q_{j\sigma_{m+1}} \beta_{m+1}(j) \qquad (3.277)$$

Proof Write

$$\mathbf{L}(\sigma; Z) = \mathbb{P}_Z(\mathbf{b}(\mathbf{X}) = \sigma; Z) = \sum_{j=1}^{s} \alpha_n(j) \beta_n(j);$$

as $\beta_n(j) = 1$, this factor will be omitted. Next, use (3.263) for $m = n - 1$:

$$\alpha_{m+1}(j) = \left(\sum_{i=1}^{s} \alpha_m(i) p_{ij} \right) q_{j\sigma_{m+1}}.$$

This gives

$$\mathbf{L}(\sigma; Z) = \sum_{i,j=1}^{s} \alpha_{n-1}(i) p_{ij} q_{j\sigma_n}. \qquad (3.278)$$

Hence, by direct differentiation,

$$\frac{\partial}{\partial p_{ij}} L(\sigma; Z) = \alpha_{n-1}(i) q_{j\sigma_n} + \sum_{l,m=1}^{s} \left[\frac{\partial}{\partial p_{ij}} \alpha_{n-1}(l) \right] p_{lm} q_{m\sigma_n}. \tag{3.279}$$

So we need only calculate the double sum in the RHS. We get

$$\frac{\partial}{\partial p_{ij}} \alpha_{n-1}(l) = \begin{cases} \alpha_{n-2}(i) q_{j\sigma_{n-1}} + \sum_{m=1}^{s} \left[\frac{\partial}{\partial p_{ij}} \alpha_{n-2}(m) \right] p_{mj} q_{j\sigma_{n-1}}, & \text{if } l = j, \\ \sum_{m=1}^{s} \left[\frac{\partial}{\partial p_{ij}} \alpha_{n-2}(m) \right] p_{mj} q_{j\sigma_{n-1}}, & \text{if } l \neq j. \end{cases}$$

Substituting the partial derivative $\partial \alpha_{n-1}(m) / \partial p_{ij}$ in the second term in the RHS of (3.279) gives

$$\sum_{r,l=1}^{s} \left[\frac{\partial}{\partial p_{ij}} \alpha_{n-1}(r) \right] p_{rl} q_{l\sigma_n}$$

$$= \sum_{r=1}^{s} \alpha_{n-2}(i) q_{j\sigma_{n-1}} p_{jr} q_{r\sigma_n} + \sum_{r=1}^{s} \left[\sum_{l=1}^{s} \frac{\partial}{\partial p_{ij}} \alpha_{n-2}(l) p_{lj} q_{j\sigma_{n-1}} \right] p_{jr} q_{r\sigma_n}$$

$$+ \sum_{r,m,l=1}^{s} \mathbf{1}(m \neq j) \left[\frac{\partial}{\partial p_{ij}} \alpha_{n-2}(l) p_{lm} q_{m\sigma_{n-1}} \right] p_{mr} q_{r\sigma_n}$$

$$= \mathrm{I} + \mathrm{II} + \mathrm{III}. \tag{3.280}$$

Next, calculate the value of each of the three terms in the RHS of (3.280). First,

$$\mathrm{I} = \sum_{r=1}^{s} \alpha_{n-2}(i) q_{j\sigma_{n-1}} p_{jr} q_{r\sigma_n} = \alpha_{n-2}(i) q_{j\sigma_{n-1}} \left(\sum_{r=1}^{s} p_{jr} q_{r\sigma_n} \right)$$

from which, by virtue of (3.265), we get

$$\mathrm{I} = \alpha_{n-2}(i) q_{j\sigma_{n-1}} \beta_{n-1}(j). \tag{3.281}$$

Further, combining terms II and III together and interchanging the order of summation yields

$$\mathrm{II} + \mathrm{III} = \sum_{r,l=1}^{s} \left[\frac{\partial}{\partial p_{ij}} \alpha_{n-2}(l) \right] p_{lr} q_{r\sigma_{n-1}} \left(\sum_{m=1}^{s} p_{rm} q_{m\sigma_n} \right). \tag{3.282}$$

The recursion relation (3.265) now gives

$$\sum_{m=1}^{s} p_{rm} q_{m\sigma_n} = \beta_{n-1}(r).$$

Hence, the RHS of (3.282) becomes

$$\sum_{r,l=1}^{s} \left[\frac{\partial}{\partial p_{ij}} \alpha_{n-2}(l) \right] p_{ir} q_{r\sigma_{n-1}} \beta_{n-1}(r).$$

So, (3.279) takes the form

$$\frac{\partial}{\partial p_{ij}} \mathbf{L}(\sigma; Z) = \alpha_{n-1}(i) q_{j\sigma_n} \beta_n(j) + \alpha_{n-2}(i) q_{j\sigma_{n-1}} \beta_{n-1}(j)$$

$$+ \sum_{r,l=1}^{s} \left[\frac{\partial}{\partial p_{ij}} \alpha_{n-2}(l) \right] p_{lr} q_{r\sigma_{n-1}} \beta_{n-1}(r). \qquad (3.283)$$

The double sum in the RHS of (3.283) can be subjected to a similar procedure, substituting the derivative $\partial \alpha_{n-2}(l)/\partial p_{ij}$ and reorganizing the resulting sums as above. Since $\alpha_r(0)$ does not involve any p_{ij}, differentiation stops at $l = 0$. Finally, we obtain

$$\frac{\partial}{\partial p_{ij}} \mathbf{L}(\sigma; Z) = \sum_{m=0}^{n-1} \alpha_m(i) q_{j\sigma_{m+1}} \beta_{m+1}(j),$$

as required. $\qquad \square$

Lemma 3.8.4 leads to (yet) another form of (3.226)–(3.228), (3.255), (3.258)–(3.259) and (3.268)–(3.270). Indeed, insert (3.277) into (3.275) and interchange the order of summation in the denominator. We obtain

$$\sum_{m=1}^{n-1} \alpha_m(i) \left[\sum_{j=1}^{s} p_{ij} q_{j\sigma_{m+1}} \beta_{m+1}(j) \right],$$

where backward recursion (3.265) gives

$$\sum_{j=1}^{s} p_{ij} q_{j\sigma_{m+1}} \beta_{m+1}(j) = \beta_m(i).$$

Hence we have the equality

$$\sum_{j=1}^{s} p_{ij} \frac{\partial}{\partial p_{ij}} \mathbf{L}(\sigma; Z) = \sum_{m=1}^{n-1} \alpha_m(i) \beta_m(i)$$

which immediately gives the RHS of (3.268).

Based on (3.274)–(3.276), we rewrite the Baum–Welch transformation Φ for the HMM filtration problem as

$$\Phi: \ (\lambda, P, \mathbf{Q}) \mapsto \left(\widehat{\lambda}^*, \widehat{P}^*, \widehat{\mathbf{Q}}^* \right), \qquad (3.284)$$

where

$$\widehat{\lambda}_j^* = \left(\sum_{m=1}^{s} \lambda_m \frac{\partial}{\partial \lambda_m} \mathbf{L}(\sigma;Z) \right)^{-1} \lambda_j \frac{\partial}{\partial \lambda_j} \mathbf{L}(\sigma;Z), \tag{3.285}$$

$$\widehat{p}_{ij}^* = \left(\sum_{m=1}^{s} p_{im} \frac{\partial}{\partial p_{im}} \mathbf{L}(\sigma;Z) \right)^{-1} p_{ij} \frac{\partial}{\partial p_{ij}} \mathbf{L}(\sigma;Z), \tag{3.286}$$

and

$$\widehat{q}_{jk}^* = \left(\sum_{m=1}^{\kappa} q_{jm} \frac{\partial}{\partial q_{jm}} \mathbf{L}(\sigma;Z) \right)^{-1} q_{jk} \frac{\partial}{\partial q_{jk}} \mathbf{L}(\sigma;Z). \tag{3.287}$$

It is important to stress that, by virtue of (3.268)–(3.270), the transformation (3.284) has an essential computational advantage: it is reduced to a superposition of *local* transformations. At each step one selects a position $l = 0, \ldots, n-1$ in the string and updates the local variables with index l only. See Rabiner and Juang, 1993. It results in a great economy of calculation, requiring ns^2 operation whereas a straightforward approach results in ns^n operations.

The analysis of convergence of iterations Φ^N of map Φ follows basic geometric ideas going back to the first half of the 20th Century. It is carried out in Section 3.9, and the outcome is stated as Theorem 3.9.9 (in the case of the HMM filtration problem). A similar result holds for the interpolation problem.

A considerable part of this section has been an illustration of the importance of mathematics aimed at enabling simulations on computers. We finish it with a story about computers.

Recently, the authors of this book came across an account by A. Samarskii, a prominent Russian applied mathematician and a full member of the Russian Academy of Sciences, about an early stage of parallel computations in the USSR. (See V. Gubarev, *Stalin's White Archipelago*, Moscow: 'Molodaya Gvardia' Publishing House, 2004 (in Russian).) In the late 1940s, the Soviets were busy constructing their own nuclear bomb, which required a lot of calculations. In particular, it was necessary to solve numerically systems of hundreds of linear equations per day. The Soviet computer industry of the period was limited to fleet of mechanical comptometers, but they still managed to perform calculations rapidly and reliably. The Soviet solution was quite elegant: Samarskii, who was the head of a computational group, presided over about 30 young female 'computors', recently graduated from a Moscow Institute of Geodesy. Each girl had to solve, on her personal comptometer, a dozen equations and to give her results to another girl, for comparison and further use, in accordance with a specially designed algorithm of parallel computation. By the time of their first test explosion (August, 1949), the

Soviets managed to numerically predict the results of the test with a margin of error of 30%, which, according to Samarskii, was better than the level of accuracy achieved by the Americans (who already used electronic computer prototypes).

A factor that might have contributed to Soviet efficiency of that period was that a failure to perform a correct calculation might be treated as an act of sabotage with dire consequences. A brilliant physicist or engineer toiling over radioactive material could easily be transformed into a prisoner sent to mine the very same material without any protection.

At the time, computors were already widely employed worldwide. For example, many British Universities had a post of a computor attached to a mathematics professor: the operator's duties were to do calculations following the professor's instructions, by assembling a set of specialised calculators suitable for a given problem.

3.9 Generalisations of the Baum–Welch algorithm. Global convergence of iterations

I felt like the old minstrel who has been singing his song
for 18 years and now finds, with considerable satisfaction,
that his folklore is the theme of an overpowering symphony.
H.O. Hartley (1912–1980) American statistician

As was determined above, the Baum–Welch transformation for the HMM filtration problem can been defined by the (equivalent) formulas (3.232) (3.271), (3.284), and for the interpolation problem in (3.237). For definiteness, we will talk about unrestricted problems only. In the case of the filtration problem, the characteristic feature of the transformation is that it takes a current model Z to a re-estimated, or updated, model $\widehat{Z}^*$, where $\widehat{Z}^* = \Phi(Z)$ lies in a domain

$$\mathcal{D}(Z) = \{\widehat{Z} \in \mathcal{U} : \mathbf{L}(\sigma; \widehat{Z}) \geq \mathbf{L}(\sigma; Z)\}; \tag{3.288}$$

cf. Theorem 3.7.12. Similarly, for the interpolation problem, $\widehat{P}^* = \Pi(P) \in \mathcal{D}(P)$ where

$$\mathcal{D}(P) = \{\widehat{P} \in \mathcal{P} : \mathbf{L}(\widehat{P}|\mathbf{X_T} = \mathbf{x_T}) \geq \mathbf{L}(P|\mathbf{X_T} = \mathbf{x_T})\}; \tag{3.289}$$

cf. Theorem 3.7.8.

The proof of convergence of Baum–Welch does not use its specific character. In particular, the structure of the Markov chain is used only in establishing monotonicity in (3.213) and (3.234). This allows us to cover the issue of convergence for a larger class formed by *expectation–modification* (EM) algorithms and their generalisations called *generalised expectation–modification* (GEM) algorithms.

The epigraph to this section is taken from X.-L. Meng, D. van Dyk, "The EM algorithm – an old folk-song sung to fast new time", *J. Roy. Stat. Soc. B*, **59** (1997), 511–567. It was followed by the following text: "Just as a folk-song typically evolves many years before its tune is well recognised, various EM-type methods or ideas ... can be found in the [early] literature. ... The folk-song analogy is also accurate in the sense that it signifies the collective effort in developing the EM algorithm. ... Baum et al. (1970) is perhaps the most sophisticated." By 1992, the number of EM-related publications exceeded 1,000 (by now it is much larger). Curiously, among the 300 journals where the above 1,000 papers were published, the leading place was occupied, unsurprisingly, by the *Journal of the American Statistical Association*, but the *Journal of the Royal Statistical Society*, Ser. B, was pipped into the fifth place by the *Journal of Dairy Science*.

The EM algorithm works as follows. We have two sample spaces $\mathscr{X}$ (inaccessible 'full data' samples) and $\mathscr{Y}$ (observed 'incomplete data' samples) and a many-to-one map Ψ from $\mathscr{X}$ to $\mathscr{Y}$ (an observation mechanism that is inaccurate). That is, instead of observing a sample vector $\mathbf{x} \in \mathscr{X}$, we observe an 'inaccurate' vector $\mathbf{y} = \Psi(\mathbf{x}) \in \mathscr{Y}$. This situation often occurs in statistics when data can be grouped, censored, truncated or missing. Let $f(\mathbf{x};\theta)(= f_{\mathbf{X}}(\mathbf{x};\theta))$, $\mathbf{x} \in \mathscr{X}$, be the PDF of the random sample vector $\mathbf{X}$ depending on a parameter $\theta \in \Theta$. Then the PDF $g(\mathbf{y};\theta)(= g_{\mathbf{Y}}(\mathbf{y};\theta))$ of the random vector $\mathbf{Y} = \Psi(\mathbf{X})$ is given by

$$g(\mathbf{y};\theta) = \int_{\mathscr{X}(\mathbf{y})} f(\mathbf{x};\theta)\,d\mathbf{x}, \quad \mathbf{y} \in \mathscr{Y}, \tag{3.290}$$

where $\mathscr{X}(\mathbf{y})$ is the inverse image $\{\mathbf{x} : \Psi(\mathbf{x}) = \mathbf{y}\}$. The parameter θ is unknown and should be estimated by the method of maximum likelihood, i.e. by maximizing $g(\mathbf{y};\theta)$, or, equivalently, $\ln g(\mathbf{y};\theta)$, over $\theta \in \Theta$. Since the sample $\mathbf{x}$ is unavailable, we replace the log-likelihood $\ln f(\mathbf{x};\theta)$ by its conditional expectation given $\mathbf{y}$. In the description of the algorithm, this is done for an arbitrary $\theta \in \Theta$, but in subsequent iterations, θ will be chosen to be $\theta^{(N)}$, the value obtained after the Nth iteration. See below.

Equation (3.290) is the starting point for the so-called E-step of the EM algorithm. To perform the iteration of the E-step, set

$$h(\mathbf{x}|\mathbf{y};\theta) = \frac{f(\mathbf{x};\theta)}{g(\mathbf{y};\theta)}, \quad \mathbf{x} \in \mathscr{X}, \, \mathbf{y} = \Psi(\mathbf{x}) \in \mathscr{Y}, \tag{3.291}$$

$h(\mathbf{x}|\mathbf{y};\theta)$ being the conditional density of $\mathbf{X}$ given that $\Psi(\mathbf{X}) = \mathbf{y}$. To maximise $\ln g(\mathbf{y};\theta)$, we write down the log-likelihood $\ell(\theta')(= \ell(\mathbf{y};\theta'))$, in the form

$$\ln g(\mathbf{y}|\theta') = \ln f(\mathbf{x}|\theta') - \ln h(\mathbf{x}|\mathbf{y},\theta'),$$

and take the conditional expectation $\mathbb{E}_\theta[\,\cdot\,|\Psi(\mathbf{X}) = \mathbf{y}]$. So, ℓ is considered as a function of the variable $\theta' \in \Theta$:

$$\ell(\theta') = \ln g(\mathbf{y}|\theta') = Q(\theta'|\theta) - H(\theta'|\theta). \tag{3.292}$$

Here $Q(\theta'|\theta)(= Q(\mathbf{y}, \theta'|\theta))$ and $H(\theta'|\theta)(= H(\mathbf{y}, \theta'|\theta))$ stand for the conditional expectations

$$
\begin{aligned}
Q(\theta'|\theta) &= \mathbb{E}_\theta\left[\ln f(\mathbf{X}|\theta')|\Psi(\mathbf{X}) = \mathbf{y}\right], \\
H(\theta'|\theta) &= \mathbb{E}_\theta\left[\ln h(\mathbf{X}|\mathbf{y}; \theta')|\Psi(\mathbf{X}) = \mathbf{y}\right],
\end{aligned}
\tag{3.293}
$$

which are assumed to exist for all pairs $\theta', \theta \in \Theta$.

We now define the Nth iteration of the EM algorithm as a map $A\colon \theta^{(N)} \in \Theta \mapsto \theta^{(N+1)} = A(\theta^{(N)}) \in \Theta$ as follows.

The E-step. Given $\theta^{(N)}$, determine $Q(\theta|\theta^{(N)})$.

The M-step. Choose $\theta^{(N+1)}$ to be a value which maximises the function $\theta \in \Theta \mapsto Q(\theta|\theta^{(N)})$:

$$\theta^{(N+1)} = \operatorname{argmax}\left[Q(\theta|\theta^{(N)})\colon \theta \in \Theta\right]. \tag{3.294}$$

The value $\theta^{(N+1)}$ depends on $\theta^{(N)}$ and the sample vector $\mathbf{y}$.

The hope here would be (and perhaps initially was) that the sequence of subsequent values $\theta^{(N)}, N = 0, 1, \ldots$ obtained by iterating the algorithm (it is often called an EM-sequence) will be 'nice'. Ideally, $\theta^{(N)}$ might converge (rather quickly) to θ^*, the MLE maximising likelihood $g(\mathbf{y}; \theta)$ in (3.290). Unfortunately, this is not always the case, and the large part of the forthcoming discussion aims to clarify this issue. In view of applications, we use in examples the alternative notation $L^{\text{full}}(\theta) = L^{\text{full}}(\mathbf{x}; \theta)$ for likelihood $f(\mathbf{x}; \theta)$ and $L^{\text{obs}}(\theta) = L^{\text{obs}}(\mathbf{y}; \theta)$ for $g(\mathbf{y}; \theta)$, with $\ell(\theta') = L^{\text{obs}}(\theta')$. One good bit of news here is that in the course of the iterations, the value $L^{\text{obs}}(\theta^{(N+1)})$ is no less than $L^{\text{obs}}(\theta^{(N)})$:

$$L^{\text{obs}}(\theta^{(N+1)}) \geq L^{\text{obs}}(\theta^{(N)}). \tag{3.295}$$

We see that to make M a point-to-point map, we have to specify $\theta^{(N+1)}$ among maximisers. This choice may influence various aspects of the implementation of the EM algorithm, both theoretical and practical. It is also clear that the choice of an initial value $\theta^{(0)}$ has to be judicial.

Example 3.9.1 In this example, ϕ stands for the standard normal PDF $N(0, 1)$:

$$\phi(x) = \frac{1}{\sqrt{2\pi}}e^{-x^2/2}, \; x \in \mathbb{R}. \tag{3.296}$$

We observe an IID random sample $\mathbf{Y} = \begin{pmatrix} Y_1 \\ \vdots \\ Y_n \end{pmatrix}$ from a distribution with PDF

$$g(y;\theta) = \frac{1}{2}(\phi(y) + \phi(y-\theta)), y \in \mathbb{R}, \tag{3.297}$$

representing an equal mixture of a standard normal PDF and a normal PDF with mean $\theta \in \mathbb{R}$ and unit variance. In this example, the complete data $\mathbf{x}$ consist of pairs $x_1 = (y_1, \alpha_1), \ldots, x_n = (y_n, \alpha_n)$ where $y_j \in \mathbb{R}$ and $\alpha_j = 0$ or 1, $j = 1, \ldots, n$; α_j specifies by which PDF, $\phi(x)$ or $\phi(x-\theta)$, the jth observation point y_j is generated. The function $\Psi(\mathbf{x}) = \mathbf{y}$ erases the α_js and leaves only points $y_1, \ldots, y_n$. The same is applicable to random samples $\mathbf{Y}$ and $\mathbf{X}$.

In this example, the log-likelihood $\ln L^{\text{obs}}(\theta)$ is given by

$$\ln L^{\text{obs}}(\theta) = \sum_{i=1}^{n} \ln\left(\frac{1}{2}\phi(y_i) + \frac{1}{2}\phi(y_i - \theta)\right), \tag{3.298}$$

and the log-likelihood $\ln L^{\text{full}}(\theta)$ by

$$\ln L^{\text{full}}(\theta) = \sum_{i=1}^{n} \mathbf{1}(\alpha_i = 0)\ln\phi(y_i) + \mathbf{1}(\alpha_i = 1)\ln\phi(y_i - \theta). \tag{3.299}$$

The above description of the EM algorithm in this case leads to the following formula:

$$Q(\theta', \theta) = \sum_{i=1}^{n} (1 - w_i(\theta))\ln\phi(y_i) + w_i(\theta)\ln\phi(y_i - \theta') + c \tag{3.300}$$

where $w_i(\theta) = \dfrac{\phi(y_i - \theta)}{\phi(y_i) + \phi(y_i - \theta)}$ for $i = 1, \ldots, n$, and c is a constant not depending on θ and θ'. Clearly, $Q(\theta', \theta)$ is a quadratic form in θ' and can be easily maximised.

Then, by the above description of the M-step, the value $\theta^{(N+1)}$ obtained after the $(N+1)$st iteration of the algorithm, is given by

$$\theta^{(N+1)} = \theta^{(N+1)}(\theta^{(N)}, \mathbf{y}) = \frac{\sum\limits_{i=1}^{n} y_i w_i(\theta^{(N)})}{\sum\limits_{i=1}^{n} w_i(\theta^{(N)})}. \tag{3.301}$$

Example 3.9.2 (Bivariate normal data with missing values) Let $\mathbf{X} = \begin{pmatrix} X_1 \\ X_2 \end{pmatrix}$ be a random vector with a bivariate normal distribution $N(\mu, \Sigma)$ where μ is a two-dimensional real vector $\begin{pmatrix} \mu_1 \\ \mu_2 \end{pmatrix}$ of mean values $\mu_i = \mathbb{E}X_i$ and Σ a positive-definite

2×2 real covariance matrix $\begin{pmatrix} \sigma_{11} & \sigma_{12} \\ \sigma_{21} & \sigma_{22} \end{pmatrix}$, with $\sigma_{ii} = \text{Var} X_i$ and $\sigma_{21} = \sigma_{12} = \text{Cov}(X_1, X_2)$, $i = 1, 2$. Suppose we want to estimate the collection of parameters $\theta = \{\mu_i, \sigma_{ij}, i, j = 1, 2\}$ from a random sample of size n taken from independent copies of $\mathbf{X}$, where the data on the first variate, X_1, are missing in m_1 places and the data on the second variate, X_2, are missing in m_2 places, where $m_1 + m_2 \leq n$ and positions missing entry X_1 are disjoint from those missing X_2. In this example, $\Theta \subset \mathbb{R}^5$; in fact Θ lies in $\mathbb{R} \times \mathbb{R} \times \mathbb{R}_+ \times \mathbb{R}_+ \times \mathbb{R}$. (The first two Cartesian factors $\mathbb{R}$ provide space for μ_1 and μ_2, the two copies of $\mathbb{R}_+$ do so for σ_{11} and σ_{22} and the last factor $\mathbb{R}$ for σ_{12}; the Cauchy–Schwarz inequality $|\sigma_{12}^2| \leq \sigma_{11}\sigma_{22}$ indicates that Θ is indeed a proper subset of $\mathbb{R} \times \mathbb{R} \times \mathbb{R}_+ \times \mathbb{R}_+ \times \mathbb{R}$.)

The observed data array is denoted by $\underline{y}$. For definiteness, we label the data so that (i) $\mathbf{y}^{(j)} = \begin{pmatrix} y_1^{(j)} \\ y_2^{(j)} \end{pmatrix}$, $j = 1, \ldots, m$, stand for the fully observed data points, where $m = n - m_1 - m_2$, (ii) $\begin{pmatrix} * \\ y_2^{(j)} \end{pmatrix}$, $j = m+1, \ldots, m+m_1$, denote the m_1 observation with the first component missing, and $\begin{pmatrix} y_1^{(j)} \\ * \end{pmatrix}$, $j = m+m_1+1, \ldots, n$, the m_2 observations with the the second component missing. Then $\underline{y}$ is associated with the array

$$\left\{ \begin{pmatrix} y_1^{(1)} \\ y_2^{(1)} \end{pmatrix} \cdots \begin{pmatrix} y_1^{(m)} \\ y_2^{(m)} \end{pmatrix} \begin{pmatrix} * \\ y_2^{(m+1)} \end{pmatrix} \cdots \right.$$
$$\left. \begin{pmatrix} * \\ y_2^{(m+m_1)} \end{pmatrix} \begin{pmatrix} y_1^{(m+m_1+1)} \\ * \end{pmatrix} \cdots \begin{pmatrix} y_1^{(N)} \\ * \end{pmatrix} \right\}. \qquad (3.302)$$

The full data of course would be represented by the array of vectors $\mathbf{y}^{(j)} = \begin{pmatrix} y_1^{(j)} \\ y_2^{(j)} \end{pmatrix}$, $j = 1, \ldots, n$:

$$\underline{x} = \left\{ \begin{pmatrix} y_1^{(1)} \\ y_2^{(1)} \end{pmatrix} \cdots \begin{pmatrix} y_1^{(n)} \\ y_2^{(n)} \end{pmatrix} \right\}. \qquad (3.303)$$

We use the parallel notation $\underline{Y}$ and $\underline{X}$ for random samples and identify $\underline{Y}$ as a part of $\underline{X}$ as has been indicated.

The log-likelihood $\ln L^{\text{obs}}(\theta) = \ln L^{\text{obs}}(\underline{y}; \theta)$ for θ based on the observed data $\underline{y}$ is

$$
\begin{aligned}
\ln L^{\text{obs}}(\theta) \\
= -n\ln(2\pi) - \frac{1}{2}m\ln(\det\Sigma) \\
-\frac{1}{2}\sum_{j=1}^{m}(\mathbf{y}^{(j)} - \mu)^{\mathrm{T}}\Sigma^{-1}(\mathbf{y}^{(j)} - \mu) - \frac{1}{2}\sum_{i=1}^{2}m_i\ln\sigma_{ii} \\
-\frac{1}{2}\left[\sigma_{11}^{-1}\sum_{j=m+m_1+1}^{n}(y_1^{(j)} - \mu_1)^2 + \sigma_{22}^{-1}\sum_{j=m+1}^{m+m_1}(y_2^{(j)} - \mu_2)^2\right].
\end{aligned}
\tag{3.304}
$$

The full data log-likelihood $\ln L^{\text{full}}(\theta) = \ln L^{\text{full}}(\mathbf{x}; \theta)$ equals

$$
\begin{aligned}
\ln L^{\text{full}}(\theta) &= -n\ln(2\pi) - \frac{1}{2}n\ln(\det\Sigma) - \frac{1}{2}\sum_{j=1}^{n}(\mathbf{y}^{(j)} - \mu)^{\mathrm{T}}\Sigma^{-1}(\mathbf{y}^{(j)} - \mu) \\
&= -n\ln(2\pi) - \frac{1}{2}n\ln(\sigma_{11}\sigma_{22} - \sigma_{12}^2) \\
&\quad -\frac{1}{2}(\sigma_{11}\sigma_{22} - \sigma_{12}^2)^{-1}\left[\sigma_{22}S_{11} + \sigma_{11}S_{22} - 2\sigma_{12}S_{12}\right. \\
&\quad -2T_1(\mu_1\sigma_{22} - \mu_2\sigma_{12}) + 2T_2(\mu_2\sigma_{11} - \mu_1\sigma_{12}) \\
&\quad \left.+n(\mu_1^2\sigma_{22} + \mu_2^2\sigma_{11} - 2\mu_1\mu_2\sigma_{12})\right],
\end{aligned}
\tag{3.305}
$$

where

$$
T_i = \sum_{j=1}^{n}y_i^{(j)}, \quad S_{il} = \sum_{j=1}^{n}y_i^{(j)}y_l^{(j)}, \quad i,l = 1,2.
\tag{3.306}
$$

We see that the likelihood $L^{\text{full}}(\theta)$ belongs to the exponential family, with a sufficient statistic formed by the collection $\{T_1, T_2, S_{11}, S_{12}, S_{22}\}$. Had the full-data array $\mathbf{x}$ been available, the (full-data) maximum likelihood estimator $\hat{\theta}$ of parameter θ would be given by

$$
\hat{\mu}_i = T_i/n, \quad \hat{\sigma}_{il} = (S_{il} - n^{-1}T_iT_l)/n, \quad i,l = 1,2.
\tag{3.307}
$$

This observation suggests the form of the EM algorithm in the current example. Again we assume that the value $\theta^{(N)} = \{\mu_i^{(N)}, \sigma_{ij}^{(N)}, i,j = 1,2\}$ was obtained at the Nth iteration and denote by $\mathbb{E}_{\theta^{(N)}}$ the expectation relative to the bivariate normal distribution determined by $\theta^{(N)}$. The E-step on the $(N+1)$th iteration of the algorithm is as follows. We need to calculate the conditional expectation

$$
Q(\theta, \theta^{(N)}) = \mathbb{E}_{\theta^{(N)}}\left[\ln L^{\text{full}}(\theta)|\underline{y}\right]
$$

of the full-data log-likelihood $\ln L^{full}(\underline{\mathbf{X}}; \theta)$, given that $\underline{\mathbf{Y}} = \underline{\mathbf{y}}$. It can be seen from (3.304) that this is reduced to computing the conditional expected values

$$\mathbb{E}_{\theta^{(N)}}\left[Y_1^{(j)}|\underline{\mathbf{y}}\right] \text{ and } \mathbb{E}_{\theta^{(N)}}\left[(Y_1^{(j)})^2|\underline{\mathbf{y}}\right], \quad j = m+1, \ldots, m+m_1,$$

and

$$\mathbb{E}_{\theta^{(N)}}\left[Y_2^{(j)}|\underline{\mathbf{y}}\right] \text{ and } \mathbb{E}_{\theta^{(N)}}\left[(Y_2^{(j)})^2|\underline{\mathbf{y}}\right], \quad j = m+m_1+1, \ldots, n.$$

By independence of the sample, in the first line the condition $\underline{\mathbf{Y}} = \underline{\mathbf{y}}$ can be replaced with $Y_2^{(j)} = y_2^{(j)}$ and in the second line with $Y_1^{(j)} = y_1^{(j)}$. Correspondingly, one can use the notation

$$\mathbb{E}_{\theta^{(N)}}\left[Y_1^{(j)}|y_2^{(j)}\right] \text{ and } \mathbb{E}_{\theta^{(N)}}\left[(Y_1^{(j)})^2|y_2^{(j)}\right], \quad j = m+1, \ldots, m+m_1, \quad (3.308)$$

and its mirror versions

$$\mathbb{E}_{\theta^{(N)}}\left[Y_2^{(j)}|y_1^{(j)}\right] \text{ and } \mathbb{E}_{\theta^{(N)}}\left[(Y_2^{(j)})^2|y_1^{(j)}\right], \quad j = m+m_1+1, \ldots, n. \quad (3.309)$$

Next, if a vector $\begin{pmatrix} Y_1 \\ Y_2 \end{pmatrix}$ has a bivariate normal distribution $\mathrm{N}(\mu, \Sigma)$, then the distribution of Y_2 conditional on $Y_1 = y_1$ is normal $\mathrm{N}(\mu_2^*, \sigma_{22}^*)$, with the mean

$$\mu_2^* = \mu_2 + \sigma_{12}\sigma_{11}^{-1}(y_1 - \mu_1)$$

and the variance

$$\sigma_{22}^* = \sigma_{22}\left(1 - \frac{\sigma_{12}^2}{\sigma_{11}\sigma_{22}}\right).$$

Thus, the above expected values (3.308) take the form

$$\mathbb{E}_{\theta^{(N)}}\left[Y_2^{(j)}|y_1^{(j)}\right] = \mu_2^{(N)} + \frac{\sigma_{12}^{(N)}}{\sigma_{11}^{(N)}}\left(y_1^{(j)} - \mu_1^{(N)}\right),$$

and

$$\mathbb{E}_{\theta^{(N)}}\left[(Y_2^{(j)})^2|Y_1^{(j)} = y_1^{(j)}\right] = \left(z^{(j)}(N)\right)^2 + \sigma_{22}^*(N).$$

Here

$$z^{(j)}(N) = \mu_2^{(N)} + \frac{\sigma_{12}^{(N)}}{\sigma_{11}^{(N)}}\left(y_1^{(j)} - \mu_1^{(N)}\right)$$

and

$$\sigma_{22}^*(N) = \sigma_{22}^{(N)}\left(1 - \frac{(\sigma_{12}^{(N)})^2}{\sigma_{11}^{(N)}\sigma_{22}^{(N)}}\right)$$

are values calculated at the Nth iteration (cf. (3.310)). The formulas for $\mathbb{E}_{\theta^{(N)}}\left[Y_2^{(j)}|y_1^{(j)}\right]$ and $\mathbb{E}_{\theta^{(N)}}\left[(Y_2^{(j)})^2|y_1^{(j)}\right]$ in (3.309) are obtained by interchanging the subscripts 1 and 2.

The M-step at the $(N+1)$th iteration is implemented simply by replacing the statistics T_i and S_{il} by $T_i^{(N)}$ and $S_{il}^{(N)}$, respectively, where the latter are defined by substituting, into (3.305), the missing values $y_i^{(j)}$ and $(y_i^{(j)})^2$, $i=1,2$ with their current conditional expectations (3.308) and (3.309). Accordingly, the value $\theta^{(N+1)} = \{\mu_i^{(N+1)}, \sigma_{ij}^{(N+1)}, i,j=1,2\}$ is given by

$$\mu_i^{(N+1)} = T_i^{(N)}/n, \ \ \sigma_{il}^{(N+1)} = \left(S_{il}^{(N)} - n^{-1}T_i^{(N)}T_l^{(N)}\right)/n. \tag{3.310}$$

Example 3.9.3 (Parameter estimation for exponential families) In this example, $\Theta = \mathbb{R}^n$. Recall, an exponential family of PDFs $f(\mathbf{x};\theta)$, $\theta = \begin{pmatrix} \theta_1 \\ \vdots \\ \theta_n \end{pmatrix} \in \mathbb{R}^n$, is

given by

$$f(\mathbf{x};\theta) = \exp\left[(\mathrm{grad}_\theta B(\theta))^{\mathrm{T}}[C(\mathbf{x}) - \theta] + B(\theta) + H(\mathbf{x})\right] \tag{3.311}$$

where

$$(\mathrm{grad}_\theta B(\theta))^{\mathrm{T}}[C(\mathbf{x}) - \theta] = \sum_{j=1}^n \frac{\partial}{\partial\theta_j}B(\theta)[C_l(\mathbf{x}) - \theta_j].$$

The vector-function $C(\mathbf{x})$ is a sufficient statistic. Exponential families include many popular examples: multivariate normal, Poisson, multinomial, hypergeometric (see Volume 1, Section 3.6).

Assume that a function $\Psi: \mathbf{x} \to \mathbf{y}$ is given, and we have access to the sample $\mathbf{y} = \Psi(\mathbf{x})$. The $(N+1)$st iteration of the EM algorithm works here as follows.

The E-step: write down the function

$$\begin{aligned} Q(\mathbf{y};\theta|\theta^{(N)}) &= \mathbb{E}_{\theta^{(N)}}\left(H(\mathbf{X})|\Psi(\mathbf{X}) = \mathbf{y}\right) + (\mathrm{grad}_\theta B(\theta))^{\mathrm{T}}\widetilde{C}(\mathbf{y}) \\ &\quad - (\mathrm{grad}_\theta B(\theta))^{\mathrm{T}}\theta + B(\theta). \end{aligned} \tag{3.312}$$

Here

$$\widetilde{C}(\mathbf{y})_l = \mathbb{E}_{\theta^{(N)}}\left[C(\mathbf{X})_l|\Psi(\mathbf{X}) = \mathbf{y}\right]. \tag{3.313}$$

Observe that the term in the first line of the RHS in (3.312) does not depend on θ and hence does not take part in maximisation in θ. The third and the fourth terms, in contrast, depend on θ but not on $\theta^{(N)}$. It is the term $(\mathrm{grad}_\theta B(\theta))^{\mathrm{T}}\widetilde{C}(\mathbf{y})$ that depends on both θ and $\theta^{(N)}$, where $\widetilde{C}(\mathbf{y})$ is defined in (3.313).

The M-step: given the value of the parameter $\theta^{(N)}$, you aim to find the maximum of $Q(\mathbf{y};\theta,\theta^{(N)})$ in θ (for fixed $\mathbf{y}$). When found, the maximiser is identified as $\theta^{(N+1)}$ $(=\theta^{(N+1)}(\mathbf{y}))$. Unfortunately, a straighforward maximisation, as, e.g., in Example 3.9.2, is rarely possible.

As was said above, the question of convergence of values $\theta^{(N)}$ obtained in the course of iterations is delicate and requres a detailed analysis. A relatively simple model case is where the complete data $\mathbf{X} = \begin{pmatrix} \mathbf{X}_1 \\ \vdots \\ \mathbf{X}_n \end{pmatrix}$ and incomplete data $\mathbf{Y} = \begin{pmatrix} \mathbf{Y}_1 \\ \vdots \\ \mathbf{Y}_n \end{pmatrix}$ coincide and are formed by IID vectors $\mathbf{X}_j = \begin{pmatrix} X_j^{(1)} \\ \vdots \\ X_j^{(d)} \end{pmatrix}, j = 1,\ldots,n,$ which are multivariate normal vectors with a known $d \times d$ covariance matrix Σ and an unknown mean vector $\theta = \mu \in \mathbb{R}^d$. Here, the joint PDFs $f_{\mathbf{X}}(\,\cdot\,;\theta)$ and $g(\,\cdot\,;;\theta)$ coincide and are given by the expression

$$\left(\frac{1}{\sqrt{(2\pi)^n \det \Sigma}}\right)^n \exp\left[-\frac{1}{2}\sum_{j=1}^n (\mathbf{x}_j - \mu)^{\mathsf{T}} \Sigma^{-1}(\mathbf{x}_j - \mu)\right], \tag{3.314}$$
$$\mathbf{x} \in \mathbb{R}^d, \ \theta = \mu \in \mathbb{R}^d.$$

In this case the log-likelihood $\ln L^{\mathrm{obs}}(\theta) = \ln L^{\mathrm{full}}(\theta)$ is a negative quadratic function of μ (with coefficients depending on sample $\mathbf{x}$):

$$\mu \in \mathbb{R}^n \mapsto -\frac{1}{2}\sum_{j=1}^n (\mathbf{x}_j - \mu)^{\mathsf{T}} \Sigma^{-1}(\mathbf{x}_j - \mu).$$

This function is concave and has a unique maximum, giving the MLE

$$\widehat{\mu}^* = \frac{1}{n}\sum_{i=1}^n \mathbf{x}_j.$$

The value $\widehat{\mu}^*$ can also be obtained as the limit in various approximations, which provides good practice for the EM algorithm implementation. A standard approximation technique is the steepest descent method (used to minimise the convex quadratic function). But to estimate the rate of convergence one needs the following inequality.

Worked Example 3.9.4 (Kantorovich's inequality) Let Σ be a positive definite real $n \times n$ matrix. For any vector $\mathbf{x} = \begin{pmatrix} x_1 \\ \vdots \\ x_n \end{pmatrix} \in \mathbb{R}^n$ the following bound holds:

$$\frac{\|\mathbf{x}\|^4}{(\mathbf{x}^{\mathrm{T}} \Sigma \mathbf{x})(\mathbf{x}^{\mathrm{T}} \Sigma^{-1} \mathbf{x})} \geq \frac{4\mu_- \mu_+}{(\mu_- + \mu_+)^2} \tag{3.315}$$

where μ_- and μ_+ are, respectively, the smallest and largest eigenvalues of Σ.

Solution Let the eigenvalues μ_i of Σ satisfy

$$0 < \mu_- = \mu_1 \leq \cdots \leq \mu_n = \mu_+.$$

By an appropriate change of coordinates the matrix Σ becomes diagonal, and

$$\frac{\left(\sum_{i=1}^n x_i^2\right)^2}{\left(\sum_{i=1}^n \mu_i x_i^2\right)\left(\sum_{i=1}^n x_i^2/\mu_i\right)} = \frac{1/\sum_{i=1}^n \xi_i \mu_i}{\sum_{i=1}^n \xi_i/\mu_i} := \frac{\phi(\xi)}{\psi(\xi)}$$

where $\xi_i = x_i^2/\sum_{i=1}^n x_i^2$. The function $y \mapsto 1/y$ is convex for $y > 0$, the point $\phi(\xi)$ lies on the curve and the point $\psi(\xi)$ is the linear combination of the points on the curve. Hence the minimal value of the ratio is achieved for some $\mu = \xi_1 \mu_1 + \xi_n \mu_n$ with $\xi_1 + \xi_n = 1$. In this case, $\xi_1/\mu_1 + \xi_2/\mu_n = (\mu_1 + \mu_n - \mu)/\mu_1 \mu_n$, and one obtains

$$\frac{\phi(\xi)}{\psi(\xi)} \geq \inf_{\mu_1 \leq \mu \leq \mu_n} \frac{1/y}{(\mu_1 + \mu_n - \mu)/\mu_1 \mu_n} = \frac{4\mu_1 \mu_n}{(\mu_1 + \mu_n)^2}$$

as the mimimum is achieved at the point $\mu = (\mu_1 + \mu_n)/2$. $\qquad\square$

Worked Example 3.9.5 (Steepest descent for quadratic functions.) Given a positive definite real $n \times n$ matrix Σ and vectors $\mathbf{b}, \mathbf{x}_0 \in \mathbb{R}^n$, set

$$\mathbf{g}_k = \Sigma \mathbf{x}_k - \mathbf{b} \quad \text{and} \quad \mathbf{x}_{k+1} = \mathbf{x}_k - \frac{\mathbf{g}_k^{\mathrm{T}} \mathbf{g}_k}{\mathbf{g}_k^{\mathrm{T}} \Sigma \mathbf{g}_k} \mathbf{g}_k, \quad k = 0, 1, \dots. \tag{3.316}$$

Then, for all $\mathbf{x}_0 \in \mathbb{R}^n$, $\mathbf{x}_k$ converges, as $k \to \infty$, to the unique minimum point $\mathbf{x}_*$ of the function

$$f(\mathbf{x}) = \frac{1}{2} \mathbf{x}^{\mathrm{T}} \Sigma \mathbf{x} - \mathbf{x}^{\mathrm{T}} \mathbf{b}.$$

Furthermore, with $D(\mathbf{x}) = \frac{1}{2}(\mathbf{x} - \mathbf{x}^*)^{\mathrm{T}} \Sigma (\mathbf{x} - \mathbf{x}^*)$, we have the following bound: for all k,

$$D(\mathbf{x}_{k+1}) \leq \left(\frac{\mu_+ - \mu_-}{\mu_+ + \mu_-}\right)^2 D(\mathbf{x}_k),$$

where μ_- and μ_+, as before, are the minimal and the maximal eigenvalues of matrix Σ.

Solution (Sketch) Apply Kantorovich's inequality to

$$\frac{D(\mathbf{x}_k) - D(\mathbf{x}_{k+1})}{D(\mathbf{x}_k)} = \frac{(g_k^T g_k)^2}{(g_k^T \Sigma g_k)(g_k^T \Sigma^{-1}) g_k}.$$

$\square$

Quite often it is not numerically feasible to perform the maximisation procedure in the M-step. We already spotted it in Example 3.9.3; in the case of Markov chains this is true more often than not. To this end the *generalised expectation-modification* (GEM) algorithm has been proposed. In the GEM algorithm, we simply choose $\theta^{(N+1)}$ in such a way that

$$Q(\theta^{(N+1)}|\theta^{(N)}) \geq Q(\theta^{(N)}|\theta^{(N)}). \tag{3.317}$$

Because the expectation $H(\theta|\theta)$ obeys, by Gibbs' inequality,

$$H(\theta|\theta) \geq H(\theta'|\theta), \quad \theta, \theta' \in \Theta,$$

bound (3.296) leads to monotonicity property (3.295) (which is the crucial feature of the EM and GEM algorithms).

That is, in the GEM algorithm we are looking at the inequality

$$Q(\theta'|\theta) \geq Q(\theta|\theta), \quad \theta, \theta' \in \Theta. \tag{3.318}$$

Given $\theta \in \Theta$, this determines the set

$$M(\theta) = \{\theta' \in \Theta : Q(\theta'|\theta) \geq Q(\theta|\theta)\}, \tag{3.319}$$

and we are forced to think in terms of a point-to-set map

$$\theta \in \Theta \mapsto M(\theta),$$

guaranteeing that

$$\theta^{(N+1)} \in M(\theta^{(N)}), \quad n = 0, 1, \dots.$$

A sequence $\theta^{(N)}$ with the last property is called a GEM-sequence.

Example 3.9.6 Here we give an example of a GEM sequence $\{\theta^{(N)}\}$ for which $L(\theta^{(N)})$ converges monotonically, whereas the sequence $\{\theta^{(N)}\}$ does not converge but has a unit circle as the set of limit points. Consider the bivariate normal PDF, with an unknown mean $\mu = \begin{pmatrix} \mu_1 \\ \mu_2 \end{pmatrix}$ and the unit covariance 2×2 matrix. In this example, the unknown parameter $\theta = \mu$, and the complete and incomplete data coincide: $\mathbf{x} = \mathbf{y} = \begin{pmatrix} y_1 \\ y_2 \end{pmatrix}$.

The GEM sequence $\{\mu^{(N)}\}$, where $\mu^{(N)} = \begin{pmatrix} \mu_1^{(N)} \\ \mu_2^{(N)} \end{pmatrix}$, is given by

$$\mu_1^{(N)} = y_1 + r^{(N)} \cos \vartheta^{(N)},$$
$$\mu_2^{(N)} = y_2 + r^{(N)} \sin \vartheta^{(N)},$$

where $r^{(0)} = 2$, $\vartheta^{(0)} = 0$, and

$$r^{(N)} = 1 + (N+1)^{-1}, \quad \vartheta^{(N)} = \sum_{i=1}^{N} (i+1)^{-1}, \quad k = 1, 2, \ldots.$$

Here, in plain words, r and ϑ are the polar coordinates centred at the observed vector $\mathbf{y}$. For the log-likelihood $\ln L(\mu^{(N)}) = \ln L(\mathbf{y}|\mu^{(N)})$ we have that

$$\ln L(\mu^{(N+1)}) - \ln L(\mu^{(N)}) = \frac{1}{2}\left((r^{(N)})^2 - (r^{(N+1)})^2\right)$$

$$= \frac{1}{2}[(r^{(N)})^2 - (2 - (r^{(N)})^{-1})^2],$$

since $r^{(N+1)} = 2 - (r^{(N)})^{-1}$. Now we use the elementary bound $0 < 2 - u^{-1} \le u$, for $u \ge 1$. As $r^{(N)} \ge 1$ for each k, we obtain that

$$\ln L(\mu^{(N+1)}) - \ln L(\mu^{(N)}) \ge 0.$$

Hence, the sequence $\mu^{(N)}$ is indeed GEM.

Since $r^{(N)} \to 1$ as $N \to \infty$, the sequence of likelihood values $\{L(\mu^{(N)})\}$ converges to the value

$$(2\pi)^{-1} e^{-1}.$$

But for the sequence $\{\mu^{(N)}\}$, any point of the circle of unit radius and centred at $\mathbf{y}$ is a limiting point.

We now pass to a general set-up, motivated by the background from above, to which we will periodically refer. It is convenient to work with a transformation M taking points to sets (a point-to-set map). In general, such a map will send a point $\theta \in \Theta$ to a subset $M(\theta) \subset \Theta$ where Θ is a given domain in a Euclidean space $\mathbb{R}^n$.

$$M: \theta \in \Theta \mapsto M(\theta) \subset \Theta. \tag{3.320}$$

The examples we will have in mind emerge from (3.284) and (3.245). In these examples $\theta = Z = (\lambda, P, \mathbf{Q})$ for the HMM filtration problem and $\theta = P$ for the

HMM interpolation problem. Here, the map M is actually point-to-point (i.e., one-to-one), and coincides with the map Φ in the filtration problem and Π in the interpolation problem.

From now on we will assume that $M(\theta)$ is a compact set for all $\theta \in \Theta$. This will cover the above-mentioned case of transformations Φ and Π. In fact, in the unrestricted filtration HMM problem, the transformation Φ acts on the set $\mho = \mathcal{U}$, while in the unrestricted interpolation HMM problem, the transformation Π acts on $\Theta = \mathcal{P}$, where both $\mathcal{U}$ and $\mathcal{P}$ are compact sets in the corresponding Euclidean spaces (see. (3.215) and (3.5)–(3.7)).

Definition 3.9.7 We say that M is *closed* at point $\theta \in \Theta$ if the convergence $\theta(k) \to \theta$, where $\theta(k) \in \Theta$, and $v(k) \to v$, and where $v(k) \in M(\theta(k))$ implies that $v \in M(\theta)$. If the map M is closed at each point $\theta \in \Theta$, we say M is *closed over* Θ.

For the rest of this section, we assume that all maps under consideration are closed over Θ. Clearly, if $M : \Theta \to \Theta$ is a point-to-point map, then M is closed at θ when M is continuous at the point θ. In the general case of a point-to-set map M, we will continue speaking of an *algorithm A* generating a sequence of points $\theta^{(N+1)} = A(\theta^{(N)})$, with $\theta^{(N+1)} \in M(\theta^{(N)})$, $N \geq 0$, starting from an initial point $\theta^{(0)} \in \Theta$. Such an algorithm is merely given by a point-to-point map which we again denote by A, specifying a unique choice of the point $A(\theta)$ from $M(\theta)$.

Definition 3.9.8 A function $F : \Theta \to \mathbb{R}$ is called an *ascent* (or *Lyapunov*) function for a closed point-to-set map M, with a *solution set* $\Gamma \subset \Theta$, if:

(1) F is continuous and bounded on Θ;
(2) $F(\widehat{\theta}) \geq F(\theta)$ for all $\theta \in \Theta$ and $\widehat{\theta} \in M(\theta)$;
(3) $F(\widehat{\theta}) = F(\theta)$ for some $\widehat{\theta} \in M(\theta)$ then $\theta \in \Gamma$.

An example of an ascent function arises in the context of unrestricted HMM problems. As was said above, for the filtration problem the set $\Theta = \mathcal{U}$, map M, coincides with the (point-to-point) Baum–Welch transformation $Z \in \mathcal{U} \mapsto \Phi(Z)$ (see (3.284)) and the solution set Γ coincides with $\mathcal{F}_\Phi$, the set of fixed points of transformation Φ:

$$\mathcal{F}_\Phi = \{Z \in \mathcal{U} : \Phi(Z) = Z\}. \tag{3.321}$$

In this case, we can set

$$F(Z) = \mathbf{L}(\sigma; Z), \quad Z = (\lambda, P, Q) \in \mathcal{U}, \tag{3.322}$$

for any given training sequence σ. See (3.213).

Similarly, for an unrestricted HMM interpolation problem, the set $\Theta = \mathscr{P}$ (see (3.216)), and a natural ascent function is

$$F(P) = \mathbf{L}(P|\mathbf{X}_T = \mathbf{x}_T), \ \ P \in \mathscr{P}, \tag{3.323}$$

see (3.234). The solution set here coincides with $\mathscr{F}_\Pi$, the set of fixed points of the transformation Π:

$$\mathscr{F}_\Pi = \{P \in \mathscr{P} : \Pi(P) = P\}. \tag{3.324}$$

Note that both $\mathscr{F}_\Phi$ and $\mathscr{F}_\Pi$ are closed subsets in $\mathscr{U}$ and $\mathscr{P}$, respectively.

We are interested in proving that sequences of models $(Z^{(N)})$ and $(P^{(N)})$ converge or have limit points as $N \to \infty$. Recall, geometrically $Z^{(N)}$ and $P^{(N)}$ are images of initial models, $Z^{(0)}$ and $P^{(0)}$, under transformations Φ^N and Π^N (that is, the N-fold iterations of transformations Φ and Π). Thus, we want to analyse the limit

$$\lim_{N\to\infty} \Phi^N(Z) = Z^*, \ \ \lim_{N\to\infty} \Pi^N(P) = P^*. \tag{3.325}$$

For brevity, we refer below to transformation Φ, as the argument for Π is completely analogous. It is clear that limiting models Z^* will give fixed points of the transformation Φ,

$$\Phi(Z^*) = Z^*, \tag{3.326}$$

which is in agreement with (3.274)–(3.276).

From general results established below, it will be possible to deduce the following

Theorem 3.9.9 *Let a sequence $\{Z^{(N)}\}$ be generated by iterations of transformation Φ from an initial model $Z^{(0)}$:*

$$Z^{(N+1)} = \Phi(Z^{(N)}), \ \ N = 0, 1, \dots.$$

Suppose that $F(Z) = \mathbf{L}(\sigma; Z), Z \in \mathscr{U}$, (cf. (3.322)) is an ascent function satisfying properties (1) and (2) from Definition 3.9.13, where the solution set Γ is the set $\mathscr{F}_\Phi$ of fixed points of Φ. Then:

(i) *Any limiting point Z^* of the sequence $\{Z^{(N)}\}$ lies in $\mathscr{F}_\Phi$, i.e. is a fixed point of Φ.*

(ii) *The values $F(Z^{(N)})$ monotonically increase and hence*

$$\lim_{N\to\infty} F(Z^{(N)}) = F(Z^*)$$

(which implies that the value $F(Z^*)$ is the same for all limiting points Z^*).

(iii) *Suppose in addition that the norm $||Z^{(N+1)} - Z^{(N)}|| \to 0$ as $N \to \infty$, and the limiting points Z^* form a closed compact connected set in $\mathcal{U}$. Therefore, either there exists a unique limiting point or the limiting points form a closed compact continuum.*

(iv) *Under assumption (iii), suppose that Γ has finitely or countably many points of global maximum (that is, the likelihood $L(\sigma; Z)$, possesses not more than countably many MLEs), and that $F(Z^{(N)})$ converges to the (globally) maximal value of F. Then, in assertion (iii), the limiting point θ^* is unique, and hence there is the convergence*

$$\lim_{N \to \infty} \theta^{(N)} = \theta^*.$$

Despite its assuring appearance, Theorem 3.9.9 requires strong assumptions and leaves open the important question of how to verify conditions (iii) and (iv). This is an area of intensive research, both analytic and computational. See Remark 3.9.14 below.

Our first example in the adopted general set-up is

Worked Example 3.9.10 (Lyapunov's theorem) Let F be a bounded continuous function $\Theta \to \mathbb{R}$, and A a continuous point-to-point map $\Theta \to \Theta$. Assume that F is an ascent function with a solution set Γ. That is

$$F(A(\theta)) \geq F(\theta), \tag{3.327}$$

and

$$\text{if } F(A(\theta)) = F(\theta) \text{ then } \theta \in \Gamma. \tag{3.328}$$

Let θ^* be a limit point for a sequence $\{\theta^{(N)}\}$ where $\theta^{(N+1)} = A(\theta^{(N)})$, starting from some initial point $\theta^{(0)} \in \Theta$:

$$\theta^* = \lim_{k \to \infty} \theta^{(N_k)}.$$

Prove that $\theta^* \in \Gamma$.

Solution Since the sequence $F(\theta^{(N)})$ is monotone non-decreasing in N and bounded from above, there exists the limit $\lim_{N \to \infty} F(\theta^{(N)})$. By continuity of F, this limit coincides with $F(\theta^*)$. On the other hand, by continuity of A and the above monotonicity, it also coincides with $F(A(\theta^*))$. Thus, $F(A(\theta^*)) = F(\theta^*)$, and owing to condition (3.319), $\theta^* \in \Gamma$. $\qquad \square$

Worked Example 3.9.11 (Ostrowski's theorem) Consider a sequence of points $\theta(k) \in \mathbb{R}^n$ for which the norm $||\theta(k+1) - \theta(k)|| \to 0$. Then either this sequence converges (i.e., there exists the limit $\lim_{k \to \infty} \theta(k)$), or the set of its limiting points is a closed connected continuum.

Solution A closed connected set in $\mathbb{R}^n$ is one that cannot be represented as a union of two disjoint non-empty closed sets; if such a set contains more than one point, it is a continuum. It is easy to see that the limiting points for the sequence under consideration form a closed set. Suppose this set contains more than one point, and that we have represented this set as a disjoint union $C_1 \cup C_2$ where C_1 and C_2 are both closed. Then there exists a $\delta > 0$ such that the distance of any point of C_1 from any point of C_2 does exceed δ. By hypothesis we have that for certain k_0

$$\|\theta(k+1) - \theta(k)\| \leq \frac{\delta}{3}, \text{ whenever } k > k_0. \tag{3.329}$$

Take a point $\theta_1^* \in C_1$. Then there exist arbitrarily large $k' > k_0$ such that $\|\theta(k') - \theta_1^*\| < \delta/3$. As the points $\theta(k)$ with $k > k'$ have an accumulation point in C_2, there exists $k'' > k'$ such that $\mathrm{dist}(\theta(k''), C_2) \leq 2\delta/3$. Here and below, the distance between a point and a set is understood, as usually, as the infimum:

$$\mathrm{dist}(\overline{\theta}, \overline{C}) = \inf\left[\|\overline{\theta} - \widetilde{\theta}\| : \widetilde{\theta} \in \overline{C}\right].$$

Assume that k_1 is the smallest number k'' with this property. Then we have that

$$\mathrm{dist}(\theta(k_1 - 1), C_2) > \frac{2\delta}{3},$$

and therefore by (3.329)

$$\mathrm{dist}(\theta(k_1), C_2) > \frac{\delta}{3}.$$

We see that

$$\frac{\delta}{3} < \mathrm{dist}(\theta(k_1), C_2) \leq \frac{2\delta}{3}. \tag{3.330}$$

Therefore there exists an infinite sequence of indices $k^{(1)} < k^{(2)} < \cdots$, for which (3.330) holds. An accumulation point θ_* for the sequence $\{\theta(k^{(N)})\}$ is a limiting point for the original sequence $\{\theta(k)\}$ and satisfies the relation

$$\frac{\delta}{3} \leq \mathrm{dist}(\theta_*, C_2) \leq \frac{2\delta}{3}. \tag{3.331}$$

So, θ_* does not belong to C_2. Then the point θ_* must lie in C_1; at the same time its distance from C_2 is less then δ. This yields a contradiction, and therefore the set of limiting points is a connected continuum. Compare with Theorem 28.1 in Ostrowski, 1966.

$\square$

Worked Example 3.9.12 Assume that function $F : \Theta \to \mathbb{R}$ and a point-to-point map $A : \Theta \dashrightarrow \Theta$ are as in Worked Example 3.9.10, with the solution set Γ coinciding with $\mathscr{F}_A$, the set of fixed points of A:

$$\Gamma = \mathscr{F}_A = \{\theta \in \Theta : A(\theta) = \theta\}. \tag{3.332}$$

Show that for any initial point $\theta^{(0)}$, the set of limiting points for the sequence $\{\theta^{(N)}\}$, where $\theta^{(N+1)} = A(\theta^{(N)})$, is compact and connected.

Solution The limiting points form a closed subset of the compact set $\{\theta \in \Theta : F(\theta) \geq F(\theta^{(0)})\}$; therefore they form a closed compact set. It has been proved in Worked Example 3.9.11 that the condition

$$\lim_{m \to \infty} \|\theta^{(N+1)} - \theta^{(N)}\| \to 0 \tag{3.333}$$

suffices to conclude that either there is a single limiting point (that is, the points $\theta^{(N)}$ converge to a limit) or the set of limiting points is a closed connected continuum. Suppose that condition (3.333) fails. Then it is possible to extract a subsequence $\theta^{(N_k)}$ such that there exist the limits $\lim_{k \to \infty} \theta^{(N_k)} = \theta'$ and $\lim_{k \to \infty} \theta^{(N_k+1)} = \theta''$, but $\theta' \neq \theta''$. Now the continuity of A implies that $\theta'' = A(\theta')$, and the monotonicity of F implies that

$$F(\theta'') = F(\theta') = \lim_{N \to \infty} F(\theta^{(N)}).$$

By virtue of (3.328), the equality $F(\theta'') = F(\theta')$ implies that θ' is a fixed point for A. Hence, $\theta'' = \theta'$ which yields a contradiction. Hence the condition (3.332) holds, and the set of the limiting points for $\{\theta^{(N)}\}$ is connected. $\qquad \square$

Worked Examples 3.9.10–3.9.12 are summarised in

Theorem 3.9.13 (A global convergence theorem) *Let M be a point-to-set map $\theta \in \Theta \mapsto M(\theta) \subset \Theta$ and F be a continuously differentiable ascent function F on Θ with a solution set $\Gamma \subset \Theta$. Fix an algorithm generating a sequence of points $\theta^{(N+1)} = A(\theta^{(N)}) \in M(\theta^{(N)})$ from a point $\theta^{(0)}$. Then any limiting point θ^* for sequence $\{\theta^{(N)}\}$ lies in the set Γ, and the values $F(\theta^{(N)})$ converge monotonically to the limit equal to $F(\theta^*)$ (which implies that value $F(\theta^*)$ is the same for all limiting points).*

Assume in addition, that (i) the map A can be extended by continuity to Θ, and the set Γ has been specified as in (3.332), and (ii) equation (3.333) holds true. Then the set of limiting points for $\{\theta^{(N)}\}$ is compact and connected. Suppose that we know in addition that (iii) the set of limiting points for sequence $\{\theta^{(N)}\}$ is finite or countable. Then this set is actually reduced to a single point, and therefore the sequence converges to a limit:

$$\lim_{N \to \infty} \theta^{(N)} = \theta^{(\infty)}. \tag{3.334}$$

Remark 3.9.14 A possible way to check the condition (iii) in Theorem 3.9.13 is to verify that (a) the ascent function F has a unique global maximum $\theta_{max} \in \Theta$, and (b) the values $F(\theta^{(N)})$ converge to the maximal value $F(\theta_{max})$. For the case of the HMM filtration problem, this has been epitomised in Theorem 3.9.9.

In practice, conditions (i)–(iii) of Theorem 3.9.13 are expected to be fulfilled for a broad range of cases. However, their rigorous verification is not straightforward. This forced several authors to discuss various palliative measures which may be helpful in some situations. See the monograph by McLachlan and Krishnan, 1997, and also the article C.F. Jeff Wu. "On the convergence properties of the EM algorithm",

Annals Stat., **11**, 1983, 95–103.

Epilogue: Andrei Markov and his Time

The topic of Markov chains occupies a special place in teaching probability theory. It is named after the Russian mathematician who introduced and developed this elegant concept in the 1900s, 30 years before the notion of probability was shaped in the manner we use it today.

Andrei Andreevich Markov (1856–1922) was born into the family of a Russian civil servant. His father, following the family tradition, began his career by studying at a local seminary, but then moved into a forestry inspection office and later became a private solicitor. Markov's father was well known for his frankness and high principles, qualities inherited by his son, but was also inclined to gamble at card games. Once he lost all the family's possessions, but luckily his opponent was unmasked as a cheat, and the loss was declared void. His son by contrast loved chess and was considered one of the best amateur players of the time. When Mikhail Chigorin, a Russian chess master, was preparing for his 1892 match for the World Chess Championship with the Austrian Wilhelm Steinitz, the reigning World champion, he played a sparring series of four games with Markov; Markov won one and drew another. (Chigorin was later dramatically defeated in the decisive game by Steinitz, to the deep disappointment of numerous chess enthusiasts in Russia who still deplore this loss). Markov had already defeated Chigorin in 1890, in a beautiful game recorded in a number of chess textbooks. In an Oxford vs Moscow match played by telegraph in 1916, in the middle of World War I, Markov, representing Moscow, gave another beautiful example, this time against Paul Vinogradov, a social scientist of Russian origin and a professor at Oxford.

In childhood Markov suffered from tuberculosis of the bone, particularly afflicting one leg so that he had to use crutches. However, he was very active and managed to play with other boys by jumping on his healthy leg. After the family moved to St Petersburg, before his tenth birthday, Markov had successful surgery and thereafter walked normally, with only a slight limp. He loved walking, and his

479

favourite saying was 'You must walk if you're still alive.' He was not a brilliant pupil in the gymnasium (a high school in Imperial Russia and Germany) but showed extraordinary abilities in mathematics. It was probably in the genes, as his younger brother also became a prominent mathematician. During his final year at the gymnasium he invented a method of solving linear differential equations and wrote a long letter to a number of prominent Russian mathematicians. Their responses, that his method was not as good as the standard method (which is taught in modern differential equations courses and was already well known by then), only encouraged his interest in mathematics, and he enrolled in 1874 at the Department of Physics and Mathematics of St Petersburg University.

At university, Markov was an outstanding student and was awarded numerous prizes and grants, including an Imperial stipend. His marks were always the highest, except in theology (then a part of the syllabus) and inorganic chemistry (where his examiner was Mendeleev, the father of the Periodic Table and inventor of the 40% standard of alcohol in vodka). His favourite professor was Chebyshev with whom Markov became particularly close and had long conversations after lectures. He graduated in 1878 and earned his Magister's degree in 1880 (roughly corresponding to the modern PhD). The DSci. (Doctorate degree) was awarded to him in 1885, and in 1896 he was elected a member of the Russian Academy of Sciences. In his lifetime he published more than 120 papers and monographs, about a third of them addressing topics from probability theory.

The idea of the Markov chain emerged in his paper of 1907. It is remarkable that Markov did not foresee a wide application of his theory. In his attempts at showing its use he analysed a sequence of 200,000 letters from 'Eugene Onegin', an 1820s novel in verse by Alexandre Pushkin, and another one, of 100,000 letters, from an 1850s Russian novel 'Years of Childhood', by Serguei Aksakov. ('Eugene Onegin' is still probably the most popular piece of poetry in Russia, and it is not uncommon to find people able to cite this long text by heart.) Markov checked that the succession of vowels and consonants in these texts is accurately described by a Markov chain with suitable transition probabilities. Without a computer, he had to do all the analysis by hand, which took months of diligent work, including continual error-checking.

Another example Markov had in mind was card-shuffling; he also spent time in various related calculations. (It is well-known that probability theory from its very beginning was strongly influenced by gambling.) It is worth noting that the card-shuffle example became popular in many areas of research influenced by the advent of computers.

Markov's high research standards often led him into disputes with colleagues whom he criticised for lack of rigour (one of his targets was Sonya (Sophia) Kovalevskaya). In general, at this time there was a schism in Russian mathematics, as the St Petersburg school followed much higher standards than did Moscow mathematicians; this did not always help to maintain friendly relations between the two communities.

Despite (or perhaps because of) his frankness and unwillingness to find a compromise, Markov had many devoted friends among colleagues, notably Alexandre Lyapunov (1857–1918). Lyapunov, although only a year younger, was considered as a follower of Markov (in particular, he was consulted by Markov when in the 1900s he had to give courses in probability, an emerging area of mathematics with a great potential for applications). Markov married in 1883; his wife was a former private student whom he had successfully coached in maths, but the marriage was delayed by several years as the bride's mother did not agree until the suitor was able to prove he was financially solvent. They had three adopted children and one natural son, also named Andrei (1903–1979), who later became a professor of mathematical logic at the Moscow State University and a corresponding member of the Soviet Academy of Sciences.

Markov was a confirmed liberal and leading activist in the organization of Russian science education, and in social and political life in general. In 1901, when the Russian Church Synod announced its decision to excommunicate Leo Tolstoy, Markov submitted a petition to be excommunicated with him. When a member of the Synod approached him to discuss the matter, Markov responded he was only prepared to discuss mathematical topics. In 1903 he declined offers of an Imperial medal as a sign of his disagreement with governmental policies restricting the financial and general freedoms of Russian universities. In 1905 he co-signed a letter of protest against the poor state of Russian schools; when he was reprimanded by Grand Duke Constantin Romanov, he famously replied: "I do not change my views by orders from my superiors." In 1908 the Imperial Cabinet Minister of Education ordered university professors to report the political activities of students (who were becoming more and more radical). Markov protested ("I refuse to be an agent of the government at my lectures!") and offered his resignation. The resignation was not accepted but he was deprived of various honours during the remaining period of the Romanov monarchy. In 1912 he opposed celebrations of the 300th anniversary of the House of Romanov and organized instead a scientific session dedicated to the 200th anniversary of the Law of Large Numbers. After the Bolshevik Revolution, in 1921, he, together with other colleagues, strongly protested to the Soviet authorities, arguing that candidates should be admitted to the universities on the basis of their knowledge of the subject and not on their class origins or political views. This last stance was a particularly bold step as he was already seriously ill and in need of special foods and medicines, only available through government sources at this time of general chaos and shortages in Russia.

Markov was a brilliant lecturer. Many of his courses were lithographed and later translated and printed in Germany. Among his students were Günter, Markov Jr and Voronoi, future eminent mathematicians in their own right.

Towards the end of the academic year of 1921/2, Markov was so poorly that his son Andrei had to lead his father to the lecture theatre. Markov's general state of health was undermined when he learned about the tragic end of Lyapunov who had committed suicide after his wife died of tuberculosis (aggravated by malnutrition) amid the deprivation which marked the Civil War in Russia. Markov's departure was a great loss for Russian mathematics. As Günter wrote shortly after his death: "[Markov] was the natural leader of the circle of disciples who gathered around him; he will remain our leader, as what rests in the cemetery is only what was mortal in him – his high spirit will live forever in those who surrounded him."

After Markov's fundamental works, the theory took a turn towards a general concept of random processes, which included Brownian motion, a continuous-time/space version of a homogeneous random walk. The great names here are Kolmogorov, Wiener, Lévy, Doob, Itô, Feller and, later on, Dynkin. Some of these names have been mentioned in this book.

Bibliography

Some books are undeservedly forgotten,
none are undeservedly remembered
W.H. Auden (1907–1973), British poet

Bharucha-Reid, A.T. *Elements of the Theory of Markov Processes and their Applications.* New York: McGraw-Hill, 1960.

Billingsley, P. *Statistical Inference for Markov Processes.* Chicago: University of Chicago Press, 1961.

Cappé, O., Moulines, E., Ryden, T. *Inference in Hidden Markov Models.* New York: Springer, 2005.

Daley, D.J., Gani, J. *Epidemic Modelling: an Introduction.* Cambridge: Cambridge University Press, 1999.

Dembo, A., Zeitouni, O. *Large Deviations Techniques and Applications.* New York: Springer, 1998.

Doob, J. *Stochastic Processes.* New York: Wiley, 1953.

Durbin, R., Eddy, S., Krogh, A., Mitchison, G. *Biological Sequence Analysis. Probabilistic Models of Proteins and Nucleic Acids.* Cambridge: Cambridge University Press, 1998.

Durrett, R. *Essentials of Stochastic Processes.* New York: Springer, 1999.

Deuschel, J.D., Stroock, D.W. *Large Deviations.* Boston: Academic Press, 1989.

Dynkin, E.B. *Foundations of the Theory of Markov Processes* [in Russian]. Moscow: Fizmatgiz, 1959. English translation: *Theory of Markov Processes.* Oxford: Pergamon, 1960.

Dynkin, E.B. *Markov Processes* [in Russian]. Moscow: Fizmatgiz, 1963. English translation: *Markov Processes,* Vols. 1, 2. Berlin: Springer, 1965.

Dynkin, E.B. *Markov Processes and Related Problems of Analysis.* Cambridge: Cambridge University Press, 1982.

Dynkin, E.B., Yushkevich, A.A. *Controlled Markov Processes and their Applications* [in Russian]. Moscow: Nauka, 1975. English translation: *Controlled Markov Processes.* New York: Springer-Verlag, 1979.

Feller, W. *An Introduction to Probability Theory and its Applications,* Vol. 1. New York: Wiley, 1950. [Latest edition: 1968].

Feller, W. *An Introduction to Probability Theory and its Applications,* Vol 2. New York: Wiley, 1966. [Latest edition: 1971].

Grimmett, G., Stirzaker, D. *Probability and Stochastic Processes.* Oxford: Clarendon Press, 1982. [3rd edition: 2001].

Grimmett, G., Stirzaker, D. *Probability and Random Processes: Problems and Solutions*. Oxford: Clarendon Press, 1992.

Karlin, S. *A First Course in Stochastic Processes*. New York: Academic Press, 1966.

Karlin S., Taylor, H.M. *A Second Course in Stochastic Processes*. New York: Academic Press, 1981.

Kelly, F.P. *Reversibility and Stochastic Networks*. New York: Wiley, 1979.

Koski, T. *Hidden Markov Models for Bioinformatics*. Dordrecht: Kluwer, 2001.

Kingman, J.F.C. *Poisson Processes*. Oxford: Oxford University Press, 1993.

Kullback, S. *Information Theory and Statistics*. New York: Wiley, 1959. [Latest edition: Mineola, N.Y.: Dover; London: Constable, 1997].

Lehmann, E. L. *Testing Statistical Hypotheses*. New York: Wiley; London: Chapman and Hall, 1959. [Latest edition: Lehmann, E. L., Romano, E.P. *Testing Statistical Hypotheses*. New York: Springer, 2005].

MacDonald, I.L., Zucchini, W. *Hidden Markov and other Models for Discrete-valued Time Series*. London: Chapman and Hall, 1997.

Martin, J.J. *Bayesian Decision Problems and Markov Chains*. New York: Wiley, 1967.

McLachlan, G.J., Krishnan, T. *The EM Algorithm and Extensions*. New York: Wiley, 1997.

Norris, J.R. *Markov Chains*. Cambridge: Cambridge University Press, 1997.

Ostrowski, A.M. *Solution of Equations and Systems of Equations*. New York: Academic Press, 1960. [Latest edition: 1973].

Rabiner, L.R., Juang, B.H. *Fundamentals of Speech Recognition*. Englewood Cliffs, NJ: Prentice Hall, 1993.

Ross, S.M. *Stochastic Processes*. New York: Wiley, 1983. [Latest edition: 1996].

Shwartz, A., Weiss, A. *Large Deviations for Performance Analysis. Queues, Communications, and Computing*. London: Chapman and Hall, 1995.

Stirzaker, D. *Stochastic Processes and Models*. Oxford: Oxford University Press, 2005.

Stroock, D.W. *An Introduction to the Theory of Large Deviations*. New York: Springer, 1984.

Stroock, D.W. *An Introduction to Markov Processes*. Berlin: Springer, 2005.

Zangwill, W.I. *Nonlinear Programming: a Unified Approach*. Englewood Cliffs, NJ: Prentice Hall, 1969.

Index

When a term is encountered on numerous occasions, the page number of only three of its appearances are given; the number in italics indicates the page where the term was defined.

Printed in the United States
by Baker & Taylor Publisher Services

Printed in the United States
by Baker & Taylor Publisher Services